TECHNICAL
COMMUNICATION

TECHNICAL COMMUNICATION

2nd Edition

Mary M. Lay
University of Minnesota

Billie J. Wahlstrom
University of Minnesota

Cynthia L. Selfe
Michigan Technological University

Jack Selzer
Penn State University

Carolyn D. Rude
Texas Tech University

 **Irwin McGraw-Hill**

Boston   Burr Ridge, IL   Dubuque, IA   Madison, WI   New York   San Francisco
St. Louis   Bangkok   Bogotá   Caracas   Lisbon   London   Madrid   Mexico City
Milan   New Delhi   Seoul   Singapore   Sydney   Taipei   Toronto

McGraw-Hill Higher Education

*A Division of The **McGraw-Hill** Companies*

TECHNICAL COMMUNICATION

Copyright © 2000, 1995, by The McGraw-Hill Companies, Inc. All rights reserved. Printed in the United States of America. Except as permitted under the United States Copyright Act of 1976, no part of this publication may be reproduced or distributed in any form or by any means, or stored in a data base or retrieval system, without the prior written permission of the publisher.

This book is printed on acid-free paper.

1 2 3 4 5 6 7 8 9 0 DOC/DOC 9 0 9 8 7 6 5 3 2 1 0 9

ISBN 0-256-22058-1

Vice president/Editor-in-chief: *Michael W. Junior*
Sponsoring editor: *Andy Winston*
Development editor: *Sarah Reed*
Marketing manager: *Ellen Cleary*
Senior project manager: *Jean Lou Hess*
Senior production supervisor: *Lori Koetters*
Freelance design coordinator: *Laurie J. Entringer*
Cover photo: *© 1997 William Whitehurst, The Stock Market*
Supplement coordinator: *Carol A. Bielski*
Compositor: *Shepherd Incorporated*
Typeface: *10/12 Garamond*
Printer: *R. R. Donnelley & Sons Company*

Library of Congress Cataloging-in-Publication Data

Technical communication / Mary M. Lay ... [et al.]. —2nd ed.
    p. cm.
  Includes bibliographical references and index.
  ISBN 0-256-22058-1
  1. Communication of technical information. I. Lay, Mary M.
T10.5.T413 2000
601.4—dc21   99-067377

http://www.mhhe.com

Preface

The first question you will probably have about this text is "why five authors?" The answer to that question illustrates the changing nature of technical communication.

Technical communication—the specialized communication that helps readers, viewers, or listeners respond to the challenges of a technological world—is complex, must be socially and legally responsible, and has important impact on everyday life. More and more often, technical communication is part of a team effort. Products and processes in business, industry, and education are too complex for a single person to know everything that is essential to the success of a project.

Each of the five authors of this textbook is an expert in some aspect of technical communication. That expertise comes from many years in the classroom and many years in industrial, research, business, and scholarly settings. We believe that by combining our voices, our experiences, and our knowledge we can bring you the latest communication strategies that will enable you to function well in this complex technological world.

All of you will become technical professionals who communicate. As a professional, you will need to know how to conduct research about technology and how to help your readers, viewers, and listeners—your audiences—use your information to solve problems. You will need to display technical data visually, present information orally, design and package effective documents, and use communication technologies to accomplish these tasks. In addition, you need to know what forms and formats are traditionally used on the job, how to choose an appropriate style for your communication, and how to analyze your audience's needs and interests. In order to be effective in your job, you will need to know how to establish your credibility as an author, how to write as part of a collaborative team, how to help your audience make decisions, how to avoid endangering your audience, and how to meet many more communication challenges.

In this text, then, you receive the advice of five teachers, scholars, and practitioners, and we have worked hard to make that advice reflect a growing body of knowledge about technology and communication.

Organization   *Technical Communication* has four parts. Part One, "Understanding Technical Communication," focuses on communication in today's workplace and helps you understand the workplace's uses, creators, and audiences. Chapter 1 looks at the nature of technical communication and how it is affected by increasing use of technology, by growing regard for how society affects and is affected by technical communication, by the global nature of the workplace, and by increasing awareness of cultural diversity. This chapter, also provides you with some of the history and special characteristics of technical communication. In Chapter 2 you will learn about writing and writing processes as they occur in the workplace, and Chapter 3 teaches you how to address your audiences effectively by thinking rhetorically and determining your audience's needs, attitudes, and knowledge. The focus of Chapter 4 is on the persuasive nature of technical communication and how to establish your credibility, how to appeal to your audience's values, and how to provide good reasons for your argument.

The chapters in Part Two, "Acquiring the Tools of Technical Communication," give you the strategies for writing effective technical communication, in whatever situation you are called upon to do so. Chapter 5 introduces you to the nature of collaborative writing—producing documents as part of a writing team. It provides practical advice for maintaining effective interpersonal relationships in that team as well as for employing the new and emerging technologies that support collaboration. In Chapters 6 and 7, you'll learn research strategies for collecting and generating information, whether you use the library, conduct a survey, or navigate the Internet. In Chapter 8, you'll look at editing and style issues, and you'll learn what to check, change, or correct as you revise, edit, and proofread your documents. Finally, Chapter 9 covers document design and packaging—everything from type styles to binding—and Chapter 10 demonstrates how to display data visually in graphs, tables, charts, and drawings. Chapter 10 also introduces you to the software and hardware used to create effective visual displays in print, multimedia, and oral presentations.

Part Three, "Creating Effective Documents," focuses on the typical forms and formats used to organize technical documents. Chapter 11 covers definitions and descriptions, and Chapter 12 looks at creating effective instructions, specifications, and procedures—all traditional systems for organizing technical information. Chapters 13 and 14 focus on the longest and often most challenging type of technical communication—the report. These chapters will show you how to recognize and create the traditional types and parts of reports and how reports help readers make decisions, based on technical, managerial, and social criteria. Finally, Chapter 15 demonstrates how effective proposals, which set forth solutions to problems, can win not only approval but also funding for your ideas.

The final part of this text, "Developing and Maintaining a Professional Edge," concentrates on some common and challenging communication situations. Chapter 16 introduces you to what is the most personal and frequent type of workplace communication, correspondence, and it shows you how to apply what you've learned so far to the creation of a successful job search. The last chapter shows you how to apply much of what you have learned about written documents and visual displays to oral presentations, including the job interview.

Several features in this text make it useful and interesting for you.

Special Features

■ SOCIAL CONSTRUCTION

Many of the examples, assignments, and exercises in the text are *socially situated,* in that they address social as well as technical issues. We don't believe that either communication or technology are isolated phenomena; instead, we believe that they occur in a rich setting that is shaped by economics, politics, ethical and legal considerations, and social and cultural forces. Therefore, the examples, assignments, and exercises we offer here assume that the technology is created and used by people with various values, interests, and needs—that writers can and should help audiences understand and use technology, make decisions about technology, and solve problems with technology. When writers fail at their jobs, there are consequences to those failures; writers can alienate, frustrate, or endanger their audiences.

■ COLLABORATION

One of our goals is to help you see that the need to communicate springs from unique and describable workplace problems, values, and goals. In fact, in this text, we assume that you never write alone—whether you *collaborate* within a writing team or listen and respond to the needs of people in your organization, your communications are socially constructed.

■ CASE STUDIES

This text also contains a **set of documents,** upon which many exercises are based. These documents, collected in Appendix A, address both technical and social issues: the AIDS epidemic, Y2K computer programming issues, environmental and public relations concerns. You'll read how community and medical leaders have responded, through technical communication, to health care workers and the general public's need to know about AIDS and the HIV virus. You'll read about bioremediation, a process used in the *Exxon Valdez* oil spill. You'll see how one university responded to community concern about the number of birds trapped and killed in experimental agricultural fields. And, you'll see how Web documents are being used to address issues of interest to a broad segment of society.

■ TECHNOLOGY

To be successful in the workplace, people must use a variety of media to communicate their ideas effectively. We know that documents in today's workplace can take a variety of forms: print, Web pages, CD-ROMs, e-mail, and multimedia, to name just a few of the most common. Therefore in this text we try to familiarize you with software and hardware as well as the types of communications you will be called upon to create using this technology.

Through this text, we'll teach you **how to use technology to communicate.** You'll see not only how to find information within computer databases and in the electronic environment of the Internet and how to use the computer to draft and revise your documents, but also how to use the computer software to display data, organize oral presentations, and even collaborate within a writing team.

■ PERSUASION

This text emphasizes the **persuasive nature of technical communication.** Thus, you'll not only learn how to convey technical information in a clear and concise manner, but you'll also learn how to persuade your audience that you have found solutions for problems, help an audience make decisions, and demonstrate that you deserve to be hired or funded.

■ DIVERSITY

Within this text, we also remind you about the **culturally diverse and international nature** of technical communication. We point out when to think about your audience's level of literacy, how to meet the needs of color-blind or hearing impaired audiences, and when to consider an international audience's response to your words and symbols.

■ LEARNING AIDS

Finally, **each chapter begins with a quotation** from a technical communication scholar, teacher, or practitioner. These statements should help you think and discuss some of the main principles within each chapter. Throughout the text, we have **highlighted important or new words and defined them in the margin,** so that you can easily review them. There is also a glossary that has longer definitions of some terms and that allows you to review quickly any points you might want to check as you read. And, we have created **Writing Strategy** checklists and worksheets for you to use in creating documents now and in the future.

The Instructor's Resource Guide

The Instructor's Resource Guide (IRG) to this textbook offers a range of supplemental activities, readings, and information to new and experienced teachers. We provide sample syllabi for different length terms, and we offer suggestions about how to integrate the case studies into the classroom. With each chapter, we give an overview, set of goals, definitions of important terms and concepts, teaching strategies, and supplementary activities and assignments. When appropriate, we provide overhead transparency masters and handouts. Finally, we have reprinted several landmark and recent articles that inform different aspects of teaching technical communication.

Acknowledgments

We want to thank some special people who contributed to either our collective or individual efforts. Ann Hill Duin, Steve Doheny-Farina, and Sherry Little contributed much to the first edition of this textbook, and their words and thoughts still have great impact on this second edition. Linda Jorn, University of Minnesota, shared with us the AIDS documents that appear in this book, and Linda Van Buskirk, Cornell University, alerted us to and helped us gather the *Exxon Valdez* documents. Sam Racine and Linda Clemens, University of Minnesota, helped gather and document research sources, and Linda developed the Instructor's Resource Guide, using the excellent foundation provided by Laurie Gardener in the manual that accompanied the first edition. Paul Brady, University of Minnesota, provided examples of Web page design from his courses. Susan Wakefield and Linda Clemens, University of Minnesota, worked on permissions. Patricia Goubil-Gambrell, formerly of Texas Tech University, alerted us to some of the examples that appear in chapters 13 and 14. Andrew Stephenson, at Penn State, offered information on scientific writing and gave permission to use and revise his work, and Gay Gragson enhanced our conversations about audience.

The teachers and scholars who reviewed the text in its various stages provided valuable suggestions and reactions, and at times they even generously offered some of the words, ideas, and opinions that appear in this text. We want to thank the entire staff at Irwin/McGraw-Hill for their help in completing this challenging project. In particular we would like to thank Craig Beytien for getting us all together in the first place, Sarah Reed for helping us maintain a clear vision and a positive attitude, Jean Lou Hess for bringing the entire project to completion, and Ellen Cleary for her creative contributions. The teachers and scholars who reviewed the text in its various stages provided valuable suggestions and reactions, and at times their suggestions appear in the text:

Tamara Aldin, University of Washington

Rance Baker, San Antonio College

Carol Barnum, Southern Polytechnical State University

Ron Bishop, Drexel University

Lee Brasseur, Illinois State University

Sandra Corse, Georgia Institute of Technology

Judy Hakola, University of Maine

June Hankins, Southwest Texas State University

TyAnna Herrington, Georgia Institute of Technology

Dan Jones, University of Central Florida

C. Hugh Marsh, University of California, Santa Barbara

Eric Pappas, Virginia Polytechnic Institute

Sabrina Peters, Texas Tech University

Carolyn Plumb, University of Washington

Rita Reeves, East Carolina University

Randall Sadler, University of Arizona, Tucson

Brenda R. Sims, University of North Texas

Zach Thundy, Northern Michigan University

Thomas L. Warren, Oklahoma State University

Harriet Wilkins, IUPUI

Carole Yee, New Mexico Institute of Mining and Technology

Donald E. Zimmerman, Colorado State University

We would also like to extend a special thanks to the professors who helped us prepare the 1st edition by providing their suggestions and feedback:

Jo Allen, East Carolina University

Dennis Barbour, Purdue University, Calumet

Steven Bernhardt, New Mexico State University

Virginia Book, University of Nebraska, Lincoln

Sam Geonetta, University of Cincinnati

Hillary Hart, University of Texas, Austin

David Helgeson, British Columbia Institute of Technology

Dan Jones, University of Central Florida

David Kaufer, Carnegie Mellon University

Charles Kostelnick, Iowa State University

Marilee Long, Colorado State University

Carolyn Miller, North Carolina State University

Carolyn Plumb, University of Washington

Kathy Underwood, University of Washington

We are thankful to these people for their assistance. Any errors that remain are ours alone.

Finally, we thank our students, our colleagues, our friends, and our families who continue to give us their insights, patience, support, and love.

Mary M. Lay
Billie J. Wahlstrom
Carolyn Rude
Cynthia Selfe
Jack Selzer

Brief Contents

Contents

Understanding Technical Communication

Technical Communication in Today's Workplace

The important thing is to get started. I'll often put a piece of paper in the typewriter and decide the paper doesn't want to be written on, so I'll throw it away and take another piece of paper.

Russell Baker, *Growing Up* (New York: Plume, 1983).

A web of glass spans the globe. Through it, brief sparks of light incessantly fly, linking machines chip to chip and people face to face.

Vinton G. Cerf, "Networks," *Scientific American* (September 1991), p. 72.

Introduction

communication skills
reading, speaking, listening,
writing, and designing
visual displays

Given the complex world you are preparing to enter, good **communication skills**, familiarity with current and emerging communication technologies, and a deepening awareness of the nature of intercultural communication are critical to your success. We hope the communication skills and insights you gain here will make it easier for you to succeed at work and to be a knowledgeable, ethical, and effective citizen and employee.

This book is about communicating in a rapidly changing workplace. To be successful in your career or job, you will need to create the specialized kinds of communication that people use in business, industry, government, and education. Knowing how to use communication technologies appropriately to create these specialized documents is also critical. And while it is true that technology has changed the workplace, the workplace you will enter has also been changed by the increasingly multinational nature of the economy and the changing demographics of American society. To be effective at work, you will need to communicate effectively with a diverse population.

Some of you reading this book may become technical communicators, knowledge engineers, information developers, or information architects, as such professional communicators are called today. Most of you, however, will be technical professionals for whom communication will play a large role in your work. Right now you may be taking a technical communication class because it is required, but as you move into the work world, your ability to do your work and advance in your position will largely depend on good communication skills.

If you are planning to become an engineer, you will find that engineers are often responsible for writing feasibility studies for new buildings, electrical systems, and computer networks. If you are majoring in computer science, you will have to write specifications for hardware and software. If your goal is to manage the technical products division of an industrial corporation, you must use written documents, lectures, and videotapes to teach technicians how to install products and customers how to use those products. If you are in one of the agricultural or environmental sciences, you will have to create environmental impact studies and write grants.

Whatever your technical profession, a significant part of your time will be spent on technical communication. In a survey of 1,400 members of the American Institute of Chemists, the American Consulting Engineers Council, the International City Management Association, the Modern Language Association, the American Psychological Association, the Professional Services Management Association, and the Society for Technical Communication, respondents said they spent 44 percent of their professional time in some kind of writing activity (this includes brainstorming, note-taking, organizational planning, drafting, revising, and editing) (Lunsford and Ede). If you intend to be one of the more successful practitioners of your profession, count on spending even more time than the average practitioner engaged in communication activities.

If instead you choose technical communication as a profession, you will join a growing number of people employed in business, government, and industry. In 1990 the U.S. Bureau of Labor Statistics classified approximately 70,000 workers as technical writers, practitioners in a field that has only been recognized officially as a career since the 1950s. The 1998-99 *Occupational Outlook Handbook* published by the U.S. Bureau of Labor Statistics identified about 286,000 jobs held by writers and editors in 1996, a significant portion of which are in technical communication. The Society for Technical Communication, the largest professional organization of technical communicators, has more than 20,000 members worldwide. The Bureau of Labor Statistics has indicated that through the year 2006, opportunities will be good for technical communicators because of society's need to deal with technical information. In its 1999 Annual Guide to the Best Jobs in the Future, *U.S. News and World Report* listed technical communication as one of the 20 hottest jobs.

Whatever your position in the science and technology workplace, you will be called upon to show competence in the field of technical communication. Our goal in this book is to make sure you enter the workforce with the communication skills and knowledge you need. For that reason, when we speak of technical communicators, we are thinking not only of those of you who may choose that profession, but also of everyone who uses communication in the workplace.

The History of Technical Communication

Societies have always needed communicators to serve their information needs. These communicators described the functioning of the natural world so that nonspecialists could understand it, and explained processes so that others could perform them. Communicators have been essential to the development and distribution of innovations in communication technology and in tracking resource allocation and use.

For nearly 2,000 years, teachers and philosophers have addressed problems associated with communicating complicated information accurately and effectively. Some of the theories behind modern technical communication come from Aristotle, the most famous of communication philosophers, who lived in ancient Greece. In *Rhetoric,* Aristotle discussed various ways to create effective messages (Example 1-1). He distinguished between **persuasion,** ways to motivate listeners and readers, and **argument,** the arrangement of persuasive points. His emphasis on persuasion and argument still influences how we view technical communication and how we structure effective documents.

Aristotle was concerned with creating effective messages with whatever means of persuasion were available. Aristotle was working in an oral medium, and his communicators were orators. Today, technical communicators have a range of means—video, print, slides, audio, multimedia—that

persuasion
motivation of a reader or listener

argument
the arrangement of persuasive points

Aristotle
(384-322 B.C.) author of *Rhetoric,* which describes the art of creating the most persuasive message on any subject

EXAMPLE 1–1   Aristotle from *Rhetoric*

Rhetoric may be defined as the faculty of observing in any given case the available means of persuasion. This is not a function of any other art. Every other art can instruct or persuade about its own particular subject matter; for instance, medicine about what is healthy and unhealthy, geometry about the properties of magnitudes, arithmetic about numbers, and the same is true of the other arts and sciences. But rhetoric we look upon as the power of observing the means of persuasion on almost any subject presented to us; and that is why we say that, in its technical character it is not concerned with any special or definite class of subjects.

Source: Aristotle. *Rhetoric,* Book I. *The Rhetorical Tradition: Readings from Classical Times to the Present.* Ed. Patricia Bizzell and Bruce Herzberg (Boston: Bedford Books, 1990), p.153.

EXAMPLE 1–2   Frontinus from *Aqueducts of Rome, II*

[Aqueduct] Claudia, flowing more abundantly than the others, is especially exposed to depredation. In the records it is credited with only 2,855 quinariae, although I found at the intake 4,607 quinariae—1,752 quinariae more than recorded. Our gauging, however, is confirmed by the fact that at the seventh mile-stone from the City, at the settling reservoir, where the gauging is without question, we find 3,312 quinariae—457 more than are recorded, although, before reaching the reservoir, not only are deliveries made, to satisfy private grants, but also, as we detected, a great deal is taken secretly, and therefore 1,295 quinariae less are found than there really ought to be.

Source: Frontinus. *The Stratagems and the Aqueducts of Rome,* vol. II. Ed. Mary B. McElwain. Transl. Charles E. Bennett (Cambridge: Harvard University Press, 1930), pp. 70–72.

Aristotle could not imagine, but their primary concern is still the creation of effective messages. What Aristotle says about subject matter is of particular interest to technical communicators today. Because of their rhetorical training, technical communicators are not bound by subject matter but can look for the "means of persuasion" on almost any subject. Nevertheless, it is essential that we develop in our work environment an understanding of the subject matter with which we deal.

Scholars have traced the practice of technical writing back to the Sumerians, but **Sextus Julius Frontinus** is generally recognized as the author of one of the first extended pieces of technical communication. In approximately A.D. 97, Frontinus wrote a manual for building and maintaining aqueducts while working as the water commissioner in Rome (Example 1-2).

Sextus Julius Frontinus (ca. A.D. 35–103) a Roman author and soldier whose writing about aqueducts is an example of early technical communication

He was asked to account for the distribution of natural resources in much the same way technical communicators working for the government today are asked to describe resource use and allocation. Then, as now, government relied on the work of technical communicators to provide the information necessary to maintain civilization's infrastructure.

Technical communication also has been deeply connected to the development of new communication technologies and the distribution of information about how to use the technology throughout the population. Technical communicators, for example, have worked in computer fields since they began and now work with rapidly emerging communication technologies. The writing of Pliny the Elder illustrates technical communicators' close ties to emerging communication technologies, right from the very beginning.

Pliny the Elder (A.D. 23–79) was a Roman soldier and administrator who held a number of offices in Gaul, Africa, and Spain. His only surviving work, the *Natural History,* is a compilation of natural and scientific knowledge (more than 20,000 "facts") from more than 2,000 ancient and contemporary (to him) texts. In the true spirit of a technical communicator who wants to obtain the most accurate information possible, Pliny died of asphyxiation from the sulfurous fumes after getting too close to the eruption of Mt. Vesuvius.

> **Pliny the Elder** (23–79) a Roman soldier and administrator who wrote and compiled information about natural history

In Example 1–3 you can read a portion of one of the earliest documents describing the development of a new communication technology, paper. Pliny explains the process by which paper is made from papyrus. This example is also interesting for the ways in which Pliny defines terms and the evidence of the impact of politics even in these early examples of technical communication. Technical communication has changed since the Greeks and the Romans, but in every age, it has aided people in making sense of technology, and it has helped them apply scientific principles to practical problems.

During the **English Renaissance** (1475–1640), technical manuals helped people do everything from preparing the ground for planting to curing diseases. In Example 1–4, you can see how **Reginald Scot** described how to cultivate a hop garden in 1574. In every age, people have needed instructional manuals to help them accomplish tasks. Technical documents from the Renaissance were directed chiefly toward helping people "perform tasks—explaining the process, providing instructions, describing tools needed to perform these tasks" (Tebeaux and Killingsworth, 9).

> **English Renaissance** (1475–1640) a period when technical writing taught people everything from surgery to planting
>
> **Reginald Scot** (1538–1599) a technical communicator of the Renaissance who described how to cultivate a hop garden

Notice especially how technical illustrations are used in Example 1–4. They show how to tie hops to poles, indicating that this document was designed for a lay rather than an expert audience who would already have known how to perform some of these tasks. These technical illustrations have their counterparts in today's directions for completing a variety of tasks such as assembling furniture, connecting peripherals to a computer, and designing Web pages.

EXAMPLE 1–3   An Ancient Process: Making Paper from Papyrus

68. Before I leave Egypt I shall also describe the nature of the papyrus plant since our civilization—or at any rate our written records—depends especially on the use of paper . . .

74. Paper is made from papyrus by splitting it with a needle into strips that are very thin but as long as possible. The quality of the papyrus is best at the center of the plant and decreases progressively towards the outside. The first quality used to be called "hieratic" paper and in early times was devoted solely to books connected with religion, but to flatter the emperor, was given the name "Augustus"; the second quality was called "Livia" after his wife, and so the term "hieratic" was relegated to the third category.

75. The next quality had originally been called "amphitheatre" after the place where it was made. This paper was taken over by the ingenious workshop of Fannius at Rome; they made its texture finer by careful dressing, and so upgraded ordinary paper to rank with the first quality—this paper was known by the name of its maker . . . .

76. Next is "Saitic" paper, called after the town where it was produced in the greatest quantity, and made from low-grade scrapings; then comes "Taeneotic" from a neighboring place, made from fibre still closer to the outside covering of the papyrus. (This paper is sold by weight, not quality). Finally there is *emporitica,* "packing" paper, which is no good for writing on but is used to cover documents and a wrapping for merchandise; for this reason it takes its name from the Greek word for a merchant. After this is the outer layer—the actual papyrus—which is like rush and of no use for ropes, except for those used in water.

77. All paper is "woven" on a board dampened with water from the Nile; the muddy liquid acts as glue. First an upright layer is smeared on the table—the whole length of the papyrus is used and both its ends are trimmed; then strips are laid across and complete a criss-cross pattern, which is then squeezed in presses. The sheets are dried in the sun and joined together; each succeeding sheet decreases in quality down to the worst. There are never more than twenty sheets in a roll.

78. There is a big difference in the width of the various types of paper. The best paper is 13 inches wide, and the "hieratic" 11 inches; Fannian is 10 inches, and amphitheatre 9 inches, which Saitic is not as wide as the mallet used in the manufacturing process. Other criteria of quality in paper are its fineness, thickness, whiteness, and smoothness.

Source: *Natural History: A Selection* by Pliny the Elder, trans. John F. Healy (Penguin Classics, 1991) copyright © John F. Healy, 1991. pp. 175, 177–78.

EXAMPLE 1–4 A Perfite Platforme of a Hoppe Garden

22 *A perfite platforme*

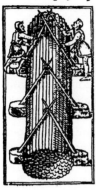

This figure or rounc is moje than halfe filled with Poales, and not filled vp to the top.

one ende of the roume woulde be full before the other , whereas nowe they shall lye euen and sharpe aboue , lyke an Hayestacke , oz the ridge of an house , and sufficiently defende themselues from the weather.

If you thinke that you haue not Poales ynowe to fryll the roume , pull downe the Wyths oz bandes lower, & your roume will be lesse.

¶ *Of tying of Hoppes to the* Poales.

When your Hoppes are growne about one oz two foote high, bynde vp (with a Rushe oz a Grasse) such as declyne from the Poales, wynding them as often about the same Poales as you can, and directing them alwayes according to the course of the Sunne, but doe it not in the morning when the dewe remayneth vpon them, if your leysure may serue to doe it at any other time of the day.

If

of a Hoppegarden 23

If you laye softe grœne Rushes abroade in the dewe and the Sunne, within two oz thrœ dayes, they will be lythie, tough, and handsome foz this purpose of tying, which may not be foze

slowed,foz it is most certaine that the Hoppe that lyeth long vpon the grounde before he be tyed to the Poale,prospereth nothing so wel as it,which sooner attayneth therebnto.

¶ *Of hylling and hylles.*

Nowe you must begyn to make your hils, and foz the better doing therof,you must prepare a toole of Iron fashioned somewhat lyke to a Cowpers Iddes , but not so much bowing, and therfoze lykest to the netherpart of a Showell,the powle wherof must be made with a round hole to receyue a helue,lyke to the helue of a Mattocke , and in the powle also a naple hole must be made, to fasten it to the helue.

This

This toole is here aboue propoztioned, better propoztioned, th.n that on the other side following.

Source: Reginald Scot. *A Perfite Platforme of a Hoppe Garden.* 1574. (Amsterdam: Theatrvm Orbis Terrarvm, Ltd., 1973). The English Experience 620. Short Title Catalog 21865. As discussed in Elizabeth Tebeaux and M. Jimmie Killingsworth. "Expanding and Redirecting Historical Research in Technical Writing: In Search of Our Past," *Technical Communication Quarterly* 1.2 (1992), 5–32.

Technical communication really came into its own as an academic field and as a profession because of the needs of people during World War II. This war relied heavily on technology, and the awful and complex weapons used by thousands of average individuals, or nonexperts, required a variety of technical documents to explain their use and maintenance. Airplanes, guns, bombs, machines, and electronic equipment all needed user and repair manuals as well as specification sheets; after the war, the large technology companies—General Electric, Westinghouse, and General Motors— opened technical writing departments after discovering that having technical communicators on the payroll saved money (Connors).

Technical communication has always provided essential information. Technical communicators are critical to society. Like their ancient predecessors, contemporary technical communicators create the documents that help us run everything from nuclear submarines to home computers, tell us how new technology works, and document the allocation of natural resources. Today, as in the past, the ability of technical communicators to do their job well is critical, for we live in a world where the malfunction of such a small part as an O ring can lead to the explosion of a space shuttle and the death of its crew.

The Definition of Technical Communication

audience
the listeners, viewers, and readers of documents

purpose
why a communication is needed and what it is supposed to accomplish

style
the communicator's choices about tone, point of view, level of formality, use of figures of speech, voice, and emphasis

technical communication
applied communication designed to perform specific tasks or help the audience solve specific problems

genre
a category of document characterized by a particular style, form, and content (e.g., instructions, reports, warning labels)

clarity
a cluster of elements that includes freedom from ambiguity, logical development of ideas, unity, coherence, appropriate style and usage, clear transitions, and accurate word choice

Technical communication is a specialized form of communication. It shares many characteristics of other forms of communication, but it differs from these other forms in its **audience** (its listeners, viewers, or readers), **purpose** (why a communication is needed), **style** (tone, voice, point of view, etc.), format, situation, content, and use. **Technical communication** is applied communication, communication designed to perform specific tasks or help the audience solve specific problems. That task may be to inform users about an update in computer software, teach line workers how to produce a new product, or warn customers about unsafe ways to operate a machine. Or the task might be to inform users how to repair a copying machine, select a home computer, or choose a method to clean up an oil spill. To help you understand the special nature of technical communication, we want first to place it on a continuum with other kinds of communication.

TECHNICAL COMMUNICATION AND OTHER KINDS OF COMMUNICATION

In high school and college, you have read and written a wide variety of materials, and the breadth of your experience will prepare you to understand the special role technical communication plays in business and industry. Imagine all the kinds of writing you have done or read on a spectrum of the kind shown in Example 1–5.

Although all kinds of writing communicate, notice how these kinds differ from each other in audiences, **genres,** purposes, and reasons people read them. Technical communication also differs from other kinds of communication primarily because it is designed to help others make decisions and complete tasks. To perform these functions, technical communication emphasizes **clarity**, **accuracy**, **conciseness**, **readability**, and **legibility.**

Technical communication is usually put to some use rather than enjoyed for its own sake. In this way it differs from creative writing. Technical communication also differs from academic writing. When you write an essay exam or a term paper, you are trying to prove to your instructor that you have mastered the material and can organize and present it intelligently when asked. When you do technical communication, however, your goal is not to prove to the user you know how something works but to help the reader master the task at hand. People are unlikely to curl up

EXAMPLE 1–5 Communication Spectrum

	Creative Writing	**Academic Writing**	**Technical Writing**
Purpose	■ Writing to express ■ Writing to demonstrate creativity	■ Writing to learn ■ Writing to demonstrate competency ■ Writing to think critically	■ Writing to persuade ■ Writing to inform ■ Writing to help others complete tasks
Genres	■ Nonfiction ■ Novels ■ Poetry ■ Short stories ■ Screenplays	■ Lab reports ■ Book reports ■ Themes/essays ■ Essay exams ■ Journals	■ Proposals ■ Correspondence ■ Instructions ■ Descriptions ■ Feasibility reports ■ Grants
Reading	■ Texts are read from beginning to end. ■ Texts are read at school or home. ■ Texts are read by teacher, student, or audience. ■ Texts are read for enjoyment.	■ Texts are read from beginning to end. ■ Texts are read at school or home. ■ Texts are read by teacher. ■ Texts are read for assessment.	■ Texts are read as needed. ■ Texts are read where needed. ■ Texts are read to solve a problem. ■ Texts are read to gain specific data.

accuracy
the validity and precision of the information in a technical document

conciseness
writing as clear and as brief as it can be and still be effective

readability
how easily a document can be understood

legibility
the design and formatting characteristics of a document, including typeface, type size, use of white space, and kerning (the space between letters)

on the couch with a good book of instructions. Instead, they are more likely to have those instructions at hand as they try to accomplish something: follow instructions on how to download the newest version of Netscape, consult a technical report to help them choose a new microwave, or read a governmental report to decide how to vote on the creation of a new incinerator.

Examine the three passages in Example 1-6. They represent short examples of creative, academic, and technical writing. What differences do you see between them in the type of audience, the purpose, the approach, the vocabulary and sentence structure, the organization, formatting, and where and why they might be used? Who reads poetry, planting instructions, or student essays and for what reason? For what reasons did people write these three examples?

Technical communication also differs from other forms of communication in the ways it is practiced. For example, technical communication is far more often a collaborative activity than are other kinds of communication.

EXAMPLE 1–6   Technical and Nontechnical Communication

Creative Writing—Poetry

Revolution is the Pod
Systems rattle from
When the Winds of Will are
    stirred
Excellent is Bloom

But except its Russet Base
Every Summer be
The Entomber of itself,
So of Liberty—

Left inactive on the Stalk
As its Purple fled
Revolution shakes it for
Test if it be dead.

Emily Dickinson, 1866.

Source: Reprinted by permission of the publishers and the Trustees of Amherst College from THE POEMS OF EMILY DICKINSON, Thomas H. Johnson, ed., Cambridge, Mass.: The Belknap Press of Harvard Univeristy Press, Copyright © 1951, 1955, 1979, 1983 by the President and Fellows of Harvard College.

Academic Writing—A Student Answer to an Essay Question

Don't Treat Our Soil Like Dirt

In gardening, everything starts with the soil. Healthy soil produces healthy plants, which give rise to higher yields and higher resistance to damage from insects and disease. Through soil building a gardener becomes a true steward of the earth, leaving the land in better condition than when it was originally tilled. Soil building or soil conditioning is best done through the addition of organic matter.

The most common forms of organic matter for the garden are manure; homemade compost; peat moss; leaves, grass clippings, and vegetable waste; and crops that can be turned back into the soil. Organic matter begins to improve the soil immediately. In sandy soils, it acts as a binder, increasing water- and nutrient-holding capabilities. In clay, organic matter acts as a bunch of tiny wedges separating tightly clumped clay particles, allowing them to breathe. And every cubic inch of good garden soil contains millions of microorganisms too small to see which help cause the organic materials to release valuable minerals and trace elements . . .

Technical Writing—Directions

Flower bulbs will do well in any good, well-drained soil. Plant them, with top up, at the depth appropriate to their type as indicated in the chart below.

Tulips	15 cm
Daffodils and hyacinths	10 cm
Crocuses	5 cm

In other words, several people—writers, editors, technical experts, graphic artists—may work together to create one technical document.

Finally, technical communication is usually produced within an organization that has identifiable goals, customs, and values, and each technical document that an organization creates potentially reinforces or changes its corporate image and practices. For example, technical communicators in banks not only research and analyze specific financial questions but also help formulate monetary policy by making recommendations. Technical communicators in advertising firms not only write copy but help establish customer research protocols and influence marketing strategy. Because of the global nature of business, technical communicators increasingly help to develop strategies for designing the documents used in large engineering projects for which parts are made in one country, assembled and installed in a second country, and maintained in a third.

TECHNICAL COMMUNICATION AND TECHNICAL WRITING

The class you are taking may be called "Technical Writing" or "Technical Communication," and you may be confused by how we have used those terms in this book. **Technical writing** is a subset of technical communication. When communicating about complex scientific or technical information, people use more than written documents, however. They may work with PowerPoint™ presentations, video, Web pages, or CD-ROMs. The field of technical communication covers this whole range of communication activities.

technical writing
a subset of technical communication that includes written documents

We focus primarily on written communication in this book because that makes up a substantial amount of the kind of communication you will be responsible for at work, but we look at other media as well.

Because we know that technical communicators need skills in a variety of media—writing and speaking are the two most obvious—we generally refer to technical communication when we discuss principles and practices that help you create documents in any medium, for any audience. However, when we discuss specific types of technical communication—for example, writing reports or letters—we'll narrow our terms to technical writing.

Defining technical communication is difficult, and some experts in the field resist broad definitions that include practically every type of manual or process description, including cookbooks. Other experts resist narrow definitions that exclude any communication whose main topic is not technology. For our purposes, technical communication conveys complex technical information in easily understandable forms, often to inexperienced users or to lay audiences. Although at times users might be experienced and educated in the subject of the technical document, they cannot share completely the writer's or speaker's education and knowledge of the subject of the document.

EXAMPLE 1–7   Samples of Technical Communication

Sample 1

Instructions for Printing a File Using a Macintosh Computer

To print a file that you have edited, use the Print command from the File menu. When you use this command, you will see a dialog box. Simply click on the Print button to make a single copy of your file.

   If you wish to make more than one copy, delete the "1" in the box to the right of "Copies," and type in the desired number.

   If you wish to make a copy of just a few pages of your file, in the box to the right of "From," type the first page number of the section to be printed. In the box to the right of "To," type the last page number of the section to be printed. If you wish to print just one page of a file, type that page number in both the "From" and the "To" box.

Sample 2

Definitions of an HTML Web Page Table

Tables are defined using the **TABLE** element, while the content of the table is laid out as a sequence of table rows (**TR**), which in turn contain table headers (**TH**) and/or table data (**TD**). A table can also have a caption, defined by the **CAPTION** element. A caption can contain all forms of character formatting markup, including hypertext anchors . . . Tables can have borders and dividing lines, or can be borderless. The start tag <**TABLE BORDER**> ensures that the table is drawn with borders and dividers—you can adjust the thickness of the border by assigning a value (in pixels) to the **BORDER** attribute.

Source: Ian S. Graham, *HTML Sourcebook*, 3rd ed. (New York: John Wiley, 1997), p. 73. Reprinted by permission of John Wiley & Sons, Inc.

HTML
an acronym for HyperText Markup Language, which is a programming language used for the creation of Web pages

   Technical communication conveys information and creates understanding—it does not convey meaning. To a large extent, meaning rests in each person. Whether or not the material written is meaningful depends on the needs of the person reading it. As we see in Example 1-7, Sample 1, the writer helps the reader understand how to print a file on the Mac, but if the reader doesn't have a printer or doesn't wish to print a file, then the passage lacks meaning. Likewise in Sample 2, if the reader does not know what **HTML** (HyperText Markup Language) is or has no idea of what creating a Web page table means, the passage lacks meaning.

   To say that a text lacks meaning doesn't imply that a reader doesn't understand the words that have been written. Understanding the words and knowing what they mean are not always synonymous. For a communication to have meaning depends on whether the reader has the background necessary to make sense of the document. Meaning cannot be controlled by the writer. Therefore, you will need to concentrate your energy on creating documents that are understandable. You must work to establish a common basis of understanding with your audience based on vocabulary, experience, values, and education, so that your information is accessible.

Technical communication doesn't occur in a vacuum. It often requires a sophisticated understanding of the **social contexts** or the political, economic, cultural, and organizational settings within which communication takes place. These contexts involve legal, ethical, and social responsibilities involved in the communication of technical information. You probably have heard about a number of examples of technical communication in social contexts: the reports and memoranda involved in the *Challenger*[1] disaster, the reports and news stories documenting the cleanup efforts of the *Exxon Valdez*[2] Alaskan oil spill, the policies and news stories growing out of research on cloning, and the descriptions and solutions offered for the Year 2000 problem on computers.

One way in which technical communication is a socially situated activity is its relationship to issues of public responsibility and ethics. Sometimes people who do technical communication think that the documents they create have no social impact. That's not the case. Increasingly, communicators need not only clear competencies with language, but also a sense of what could be called "public responsibility" and "civic courage" (Giroux).

With growing frequency, technical communicators must consider the legal and moral dimensions of their work. **Product liability law** regards written and online documentation as an integral component of a given product. Consequently, if the documentation for a product or software is faulty and someone is hurt or loses money as a result, then the corporation (and ultimately the person who prepared the documents) is legally responsible. Additionally, First Amendment considerations of privacy and ownership must be factored into almost all kinds of online communication. In cases such as the *Challenger* disaster, technical communicators must ask to what extent they were responsible for the communication failures that permitted the launch of a spacecraft under dangerous conditions.

Our growing understanding of the social contexts within which technical communication takes place has changed the ways we think about this subject. Two decades ago, you would have learned about technical communication only as a series of formats that could be memorized and mastered—memoranda, feasibility reports, lab reports. Now we know that having that information alone is not enough. You also must understand technical

Technical Communication as a Socially Situated Activity

social contexts
political, economic, cultural, and organizational settings with particular values, goals, and communication styles

product liability laws
laws defining the responsibility of companies (including their writers) to protect consumers

---

[1]*Challenger* was the U.S. space shuttle that exploded 73 seconds after takeoff on January 28, 1986, killing all astronauts on board.

[2]*Exxon Valdez* was the tanker that ran aground on March 24, 1989, releasing about 11 million gallons of crude oil into Prince William Sound and the Gulf of Alaska.

communication as it functions within complex organizational and cultural contexts. You need to understand the values and goals of the organization in which you work, for example, and how your communications reflect or can influence those values and goals. You need to understand the cultural issues that can affect global communications: different levels of politeness and directness, different ideas for layout and design, different ideas on appropriate diagrams and illustrations, and even different understandings of what color signifies.

Technical documents are written and read for a wide range of purposes: for informing experts within an organization and persuading nonexperts in cooperating firms; for passing along information to managers and for educating line workers; for describing products to clients and for limiting the liability of a corporate entity; for communicating with regulatory groups; and for representing an organization's work to governmental agencies. Technical communicators work within organizations that have a variety of goals and values that shape both what is communicated and how it is communicated.

Special Characteristics of Technical Communication

The communication you will do on the job differs significantly from much of what you learn during your college career. Some of the differences result from having different purposes and different audiences for the communications you create. Other differences arise from the nature of technical communication itself and from the conditions under which you are expected to create effective communications.

AUDIENCE-BASED ORIENTATION

audience-based orientation
writing that is designed from the point of view of the audience and what it needs rather than from the point of view of the writer

One characteristic of effective technical communication is its **audience-based orientation.** In school you have probably written papers with a keen awareness that your teacher would be reading them. Both what and how you wrote were shaped by the awareness of your audience. In other words, what you wrote was audience-based in its orientation. In contrast, writer-based prose, such as journal and creative writing, is characterized by self-expression of feelings, experiences, and emotions, much of which is not intended to be read at all.

If you have worked or interned, you know that when you communicate technically, you are not addressing the teacher and often you are not even addressing the boss. Technical writing addresses an audience of users whom you do not know personally. The audience-based orientation of technical communication helps you to write, speak, and design not from the point of view of what you know but from the point of view of what inexperienced or beginning users need to know. Notice how the instructions for planting bulbs in Example 1–6 are aimed directly at the readers, telling them what they need to know. In the two examples of technical writing in

Example 1–7 notice how the user is addressed directly. In Example 1–8, look at the ways that students in beginning technical writing classes are developing an audience-based orientation in their work.

An emphasis on audience-based writing is essential to working successfully in the diverse and multinational arena in which you will find yourself. Look again at the instructions in Example 1–6. Notice that the directions for planting bulbs give depths for planting in centimeters (cm). It's likely that a European audience, familiar with the metric system, would have no difficulty planting bulbs at depths of 5, 10, and 15 cm, but for an American audience raised on inches instead of centimeters, these directions are problematic. Approximating depth when planting bulbs is one

---

EXAMPLE 1–8   Student Examples of Audience-Based Writing

Sample 1—The Letter of Application

Dear Mr. Smith:

This letter is in response to the ad you placed in the *Minneapolis Star Tribune* for the Activities Director position. My experience in planning activities for children, adults with disabilities, and seniors would make me a successful candidate for the position. Let me explain why I would be a contributing leader of your staff.

Sample 2—Food Safety Literature Review

Introduction

Recent food outbreaks have made people concerned about food safety. This review discusses the new food safety techniques and opinions of experts concerning the bacteria and protozoa responsible for the outbreaks. *Clostridium Botulinum* is a concern when sterilized products are contaminated and packaged in modified atmosphere packaging. *Escherichia coli* is a common food contaminate. New systems and tests for eggshells, meat carcasses, and sausage are being researched to better protect consumers. Tests and systems used for *E. coli* can often be used to protect consumers from *Salmonella. Crytosporidium,* a protozoan, does not contaminate food, but the water in the food. No test can detect it in food.

Science is making progress to help prevent outbreaks but cannot offer you a perfect guarantee against outbreaks. Professionals need to communicate to consumers the better advice available to best protect consumers.

Beef, ice cream, water, cheese, and apple juice all have more in common than just being food items. Each also has had an outbreak of disease associated with it. These outbreaks show that common food items can be a danger. Media have focused on these outbreaks by informing you of the callbacks and illness statistics; in doing so, they have caused public concern.

If you are worried about these outbreaks, you need to go to the source of the research and not rely entirely on the media for the expert information . . .

thing, but approximating sizes when engaged in metal fabrication is quite another. Technical communicators need to know what their audience requires to carry out its tasks effectively, and they must provide that information in a usable form for that audience.

FOCUS ON SUBJECT

focus on the subject
the main subject
determines the kind of
information and the form in
which it must be conveyed

Technical communication also differs from other forms of communication in its **focus on the subject**. In technical communication, the main subject determines the kind of information and the form in which it must be conveyed. If the task is to provide instructions on printing a file, the document will be limited to that subject and organized to depict the steps in the printing. In creative writing, one focuses on an object, of course, but that focus doesn't necessarily shape the piece. A poet writing about a flower would not have to organize the poem by having a stanza on petals, another on leaves, and a third on color, but a technical communicator charged with describing flowers might well have a section on petals and one on color.

The intent of the poem is to evoke an emotional response rather than to describe function or organization. With this distinction in mind, look again at creative writing in Example 1–6. Although the poem is about a flower, the poet's focus is the flower's symbolic meaning. In the technical writing example, however, the writer's intent is to instruct a person in the planting process. In the academic writing example, the student's intent is to describe the importance of organic material in soil. Technical communication focuses on the subject in this way because its purpose is to instruct or describe rather than to evoke images or emotions in the reader.

CORPORATE REPRESENTATION

organizational culture
the corporate sense of
identity and value systems

When you are creating documents at work, you do so as a representative of the business or corporation that has hired you. All corporations have their own **organizational culture**, their sense of who they are and what is important to them. Corporations have value systems that cover everything from dress codes to requiring all managers to use the same time management system and to carry the same kind of planning calendar.

Communications within organizations also reflect corporate philosophies, such as safety, open-door management, outreach to the public, environmental awareness, and so on. Corporations try to create a corporate "feel" to their communications so that they convey corporate philosophy uniformly and can be readily identified as having the same source. All communications are expected to reflect corporationwide design considerations. A corporation may decide that all its publications must be in the same color, its illustrations placed in the center of the page, or its logo or watermark appear in the same place and in the same format on all documents, including stationery, presentation overheads, and Web sites.

Corporations, as we will see in later examples, decide how they want to appear to their customers and how their customers will appear to them. This affects the tone and style used in all of their publications. One company may adopt a friendly, down-home approach to its readers while a second company may choose a more formal, respectful approach. Working toward a uniform corporate identity is often called branding or **brand identification,** which is designed to convey specific information to consumers at a glance. Because what you create is read or viewed by people inside and outside the company, it will need to represent the company in keeping with its corporate culture.

brand identification
a way businesses try to project their corporate identity in all their products and documents

Generally, developing an awareness of yourself as representing the corporation through your documents takes awhile, and employers will wait and watch to see if new hires make the transition from student to organizational member. This transition is reflected in your awareness of corporate culture and your ability to make your documents fit that company's vision of itself.

A word of caution is needed, however. This book is not trying to turn you into a good corporate clone when it points out the importance of understanding corporate identity in the creation of successful documents. Sometimes technical communicators believe that seeing themselves as corporate representatives will stifle all of their individuality. You do need to understand a corporation's culture to be an effective member in it, but as an effective communicator, you also need to remember that your writing has the potential to change the company just as the company has the potential to affect your writing. Your ability to create change within your corporate workplace is one of the things that makes developing your communication skills so important.

COLLABORATION

As we noted earlier, much communication, particularly writing, in business, industry, government, and education is collaborative. Because people who convey information about technology have special skills, specific education and experience, or special knowledge, the most effective documents are often produced by people working in **collaboration.** For example, a computer manual might require the input of the software developer, the systems analyst, the technical communicator, the document manager, and the graphics specialist. One person cannot know it all and write everything, and many people in an organization will review your work. In fact, your organization's name rather than your own may appear on the final product.

collaboration
many people with different skills working together on a task or document

Effective collaboration means listening as well as stating your point, understanding others' points of view as well as expressing your own, and working and writing together as well as alone. Technology, particularly computers linked to the Internet and to intranets, makes collaboration easier by facilitating the sending and editing of documents, but the computer can make it harder to ensure that interpersonal relations build and grow as they can during face-to-face communication. We will address such relations as we go along.

STYLE AND ORGANIZATION IN TECHNICAL COMMUNICATION

Technical communication has special stylistic and organizational features that distinguish it from other forms of communication. For example, technical writing often reveals the topic and a position on that topic at the very beginning of a document. Suspense is a characteristic of some creative writing, but it's definitely not part of technical communication style. Your reader will want a preview of what is to come in the document, not a surprise to be discovered after ten pages of reading. Previews are important because some readers may elect not to read the document but pass it on to others. Much of the writing that you have done in the past relied on **inductive organization**—you described or listed ideas and evidence and then came to a conclusion. Technical communication uses **deductive organization** a great deal; you reveal your conclusions and then demonstrate how you reached them.

inductive organization
an organizational pattern in which you give evidence and then come to a conclusion

deductive organization
an organizational pattern that begins with conclusions and then demonstrates how they were reached

As a technical communicator, you'll also need to know how to communicate directly and concisely so that your audience can easily and safely make use of your ideas. In the past many people found technical writing very formal and distant, an impression created in part by an excessive use of the passive voice. Readers of technical documents encountered passages like the following:

> Explanation was needed for the striking performance between KN-211 subsurface and KN-135 subsurface. There was a need to find the cause of such difference in performance when the starting oil compositions appeared similar.

Today, most technical communicators try to write more directly and personally, as in the rewritten example below:

> The striking difference in performance between KN-211 subsurface and KN-135 subsurface clearly needed explanation. What caused such difference in performance when the starting oil compositions appeared similar? (Bragg et al., 59)

These rewritten sentences are much easier to read, rely more on the active voice, and engage the reader more effectively.

Even when describing a product or process or when recording data, you need to persuade your audience that the description or record is complete and appropriate. As a technical communicator, you'll begin and end with conclusions and recommendations, and you will use the discussion segments of your document to convince your reader to accept and act upon your conclusions and recommendations.

VISUAL DISPLAY IN TECHNICAL COMMUNICATION

To an extent that you may find surprising, technical communicators rely on **visual display** of words and data to help their audience understand the subject. By collaborating with a graphic artist or designer and using computer software to generate displays, you can help your audience follow

visual display
the design or presentation of words or data on a page

your communication. To see how visual display helps a reader to navigate a text, see Examples 1-9 and 1-10. The first example forces a reader to sort through information. Example 1-10 uses a visual display of information to help a reader understand and act upon what he or she has read.

A reader relies on visual cues to process information, and part of becoming an effective technical communicator is to master the repertoire of visual display tools available as you create documents. For example, in creating your own technical documents, you can use visual display to group information into different units. These groups are often separated by white space or marked by visual cues such as bullets or numbers. In Example 1-10, types of material—poetry, prose, illustration, and special works—are grouped in bulleted lists. You can also use visual display to queue information or order groups hierarchically to convey order and importance. In Example 1-10, for instance, Fair Use Guidelines and Limitations appear as major headings because they appear in boldface and are less indented than the lists that follow them.

EXAMPLE 1–9 A Document Lacking Attention to Visual Display

Subject: Potential Copyrighting Infringement Relative to Course Materials

I am sure that you have received some word that we need to update our copy guidelines for course material. A number of faculty members require that students purchase copies of essays that aren't available in textbooks. Sometimes faculty members just want students to read one essay or example out of a book and don't want to require that students buy the entire book. Ever since the much publicized Kinko's Graphic Corporation court case, we have worried about our own on-campus copying service violating copyright laws. We need to make sure that all faculty members who have students buy duplicated or copied articles do not violate the "fair-use" guidelines. Under those "fair-use" guidelines, you can copy a portion of the copyrighted material without copyright infringement if you are using the copies of the work for a nonprofit educational purpose, or if the time at which you were inspired to use the material and the time when you needed it were so close that you did not have time to request permission to copy it.

However, there are some limitations on these guidelines. For example, if you are copying poetry, you can't copy more than 250 words or two pages. But if you are copying prose, you can't copy a complete article, story, or essay if it is over 2,500 words. If it is more than 2,500 words, you can copy up to 10 percent or a maximum of 1,000 words. If you are copying illustrations, such as a chart, diagram, drawing, and such, you can copy only one per book or per periodical issue. Finally, if you are copying from a work that combines text and illustrations (like a children's book or poetry), you can't copy more than 10 percent of the words of the text. If you want to copy more, then you have to ask for permission.

Please learn these guidelines because, as of December 15, our campus copy center will be following them.

EXAMPLE 1–10   Document with Attention to Visual Display

DATE: October 20, 1999
  TO: All Faculty Members
FROM: Vice President Kofflin
  RE: Copyright and Fair-Use Guidelines for Course Material—Effective December 15, 1999

On December 15, 1999, the following guidelines will be in effect for the campus copying center. All faculty members must follow these guidelines when requesting that the campus copying center duplicate and sell course material to students. Failure to follow these guidelines could result in litigation between the University, faculty members, and copyright holders (for example, Kinko's Graphic Corporation court case).
Fair-Use Guidelines: You may copy a portion of a copyrighted work without copyright infringement if:

- You are using your copies of the work for a nonprofit educational purpose.
- The time at which you were inspired to use that copyrighted work and the time at which you will actually use it are so close that a permission request could not be processed (usually two months' time or less).

Limitations: The following limitations on the portion of the copyrighted work must be applied to the guidelines above:

- *Poetry*—No more than 250 words or two pages.
- *Prose*—Complete articles, stories, or essays that are less than 2,500 words may be copied. If the work is longer, no more than 10 percent or a maximum of 1,000 words.
- *Illustrations*—No more than one chart, diagram, drawing, graph, cartoon, or picture per book or per periodical issue.
- *Special works*—No more than 10 percent of the words in the text in works that combine text with illustration, such as children's books or poetry.

If you have any questions about these guidelines, please contact the campus copy center at 555-1224. Again, these guidelines will be in effect as of December 15, 1999.

Technical communicators help readers filter information by seeing differences and similarities between units or queues. For example, readers see that elements with the same type size, darkness, or placement are similar or in the same category. In Example 1–10, the reader sees that the categories Fair-Use Guidelines and Limitations are of equal importance. Finally, your readers will visually abstract the purpose of the elements in a document based on the elements' relationship to the entire document. In Example 1–10, readers will know that because it comes at the top of the document, following the RE: (for regarding) cue, Copyright and Fair-Use Guidelines for Course Material is the title or subject of the entire document (Martin).

Your ability to work with visual display is becoming increasingly important because so many documents today are designed to be displayed online or in Web pages. Although the style requirements of a Web page differ from those of a printed page, as we will see later, neither document can be effective unless its creator attends to how the information is displayed visually on the page or on the screen. Example 1–11 is a screen

EXAMPLE 1–11   Web Page for an Online Course

Rhet 3400 Managing Information on the Internet

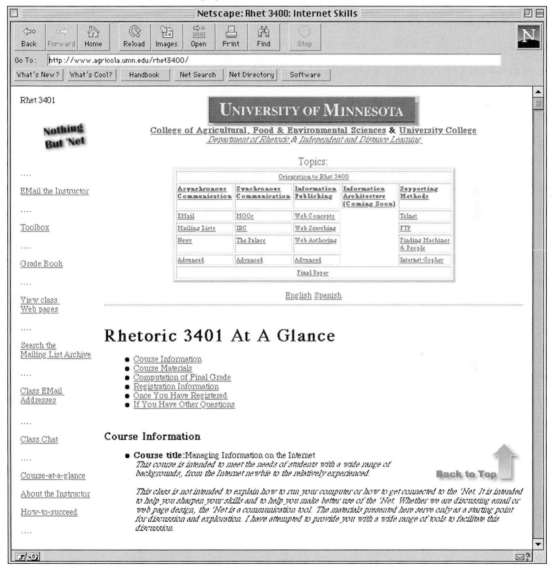

If you would like to see this site in color, you can visit it at <http://www.agricola.umn.edu/rhet3400>

shot of the introductory page of an online course, Managing Information on the Internet, taught at the University of Minnesota. Notice how technical communicators working on the Web must juggle functionality, content, and visual display.

**The Changing
Nature of the
Workplace**

As we mentioned earlier, because technical communication involves people—designers, engineers, managers, advertisers, clients, customers, and users—we say it is socially situated. As a consequence, you must understand the nature of the workplace to be an effective technical communicator.

Constant change marks life in the last part of the 20th century, and nowhere is that change more apparent than in the corporate world. One obvious change is the way information keeps increasing in the workplace. Think about the epigraph at the beginning of this chapter: "A web of glass spans the globe. Through it, brief sparks of light incessantly fly, linking machines chip to chip and people face to face" (Cerf). Until 125 years ago, there wasn't even a single bridge crossing the Mississippi River. Goods going from the East Coast to the West Coast had to stop at the river until they were ferried across and could continue on their way. The web of glass Cerf speaks of has done much to obliterate time and space for individuals and for businesses. This era is marked by enormous transnational data flows.

By 1993 worldwide communication on the Internet alone included 39,000 networks in 100 countries, with 5,000,000 users worldwide (Marine et al., 1). By 1998 there were between 35 and 37 million users of the Internet worldwide (with 83 percent in the United States), and 14.7 million U.S. households had access to the Web directly and another 8.9 million had access through online service providers. Currently 64 percent of Americans have access to the Web at home, and 37 percent of households in the United States have personal computers. Eighty percent of people with access to the Internet access it daily and 50 percent spend 10–20 hours a week online. In 1997, 1.8 million U.S. businesses were online, up from 1.1 million in 1996. By 2001 it is projected that 92 percent, or 5.9 million, of U.S. businesses will be online. There were only 23,500 commercial sites at the end of 1995 but by the end of 1996 there were 95,000 sites. Internet business is truly global—more than 10,000 commercial Web sites are offered in Japanese (Global Sight @www.global-sight.com, August 1998).

New communication technologies affect the ways in which we communicate and how we interact with each other. The Internet, combined with cellular phone service, fax machines, and fax modems, has altered how and where we send and receive information. Desktop videoconferencing is spreading as the price of desktop minicams has come down to about $50 for black and white and $200 for color. Computers have given many people the opportunity to design their own newsletters and brochures, to compile and send complex information, and to edit and revise documents fast and efficiently. Word processing and HTML editing programs have made it relatively easy for the average office worker to create Web pages for display on the Internet.

Another important change is the increasing diversity in the workplace in the United States. According to *Workforce 2000*, nearly two-thirds of all people entering the workforce between 1990 and the year 2000 will be women. The workforce is also becoming more varied in terms of race, ethnicity, and cultural backgrounds, and many businesses have both employees and customers who do not speak English as their first language or who have difficulty reading English. In the near future, immigrants will represent the largest share of the increase in the workforce since World War I. In 1965 America was about 80 percent white and 10 percent black. By 2050 America will be 26 percent Hispanic, 14 percent black, 8 percent Asian, and 52 percent non-Hispanic white. These changes have come about as a result of the 1965 Immigration and Nationality Act that ended quotas based on national origin that were in place since 1924 to keep the country from changing its ethnic and racial makeup (Tilove) and as a result of differing birthrates among ethnic and racial populations in the United States.

Throughout the 1990s approximately 600,000 immigrants annually will have entered the United States and two-thirds of them will enter the workforce. By the year 2000, nonwhites, women, and immigrants will make up more than 83 percent of the additions to the workforce (Johnston and Packer). Where the immigrants come from documents the increasing diversity of the U.S. population and workforce. According to the *Statistical Abstract of the United States: 1998*, the 915,900 immigrants to the United States in 1996 came from the following continents:

Europe	147,600	16.1%
Asia	307,800	33.6%
North America (including Mexico and Central America)	340,500	37.1%
South America	61,800	6.7%
Africa	52,900	5.8%

This diversity in the workplace is augmented by the increasingly global nature of business. Doing business in the global marketplace means dealing with people who have different cultures, languages, and business practices, who live in different time zones, but who all have need for the latest technology to improve their environment and lives and to compete in their own marketplaces. It also means learning to navigate through a sea of legal, ethical, and environmental considerations. Therefore, technical communicators must develop an international perspective to perform their work effectively.

THE GLOBAL CORPORATE WORLD

Many corporations are already multinational and global in nature and have offices in numerous countries, with suppliers and customers from around the world. This change profoundly affects the way these organizations carry out their communication. New trade agreements and free trade zones mean that businesses must become global. The globalization of business is easily seen in the interconnectedness of the world's stock markets. Bad economic news from one country causes worldwide financial reactions. Although short lived, in the late 1990s, for example, as the economies of the Pacific Rim nations experienced financial crises, the stock markets of the United States and Europe reacted in a volatile way.

In addition to considering international monetary conditions and developing global communication systems, corporations must consider cultural diversity in new ways. For example, much has been made of the difference in work habits between Japanese and U.S. employees and managers. These cultural differences, as well as those in verbal and nonverbal communication, views of time and promptness, and cultural and religious observances, cannot be ignored if companies are to do business on a global scale.

For example, differences in the calendar and how the work week is divided are significant issues with which businesses must contend. For example, May 23rd may not mean much to most American workers, but to Canadian citizens it's Victoria Day, and businesses are closed for the holiday. In Israel, the Sabbath is Saturday, and all businesses close by sundown Friday night. No public transportation is available from Friday night until Saturday night except taxis, which are driven by Arabs for whom Saturday is not the Sabbath. As a technical communicator, you'll need to be sensitive to cultural diversity, and that means sometimes adopting concepts and processes as appropriate.

In the global marketplace, businesses are affected by changing practices. Products are less likely to be created from parts manufactured on the premises or even in the nation. As a result, national and global standards for manufacturing and for communicating exist both within countries and across them. The **American National Standards Institute (ANSI)** makes available standards that are adopted by business and industry in the United States. Even though businesses are not required to follow these standards, the standards represent what most businesses think desirable. As a technical communicator, you'll need to know these standards because they cover such conventions as the design of safety signs and labels.

American National Standards Institute (ANSI) sponsor of quality and reliability standards

Standards provided by groups such as the **International Standards Organization (ISO)** also affect how you'll create documents and how your company will do business. The ISO 9000 series, for example, makes quality control of supplies and manufacturing processes possible. These standards also include guidelines for the content and visual display of the documents you'll create for your company.

International Standards Organization (ISO) sponsor of international quality and reliability standards

In general, ISO and ANSI standards improve quality control and communication because they allow for the production of generic or uniform processed materials. These materials can be incorporated into manufactured parts or hardware assemblies. Again, these standards cover everything from units of measure, pipe fittings and valves, fasteners and screw threads, to the content of documents. There are, for example, ISO standards for vocabulary and for common symbols. They provide guidelines for how documents are to be formatted (paper type and size, holes, representation of dates and times) and standards for document reproduction (typeface, paper, required signature forms).

Technical writers have reported that globalization has profoundly changed the way they think of their documents. For example, Steve Gillispie, a technical writer for a U.S. elevator company, reports that

> his company recently found itself installing Korean-made escalators in Mexico. U.S. engineers were doing the installation; Mexican engineers and technicians would later be doing the maintenance and repair. In the contract for purchasing the escalators, the elevator company had carefully included a clause requiring the Korean manufacturer to provide an English translation of the installation and maintenance instructions. These instructions were *organized* so differently from those the engineers were used to that they were virtually useless. Technical writers in the home office hastily rewrote the manuals, faxing each page as they finished it to Mexico so that the engineers on-site could proceed with the installation. However, the writers also had to consider whether even these rewritten instructions would be suitable for use by the Mexican engineers who would subsequently maintain the equipment, or if yet another version of the manual would be necessary. (Thrush, 273)

Communicating with Other Cultures

The word **culture** refers to socially constructed and transmitted behaviors, beliefs, values, and institutions. Culture affects all aspects of human life from telling people what time to eat the big meal of the day to acceptable dress for a job interview. The rules of a culture are known to its people in an almost unconscious way, so it's hard for someone entering into a new culture to grasp the rules and their significance. In Japan, for example, people arrange themselves around a business table in a hierarchical fashion, derived from the way people would be seated at the Japanese tea ceremony, a central ceremony in cultural life. In the United States we save a spot at the head of the table for the boss, much as it is often reserved at our dinner table for the head of the family.

culture
all socially contructed behaviors, beliefs, values, and institutions

In turn, communication codes do not easily transfer from one culture to another; in the United States, we associate death with the color black, but in some Asian cultures mourners wear white instead. Technical communicators who are not aware of cultural rules or who work from their own set of unconscious rules invite insult and misunderstanding. Cultural awareness is of critical importance in the world market and in our electronic communications that reach a global audience.

Diversity and Levels of Literacy

Although some people tend to think of corporate employees in terms of the corporate white *man,* such images are out of date. In the United States white men have been a minority in the workforce since 1980. By the year 2000, white men will make up less than one-fifth of the new entrants to the labor force (Hage). The labor force that you will be entering will have more women, African Americans, Asian Americans, Hispanics, and other groups than ever before. Moreover, these groups are going to bring with them new ideas, values, and perspectives on how things can be done.

This diversity represents great challenges and opportunities to people in the business world and the way they communicate. Sensitivity to language, a broader repertoire of problem-solving techniques, and an awareness of gender, race, and cultural differences in communication are essential for success as a communicator. The United States is a remarkably diverse and rich environment in which to work. Japan, in contrast, is homogeneous, with more than 99 percent of its population of Japanese origin and therefore with almost identical education and customs. The United States has one of the least homogeneous populations in the world, having brought together people from all over. This diversity may make life richer and more interesting, but it also means that to communicate successfully, one has to consider a wider range of variables.

Diversity isn't the only variable that affects communicators. Literacy levels in the United States are quite low. The United States Department of Education analyzes to what extent people in the United States possess document literacy: the knowledge and skills that enable a person to locate and use information that exists in such places as indexes and on the stubs of our paychecks. The *Digest of Education Statistics* indicated that although 95.5 percent of the population in 1991 could perform simple reading tasks such as matching coupons with a grocery list, only 57.2 percent could follow directions using a map and only 20.2 percent could decipher a bus schedule.

Our declining ability to use practical information, coupled with the increasing number of U.S. citizens who speak English as a second language or have difficulty understanding English at all, creates special problems as we seek to communicate technical information to the population at large. As we move through this book, we examine a number of ways you will be able to improve your communication skills to reach this increasingly diverse population.

TECHNOLOGICAL IMPACT

One focus of this book is on changing technologies and their impact on communicative processes. Knowledge of both traditional and emerging technologies is essential for success in your field. For example, many large

EXAMPLE 1–12 How People Communicate in the Workplace

Method of Communication	Percentage of Time
Communicating face to face	37%
Using new media	23%
Using conventional media	12%
Conducting other activities	28%

Source: Jeanne W. Halpern. "An Electronic Odyssey." *Writing in Nonacademic Settings*. Ed. Lee Odell and Dixie Goswami (New York: Guilford Press, 1985), p. 163. Adapted by permission of the Guilford Press.

and small businesses and industries not only use communication technologies to plan activities, communicate with far-flung staff, and create print, multimedia, and online materials, but also market these communication technologies and train workers to use them. As Example 1–12 shows, people generally spend nearly twice as much time using electronic media as using conventional media such as pens, pencils, and typewriters (Halpern).

Corporations are investing heavily in new communication media because they help in decision making, in collaboration, and in access, storage, and transfer of information among distant sites. Because the business world relies increasingly on high-technology products and processes, you will be surrounded by these technologies and expected to use them in your work.

Technological changes in communication practice usually affect the business environment first. The telephone, for example, was designed for business communication. No one predicted that it would move into the home and have such a prominent role there. Similarly, the computer, the fax, and the laser printer are all moving from the business environment into the home. Thus, you will be exposed to an ongoing array of changes in communication technology with which you have had no prior experience at home. You will be expected to incorporate these devices and processes into your communication procedures quickly. And you will be called upon to explain to others how to use them.

To help you develop the skills you need, we will examine how communication technologies will affect your work and the kind of documents you will be expected to create. You will be expected to understand the computer and its use for collaborative decision making and collaborative writing; you will be expected to have at least minimal expertise with word processing and document design, including some skill with graphics programs used for creating print documents and programs for presentation graphics. You will be expected to make judgments about what medium would best present a particular message.

The widespread use of the Internet and intranets means that as a communicator you will be expected to know how to create electronic communication such as e-mail and to send documents as enclosures. As you conduct research for your company, you will need to rely on your Websearching techniques. Your company will expect you to know and use effectively a variety of presentation software and to have some skill in videoconferencing. Technology is always changing, and you will be expected to keep up with the new and emerging technologies in the workplace.

SOCIAL IMPACT

Because of the complexity of current life and the rate at which change occurs in it, people often feel that they cannot make a difference. Many people who work for large corporations feel that they are basically anonymous workers whose contributions to the corporation are not of a personal nature. Yet, as we understand more about the role communicators and communication play in an organization, we realize that people can and do make a difference and that they can initiate and carry out change within organizations. Sometimes those changes are not good because writers and other communicators fail to see the connection between their acts and the larger world. The human loss and pain associated with the *Challenger* accident, the Johns Manville asbestos case, or the Dow-Corning silicon breast implant case, among others, underline how important communication can be.

What we write—or don't write—and how effective our communications are profoundly affect others. Engineers involved with the *Challenger* wrote memoranda explaining their reservations about a cold weather launch, but found themselves overridden by other managers. The result was an explosion, loss of life, ruined careers, and a horrified nation who watched the explosion replayed on television. If we do not write or communicate effectively, lives may be lost and profits may diminish.

SUMMARY

- This chapter provided a brief introduction to technical communication and its importance in the changing work environment. This environment is affected by rapid technological change, increasing numbers of women and minorities in the workplace, and the global nature of the economy. As a technical communicator, your task is to help people cope with technological change and make the most of technology itself in ethically and socially responsible ways.

- Technical communication has a long history of serving people by providing useful information. Technical communicators in the past, like those today, have worked to make technology accessible, account for the use of natural resources, and promote and explain new communication technologies. In current times technical communicators work in a variety of fields from computer science, to natural resources, to biomedical technology, to technical sales and marketing. They are knowledgeable about and competent in working with traditional and emerging communication technologies as they create a variety of print, oral, visual, and electronic documents.

- Technical communication differs from creative and academic writing because it is designed to help others make decisions and complete tasks. It is distinguished from other forms of writing by being
  — Audience-based
  — Focused on the subject
  — Dependent upon technology (both for its subject matter and its means of production)
  — Representative of the corporate culture that produces it
  — Collaborative
  — Marked by its technical style and visual display
- Because technical communication is a socially situated activity, practitioners need to be aware of the legal, ethical, and social dimensions of their work. They need to know about product liability law and how to create effective communications for a diverse workplace and a global marketplace.
- In addition, technical communicators must be at home in a world of changing technologies. The Internet alone, with its 37 million users worldwide, is profoundly affecting the documents technical communicators must create and the way the texts they create are distributed, displayed, and used.
- No longer can technical communicators rely simply on knowing a format or two used in business communications. Instead, they must have a whole repertoire of tools for creating effective communications to reach an increasingly diverse and multicultural audience.

---

1. Example 1–13 is a section of directions on how to use a self-injection system to combat migraines. Analyze this example for ways it is audience-based communication and ways in which it focuses on the subject. Be prepared to discuss your findings in class.

2. Example 1–6 showed three different kinds of writing—creative, academic, and technical. Technical communication is supposed to be clear and concise so that it can be easily used. Reread the instructions for planting flower bulbs (reprinted below). These are the actual instructions that came with a bag of mixed bulbs packaged in Holland that were intended for planting in a Minnesota garden.

ACTIVITIES AND EXERCISES

EXAMPLE 1–13   Sample Directions

How to Use the Self-Injecting Unit

- BEFORE INJECTING—Clean the skin area to be injected (top of the thigh or the outside of the upper arm).
- INJECTING—Pick up the self-injecting unit and pull off the blue cap. This will remove the gray needle guard.
- Pull black safety lock away from the red release button. DO NOT TOUCH THE RED RELEASE BUTTON. The safety lock is now OFF.
- Place the gray tip of the self-injecting unit securely on the skin area to be injected.
- Hold the self-injecting unit steady. Push in the red release button until you hear a click. Hold the unit against the skin at least 5 seconds before removing the unit from the skin. If the unit is removed from the skin too soon, not all the medicine will be released.
- Slowly remove the self-injecting unit. Do not touch the needle.

Source: Cerenex Pharmaceuticals. Division of Glaxo Inc.

These directions seem to be clear, and they certainly are short, but are they as well written as they can be? What problems do you see with these instructions? Where are they ambiguous? Where would a naive user run into trouble? How could you make them better without making them long and complex?

Technical Writing—Directions

Flower bulbs will do well in any good, well-drained soil. Plant them, with top up, at the depth appropriate to their type as indicated in the chart below.

Tulips	15 cm
Daffodils and hyacinths	10 cm
Crocuses	5 cm

---

3. Corporations and businesses are concerned about having a well-defined identity or "brand." Choose a large organization in which you are interested—for example, a car or computer company, a cruise line, a credit card or airline company, a bank or a cereal manufacturer. Gather several examples of their documents from magazines and newspapers, watch for their ads on television, and visit their Web sites. How does the organization describe itself? What are the identifying characteristics of its documents? How do those identifying characteristics convey the corporate identity or brand? What can you find in the documents that differentiate your organization from one of its competitors (Ford's corporate identity from GM's, for example)?

   In a brief oral presentation to your class, describe what you have found out about corporate culture as shown in the corporation's documents.

EXAMPLE 1–14   Adding the Features of Technical Communication

Some people call the photoreceptor the photoconductor—it's all the same. The photoreceptor is covered with photoconductive material that will conduct electricity in the light but not in the dark. The photoreceptor accepts the latent electrostatic image and provides the surface where this image can be developed and then the image is transferred to paper. The whole process begins with the charge—which happens when the photoreceptor is exposed to an electrostatic charge. This charge goes across its entire surface and is uniform. And then the photoreceptor is exposed to the image of a document—this image is projected onto the photoreceptor. The light that comes from the white background areas of that document dissipates the electrostatic charge on the photoreceptor surface. The charge remains on the photoreceptor in the dark areas where no light has been reflected from the original. A sheet of paper is placed in contact with the photoreceptor and the image is transferred. But before that happens, a developing substance is spread over the photoreceptor surface. The particles of this substance adhere to the projected image areas because of electrostatic attraction. Then again during transfer, the toner particles are transferred from the photoreceptor to the paper by imparting an electrical charge of opposite polarity to the paper surface. Finally, the surface of the photoreceptor is cleaned of residual particles after the image is fused to the paper by heat. If it is not fused by heat, it can be fused by pressure. To understand this process, you have to remember that the developing substance used here is often called the toner and that opposite electrical charges attract, while similar ones repel. Now you know all about the photoreceptor in the copying process. Remember that the photoreceptor provides the medium for forming an image that is developed and transferred onto paper.

---

4. Test the communication requirements of your future profession by interviewing a professor who is active in consulting or research as well as teaching, a graduate student who also has some workplace experience, or a professional in your community. Ask what media (oral, graphic, online, or written) they use most frequently. Ask whether the person ever collaborates with others in planning, writing, or revising. Ask whether the organization in which the person communicates has a prescribed style or format (formal or informal; letters, phone calls, memos, reports, and such).

   Display the information you gather during your interview in a table or chart. Be prepared to describe your graphic display orally to your instructor and the students in your class.

5. Example 1–14 contains a piece of writing that was written much like an essay. Working in pairs during class time, revise the passage so it has the necessary features of technical communication.

6. In Appendix A, Case Documents 2, you will find five documents written by the Exxon Company and their associates to explain the bioremediation process used to clean up the 1989 *Exxon Valdez* oil spill in Prince William Sound. The documents are (*a*) "How Tiny Organisms Helped Clean Prince William Sound" (look at the first four paragraphs), (*b*) the Executive Summary of "A Report Detailing the Effectiveness and Safety of Bioremediation in Prince William Sound," (*c*) the introduction to "Bioremediation Technology Development and

Application to the Alaskan Spill," (*d*) "The Prince William Sound Bioremediation Story," and (*e*) "Bioremediation for Shoreline Cleanup Following the 1989 Alaskan Oil Spill" (look at the first section, "The Bioremediation Process"). Choose two of these documents that interest you and read them thoroughly. Then in Appendix A, Case Documents 4, you will find Web sites designed to address the Y2K (Year 2000) problem. Choose one of these Web sites to examine and study it thoroughly.

After you have examined these sources, in a brief report to your classmates, answer the following questions about Case Documents 4 and the Web site you selected:

- What are the audience, purpose, and situation for each communication? How are they similar and how are they different?
- What aspects of a document differ when it is online or in print? What role do color and visual display play in these documents?
- Do the writers use inductive and/or deductive organization?
- What characteristics of technical communication do you see in each communication?
- Finally, if you were to rank these communications in terms of clarity, conciseness, effective visual display, focus on the subject, reader-based orientation, and so on, is one communication more effective than the others?

7. Visit two or three of the following Web sites devoted to different aspects of technical communication:

- Society for Technical Communication (http://www.stc-va.org)
- Association of Teachers of Technical Writing (http://english.ttu.edu/ATTW)
- Bibliographic and WWW Resources in International Technical Communication (http://www.nmsu.edu/techprof/attwrsrc/title.html)
- Technical Communication Quarterly (http://rhet.agri.umn.edu/~tcq)

In a memo to your instructor, describe what you have learned about the field of technical communication, its practitioners, salaries, job expectations, and publications.

WORKS CITED

Aristotle. *Rhetoric,* Book I. In *The Rhetorical Tradition: Readings from Classical Times to the Present.* Eds. Patricia Bizzell and Bruce Herzberg. Boston: Bedford Books, 1990.

Baker, Russell. *Growing Up.* New York: Plume, 1983.

Bragg, James R., et al. *Bioremediation for Shoreline Cleanup Following the 1989 Alaskan Oil Spill.* Florham Park, NJ: Exxon Research and Engineering Company, 1992.

Cerf, Vinton G. "Networks." *Scientific American,* September 1991, p. 72.

Connors, Robert J. "The Rise of Technical Writing Instruction in America." *Journal of Technical Writing and Communication* 12 (1982), pp. 329–51.

Digest of Education Statistics. 1991.

Frontinus. *The Stratagems and the Aqueducts of Rome*, vol. II. Ed. Mary B. McElwain. Transl. Charles E. Bennett. Cambridge: Harvard University Press, 1930, pp. 70–72.

Giroux, Henry. *Border Crossings: Cultural Workers and the Politics of Education.* New York: Routledge, 1992.

Global Sight Web page: <http://www.global-sight.com>.

Hage, Dave. "White Male Influence Waning?" *Minneapolis Star Tribune*, February 23, 1988, sec. D, pp. 1–4.

Halpern, Jeanne W. "An Electronic Odyssey." *Writing in Nonacademic Settings.* Ed. Lee Odell and Dixie Goswami. New York: Guilford Press, 1985, pp. 157–89.

Johnston, William B., and Arnold H. Packer. *Workforce 2000: Work and Workers for the 21st Century.* Indianapolis: Hudson Institute, 1987.

Lunsford, Andrea, and Lisa Ede. *Singular Texts/Plural Authors: Perspectives on Collaborative Writing.* Carbondale: Southern Illinois University Press, 1990.

Marine, April; Susan Kirkpatrick; Vivian Neou; and Carol Ward. *Internet: Getting Started.* Englewood Cliffs, NJ: Prentice Hall, 1993.

Martin, Marilyn. "The Visual Hierarchy of Documents." *Proceedings from the International Technical Communication Conference*, VC-32-35. 1989.

Pliny the Elder. *Natural History*, 13. Transl. John F. Healy. New York: Penguin Books, 1991, pp. 175–76.

Statistical Abstract of the United States: 1998.

Tebeaux, Elizabeth, and M. Jimmie Killingsworth. "Expanding and Redirecting Historical Research in Technical Writing: In Search of Our Past." *Technical Communication Quarterly* 1, no. 2 (1992), pp. 5–32.

Thrush, Emily A. "Bridging the Gaps: Technical Communication in an International and Multicultural Society." *Technical Communication Quarterly* 2, no. 3 (1993): pp. 271–83.

Tilove, Jonathan. "The Coming White Minority." *Minneapolis Star Tribune,* July 13, 1998, p. A6.

U.S. News and World Report. *1999 Annual Guide: Best Jobs for the Future.* "20 Hot Job Tracks." p. 85.

2

The Writer and the Writing Processes in the Workplace

Introduction ❙ Writing as Process in the Social Context
of the Workplace ❙ Writing as Problem Solving
in the Workplace ❙ An Overview of Writing Processes
in the Workplace ❙ Some Writing Process Strategies to Try

Workplace literacy includes not only the traditionally defined literacies of reading, writing, and math, but also computer skills, oral communication, teamwork, problem solving, and effective interpersonal communication. The workplace needs people who can apply these skills creatively and in combination to get things done.
Paul R. Meyer and Stephen A. Bernhardt, "Workplace Realities and the Technical Communication Curriculum: A Call for Change," in *Foundations for Teaching Technical Communication: Theory, Practice, and Program Design*, Katherine Staples and Cezar Ornatowski, eds. (Greenwich, CT: Ablex, 1997), p. 86.

Now that you understand more about the modern workplace, let's look at how technical communication can solve problems found there. By definition technical communication is *applied communication*; that is, it's designed to *accomplish specific tasks* or help an audience *solve specific problems*. And there are as many purposes for writing, audiences for written documents, and kinds of writing as there are problems in the typical workplace—in other words, plenty! **Introduction**

None of these challenges makes it easy on individuals who write as a major part of their jobs—and most employees are in this situation. Imagine yourself, for example, in your first job, a few months after your graduation from a program in mechanical engineering. After completing your initial on-the-job training, you've been meeting with your first work team, a group of mechanical and chemical engineers like you who are charged with finding a more dependable means of applying a protective coating to new automobiles. Your company is engaged in a research-and-development effort to find a coating that will protect the finish of new automobiles from stones, gravel, water, ice, and other enemies. The company has developed some promising possibilities, but it needs to make sure that those possibilities are practical: Workers on the assembly line must be able to spray this new coating easily onto finished automobiles.

That's where you and your small group of co-workers come in. All of you have been brought together to test various kinds of spray guns to see if the new coatings can be applied effectively by such devices. You've been working on the project for a few weeks now, testing various spray guns with various formulas of the new coating, when your supervisor asks you to write a progress report on how things are going. "These things are routine, so don't worry about it too much," she tells you. "I just need to know how you are coming along on the project so that I can keep my own bosses informed on how things are progressing. A few pages will be just fine. By Friday." Then she vanishes and leaves you to write the report.

You say to yourself, "What do you mean, 'don't worry'? It's my first report! And how can you say it's 'routine'? How can it be routine if I've never done one before and have no idea what you and your bosses want or need? And Friday is only three days away!"

What *do* readers want and need? Where *do* writers begin a document like this? How *do* writers make sure the job is done on time? How *can* major writing tasks be broken up into smaller, more manageable substeps or processes? Although many writers might begin by knocking out a few sentences in something of a panic, we suggest that you take a more deliberate approach—that you begin by doing some careful planning before you actually draft any sentences. Establish clear objectives, and plan the production process. No matter how pressed you are for time and no matter how sure you are of what you are doing, you will probably have better success if you begin by collecting your thoughts on what you need to accomplish and how you need to approach the task. Plan well, and you are likely to find the actual task of writing going more smoothly. Your decision making throughout the process of developing your document is best guided by a clear plan with a clear set of objectives and a vision of the writing processes that you will use.

This chapter is designed to help you both *think* through and *work* through the major writing tasks in systematic ways. In the first part of this chapter, we offer you two important conceptual tools for thinking about major workplace writing projects. We encourage you to think about such projects as

- A set of processes that occurs in complex social contexts and that typically involves the efforts of many people.
- A problem-solving activity that demands a great deal of thinking about purpose and audience.

In the second part of the chapter, we provide two resources to help you work through major writing projects:

- An overview of the typical components involved in the writing process (although, we add with caution, there are many such processes, not one).
- A number of practical, process-based writing strategies for undertaking major projects. These strategies should help you think of such projects in smaller, more manageable increments.

Writing as Process in the Social Context of the Workplace

writing process
the activities involved in the writing task, including exploring, generating, and representing information; planning documents and arriving at goals; structuring and organizing; drafting; revising; and editing

Within the workplace, you will probably produce a wide range of drafts for major writing projects. Technical communicators produce these drafts because the **writing process** in the workplace is a **social process**. By "social," we mean many people are involved in helping a writer plan, draft, and revise. At times you may sit at your desk or in front of your computer alone and write, but the opinions, requirements, and needs of your supervisors, co-workers, and clients will influence your writing.

Depending on the organization where you work, you might well submit plans and drafts to other people for reading and reaction. You may seek additional information from other people that may figure into the revision of documents. You may test your documents on multiple users who are representative of the various audiences your communication will eventually have. And you may design your document using procedures prescribed by your organization's publications group (the professional writers, editors, and graphics staff).

Moreover, you will be writing within specific circumstances. For example, let's say that you are working with your manager to plan a report on purchasing new pollution-control devices for your company's factories. You'll need to read governmental guidelines for such devices, assess local community reactions by reading newspaper articles and interviewing individuals, read previous studies done by your organization and others on the topic, check with the financial department on your budget for the project, and read scientific studies on the effectiveness of the pollution-devices you'll recommend.

You'll probably have a co-worker or editor give you an opinion on your first draft, perhaps someone who can help you decide how much technical detail should go into such a report. The graphics staff may assist you in producing technical drawings of the pollution-devices. And, finally, you will need to ask your manager about the company's position and history with pollution-control devices—your company has been praised for being a leader or criticized for lagging behind. All these people and their concerns and needs make your writing social.

In addition, you may find that the company you work for prescribes a standard **document production process** involving a specified number of stages and checkpoints that you must go through in the creation of documents. This process allows companies to control the quality of documents, to coordinate the timing of any writing that is tied to specific product releases, and to assure a consistent style and organization in all writing projects. Generally, document production processes have been negotiated over time, often in cooperation with a company's publication department, and are subject to change through the negotiation of a number of people and interests. Whether you are an engineer involved in product manufacture and design, a member of the sales force that markets products and communicates with clients, or a manager and administrator who conceives of company direction, you will have to learn your own company's document production processes.

If you work on a collaborative writing team to produce your document, your writing processes will be affected by the individuals who make up the team and the leadership and negotiations taking place within the team. Finally, your own writing processes will reflect the ways you have developed so far in your life. The ways that you plan for writing, begin writing, revise writing, and involve others in reading your writing come from the people who guided your experience and education so far. All these aspects make your writing social.

As we discuss writing processes in the social context of the workplace, remember that you'll need to select the strategies that best fit your social circumstances for any particular project. Writing Strategies 2–1 lists a number of questions to ask about the writing process of any project.

social process
the involvement of many people in writing in the workplace, such as supervisors, co-workers and clients who help a writer plan, draft, and revise

document production process
the specified number of stages and checkpoints a technical writer must go through in creating documents

---

WRITING STRATEGIES 2–1 Some Questions to Ask About Writing Processes in the Social Context of the Workplace

Social and Cultural Concerns

What is appropriate? What is polite?
What is typically done? What has been done in the past?
What is responsible writing in terms of our organizational obligations?
What is responsible writing in terms of my own professional obligations?
How does my organization typically present itself to the public?

Concerns About Language

What tone or style will best convey this information?
Does my audience have difficulty with English? Will my work be translated?
How should I address my readers? Second person or more formally?
Can I refer to myself in my document? Or do I speak for the organization as a whole?
Who can help me define technical terms?
How can I engage all my readers? Is a glossary needed for nonspecialists?

Concerns About Content, Organization, and Structure

What information is necessary? What is extraneous?
What sections are required or standard in my company? Is an abstract needed? Index? Glossary?
What organizational frameworks are useful or typical in this kind of document? This section?

Concerns About Genre

What is typical of this genre (type of document)? Is it a memo? A year-end report? A feasibility study?
Have similar documents been written?
Do we have a prescribed format for this information?
Would visual presentations (tables, charts, graphs) be typical or helpful? Where? When?

Concerns About Managing the Writing Process

What are the deadlines on the project? Are they firm?
What are the various parts of the production process? Who is in charge?
How many writers and technical experts will be involved? What is their level of skill?
Does an editor plan to review my work?
What checkpoints or review stages will be required of my work? Who is involved in these stages? My manager, editor, or co-workers?

Writing as Problem Solving in the Workplace

As we said in the introduction to this chapter, technical communication is *applied* communication; that is, it's designed to accomplish specific tasks or help an audience solve specific problems. But how do you make use of this concept of problem solving when you write? How do you learn to approach a writing task as an effective problem solver?

Three different approaches can help you begin to think about writing as problem solving—thinking rhetorically, thinking of yourself as a problem solver, and thinking of yourself as solving a problem for *some people.* Try each one out, become familiar with it, and then you will find that one or two approaches work especially well for you. Each approach involves thinking about the purpose of a writing project. When you think of purpose, imagine both what you want to find out and the intended effect of your discovery on your audience. What problem do you want to solve? What do you want your audience to do, think, or feel after reading or listening to your solution? What action do you want the audience to take after reading or listening to your communication?

THINK RHETORICALLY

Learning to think rhetorically is the first way we would like you to think about purpose. **Thinking rhetorically** means considering the elements of the rhetorical situation that make up all communications: the *communicator* (you and your collaborators), the *audience* (those who receive the communication), the *content* or the *topic* (what the communication is about), the *genre* (the form the communication takes), and the **medium** in which you make your communication (print, speech, video, graphics, online, multimedia, hypertext). One easy way of planning effectively is to make sure that all of those factors are considered in your plans and are expressed in a **statement of purpose**.

If you were the mechanical engineer working on protective coatings for new automobiles described earlier, for example, you should think of all five factors. The medium in this case is easily defined: You were told to write a progress report. And the topic seems easy to discern as well: The report will concern progress on a certain technical task—what's been done so far and what remains to be done. But it may not be so simple as that. The content may be easy to describe if all is going well, but if any difficulties have arisen, the readers will need to be reassured somehow or warned that no solution appears workable. And you will have to decide if it is enough simply to report progress or whether you should have other information or recommendations to offer to your readers. And who are those readers, anyway? Are you really writing to your supervisor or to your supervisor's supervisors, or both? Do any future readers need to be considered? Will co-workers also read the progress report?

Finally, as the engineer you will need to think about yourself as a communicator and your aim in writing this document. You will want to report progress, of course (assuming that there's some progress to report). But you may also wish to introduce yourself and your fellow workers, to reassure the readers that any difficulties are being dealt with satisfactorily, to brag a little bit, perhaps even indicate somehow that the project should go in a certain promising direction or that it should continue to get support

thinking rhetorically
considering the communicator, audience, content or topic, genre, and medium of a communication

medium
the channel through which one communicates, such as print, speech, illustrations, and the Internet

statement of purpose
all the factors of planning: communicator, audience content, genre, and medium

because it is likely to have a good outcome. Under some circumstances you would be likely to involve your team members in the report—at the minimum, you would tell them you've been asked to write it, and you might well solicit their suggestions for what to include.

It's hard to be too specific in talking about a hypothetical example, of course. The point is that by identifying and thinking about the elements of the rhetorical situation, you'll develop a good sense of purpose. After pondering these elements and before going on to the actual writing, the engineer we've been discussing might summarize the purpose this way (maybe even in writing):

> I need to write a progress report that will remind my supervisor—and especially my supervisor's supervisors—of what our team has been working on. I also want to explain the good work we've been doing and the interesting results that we've been getting. I think I'll send copies to all my coworkers so that they know about the report and so that they'll have a sense of the high regard I have for them. I want to make a good first impression, too—to indicate that I know how to write well and that I can deliver my work on time.

Notice that in this formulation the engineer attended to all the elements of the rhetorical situation: writer (and the writer's aims), audiences, topic, medium, and genre. By writing down a statement of purpose, you have a good **benchmark** to use in evaluating decisions that you will make in your developing text.

benchmarks
clearly defined expectations of goals against which writers measure their progress in the document production process

Devising statements of purpose like these is generally quite easy. Their ease and informality lend them to the composition of routine memos and correspondence and oral presentations, for instance. Even in routine circumstances or when time is short, get in the habit of taking a few minutes to think rhetorically. Example 2–1 is a memo that grew out of a need to convey an important, but brief, piece of information about a product line—a kneeling chair for clerical workers who use computers within an organization. Information about the chair line was important because it involved safety issues, production delays, and possible client satisfaction.

To practice writing a statement of purpose, use this memo. Read it and try to come up with an informal statement of purpose that the memo's author, Carla Etchison, might have used to aid her in drafting the memo.

ideal and actual situations
the way a writer thinks as a problem solver by imagining purpose as a bridge between ideal situations and actual ones

THINK OF YOURSELF AS A PROBLEM SOLVER

A second way to determine *purpose* is to think of it as the bridge between **ideal and actual situations.** In this case, what you discover and often what you recommend to your audience can help turn the actual situation into the ideal one.

EXAMPLE 2–1   Purpose—Thinking Rhetorically

The Computer Company
For Your Computing Needs

DATE: 28 December 2000
  TO: Dan Musgrove, Shipping
        John Loomis, Purchasing
        Carol Adamsson, Division Manager
FROM: Carla Etchison, Technical Sales and Customer Relations
  RE: Kneeling Pads for Computer Chair #332

  CC: Lemonia Williams, Production Manager
        Sandra Blevins, Product Engineer
        Ethel Walmera, Research and Development

As Carol requested (memo dated 12 November 2000), I have checked into the 16 cases of defective kneeling pads on Computer Chair #332. All these chairs were traced to the same production batch #332-49-7711 (dated 1 December 1999). The pads on these chairs were manufactured from a single shipment (Invoice #769-1) received from our supplier, Amos Stuffing, on 10 November 1999. The supplier at Amos has acknowledged that some pads in this shipment represented a slightly different grade of high-density foam. According to their estimate, approximately 14 percent of the shipment may have been involved.

According to Amos, the composition of this foam will degrade with heavy use, especially in situations involving high pressure demands (>523 psi) over minimal surface areas. Amos has offered to reship the affected portion of the shipment immediately, free of charge to our firm.

As Carol has requested, I have sent cards to all of our client retailers who may have purchased chairs with similarly defective pads and offered immediate replacement, without charge.

From our similar experience three years ago, we can anticipate that approximately 1,500 chairs were affected and that approximately 900 claims will be made within the next month. Lemonia Williams has outlined a production plan for reconditioning the affected chairs to specification, and Dan anticipates a 2-day delay in reshipping during the next 30-day period.

Thank you all for your cooperation on this project.

The following examples will help clarify this way to think about purpose:

- Your company is on the brink of developing an effective coating for protecting the finishes of automobiles (*ideal situation*), but it cannot test the product until a testing protocol has been written and approved (*actual situation*); consequently, you must test and respond to the feasibility of a number of spray guns that could be used to apply the product (*purpose*).
- Your company and its customers would like a reliable and inexpensive adhesive that can be applied under water (*ideal*

situation), but so far no one has even written a preliminary description of the desired product (*actual situation*); therefore, your team must write a preliminary description of the concept and the product for management to approve before actual research can begin (*purpose*).

- Domestic animals like cows sometimes develop a disease called malignant catarrhal fever, so it would be a good idea to have a treatment for the disease handy at your farm (*ideal situation*), but you currently have no idea how to treat the condition (*actual situation*); so you need to compose a literature review to summarize effective treatments (*solution*).

- People who fight forest fires should be able to count on absolutely reliable equipment, including chain saws (*ideal situation*); however, in real life the chain saws often break down during use (*actual situation*). Consequently, you need to write a document that will enable firefighters to make routine repairs on chain saws when they are out in the field (*purpose*).

- Your microbiology professor would like to have absolutely brilliant lab experiences to offer students (*ideal situation*), but in recent years her lab materials have been getting stale (*actual situation*); so she asks you to research some better possibilities that she might be able to incorporate into her courses and write a first draft of the activities (*purpose*).

As Example 2-2 indicates, you can establish a good sense of purpose if you can supply three different sorts of statements. First, there should be a statement of a goal or desired state that you and your organization consider important. The goal can be a large, general goal ("We want to reduce our costs"), but the more specific you can be, the better ("We want the parents of children placed in body casts for the treatment of congenital hip dislocations to know how to care for the children and their casts at home; that way the children will heal better"). Then, after naming a connecting word—"but" or "however" usually do very nicely—offer a statement about things as they are. This statement articulates a clash between the ideal state and the actual one: there's something that keeps the ideal statement from being true ("Right now the children's parents have no idea how to care for a toddler in a body cast"). Finally, in a third statement add a connecting word such as "so" or "therefore" and then name some activity that you could do to bring the actual state in line with the ideal one ("Let's write a brochure that we can give to parents to explain how to care for the kids and casts at home"). If you can prepare a sentence that includes all three statements, you probably have a good sense of your purpose.

A disadvantage of this way of thinking about problems, however, is that it can be impersonal, focusing on organizational or institutional problems rather than **human problems**. Because communications or documents are

human problems
a writer's purpose of solving a problem for someone rather than solving technical problems

---

EXAMPLE 2–2   Purpose—Thinking of Yourself as a Problem Solver

First Statement   Describe a goal, desired state, or value that you or your organization considers important.
  Example: "A laptop computer would make it easy to collect data on site."

Add a connecting signal word: BUT or HOWEVER.

Second Statement   Describe a condition that prevents the first statement from being achieved or realized at the current time. The statement could describe the status quo or define a competing goal or value or desired state, but it should reveal a clash with the first statement.
  Example: "But we can't purchase a laptop computer until we know more about their capacities, features, and costs."

Add a connecting word: SO or THEREFORE.

Third Statement   Describe some action—a study, a research program, and so on—that could resolve the difficulty.
  Example: "So I'll do some checking into what kinds of laptops are available and what they cost."

---

directed to people, you may need a way to address the human component of problems. For this reason, we recommend that you also consider another approach to discovering purpose.

THINK OF YOURSELF AS SOLVING A PROBLEM FOR *SOME PEOPLE*

Within organizations, writing tasks generally follow from the problems and needs of real people. As a technical communicator you might think of defining problems not as technical or institutional problems but as people problems. No problem exists unless it is a problem for someone, and this understanding provides us with a third way to think about purpose.

The purposes for which writers compose documents, or for which managers assign writing projects, are tied to the needs of the people who will read a document. Consequently, you might think about problem solving in a very human way in order to have a truly robust sense of your goals. Again, think of yourself as a problem solver. But now, instead of thinking of yourself as engaged in solving a *technical* problem, see yourself as working on a *human* problem. See yourself as solving a problem for someone.

At work it is easy enough to define an audience's problem because you often write at someone's request and that person defines his or her needs for you. At other times, however, it isn't so easy to uncover your audience's needs. If your audience is available, you can ask the person to describe exactly what he or she needs, and why. You should be assertive about talking to your prospective reader about what problems or needs to address. But sometimes it isn't easy to approach your audience.

Consider once more the progress report on the protective coating for new automobiles mentioned in the first pages of this chapter. Remember how vague the boss was about defining the purpose of the report and who would be reading it? Should that report be simply a routine record of what has been done, or should it be designed to assist readers in directing future work in especially promising directions? Is this the place for the writer to request additional resources that might be handy in getting the work done? What can a technical communicator do in circumstances when it is difficult to learn an audience's needs and when a specific sense of purpose is absolutely essential? See Writing Strategies 2–2 for guidelines for planning your document.

Throughout this book, you will learn ways of thinking about your audience that can improve the quality of the document you eventually create. Advice about audience will especially aid you in developing the kind of information your audience will find useful in your communications. But in thinking about purpose in planning, you must consider audience in specific ways. When you plan, take some time to think about your goals in an audience-centered way; you'll learn more about methods for doing that when you study Chapter 3. Identify your audience and its problem, identify what you will do and what kind of communication you will produce to solve that problem, and name explicitly the action(s) your audience should take as a

---

WRITING STRATEGIES 2–2   Planning Your Document

I. *Questions about Audience*
   A. Who is my audience and what are my audience's specific problems?
   B. What will I do and what communication will I produce in order to solve my audience's problems?
   C. What will my audience do as a result of my communication?

II. *Questions about the Process*
   A. What resources will I need for this document—people, books and articles, existing reports, and so forth?
   B. What visual presentations will I need to support my document, and who will prepare them?
   C. When should I schedule the activities noted in questions II A and B?
   D. When should I schedule my writing and production activities so that I can be sure to meet my deadline?

result of your communication or document. Keep Writing Strategies 2–2 in mind as you read more about thinking of yourself as solving a problem for some people.

Once you have done some preliminary thinking about the problem-solving context within which your writing project takes place, start thinking about the specific processes you need to undertake to complete the project. Getting an accurate and complete sense of all the factors influencing writing processes is challenging, but we can start with an *overview* of writing processes, or the strategic activities that make up writing processes:

An Overview of Writing Processes in the Workplace

- Exploring, generating, and representing information for documents.
- Planning documents and arriving at goals.
- Structuring and organizing.
- Drafting.
- Revising.
- Editing and proofreading.

We'll devote entire chapters later in the book to some of these activities.

EXPLORING, GENERATING, AND REPRESENTING INFORMATION

Some exploratory and developmental activities can give you a sense of the writing task you face and help you plan your writing project. Although you can use these activities in early stages of a writing task, they may also help you refine and improve your document throughout the writing process.

One of the most difficult tasks you face is prioritizing information. But because you have not yet written your document, you might be unsure of what you actually need and what you can discard. At this point you need to select and organize your information—to make some sense of it, to decide what is important and useful and what is not. One of the best ways to do this is to outline what you have, what you want to save, and what you still need to find. In this chapter, we explore outlining—to show how outlining can help you select and organize the information you gather from your sources.

We all use a variety of ways to organize the information we collect. Sometimes we do things as simple as stack books and papers together in groups. If, for example, you were writing about engineering disasters, one pile might be for the information about the collapse of the Hartford, Connecticut, Civic Center, and another pile might cover the explosion of the space shuttle *Challenger*. This stacking method physically organizes information, but it doesn't allow for very specific organizing. What if one book or one set of notes contains information about five different topics that you may want to use in five different places? Usually we handle this problem by describing our organization. For example,when some sources overlap, we might find ourselves affixing notes to the covers of the books and papers indicating where the information contained in them goes. The need to keep track of and cross-reference data is one reason Post-it notes are so popular.

As you explore your writing task further, you may wish to brainstorm; to try some quick, informal writing sessions; or to present your tasks visually in a **flow chart** or diagram. These activities help you to flesh out your writing goals and begin to gather the information you will need. They also help you to get a good initial sense of your audiences, the potential structure of your document, your focus, and your timelines. As you explore writing tasks and begin to generate your content, you get a sense of the social factors associated with that task: the impetus for your project, the people involved, ethical and legal issues, the nature of conventions involved in a particular document, and more.

flow charts
diagrams of processes that show the sequence of stages

PLANNING

planning
the portion of the writing process in which decisions are made about a project's structure, scope, purpose, audience, strategies, tactics, and timetable

Although some people see **planning** (i.e., getting a sense of your goals and planning the writing process itself) as an activity that takes place before the real business of drafting and redrafting begins, it is better to think of planning as occurring *throughout* a writing task.

As we discussed earlier, writing a statement of purpose for a document is an important strategy to use in approaching major writing tasks. You can also plan the development, structure, ways of handling timetables and organizational constraints, and sources to tap for needed information.

Within organizations, planning is often a social, public, and collaborative activity. Because many people, with differing levels of expertise, are often involved in a communication project, all their activities must be coordinated and scheduled. For example, your manager might plan not only the timeline for product development but also the interaction of those involved in that development. Such careful planning, which encourages frequent communication and feedback, can help a group avoid conflict and misunderstanding while creating a document.

Given all of these factors, you will find yourself making and remaking plans at many levels. These plans are dynamic representations, continually shifting and constantly reforming as you think of new possibilities and respond to new ideas and your own developing text. Indeed, when you write you make plans that shape your writing decisions, and your decisions in turn affect your plans. Writing plans also are shaped by a broad range of social, historical, political, and economic factors. These factors include your position within a company, the ways that documents have been con-

EXAMPLE 2–3   Sample of a Monthly Planner Showing Schedule for Coordinating Activities for New Software Documentation for a Personal Computer

December 2000

	Week 1	Week 2	Week 3	Week 4
Monday	Manager meets with writing group to explain project timeline and assign project	Writing project begins		
Tuesday	Lead writer meets product development staff to discuss software use and audience		Writing group meets to check consistency of various sections of the document	Editor receives complete document for review
Wednesday	Lead writer meets with writing group to divide documentation into sections	Graphics staff presents recommendations for document design		Product development staff checks document for technical accuracy
Thursday	Lead writer meets with graphics staff to discuss visual displays in documentation	Writing group begins to share drafts		
Friday	Writing group, manager, and lead writer meet with product development staff to discuss any last-minute changes in the software		Final draft sent to graphics staff	

structed in the past, the relative importance of a given project within the context of an organization's larger efforts or within the context of existing cultural priorities outside the organization, and the cost of doing business in any given sector of the economy.

Example 2-3 shows how such a complex project might be planned. In addition to pencil and paper devices such as those we've shown here to help you plan your projects, many software products can assist you with planning. Inexpensive planning software (see Example 2-4) allows you to create, update, and print out plans. This planning software often provides you with a template for entering the steps that you need to accomplish in putting your project together. Once entered, the software will arrange your steps in a variety of formats to suit your needs.

EXAMPLE 2–4 Planning Software

ID	ℹ	Task Name	Nov 28, '99	Dec 5, '99	Dec 12, '99	Dec 19, '99
			M T W T F S	S M T W T F S	S M T W T F S	S M T W T
1	▦	Meetings	███████████			
2	▦	Meet with writing group to explain project time line and assign project	██			
3	▦	Lead writer meets product development staff to discuss software use and audience	██			
4	▦	Lead writer meets with writing group to divide documentation into sections	██			
5	▦	Lead writer meets with graphics staff to discuss visual displays in documentation	██			
6	▦	Writing group, manager, and lead writer meet with product development staff to discuss any last-minute changes in the product	██			
7	▦	Writing project begins		████████████████████████		
8	▦	Graphics staff presents recommendations for document design		██		
9	▦	Writing group begins to share drafts		████████		
10	▦	Writing group meets to check consistency of various sections of the document			██	
11	▦	Final draft sent to graphics staff			██	
12	▦	Editor receives completed document				██
13	▦	Product development staff checks document for technical accuracy				██

STRUCTURING AND ORGANIZING

Often, your plans may include ideas about organization and structure. For example, as you explore how a document might develop, think about its overall organizational structure, including the number and types of sections, graphics, appendixes, tables, and headings. Think as well about subsections, paragraphs, and sentences.

Structuring and organizing take place both as you prepare to draft and as you proceed throughout the writing task. Sometimes, attempts to structure and organize information are simple. Piling books up in different stacks and using Post-it notes to cross-reference them, as we mentioned earlier, is an example of a simple approach.

At other times, you might need a more formal organization. Some writers like to make detailed, formal outlines of a document before they begin. Others prefer to make outlines after their writing team produces a second or third draft of a document as a way of checking on the overall organization of a project. Indeed, many writers use the concept of outlining at several stages of a writing project.

DRAFTING

Drafting—physically producing words and graphics—may be what you picture when you think of writing, although it may actually represent a relatively small portion of everything involved in a writing task. Drafting is a framework within which all other writing activities seem to happen as you write.

drafting
the part of the writing process that involves physically producing words and graphic displays in a document

The physical behaviors that you exhibit are also unique to you. You might compose in fits and starts or in planned sessions of concentrated activity; you might swear by a one-draft method or by multiple versions; you might prefer to draft by hand or on a computer. In any case, your behaviors will differ according to various constraints associated with your writing tasks, including:

- The length of a document, its organization, and its ultimate audience(s) and purpose.
- Your familiarity with the material and your educational background.
- The strategies accepted within an organization.
- The extent of collaborative activity.
- The nature of the writing task.
- The use of writing technologies such as computers or Dictaphones.
- The availability of information.

REVISING

Although it is impossible to separate drafting activities from those of revision (often these activities happen almost simultaneously), we use the term

revision
a critical examination, reading or rereading; a revisioning of a communication

revision here to refer to a critical examination, reading, or rereading of a text by a writer and the changes that grow out of these considerations. As we have seen with all the composing activities, revising takes place at many different levels, at many different times.

Moving among various kinds of revising efforts, you might rethink written plans and notes early in the process of a writing task; recast sentences, paragraphs, and sections throughout a drafting sequence; and restructure entire documents as a piece emerges from a collaborative team effort. You might also revise in any number of ways: making notes with a pencil in the margin of a document, creating entirely new texts on a computer, or cutting and pasting sections of your draft with the drafts of other writers. You also may respond to the testing of your document by representative users.

editing
a substantive look at how well a communication meets stylistic, grammatical, spelling, punctuation, usage, and consistency conventions

proofreading
a stage in the writing process involving a careful reading for surface errors, such as missing words, spacing problems, misspellings, and typos

EDITING AND PROOFREADING

For you as an individual writer, editing and proofreading involves another kind of critical attention to writing, often the kind of attention that focuses on the surface level of text and graphics. The surface level includes proper spelling, grammar, and usage; complete headings; punctuation; use of white space and such. Editing and proofreading looks at how this surface level reflects the document's purpose, communicates to intended readers, and satisfies your conception and goals. Specifically, **editing** is the alteration or refinement of a document so that it conforms to an acceptable standard. **Proofreading** is the reading and marking of corrections in a document, usually printed, by comparing it with the final draft.

Some Writing Process Strategies to Try

After you have done some preliminary thinking about a writing project as a problem-solving activity—and you have developed a statement of purpose that can guide your writing—there comes a time when you have to begin drafting an actual document. For most writers, there is no easy way to go about this task. You just have to start putting words down on a page or typing them into a computer. Even the most complicated writing task begins with this activity that seems both simple and difficult. There are, of course, strategies that writers have found helpful in undertaking the task of drafting, but no single strategy will work for all writers. Thus, you are going to have to experiment to find which strategies work best for you or for the writing groups of which you are a member. And you'll have to be flexible and resourceful. The more strategies you have in your repertoire as a technical communicator, the more effective you can be as a problem solver.

One of the keys to success in generating early drafts is to keep yourself from being too critical about your words and ideas at the beginning of a writing project. When you are starting your project, be willing to accept imperfections in your expressions and ideas instead of stopping to search for the very best possible phrase. Eventually, of course, you must devote time within a writing project to taking long, critical, detailed looks at individual words and expressions, but generally this kind of activity comes when you have produced at least one draft of a project. Stopping to evaluate and worry about individual words or ideas before thoughts ever get down on paper or onto a computer screen can make it difficult to generate a first draft. The following activities should help as you begin your writing process.

START DRAFT OR ZERO DRAFT

You might find it useful to think of your first draft simply as a quick list of notes and ideas that you put down on paper without too much regard for order or value at this point. You will have your audience and purpose in mind, but a **start** or **zero draft** can be done in one short sitting and still be quite useful. The start draft is a quick listing of everything you know or think about a subject. These drafts often include elements of your writing plan, brainstormed lists, references to other documents that need to be checked, ideas for organization of a document, and notes for a later interview or group session—anything that will be useful in making some sense of a writing task.

start or zero drafts quickly written lists of everything the writer knows about a subject

Start drafts are especially helpful when you don't know where or how to begin. To create a start draft, sit down at your usual writing space and begin writing. Keep writing—whatever comes to mind about your topic— for 15 to 20 minutes. Write down words, phrases, or complete sentences; do whatever is best for you. You might have learned this technique, often called freewriting, in earlier writing courses and found it useful in getting started.

Let's say that your group is producing an engineering study for the Central Bank. During your brainstorming session, you describe your audience as supervisors who will check the report for an impression of your work, coworkers who may depend on your technical details for future projects, and the Central Bank (your clients) which will use the report to make a decision (see Example 2-5).

When you are done with your start draft, examine it for material that seems promising. Once you have completed this draft, rearrange any promising material it contains into some order that will help you work on later drafts. Various important points from your start draft may suggest essential sections of a document to you or may lead you to a more carefully articulated statement of your purpose, the audiences to whom your document will be directed, and the scope and format of the information that you will be presenting. You can reduce your start draft to an outline and then check for completeness and consistency. As you work with the start draft to create an outline, add or delete material as needed (see Example 2-6).

EXAMPLE 2–5   Start Draft for an Engineering Report

Our purpose was to evaluate the existing building—Central Bank. We wanted to see if the bank can move in and not look or build elsewhere. Everything looks like a "go." We decided that Central Bank can move in and get the things they want after some reconstruction. We considered their budget and schedule, but they will have to pay overtime to get into the building at the time they want. We covered electrical, plumbing, and heating as well as air-conditioning. The improvements that need to be made can be done—and not be too costly, at least not as much as it would cost to build or improve elsewhere. The building (220 Columbia St.) is close to one we worked on before at 112 Columbia so we used some of the information from that report. We did that for United Plumbing—their needs were close enough to be useful. We looked at 114 Columbia, too, but the electrical systems in that building had been improved three years ago—and were more complicated than Central Bank wants—not any good information there. Because we looked at heating and air-conditioning, we had to look at ventilation. Central Bank needed better and more centrally located elevators than United Plumbing. We made sure that handicapped clients of the bank would have access to everything, even hand-dryers in the restrooms. A major electrical challenge would be installing a handicapped-access elevator from the parking garage in the rear. Broughton, Horn, Karis, and Grant looked at this building seven years ago—not many changes since then, so we used their report, too. But our survey needed to be very detailed—both the systems that exist now and what improvements were necessary. We talked to the VIPs at Central Bank to see if they will use the same maintenance company. We came up with a construction schedule and included the overtime in it. Made sure we looked at fire protection and codes, of course. We got existing blueprints and plans from the architect—building isn't that old and hasn't been modified greatly. Made sure we really knew exactly what this client wanted—the type of facilities and arrangement.

To summarize, Example 2–5 shows a start draft for the introduction to an engineering report that evaluated the feasibility of a branch of a national bank moving into an existing building. This writer has done much of the research before beginning the start draft, but the activity revealed some gaps in that research. Example 2–6 shows what happened when the writer of Example 2–5 rearranged the start draft. In the second version, the writer grouped ideas under subheads to organize the material and focus the group's thinking.

DIAGRAM DRAFTS

diagram drafts
start drafts using designs such as circles, boxes, and lines to represent sections of a document

Representing information visually and working manually with these physical representations can be a productive way to begin drafting your documents. So, instead of writing sentences to represent information, you can create a **diagram draft**. To use this strategy, create designs (boxes, squares, circles, and so on) for each section or category of your draft. Then create designs of a different shape for as many pieces of information as you can think of. Label each of these designs with a key word or two to

---

EXAMPLE 2–6   Outline Based on Start Draft Example 2–5

Purpose
  To evaluate and recommend whether Central Bank, Albuquerque Branch, can occupy an existing building at 220 Columbia Street

Scope of the Report
  Heating, ventilation, and air-conditioning

  Electrical

  Plumbing

  Fire protection

  Elevators

  Provisions for handicapped people

Sources of Information
  Previous study of building (7 years ago) by Broughton, Horn, Karis, and Grant

  Study of similar building located at 112 Columbia Street for United Plumbing, Inc.

  Blueprints from architect

  Discussions with building personnel

  Interviews with banking officials

  Detailed survey of facility and systems

Basis for Recommendations
  Arrangement and type of facilities needed

  Central Bank's budget

  Moving-in schedule

  Construction schedule

  Overtime needed

  Code considerations

---

represent the information or the content of the draft. Then experiment with how these parts fit together.

For example, if you decide to organize the engineering report for Central Bank that you brainstormed in Example 2–5 by each facility you had to study, you could represent Heating, Ventilation, and Air-conditioning by a square; Electrical by a circle; and Plumbing by a triangle. Then you could draw connecting lines (e.g., connect general office lighting to Electrical, sump pumps to Plumbing, plumbing stacks to Plumbing) as you can see in Example 2-7. After creating the diagram, you could reorganize the information you generated into an outline form much as the writer of the start draft used that draft to create the outline shown in Example 2-6.

If you prefer, you can list these pieces of information and your key word categories on separate index cards and shift them around to find the best matches. Or you can use different colored highlighters to code your information according to categories. If you have a computer-supported

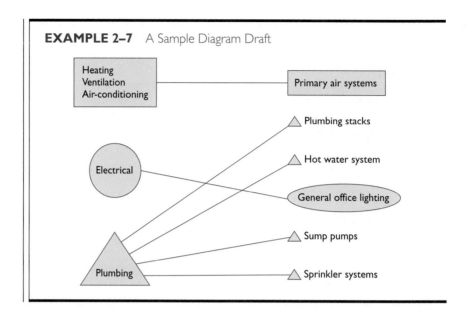

EXAMPLE 2–7   A Sample Diagram Draft

drawing program or a hypertext writing environment such as *Story-Space*™ to create these diagrams and elements, you can move categories and pieces of information around in several different ways as is shown in Example 2-8.

As you work with the categories and specific information in a diagram draft, ask yourself the questions shown in Writing Strategies 2-3 on page 58. Using Example 2-7 as your starting place, try the various approaches in Writing Strategies 2-3, and see what kinds of results you can achieve.

INTRODUCTION DRAFT

You might find it helpful and even necessary to draft the introduction of a document in considerable detail before going on to draft the rest of the document. Writers often spend a great deal of time within an **introduction draft** laying out the following items:

introduction drafts
drafts at the beginning of a document including a purpose or problem statement, a goals or outcome statement, and an information map

information maps
concise statements that map out the structure of a communication

- *Purpose* or *problem statement*: a concise statement that articulates the purpose for the document or the problem that the document is designed to address.
- *Goals* or *outcome statement*: a list of the document's major goals or the outcomes that the document is designed to produce.
- ***Information map:*** a concise statement that maps out the structure of the document that follows. This statement reveals the organizational structure of the document's presentation for readers.

Example 2-9, on page 59, represents an introduction draft.

EXAMPLE 2–8   Working with a Diagram Draft in StorySpace™

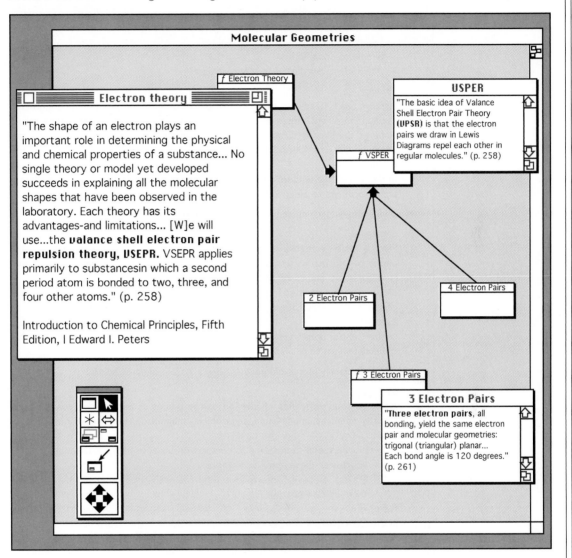

Source: Text from Edward I. Peters, *Introduction to Chemical Principles*, 5th ed. (Philadelphia: Saunders Golden Sunburst Series, Saunders College Publishing, 1990). Diagram created by authors.

WRITING STRATEGIES 2–3   Working with Diagram Drafts

■ Is there a system to these categories or pieces of information?
  —General to less general?
  —Simple to complex?
  —Are some subordinate or superordinate?

Strategy:
Arrange the subordinate categories or pieces of information inside or underneath the superordinate categories or pieces of information to show how they relate. Arrange the superordinate elements in order of importance or occurrence depending on their relationships.
Example:
In the Central Bank case, consider making Heating, Ventilation, and Air-conditioning subordinate to Electrical.

■ Are some categories or pieces of information related by association?

Strategy:
Place a major category in the center of your page. Then brainstorm as many pieces of information as possible that are related to it. Draw lines between and among related pieces of information. Examine these connections. If a cluster of pieces suggests itself as important because many lines lead to it, try placing this cluster at the center of your page and continuing your diagram.
Example:
In the Central Bank case, note that most of the problems with the building that the client wants to move into are related to Plumbing.

■ Can you classify pieces of information in a system of some sort?

Strategy:
Apply various principles of division to the pieces of information on the page. Classify them by importance, by kind, by magnitude, by success rate, and by topic.
  —Do they lend themselves to comparison because many of them are alike? If so, cluster these items by likeness.
  —Do the pieces of information lend themselves to contrast? If so, cluster them by dissimilarity.
Example:
In the Central Bank case, consider the possibility that the overtime and scheduling problems associated with changing the Plumbing systems will be much greater than that needed to change the Electrical facilities.

■ Can the pieces of information be represented in a flow chart that shows movement from one instance/decision/event to another?

Strategy:
Start with the first step in the process or narrative and work through to the end. Represent decision points in this chart with diamonds or some other design that indicates branching.
  —Do the pieces of information have a cause-and-effect relationship? Draw lines to indicate these relationships; then list evidence points beneath each relationship.

WRITING STRATEGIES 2–3 (Continued)

> —Do these pieces of information represent problem/solution relationships?
> Identify the problems with one design (bullets, squares, and such) and the
> solutions with another. Connect problems with their solutions.
> —Do the pieces of information illustrate a before-and-after relationship? If so,
> clarify the connections between the pieces of information that precede and
> those that follow.

Example:

When organizing your engineering report in the Central Bank case, note that recommending that contractors start with a major overhaul of the sewage system in
the Central Bank building might solve the other problems listed under the Plumbing category.

■ Do the categories and pieces of information represent an evaluative context of
some kind? Are some decisions? Criteria? Outcomes? Does some of the information represent evidence?

Strategy:

Place the evidence items under the appropriate claims. Brainstorm additional evidence. Brainstorm alternative, or disconfirming, evidence as well to anticipate the
audience's arguments.

Example:

If you argue that overtime is required to meet the Central Bank's move-in schedule, you might find that a number of your pieces of information support that argument. The time needed to meet code requirements and ensure proper fire protection would become solid evidence for your argument

---

EXAMPLE 2–9   Introduction Draft—"Buying Your First Home" Brochure

First-time home buyers are seldom confident home buyers—and it shows. More importantly, a lack of knowledge about the procedures involved in financing and purchasing a home may result in home buyers who pay more than they need to pay.
With today's economic pressures, few families can afford to overspend this way.

This brochure is designed to provide first-time home buyers with the information
they need in order to be more confident with lending institutions, real estate companies, and moving companies. We cannot tell you everything within these pages, and
we can't promise to make your experience problem free, but we can give you a clear
outline of the steps involved in obtaining financing for a new home, planning for your
move, and making it through moving day.

By the time you have carefully studied the information in this pamphlet, you should
be able to accomplish the following steps toward buying your home and moving into it:

■ Step 1:   Shop for mortgage financing
■ Step 2:   Apply for and obtain a mortgage loan
■ Step 3:   Plan for your move
■ Step 4:   Sail through moving day

We have organized the information into eight easy-to-follow steps, so let's get
started. Read carefully, jot down questions, and get ready to MOVE!

DRAFT OUTLINES AND REVISION OUTLINES

So far, we have discussed several writing process activities that help with getting started on your drafts, but if you're like most people, you may feel more comfortable when you have a draft to work with than when you are facing a blank page or an empty computer screen. Because the processes involved in revising or rewriting are also complex, all writers wonder at one time or another where to begin revising a document and when to stop, whom to include as readers in a revision and whom to avoid, and how complete or how skeletal a draft should be during each stage of development. These concerns are magnified when revision efforts must be coordinated among several writers. Furthermore, within an organization additional constraints of deadlines, standardized procedures for revising and editing, management supervision, and production requirements may complicate revision processes.

outlines
any of a variety of frameworks or skeletons for organizing data

Although many writers use **outlines** as a way of working toward a first draft, outlines can also be tools for revising existing drafts, even late in the writing project. Sometimes, for example, outlines can be useful as a final check on the organization of a document.

Here's how it works. Using a marking pen, highlight the purpose statement in your existing draft. Highlight the main idea of each *paragraph* by identifying one phrase or sentence. If the document is longer, highlight the main idea of each *major section* by selecting a particular sentence in the introduction to each section. Next, using the highlighted items, construct an outline of major sections and the points included in each section. Then go back to your draft and select the subpoints for each of the major points you have identified. Indent these subpoints under each appropriate main point. On a computer, you can accomplish the process of selecting, cutting, and pasting electronically.

The outline you have constructed using this strategy should communicate some very useful information; for example, how your draft is arranged organizationally, what topics are primary and what topics are subordinate, and how much detail you provide under each main topic. The outline will illustrate your information in an orderly and concise fashion that should help you get a new perspective on your work. If you use a conventional outlining approach to numbering sections and subsections (see Example 2–10), you might discover that some topics require multiple levels; that is, the decimal numbering convention reminds you to consider the levels of detail needed to cover a topic sufficiently.

Once you have constructed the outline of your draft, you will want to look closely and systematically to determine the answers to two important questions:

- What kind of organic pattern of organization does the draft currently follow?
- How does this organizational pattern match up with the purpose and the audience for your document?

EXAMPLE 2–10   Excerpt of an Outline for a Report
on Portable Computers

1. Introduction
   1.1 The benefits of portable computers
   1.2 The increasing miniaturization of computing hardware
2. Types of portable computers
   2.1 Notebook
   2.2 Sub-notebook
   2.3 Palm-top
   2.4 Wallet-sized
3. Key developments in the miniaturizing components
   3.1 CPU/chip
        3.1.1 586SX, 586DX
        3.1.2 586SL
   3.2 Power packs
        3.2.1 Ni-cad batteries
        3.2.2 Ni-mh batteries
        3.2.3 AC power adapters
   3.3 Screens
        3.3.1 B/W
        3.3.2 Color
             3.3.2.1 Active matrix
             3.3.2.2 Passive matrix . . .

When you answer these two questions, note that different purposes and audiences seem to demand—or at least suggest—different organizational arrangements. For example, if you are assigned to write a report on two different insurance plans that your company is considering, your report must compare and contrast the various benefits and drawbacks of these plans. However, if you are asked to report about the advantages of one policy, you would organize your report deductively and present and analyze each advantage as a topic.

In examining the organizational plan revealed in your draft outline, you may find it does not adequately present the needed information according to your purpose or for your particular audience. In this case, you probably need to revise the organizational structure. A new outline can help you accomplish this task in an efficient manner. After you settle on a new organizational outline that works better for your purpose and audience, for example, you can cut and paste information into the outline from your old draft.

For this purpose, keep in mind several different kinds of organizational patterns you might use to revise your draft, depending on your purpose and audience. In the examples that follow, we show excerpts from different kinds of outlines which reveal the organizational structure of various types of documents focusing on portable computers, among them the following:

- An informational report that *classifies* different types of personal computers for potential consumers (see Example 2–11).
- An internal report that *compares* the various kinds of computers so a manager can make a purchasing decision (see Example 2–12).
- Documentation written to explain the *spatial layout* of a computer keyboard for users (see Example 2–13).
- A "how-to" manual that provides a *chronological*, step-by-step strategy for using a personal computer (see Example 2–14).

Because some writing tasks are large and complex, there are times when no one of these outline patterns will suffice for representing an entire draft. For larger and more complex documents, you may need to reflect the information in one part of a report in one pattern, and another part in a completely different pattern. For example, a draft of a long and complex report on personal computers may include one section that classifies the various types of computers and another section that compares and contrasts the performance of these computers according to different criteria. Hence, the draft outline that you construct from this report may reveal one section organized around the classification/partition pattern and one around the comparison/contrast pattern.

Although these combinations seem to make sense, don't forget that when readers see the beginning of a pattern, they will expect to see that pattern played out to its conclusion. When you combine patterns, you run

---

EXAMPLE 2–11   Classification/Partition Pattern for Informational Report on Personal Computers

1.0 Personal computers
    1.1 Desktop
    1.2 Notebook
    1.3 Sub-notebook
    1.4 Palm-top
2.0 Desktop features
    2.1 CPU
    2.2 RAM
    2.3 ROM
    2.4 Ports
    2.5 Networking
    2.6 Monitor
    2.7 Input devices . . .

EXAMPLE 2–12   Comparison/Contrast Pattern for Decision Making on Internal Purchases of Personal Computers

1.0 Personal computers
    1.1 Desktop
    1.2 Notebook
    1.3 Sub-notebook
    1.4 Palm-top
2.0 Criteria for judging
    2.1 CPU speed
    2.2 RAM size
    2.3 ROM size
    2.4 Number of ports
    2.5 Networking capability . . .
3.0 Desktop
    3.1 CPU speed
    3.2 RAM size
    3.3 ROM size
    3.4 Number of ports
    3.5 Networking capability . . .
4.0 Notebook
    4.1 CPU speed
    4.2 RAM size
    4.3 ROM size
    4.4 Number of ports
    4.5 Networking capability . . .

EXAMPLE 2–13   Spatial Pattern for User Documentation

1.0 Physical layout of your new notebook computer
2.0 Screen
    2.1 Open/closed positions
    2.2 Adjustment positions
3.0 Keyboard
    3.1 Multipurpose keys
    3.2 Function keys
4.0 Ports
    4.1 Mouse
    4.2 External monitor
    4.3 External keypad
    4.4 Internal modem
    4.5 Serial
    4.6 Parallel . . .

EXAMPLE 2–14   Step-by-Step Pattern for a "How-To" Manual

1.0   Starting to use your new notebook computer
2.0   Opening the screen
3.0   Using the keyboard
    3.1   Using the multipurpose keys
        3.1.1   (step 1)
        3.1.2   (step 2) . . .
4.0   Installing software
    4.1   Word processor
        4.1.1   (step 1)
        4.1.2   (step 2) . . .

---

the risk of confusing readers. Readers can handle these changes, but you must craft the transitions carefully. The purpose of learning these patterns of information organization is not so you can construct perfect outlines but to provide you with ways of organizing information that are efficient, effective, and easily recognized by audiences in the workplace.

Writing Strategies 2–4 provides a series of questions to ask yourself about the draft outlines you use for the purpose of organizational checking and revising.

COMPUTER-SUPPORTED REVISION

groupware
computer-supported tools to aid people writing in teams in their collaboration and revision efforts

Technical communicators functioning in business and industry often use computer-supported tools, **groupware**, for collaboration and revision efforts. Groupware works on a computer network that writers use as a medium for exchanging text and graphics; for trading commentary on, or information for, drafts; for real-time, or synchronous, discussions about writing projects; or for online meetings. But computer-supported collaboration and revision activities can be supported by computers even if no networks are available.

The suggestions in Writing Strategies 2–5 will help you think in creative ways about revision activities that can be supported by computers.

Find out what computer software might help you revise. For example, you might consider using a computer-supported response program like *PREP*™ or *Prose*™ to allow various readers to flag copy electronically. A sample copy of a *PREP* commentary is shown in Example 2–15. Some computer-supported response programs also allow authors to cut and paste suggested revisions electronically. You might also consider using a program like *Daedalus' Interchange*™ that supports synchronous chatting and the printing of transcripts. Example 2–16 shows a sample *Interchange* session.

WRITING STRATEGIES 2–4   Thinking About Outlines as Guides
to Revising

- What are the major sections of the outline? What are the minor sections or subparts under each major section? Make these major and minor sections your outline headings. Are these appropriately identified in the outline or should some of the items identified as major become minor points and vice versa?
- How does the document represented in this outline meet the purpose(s) identified for this task?
- How does this order of headings and sections, points, and elements under them help to present the document's information to the audience?
- Where does the outline look thin? Which sections are overcrowded with information?
- Which sections need further information? Which needed sections are missing? Which sections are extraneous?
- Are the sections related logically? If so, how? What is the principle of organization and what are the methods of organization that provide the framework for this document and this outline? Do they suggest a useful conception of genre, or conventional form, for this task?
- Is this system of logical relations and organization articulated clearly for the reader? Where would a reader trying to move from section to section get lost?
- Are the elements at each level representing information that is equivalent in importance or status? Does further work need to be done with organizational structures to clarify subordination and superordination?

WRITING STRATEGIES 2–5   Computer-Supported Strategies for Revising

- Send a disk with the draft to several writing team members. Have readers use the "all caps" or "track changes" feature of a word-processing program to distinguish their comments from those of the author(s).
- Circulate a draft on disk among members of a writing team. Have each reader rename an electronic copy of the document with a different file name that indicates the responder's identity. Or maintain only one disk copy of a document and have each group member comment using a different type font.
- Using a computer network, circulate drafts electronically so that people do not have to coordinate busy schedules to meet face-to-face for every draft. Once an electronic draft has circulated among all the members of a writing team, send the fully annotated commentary to all group members so that everyone can see the group's comments.
- Use a synchronous network discussion to hold an online discussion of a document or section. Synchronous discussion programs like *Daedalus' Interchange*™ allow groups to print transcripts of their discussions and thus have a record of group decision making.

(continued)

WRITING STRATEGIES 2–5 (Continued)

- Use electronic mail (e-mail) to gather information to be used in redrafting a document. Try surveying potential readers or expert consultants via e-mail. Consider sending excerpted sections of a document for commentary and critique.
- Use electronic archives to find boilerplate, or already existing, material that can be used in your drafting and revision efforts.

EXAMPLE 2–15   Computer-Supported Group Commentary as a Revision Strategy Using *PREP*™

Year-End School Board Report on Instructional Computing Effort

Draft Content	Reader 1	Reader 2
Much time and attention have been paid recently to the potential of computers for addressing educational inequities in connection with poor and disadvantaged populations and women. This report examines the claims that have been made and the evidence that confirms or disconfirms each of them.	I don't really like the use of "much" here—it sets a tone that's overly academic—even for the school board.	
One of the first claims to be made about computers in educational settings when personal computers first came on the market in the early years of the past decade had to do with democratization. Indeed, researchers claimed that personal computers would provide disadvantaged school-age populations with access to information, publications, and learning approaches that have been denied them in traditional classrooms.	I wonder if we need to include some citations to actual research studies here to lend credibility. A good bibliography might help—maybe we can get some teachers to annotate one.	*Democratization* really might not go over well with this crowd politically. I think we're dealing with a pretty conservative group in this district and we might not want to be so overt in our approach.

(continued)

EXAMPLE 2–15 (Continued)

Draft Content	Reader 1	Reader 2
Even the briefest glance at the current uses of technology in our school systems belies these early claims. Computers, far from providing disadvantaged populations with access to new information, are actually being used to exacerbate the gap between those who have and those who have not. In many inner-city schools, for example, computers are being used to . . .		John, get stats here from DOE documents in the government archives in Jeff City—16 of the 22 Board members are businesspeople.

EXAMPLE 2–16   Online Synchronous Discussion as a Group Revision Strategy Using *Daedalus' Interchange*™

Revision Session: 27 July 1998

In this session, brainstorm budget considerations are offered for the report section on finances. As it now stands, the draft of this section is pretty sketchy. We've done enough site visitation at this point to be able to get down to specifics on hardware and software. You might also want to suggest school district resources that can be employed in support of individual facilities.

Eileen:

We need to know more about the size of the project we are working on! Has the Board provided any clarification about whether they are willing to contract for 6 facilities or 12?

Cornelius:

One budget consideration would be what size each site chooses to establish— they do have some choice within the specs that we provide, so how can we indicate this in a budget for the Board?

Mac:

Software choices will certainly impact on budget. What kind of things do the teachers want a lab to be able to do, what student populations will it accommodate? This section has to show some variance.

(continued)

EXAMPLE 2–16 (Continued)

Eileen:

Remember they also wanted us to consider what machines they already have at the Milton site that we can use there: 14 old Quadras. I don't think any of them can really be used. Class sizes there max about 24, so even if we could, we'd better figure on 10 new workstations, one file server, one instructor station, one backup. Cabling for all 24 stations. Positions: A director, student assistants.

Mac:

Printers, too—and what about CD ROM players, scanners, etc.?

Cornelius:

Would it help to provide budget details for one of the proposed facilities and then separate out elements that are optional or tailored to specific sites? Certainly this might help in the case of the four magnet schools, but I wonder if all of this should go in the appendixes rather than the Finances section of the report?

Mac:

Before you determine that those old Quadras can't be used, Eileen, you need to determine what kinds of things those instructors will want the lab to be able to do. Quadras can be upgraded and can work very nicely as long as you aren't asking them to do things beyond their capacity. With 16 megs of memory Quadras do OK for word processing.

Eileen:

Software purchases—How much of a budget should we set aside for software? Let's see . . . We need

Work processing—25-30 station packages.
Online reference—   "   "   "   "
Telecommunication— "   "   "   "
?

Cornelius:

A 10-pack at educational prices for Word '98 is $610. Excel 10-pack runs $700; do you need spreadsheet and data capability? But is all this really going in the Finances section? Think, folks!

SUMMARY

- In this chapter, we wanted to help you accomplish two tasks: thinking about technical writing both as an activity that takes place in the social context of a workplace environment and as a set of processes, and working through major technical writing tasks.
- First, we explored the concept of writing as applied communication—as a mechanism for solving a problem for people. The options for doing so are to think rhetorically, to think of yourself as a problem solver, and to think of yourself as solving a problem for some people.

- In describing technical writing as a social activity, we pointed out how workplace writing often involved the efforts of managers, technical experts, writers, editors, and graphics experts; how such writing was constrained by deadlines and timetables that dictate when writing must begin and come to completion; how such writing is shaped by company procedures that ensure consistency with an organization's goals; and how the styles and genres common in technical writing demand that writers listen to, work with, and seek advice from their audiences. Technical writing takes place within a context that is formed by an organization, the people in that organization, and the products, goals, values, and challenges that organization faces.

- This chapter has also provided help in working through major writing tasks. It introduced the concept of writing as a set of processes: planning, organizing and structuring, drafting, revising, and editing—adding the caution that these broad movements are often mixed in a messy and complex way.

- The chapter further provided a series of concrete strategies for you to choose from as you approached a writing task: writing plans, start drafts, outlining and drafting, purpose and audience statements, information drafts, and diagram drafts.

- Remember that although we focused on writing in this chapter, you can use many of these strategies to create oral presentations. In addition, these strategies are designed to help you not only as the sole author of a document but also as a member of a writing team.

- We suspect that you will frequently need to turn back to this chapter as you read through this book. Every time you encounter a writing task, dip into your toolbox of strategies.

ACTIVITIES AND EXERCISES

1. Review the section of this chapter in which we describe the social nature of the writing process. Recall a document that you produced or a presentation that you made as part of your summer job or as a member of some other organization to which you belong (a religious group, a service or social organization, an extracurricular group). How did the group affect the process you used to compose your presentation? For example, with whom did you discuss your plans or drafts? Why did you make the revisions you did? Why did the document or presentation take the form it did? Share your thinking with the instructor and class.

2. Read carefully the public policy paper from the Minnesota AIDS Project (Appendix A, Case Documents 1). What do you imagine the writing processes would have been for such a document? How did the writers define (or create) their audience? What purpose does the paper serve? Can you identify categories and clusters of information? Write a short memo to your instructor and classmates in which you share your observations.

3. Choose one of the Web sites in the Year 2000 Case Study in Appendix A, Case Documents 4. What do you imagine the writing processes would have been for the site you have chosen? How did the writers define (or create) their audience? What purpose does the site serve? Can you identify categories and clusters of information? Write a short memo to your instructor and classmates in which you share your observations.

4. Suppose that you are a new graduate from your university with a degree in forestry. You believe the company that has hired you ought to be doing more to reforest the federal lands it has rights to log—that it should sharply increase its investment in reforestation even though there are few immediate gains to be reaped from such an investment. By doing so, you think the company would, over the long run, gain the trust and confidence of the public and the government. What kind of document could you write in this situation? Using strategies from this chapter, how would you come up with plans for the document?

5. The following are four statements of purpose. How adequate do you think each one is? Choose one of these statements and revise it to make a more effective statement of purpose.

   a. KVC Corporation has recently announced that, "unless something unexpected happens," production at its Carr Forks Mine near Mineralton, Utah, will have to be cut by more than half because several major ore veins have been lost in a complex fault zone. As chief consultant hired to make something unexpected happen, my job is to locate the lost ore veins. I must first initiate an intensive study of the local geologic structure. Using the information I find about the individual fault zones, I should be able to locate the positions of any lost ore zones. This method has proven to be highly effective in the past, and it should permit the mine to operate once again at an economically feasible level.

   b. In the past few years, many experts and other bird lovers have observed an unusually large number of white-chested tufted titmice dying of starvation, sometimes close to regular stock bird feeders. Because our company, Bird N. Hand, Inc., is the largest producer of birdseed designed for titmice, we've been working to develop a better bird feeder so that these birds can more successfully get at the seed. I'm the chief engineer on this project, chosen because I have experience designing feeders for related species. I need to write a report that will permit our product designers to manufacture a bird feeder that allows the birds access to seeds so they won't starve to death.

   c. My manager in our Systems Department, Truman Fergerson, has assigned me to evaluate the quality of two new software packages designed to assist with inventory problems like ours: Computer Products Corporation's InvenSort 3, and Roberts Computer Software's Sort 6.3. Both of these packages are highly compatible with the computers that we currently use in our company. I will have to write a report that will permit Mr. Fergerson and other managers to determine whether one of these new packages is better than the one we now use.

   d. The purpose of our brochure is to inform citizens of Minnesota about the services offered by the Minnesota AIDS Project (MAP) and to give basic information about AIDS and HIV so that people can protect themselves. The brochure might also encourage people to consider MAP as a worthy cause and to donate time and money to it. The brochure must also make it easy for citizens to contact MAP for more information. We intend to use the brochure in many ways—in direct mailings to citizens, in response to telephone requests for information, and to distribute at events. It is possible that the brochure could have secondary uses: High school counselors and teachers could use the brochure to help educate teenagers. Newspapers could refer to the brochure in writing their stories. People who have contracted the disease could use it to take advantage of our client services.

6. Exercise 5*d* refers to a document in Appendix A, Case Documents 1. Look at two other documents from Case Documents 2, 3, or 4 at the back of this book. Then write the kinds of purpose statements that the authors of those two documents might have employed.

7. Planning is particularly useful in the designing of documents that will be translated into other languages. Many communications fail because planning of this sort is ignored. Imagine that you are designing online help screens that will be used in connection with one of the Year 2000 sites that we have included at the end of this book (Appendix A, Case Documents 4). These online help screens, moreover, will be translated ultimately into Japanese and German (or any other language of your choice). Thinking rhetorically, as a problem solver, and as a solver of problems for particular people, make a list of the kinds of questions you will want to consider as you create your online help screens.

8. Assume that you are working for a health agency in Minneapolis and you are asked by your supervisor to write a report on the status of AIDS in Minnesota. In researching this topic, you have gathered the documents concerning AIDS in Minnesota contained in Appendix A, Case Documents 1. Develop an outline that classifies all of this information. Be sure to break the information down into types and subtypes.

9. Read "AZT Step Fact Sheet" in Appendix A, Case Documents 1. Assume that you are going to draw information from this document for a report on treatments for AIDS. Are there sections of the fact sheet that could be presented in any of the following organizational patterns: chronological, spatial, classification/partition, comparison/contrast? Develop outlines for each pattern that you can see in that information.

Meyer, Paul R., and Stephen A. Bernhardt. "Workplace Realities and the Technical Communication Curriculum: A Call for Change." In *Foundations for Teaching Technical Communication: Theory, Practice, and Program Design.* Katherine Staples and Cezar Ornatowski, eds. Greenwich, CT: Ablex, 1997.

WORKS CITED

3

Audience and Technical Communication

Where is the audience located? In the text? Outside the text? Or somewhere in between? The answer is all of the above and, at the same time, none of the above. We talk about audiences in different, sometimes contradictory, ways. Indeed, we cannot help but do so, for the term is one of those, like "writer" and "style," that defies our efforts to pinpoint its meaning.

> James Porter, "Preface" to *Audience and Rhetoric.*
> Englewood Cliffs, NJ: Prentice Hall, 1992. p. x.

In Chapter 2 we advised you to spend time and effort planning—to think about your goals before you begin writing. We did so because studies have shown that on all but routine assignments, successful technical communicators spend about three-fourths of their time not writing sentences and paragraphs and words but *planning* what they want to go into those paragraphs and sentences, *developing* their information and arguments, and later *revising* their drafts into professional work. A good communication *product*—whether an engineering report or an environmental impact statement—depends on a good production *process*. And part of that process, as we indicated, involves thinking about your audience.

This chapter suggests some general ways of thinking about audience that should help you be more productive. Chapter 2 introduced you to how thinking about audience can help you plan your documents; other chapters will have more to say about this important topic. Thinking about audience is not something that you do at one point in the production of a communication and never again; rather, considering audience is an ongoing activity for good technical communicators. They think about audience as they conceive of their task; as they develop and organize content; as they draft individual sentences; as they create tables, graphs, and illustrations; and as they test the effectiveness of their document. In this chapter, we concentrate on how you can think about audience in the process of *invention*, in the act of developing your content. Our goal is to get you to think about audience in ways that will influence what you decide to put into your documents.

The Meanings of Audience

What exactly is "audience" anyway? The answer to that question is, of course, easy to supply in many cases; audience refers to the person or persons who actually hear an oral communication, actually read a written one, or actually interpret graphic displays of data. Audience denotes the real consumers of communications. The concept of a **real audience** is especially concrete for you when you are speaking because a listening audience is sitting right in front of you and is hard to ignore. Audiences who are lost or confused by an oral presentation can make their discomfort known by body language ("If this thing doesn't end soon, I'm going to scream," their bodies say), by their closed eyes and steady breathing, or by their verbal responses ("Excuse me, Michael, but I'm really confused by what you just said").

Audiences for written communications can be nearly as real (and nearly as responsive) as the ones you see when you give a presentation. If you are writing to people you know well or with whom you work closely, your audience is real and immediate indeed. For example, turn to Appendix A, Case Documents 3, where you'll see a letter that was sent to a university

real audience
the audience who actually hears, views, or reads a document

dean about trapping and killing birds in experimental fields. The letter writer addresses a real and immediate audience, the dean of the College of Agriculture. The letter writer announces a topic of concern and calls for a specific action from that audience.

When we use the term *audience* in this book, we are most often referring to that real, immediate, and concrete presence—the audience for whom you hoped to solve a problem, as you read in Chapter 2. But there is also a second way of thinking about audience: Audience can be imagined; it can be a construct in our minds that helps us to compose our documents. In this sense, an audience is an imaginative concept, rather than a flesh-and-blood presence. You probably are aware of this writer-centered sense of audience from your own writing. You have an awareness of individuals whose presence in your mind helps guide your decision making during writing. This imaginary, writer-centered concept of audience can be especially important when your sense of "real audience" is rather vague, as it is when you are writing to readers whom you know barely or not at all. In these cases, audience can be a pretty abstract notion, as abstract as the proverbial "dear sir" or "dear madam" at the top of a form letter.

Having the audience in your mind as a concrete and useful concept is important whether or not you know the real people who will ultimately encounter your work. As a writer you should think of "audience" not only as a group of real people who encounter your work, but also as a concept in your mind that will help you create effective documents.

Even if your real audience doesn't have a concrete presence for you, you can make audience a creative concept. Using audience as an element in your mind when you don't know your real audience will help you to develop more effective documents and presentations than if you created documents only to please yourself. Whether the term *audience* refers to a real presence or to something in the mind of the writer, there are questions about audience you can ask that will strengthen your communications.

Finally, before the chapter ends, you'll also be introduced to a third (and rather sophisticated) way of thinking about audience. Audience can also be a textual presence—a "character" you can actually create in your documents in order to help you meet your rhetorical aims.

imaginary audiences
audiences that are constructed in the writer's or speaker's mind

textual audiences
audiences found within the pages of documents where they help writers communicate more effectively with readers

As you can see by now, audience can be located in several different places. *Real audiences* are found wherever people use discourse—in assembly halls, in study or work cubicles, wherever people read. **Imaginary audiences** are located in writers' minds, where they can animate the writers' ability to invent effective things for real readers. **Textual audiences** are found within the pages of documents, where they help writers to communicate more effectively with readers. Let's examine each of these three meanings of audience in turn.

Ask yourself early in the composing process who will actually encounter and use the document you are preparing. It's important to know who your real audience is, and many times it is simple enough to identify that audience. If you are speaking before a group of people or sending an e-mail communication to an individual you know, your audience is easy to identify. At other times, it's not always so simple to figure out who that audience is—for example, when you write to someone you've never met or when you prepare a Web page that will be consumed by strangers far away. To get an inventory of everyone who might read your documents, you should think about whether your audience is simple and homogeneous, or complex and heterogeneous, and you should think about how your documents will be used.

Real Audiences: Determining Who Will Read Your Document

SIMPLE, HOMOGENEOUS AUDIENCES

When you write a letter to a personal friend, you know exactly who your audience is, and you know that no one else is likely to see your letter. In technical writing you will sometimes communicate with a single individual—when you write routine memos and e-mail to a colleague or collaborator, for example, or when you write personal or semipersonal letters to clients and customers. These single and well-known audiences are simple to identify, and they are often simple to address.

Other times your real audience might consist of many individuals, but you can still conceive of them as a simple audience because their needs are all roughly the same. For instance, if you give an oral presentation to a group of colleagues with similar backgrounds and needs who are gathered together to hear you for a similar purpose, you can think of them as a **simple, homogenous audience.** If you write an article for a trade journal, you can address everyone in your audience in rather similar terms because your readers can be considered homogeneous: They have the same needs and responsibilities, and they read with much the same purpose in mind. Although many individuals read the journal, all can be addressed as if you were writing to one of them.

simple, homogeneous audiences
readers, viewers, or listeners with similar backgrounds and needs who use a communication for a similar purpose

MULTIPLE AUDIENCES

In other situations, however, it is not always easy to identify the audience of a communication. There may be **multiple audiences** composed of several readers, viewers, or listeners now or in the future, and inside or outside the organization. Example 3–1 reproduces a deceptively simple memo from a university writing coordinator, Robert Topping, to his supervisor, Marianne Angell. Who else besides Professor Angell might read this document or receive a copy?

multiple audiences
several readers, viewers, or listeners in the present or future, inside or outside the organization, who receive and use a communication

EXAMPLE 3–1   A Memo to a Complex Audience

Megalithic State University
Interoffice Correspondence

DATE: October 17, 1999
   TO: Marianne Angell, Head, Department of English
FROM: Robert Topping, Director of Writing
 SUBJ.:   Recent meeting on placement of undergraduates

As you know, Marianne, last month the Writing Committee, after lengthy debate, recommended that the department should seek to improve the method used to place our students into the required first-year writing courses here at Megalithic State. I just wanted to bring you up to date on recent developments.

The committee (and hence the department) wishes to get students placed more reliably in the appropriate course. Most students complete their writing-course requirement by enrolling in English 101, Rhetoric and Composition, but about 15 percent take English 101H, the "honors" version of the course; and another 15 percent must first complete English 100, Basic Writing, before they go on to English 101. Students enroll in the three courses on the basis of the First-Year English Placement Test administered by the Academic Testing Program during the summer orientation program. The trouble is, many English faculty members (and the committee as a whole) feel that a great many placements are incorrect. Faculty report that many students placed into English 101 and 100 belong in the other course, and they feel that many students now placed into English 101 actually belong in English 101H. These incorrect placements result from the fact that the Placement Test is an indirect test—a multiple-choice, machine-graded exam—instead of a direct test that assesses students' actual writing samples. The committee, particularly Professor Sarah Maxwell, also feels that our indirect test is unprofessional: Most experts in testing agree that indirect testing is far less accurate than direct tests, particularly for nontraditional and minority students.

Consequently, the committee asked me to investigate alternatives to our current testing method. I shared our concerns informally in meetings with Stanley White, Coordinator of the Academic Testing Program, and several of his colleagues; with Gene Poritzky, who chairs the University Committee on Testing; and with Carol Thigpen, Dean of Undergraduate Studies. Dean Thigpen understands our perspective on this issue and hopes that a direct writing test can be incorporated into the summer testing program. Mr. White and his colleagues, on the other hand, are worried about changing the testing program. He wonders how time for an expanded direct test can be carved out of a busy orientation program, how these tests can be scored quickly and reliably enough to return results to students in time for registration, and who will pay to score a direct test.

I am personally convinced that all of Mr. White's objections can be answered. In my memo of September 21 to him, I tried to address his concerns. Dean Thigpen feels we all should get together to discuss the testing plan and has asked me to schedule a meeting for early November. November 8 or 9 might be good possibilities. Could you ask Brenda Chait as administrative assistant to contact everyone on this memo and arrange for a mutually convenient meeting time? Would you be interested in attending the meeting yourself?

This memo seems to be a simple report on the efforts to change the testing procedures used in the writing program at the university. The audience seems to be easy to identify: It's Professor Angell, the recipient of the memo. But with a bit of thought, you can easily imagine that others will also be likely to receive or read the memo, and the memo is a very complex attempt to address several readers at once.

Besides Professor Angell, Topping is also addressing several other people in his own department. First, there are obviously members of the Writing Committee, who want to know how efforts at reform are coming along. At least one important member, Sarah Maxwell, is addressed directly; but the other committee members are also very likely to receive copies of the memo. Second, Brenda Chait will need a copy of the memo in order to arrange the next meeting on the subject.

Then there are people outside his department that Mr. Topping is addressing both directly and indirectly. You can tell that he is planning to send copies to them because of the way he elaborately outlines the current situation and because the final paragraph prepares them to receive a phone call from Brenda Chait. The other readers include the following:

- Mr. White, who apparently opposes the proposed plan but who must be acknowledged with respect.
- Dean Thigpen, who apparently doesn't know much about the testing program (hence, all the detail in paragraph 2 in a letter ostensibly to someone who knows that information already) but who still supports the reform (and who does not wish to alienate Mr. White).
- Members of the University Testing Committee, including Mr. Poritzky, who have to agree to any changes in testing policies.

Finally, there might also be a host of future readers who will encounter the document:

- People who might be called upon to help implement the change, once it is agreed to, and read the memo as background.
- Other people who might need to have background on the matter some years in the future, such as Mr. Topping's successor.

You can probably think of other possible future readers yourself. Even a short memo like this one can have a very complex readership. The nominal audience for a document—in this case, Professor Angell—may not even be the most important audience for a communication at all.

To get a good inventory of all your readers, think carefully about all those people who might encounter your message now and in the future, both inside and outside your organization. Take note of the people to whom you'll send copies yourself, as well as the people who might be shown the document by the people who receive it, and do your best to address all of those readers in your communication. Imagine what might happen to your work after you turn it in. Who will use it? For what purposes? Where will it be sent in the organization? When you join an organization, you usually pick up quickly a sense of how things get done and a sense of who likes to be kept informed on what. Incorporate that sense into your inventory of audiences and write with them in mind. In addition, think about who might read your communications in the future. Some of your readers will discard your communication immediately, but others will file it for future use. What will those future uses be? And how can you facilitate future users in how you communicate now?

SINGLE AUDIENCES WITH MULTIPLE NEEDS

audiences with multiple needs viewers, readers, and listeners who may seem similar but have different needs for a communication

Sometimes a single or homogeneous audience may have **multiple needs.** Take, for instance, the user manual that came with your computer, CD player, or bicycle. You may be the real audience for that manual, but it will be a *very different you* who uses it each time. When you consider buying the computer, for instance, you might take a look at the manual to learn about its features and capabilities. In that sense, the manual addresses you as a potential buyer. Once you do buy the machine, you'll want to install it, and the manual must address you as an installer. Then you'll want to use it, so the manual must address you as a novice user. Later, as you use the computer and become more confident, you'll return to the manual, now as an expert, to pick up the nuances, to perform repairs, or to troubleshoot.

Or, to consider a second example, think of the various needs addressed by a booklet describing the benefits enjoyed by employees of a certain company. Because it is read by potential employees who are considering joining the company, it must address readers as a kind of recruiting device—it must convince new employees to sign on. Once the new employee joins an organization, that employee might read the booklet to learn exactly what benefits he or she possesses (the employee might want to consider purchasing additional health or life insurance, for instance). Later, if the employee experiences an illness or injury, the booklet might be consulted because the employee wants to learn whether he or she is covered or how to submit a claim for reimbursement for a medical expense.

In short, individual readers often wear several different hats. When you take inventory of your audiences, remember to account for those different hats, those different uses your document might be put to. From the perspective of the writer, individuals are complex and multiple to the extent that they have multiple needs. Each reader might, in some cases, represent a different audience at different times.

MULTIPLE AUDIENCES WITH MULTIPLE NEEDS

Finally, some documents are addressed to multiple readers who have multiple needs; not only must you address a variety of users in such documents, but you must think about their various (and sometimes conflicting) needs. Think of all the different audiences who come to a company's Web site. Look at the Case Documents 4 dealing with the Year 2000 (Y2K) problem that everyone was concerned with in 1999 and see how the writers have designed the Web sites to accommodate the many people—from expert to novice, from concerned citizen to technician—who might encounter them. Or think of the audiences of an Environmental Impact Statement: Such statements must address experts, government officials, and the general public; and they must offer conclusions, background information, and documentary evidence in a form that can serve readers over a period of some time. Obviously these complicated items are challenging to write because the team that produces them must address so many needs, some of them contradictory.

SOME ADDITIONAL ADVICE

Simple audiences. Multiple audiences. Future audiences. Possible audiences. Simple audiences with multiple needs. Identifying with any certainty the real audiences of anything you write can get pretty difficult. In fact, identifying the real audiences is often impossible because you can't control with any certainty who will actually read your work.

But identifying who actually will read your work is less important than identifying who *might* read it—and preparing for as many possible readers as you can, within practical limits. As a writer, you need to anticipate as many readers as possible so that you can design a document that is as useful as possible. In a later chapter, you'll learn how to adapt documents to your many readers—how to address heterogeneous, multiple readers in one document. For now, take inventory of your audiences by thinking about who will receive your communication directly from you, who will receive it indirectly (from someone else), and who might read it in the future. Then think about any multiple needs that your communication might have to satisfy. If you do all that, you'll have a good sense of your real audience, and you'll have begun the process of addressing those needs.

Once you've identified your readers, you can begin thinking about how to adjust your communications to them. Early and often in the writing process you should be asking three basic questions about all of the audiences that you turn up in your inventory:

Gathering Information on Real Audiences: Determining Their Needs, Attitudes, and Knowledge

1. What are the needs of my audiences?
2. What are the attitudes of my audiences toward me and toward this message?
3. What does my audience know—and not know—already?

Asking questions like these, first formulated by Richard Young, Alton Becker, and Kenneth Pike in their book *Rhetoric: Discovery and Change*, will help you develop appropriate content for your communications.

WHAT ARE YOUR AUDIENCE'S NEEDS?

We have discussed to some extent the importance of assessing your audience's needs before you formally prepare your communication. You learned to think of yourself as a problem solver and to think of your audience as someone with a problem that you need to solve. That is, you learned to address your reader's needs. You learned to think of your purpose as an attempt to alleviate problems in specific ways.

Another consideration will help you to develop appropriate content for your communications: Think of your **audience's roles.** Within those roles, what will your audience do as a result of reading or hearing your communication? Will your audience transmit the message elsewhere? Will your audience make a decision or take some action? Will your audience advise a decision maker or action taker? Will your audience see themselves as learners? Or will your audience implement a decision or action?

audience roles
the roles of transmitter, decision maker, advisor, or implementer that an audience takes after reading, viewing, or listening to a communication

Transmitters

Transmitters are people who receive a message and then simply direct the communication to the appropriate decision maker or action taker. For example, people who write grant proposals often submit them to people who simply receive the applications and pass them on to the advisors who actually rate them. Similarly, job application materials are often submitted to someone who receives the applications and passes them along to other people who do the actual hiring.

transmitters
audience members who direct a communication to a decision maker or action taker

Of course, not every message is directed to a transmitter; quite often the person addressed in a document is not a transmitter but the person who is actually involved in the action of the communication. But transmitters exist often enough that you should consider their needs, typically by making it easy for them to direct your communication to the right audience. Transmitters especially appreciate summaries, subject lines (in memos), title pages (in reports), introductions, and other references that permit them to route communications to the right people without having to read through the entire work.

decision makers and action takers
audience members who use a communication to make decisions or conduct activities

Decision Makers and Action Takers

Decision makers and action takers, by contrast, may often be considered your primary audience. For them, you should provide the information and arguments they need to make decisions or to do the activities described in your communications. Much of the information in reports, proposals, instructions, and other technical documents is included with these readers in mind. Action takers and decision makers often hold executive positions within organizations and have many things competing for their attention; they appreciate completeness, care, and thoroughness; they may also be pressed for time and appreciate overviews, concise presentations, and clear graphs and figures.

Advisors

If a decision maker lacks the time or expertise necessary to take action as a result of a communication, he or she may rely on advisors. **Advisors** are often given the role of reading communications carefully and offering guidance to decision makers. Advisors thus appreciate the details of a presentation, even particulars that are included in appendixes. If you expect that your report, proposal, or other communication will also be read by advisors, you will need to anticipate their needs as well as those of the decision makers as you draft your work.

advisors
audience members who use a communication to advise decision makers

Learners

Sometimes readers come to a document with the goal of learning. **Learners** see themselves as consuming a kind of tutorial on the subject at hand. Readers of manuals obviously are trying to learn how to perform some function, but a great many other documents are written to instruct or inform readers about something. Example 3–2 shows the type of trade publications audiences of farmers and agronomists turn to when they want to learn about effective agricultural practices.

learners
audience members who use a communication to learn how to perform a task

---

EXAMPLE 3–2   Audience as Learner

There are several key factors for getting low cost per bushel and high, or at least decent, net profits from soybeans, say Phil and Mike [McLain, winners of the North Carolina Most Efficient Yield contest]. But at the top is no-till, with crop rotation probably second overall. They started working into no-till in 1978 and have been 100 percent no-till for at least eight years.

"No-till is the one single factor over the years that has done the most to help us be more cost effective," says Phil . . . Mike agrees. "In no-till, we're making fewer trips over fields, and that alone cuts probably $15 to $20 an acre off our costs. No-till also increases moisture retention, which helps with yield."

Source: Syl Marking, "Cost-Cutting Kings," *Soybean Digest,* March 1999, pp. 16–17.

Implementers

implementers
audience members who use a communication to achieve a specific end, who initiate the action recommended in a communication

Sometimes your audiences may be in the role of **implementers**, too. Once a decision is made, executives may take action themselves or they might turn a communication over to someone else for implementation. If you anticipate that this will be the case, you must be sure to design your communications with implementers in mind. You'll have to provide the details on implementation that such readers will require.

In real life, of course, roles often shift and overlap, and one person may carry several roles. Nevertheless, thinking about your audience's roles—thinking of them as transmitters, action takers, decision makers, advisors, learners, and implementers—suggests what you ought to include in your communications.

WHAT ARE YOUR AUDIENCE'S ATTITUDES AND MOTIVATIONS?

audience attitudes
motivations or emotions that determine how an audience may respond to a communication

In addition to thinking about your readers' roles, you need to consider **audience attitudes** and motivations—their attitudes toward your message, their motivation to read or listen to your message, their attitudes in the act of reading, and their attitudes toward you.

Attitudes Toward Your Message

If readers were dispassionate people who take in messages without getting emotionally involved in them, life would be convenient. But in the real world people are emotional as well as reasonable, and they often have preformed attitudes about things that you must take into consideration as you write or speak. These attitudes are related to your readers' motivations for using a document as well as their prejudices and assumptions about its content.

Look again in Appendix A, Case Documents 2, at "How Tiny Organisms Helped Clean Prince William Sound," the public relations release concerning the cleanup of Prince William Sound in Alaska after an accident led to a major oil spill. This document includes sentences such as, "Alaska's Prince William Sound today is essentially clean" and "Repeated tests have shown that in Prince William Sound bioremediation was both effective and safe."

What are your attitudes toward those sentences? What might be the attitudes of committed environmentalists? What about the attitude of typical executives (if there is such a thing) in the oil industry? What would those audiences think of such terms as "essentially clean" and "effective and safe"? Depending on the readers' prejudices and assumptions, the phrase "major oil spill" rather than "major environmental disaster" would be met with different attitudes.

Obviously, not all audiences share the same attitudes or share yours. Therefore, even in less controversial circumstances, you must think carefully about your audience's attitudes toward what you have to say or write.

The information and arguments contained in your documents and speeches need to be considered in light of your audience's attitudes.

If possible, anticipate those various attitudes and act accordingly. Often you will have to include material in your communication that will show your awareness of differing attitudes, and you will want to include material to counter those attitudes without insulting your audience. As you will learn in the next chapter, you will frequently want to change your readers' attitudes through your writing. As a general rule, the more your audience's attitudes differ from your own, the more you will have to address and counter their attitudes in your communication. It's one thing to state a position on an issue; it's another to include everything necessary to change your audience's views on that issue. As a start, do what you can to determine your audience's attitudes before you write or speak—and then create your document accordingly.

Motivation to Read or Listen to Your Message

A particularly important attitude toward your message is the motivation of your audiences. How motivated are they to hear your presentation or read your document? With many things competing for their attention, you should not assume that your audiences see your communication as important as you do. Assess their level of interest and act accordingly. Some people may be eager to hear you out while others may be bored; some may be interested in the whole presentation, others in only certain parts. If you think for a moment how quickly you scan the contents of a Web page before deciding whether to look carefully at it, you'll have a sense of how difficult it can be to interest the general public in reading something they don't have to read. Even at work, readers are tempted to ignore or defer reading a great many things that are offered to their attention.

If your audience is unlikely to be interested in your presentation or document, you may wish to begin by **dramatizing** what you have to say in order to win their interest. An effective way to do this is to stress in your introduction the importance of the problem you have been working on so that your audience will see its significance. Unmotivated readers also might appreciate more visual presentations—illustrations, graphs, and tables in written communications; slides, photographs, and overhead projections in oral presentations; and graphics and animations in Web documents.

dramatizing
a tactic to use with unmotivated audiences to involve them in the communication, including emphasis and visual display

If your audience seems unmotivated, it usually makes sense to be brief. People don't mind paying attention for a short time to something they consider unimportant, but they do become impatient if they are asked to spend a great deal of time with what they perceive to be unimportant.

When dealing with audience attitudes, acknowledge that you are aware of a variety of motivations among them. When you give a presentation, for example, you might want to acknowledge that variation by saying, "I know that some people might find this presentation a bit dry, but if you will bear with me, I promise a good payoff in the end." If you try using a technique like this, be sure that your statement isn't an apology for your work but rather an acknowledgment of difference.

Attitudes Toward You

Your audience's attitudes toward you are as important as their attitudes toward your message. Does your audience know you at all, either by reputation or from previous work? Are you considered a colleague, a superior, or a subordinate? Are you regarded as an expert or a novice? How your audience regards you will affect the tone and presentation of your message. In many cases, particularly as you address audiences outside your organization or readers elsewhere in your organization, your audience will know little about you. When that is the case, you can build trust in your readers by following the advice about what we'll call *ethos* in Chapter 4. When you finish that chapter, you'll know how to build your credibility; for now, we want you simply to assess what your readers think about you and to begin thinking how you might counteract any negative attitudes that they might hold.

Attitudes in the Act of Reading

The preceding sentences might give the impression that an audience's attitudes are relatively firm before a communication is encountered and that attitudes change only after an audience has completely taken in a communication. Attitudes are more difficult to chart than that, and responses to communications are actually made continuously, not reserved until the end.

Communicators and their audiences are highly active, and there is a dynamic interaction between speakers and listeners, writers and readers. This active, dynamic communication is a process of intellectual negotiation that occurs in a lively (and, unfortunately, somewhat uncontrollable) social setting.

If you think about your own reading processes, you will remember that they are anything but passive. You have no obligation to passively "receive" a writer's message simply because it is presented to you. Rather, you expect writers to attend to your needs and not to express themselves without considering their audience. As you read, you actively make notations, you mentally take issue with some points and agree with others, and you may reread some passages if they are confusing or skip others in frustration. In other words, as a reader you actively create meanings relevant to your own needs and experiences, rather than simply decoding the writer's intended meanings.

Moreover, you don't necessarily accept the writer's invitation to start at the beginning of a message and proceed directly to the end. Instead you might preview the reading first by inspecting pictures and headings, maybe even the final paragraph. Sometimes you may read every word of something and go from the first word, in order, to the last. Personal correspondence and e-mail messages often get consumed that way. But more often you read selectively: Some sections get your full attention and may even get reread, others you read straight through without much reaction, still others you might skip completely, and sometimes you quit reading altogether after taking in a few sentences. Even documents that interest readers are read in ways that suggest the readers are very active indeed. For example, you read some things at one sitting with full attention, but other times you break the reading up and take it in at several sittings and read other things in between.

In keeping with this view of reading, then, we advise you to see your audiences as reading very actively and responding as they read. Anticipate possible responses and adjust your presentation accordingly. Understand that the way your audience responds to one part of your communication will affect their responses to other parts. See your reader as a dynamic presence who must be led through your documents like a respected colleague.

WHAT DOES YOUR AUDIENCE ALREADY KNOW—AND NOT KNOW?

You can also adjust your communications by asking yourself what your audience already knows and doesn't know. As you plan your document or presentation, consider your audience's organizational role, expertise and education, and familiarity with the subject. In general, size up your **audience's knowledge** before you decide what to include in a communication. Here are some things to consider that might help you to do so, particularly if you do not know your audience well.

audience knowledge
what your readers, viewers, and listeners already know about the topic of your communication

What Are Your Audience's Organizational Roles?

Your **audience's organizational roles** can provide useful clues to their needs and to what they already know. The tasks particular employees handle give them specialized knowledge. Managers, for example, know different things than technicians. You will want to consider the job titles (e.g., environmental engineer, nursing supervisor, assistant manager, group leader, systems analyst) that describe your audience and think about what those roles imply about your audience's knowledge about the topics of your communication.

audience organizational role
the task a reader, viewer, or listener is assigned, including job and professional titles

What Expertise and Education Does Your Audience Have?

People are the product of their experiences. Thus, taking into account your audience's expertise and education may help you to determine how much they know. If you know your **audience's background,** you will be able to assess with some reliability whether you are including enough information for them and whether your vocabulary needs adjustment. You will need to ask whether your audience reads or speaks English well, or at all. If you do not know much about your readers, make an educated guess about their background or, better yet, do some research (informal interviews and surveys are often easy to do) to learn more about your readers before you write.

audience background
the expertise and education readers, viewers, or listeners bring to a communication

When you think about your audience's education, think less about how much formal education they have had than about what they have studied. For after all, a reader's educational level often has little to do with his or her job-acquired special knowledge. Far more important is whether your audience has had any experience with the ideas and concepts you plan to present. If your document involves specialized knowledge—information about technology or scientific processes—it is especially important for you to consider your audience's expertise and education.

Is Your Audience Familiar with the Topics and Terminology?

audience specialty
the topics, terminology, concepts, issues, controversies, and techniques that readers, listeners, and viewers understand

Related to the question of expertise is **audience specialty**. If your audience knows your specialty well, you can count on a shared knowledge of concepts, issues, controversies, and techniques, and you may not have to provide much in the way of background information. On the other hand, if your audience comes from outside your specialized area or if you are dealing with an issue with which your readers are unfamiliar, you will need to provide explanations and accessible vocabulary throughout your communication.

The famous British evolutionary biologist John Maynard Smith once told one of the authors of this textbook that when he writes he often has a very specific image of his audience. "I usually try to keep in mind an actual person, someone who happens to be a relative of my wife, an intelligent but rather ignorant British civil servant who is bent on improving his mind." You may not be able to generate a formal portrait of your audience like this, but the more you *do* to ascertain what your audiences know and don't know, the more successful you and your communications will be.

Thinking about the Conditions Under Which the Audience Receives Your Message

reading conditions
the circumstances in which a reader encounters, consults, and uses a communication

Once you have inventoried your audience and considered their needs, attitudes, and knowledge, consider the conditions under which those readers will encounter your document or listen to your oral presentation. Most people behave as if their audience will encounter their communications under ideal circumstances. They anticipate communicating with a reader who is never interrupted, who is passionately interested, and who has unlimited time. They write as if everyone will be reading their work in a quiet study.

Your own experiences tell you that **reading conditions** are often anything but ideal. Sometimes you are hurried and can read only superficially. Sometimes you read parts of documents and skip irrelevant sections. Often you read in snippets, getting interrupted now and again as you try to navigate a text. And sometimes you read in less than ideal physical circumstances; for example, reading from a computer screen can be difficult, uncomfortable, and inconvenient. Maybe you have had the experience of trying to read a manual under the car as you do a repair, to use a document on a construction site in the wind or rain, or to read a document quickly to cope with an emergency. As a writer, you need to anticipate the conditions under which your document will be used.

Try to picture when, where, and how your documents will be consulted. As you work with this aspect of audience, do your best to plan for a document that will work wherever and however it is needed. Early recognition of what is needed on your part is essential if the final product is to be successful. In later chapters you will learn to construct tables of contents, indexes, headings, and other devices for helping readers manage more easily when reading conditions are less than ideal. You will also learn to think of other design considerations (even down to type sizes, paper styles, and cover design) as you think about ways to help your audience manage the conditions under which they read.

Two concrete examples illustrate and summarize the advice we've just given, particularly about audience needs, attitudes, and knowledge.

Audience Analysis: Sample Working Papers

SAMPLE WORKING PAPER ONE

Listen in (so to speak) as Ming Tchen thinks about the needs, attitudes, and knowledge of the audience of a manual he is writing for the director of his summer internship (Example 3–3). A prospective industrial engineer from Kansas City who has just completed his junior year of studies, Ming has been working all summer as an intern for IE Consulting, Inc., a firm that specializes in industrial engineering work. Ming's supervisor (Paula Grecco), recalling all the time it took to get Ming acquainted with the job and up to speed on everything, has given him a final task: to write a manual that Ming's successor can use to prepare for the same internship next summer. To be sure that his information will be complete, Ming considers his successor's needs; to do so, he remembers his own needs from earlier that summer.

Because he doesn't know his reader personally, he thinks of his audience as the self he was four months earlier. That reminds him to tell his successor what he or she should know about IE Consulting and about his supervisor; what he or she should do to prepare for the internship; the details about the duties of the internship; and so forth. Ming would have to think about what Paula wants the new person to know. So Paula is also an audience.

Having thought about his reader's needs, he next considers that reader's attitudes. On consideration, Ming thinks that his reader will have ideal attitudes. He assumes that reader will be eager to learn all he or she can from the document Ming will prepare, and so is likely to read the document with great interest and motivation. He also feels that his reader is likely to have positive attitudes about Ming himself: Ming can count on his reader's respect because he comes from the same university and majors in the same field, and has had the benefit of just the kind of experience that his reader is seeking.

EXAMPLE 3–3   Thoughts About an Audience's Needs, Attitudes, and Knowledge

Ming Tchen's Informal Chart

My audience is simple and homogeneous—I'm writing my successor in this internship. My audience isn't Paula—even though Paula has given me this assignment; my reader is my successor and I'm writing for that person's approval (as well as for Paula's!). Paula is just my transmitting audience. I guess I can imagine that I'm writing to the person I was three or four months ago. My goal is to prepare him or her to take my internship next summer without wasting a lot of time getting organized and oriented.

Needs: To know about IE Consulting—the company's history and specialties, key employees and company policies. What the internship involves and what you can expect to do (dwell on some examples?). A profile of Paula, the internship supervisor? Miscellaneous stuff too—parking, lunch, reimbursement policy, pay, and so forth. In general, teach everything I learned in the first two weeks. For Paula, my transmitting audience, I'll have to compose a cover note to attach to my communication.

Attitudes: My successor's attitude toward what I have to say should be ideal—he or she will be eager to read what I am offering and very curious. Highly motivated reader. As for attitudes toward me: I'll identify myself as this year's intern, mentioning that I too am an IE student at Tech. That way, the reader should be friendly to me.

Knowledge: Again, not a big problem—the person should know everything that I knew in May, so I can assume lots of IE knowledge (no problem with terms and concepts). IE 436 prepares people perfectly for this internship, so I'll build on that in detail and use comparisons to what was covered in that course. My successor will have no real knowledge of IE Consulting, so be careful there—details, details!

This shouldn't be too tough to write after all!

As for his reader's knowledge, Ming feels that he is again in a good position. Because his reader will have a similar educational background, Ming can count on him or her to know the technical processes and vocabulary that Ming plans to use, and Ming will also be able to build on his audience's familiarity with Industrial Engineering 436, a course that is a prerequisite to the internship and prepares students for exactly the kinds of experiences they will encounter at IE Consulting. At the same time, Ming's reader will also need some background information on processes and procedures unique to IE Consulting.

SAMPLE WORKING PAPER TWO

Now, listen in as Mandy O'Brien plans an oral presentation for a meeting that she has scheduled for the next afternoon (see Example 3-4). Mandy directs the employee training program at her large manufacturing company, which has a major facility in Ohio and smaller ones elsewhere in the northeastern United States. Employed by her company for two years, Mandy has been

EXAMPLE 3–4   Thoughts About an Audience's Needs, Attitudes, and Knowledge

Mandy O'Brien's Informal Chart

Tomorrow's presentation. Audience: Emilia, Sam, Harry Moore

Needs: They don't know it yet, but improvements are needed in our company's training programs. I've spoken informally to Sam about this, and he tends to agree with me that the current programs are getting old-fashioned. I feel that the ones put on by Harry Moore at our other plants aren't very good at all. They are poorly coordinated with our other facilities and not offered very professionally. In fact, Harry M shouldn't even be offering training programs—it's my job, not his. But some of these programs predate both of us so I've never complained. Still, we can be doing much better both here and at our other locations. I need to establish that early (especially for Emilia, who knows next to nothing about all this) with all that data I've collected and by indicating that I know of better alternatives. Show that our programs are broken and need to be fixed. And get myself assigned to the task of formally proposing a solution—that's my specific goal.

Attitudes: I'm not sure, but I think Emilia will be receptive to my ideas—she's given me responsibilities and she hired me because I could make our training better. She also stands to lose nothing, and she'll be very interested when I show her I might be able to deliver better training for less money. Same with Sam— we've already talked about improvements that we'd like to see at the main office, and I think he'll like some of the ideas I've been developing since we talked last. Harry will be a problem maybe. Will he see it as a personal criticism if I come on too strong about a "problem" with the training at his plants? Got to be tactful there, especially since he doesn't like to get ideas from people like me (I hear he doesn't like getting advice from younger women) and since Emilia will be there. Will she be protective of Harry's views, since he's an ass't VP? Oh well— expect the best, especially with Sam and Emilia there. Harry still sees me as a rookie, too, so I'm going to be a real pro in my presentation—don't let it get too informal; try to keep the agenda on what I've called the meeting to discuss. Sam is friendly and respects me. Emilia probably still thinks of me as inexperienced—I want her to leave this meeting very happy that I was hired; I want her to think of me as a pro.

Knowledge: Emilia has only a vague and general idea of our programs, so I'll have to make sure that my introduction goes over old ground for her. Don't bore the other two, though—Harry and Sam know most of the details of the training programs. But their expertise isn't in training, so I will need to explain my new ideas in at least some detail. Go easy on the jargon! Don't be pretentious. But show that I know my stuff, especially for Harry's sake. Maybe it would be wise to have some of my plans in short written form (but tentative!) to pass out if the occasion presents itself.

noticing for several months some weaknesses in the kind of continuing training offered by the company, especially at its facilities outside the main office.

In addition, Mandy would like to suggest some improvements of various kinds in some of the training programs traditionally offered by the company. So she has called a meeting of the company vice president (Emilia Black), the assistant vice president in charge of production at the main office (Sam Priest, who is also her boss), and the assistant vice president in charge of the other plants (Harry Moore). All of them have only a vague sense of what the meeting is for. Anticipating some reluctance from one or two members of her audience (all of whom outrank her), Mandy has called for and is planning for an informal oral presentation rather than making a written proposal. Her goal is to leave the meeting with a mandate to devise a written plan that can be considered formally at a later date. She feels that it would be unwise for her to draft a formal plan herself without creating some kind of demand for it—hence the meeting.

Mandy has informally recorded her thoughts about her audience for the presentation. Note that although the problems she is addressing are really her own responsibility, she tries hard to conceive of them in terms of her audience's needs.

Then she thinks carefully about her audience's attitudes. Her boss, she feels, is likely to think favorably about her planned changes, because she has spoken to Sam about them in the past. Indeed, some of her ideas for improvements were suggested by her boss—not a bad strategy for getting his support. The vice president (Emilia) is also likely to be cooperative, Mandy feels, because she has no real personal stake in the matter and is always interested in improvements. But the other assistant vice president (Harry) is unlikely to be receptive to her ideas, Mandy feels, because he has initiated his own employee training programs for other plants—some of them precisely the programs that Mandy wants to reform and coordinate. With Emilia at the meeting, he may be especially protective of his training programs, especially since he should have been operating those programs through Mandy's office.

Mandy also thinks about her audience's attitudes toward her. She has a satisfying professional and personal relationship with Sam, but Emilia, the company's vice president, doesn't know her well and probably views her as still somewhat inexperienced. As for the assistant vice president in charge of other plants, Mandy again anticipates trouble. Having come up from the ranks and having worked for many years at other company facilities, he tends to be protective of his territory and to view suggestions as interference. Mandy has learned that he also sometimes finds it difficult to take advice from younger women.

So Mandy is thinking of information to include in her presentation that will show a special awareness of his protectiveness toward his territory and has resolved to use a rather formal tone in her presentation in order to be as professional as possible. As for her audience's knowledge, Mandy is the expert in this case—she's the only one who knows a lot about employee training programs. She plans to provide careful explanations of the proposals that she wants to kick around; the more she thinks about it, the more she thinks it might be wise to have some written explanations to hand out to support her presentation. She'll also have to provide some background information for Emilia.

Whether you use notes like the ones illustrated in Examples 3-3 and 3-4 or whether you use other means of thinking about your audiences, the important thing is—do the thinking. Planning can make the difference between a document or presentation that speaks generally to everyone or specifically to the needs, attitudes, and knowledge of your intended audiences. The questions we have raised about audience so far in this chapter are summarized in Writing Strategies 3-1.

WRITING STRATEGIES 3–1 Questions to Ask About Audience in Preparing Documents and Presentations

Who will read this document or hear this presentation?

- Is your audience simple and homogeneous? or multiple and heterogeneous?
- Does your audience have multiple needs?
- How will all that affect the contents of your communication?

What are the needs of your audience?

- What problem of your audience are you trying to solve?
- What are the job classifications of your audience?
- What are the roles of your audience: transmitters? action takers? advisors? decision makers? learners? implementers? a combination?
- How will all that affect the contents of your communication?

What is your audience's attitude toward you and toward what you have to say?

- What does your audience think of you as a person and co-worker?
- What will your audience's response be to what you have to say or write?
- How motivated is your audience to attend to you and your communication?
- How will all that affect the contents of your communication?

What does your audience know—and not know?

- Do you need to supply any background information, explanations of terms, or other information to your audience?
- What does your audience know already that you can build on?
- What are the professional backgrounds, educational backgrounds, and fields of expertise of the various members of your audience?
- Does your audience have experience with the ideas and concepts you are presenting?
- How will all that affect the contents of your communication?

Under what conditions are the members of your audience reading?

- Is there anything unusual or less than ideal about the time and place that your audience will read your communication?
- How does that affect the contents and design of your communication?

**Creating
an Audience
or a Reader**

So far in this chapter on audience, we've encouraged you to think about your real audience—the actual individuals who will encounter your documents and presentations. But at the beginning of the chapter we also told you that we would be talking about the audience not *of* the document but *in* the document—the audience in the text.

The concept of the audience in the text is rather difficult but important to grasp. We will try to explain this concept with a few examples. Consider first the short poem in Example 3-5. It is called "Spring and Fall: To a Young Child" and was written around 1880 by Gerard Manley Hopkins, a Jesuit priest who died in Dublin in 1889. The speaker in the poem is addressing a young child, "Margaret," who is crying because she feels a sense of loss when she sees the "Goldengrove unleaving"—the beautiful autumn leaves falling off the trees in front of her and turning into late November "leafmeal" (bits of leaves).

Who is the audience of this poem? Despite its title and the presence of "Margaret" in the poem, we know that "Spring and Fall" wasn't really written for a young child after all. Children cannot understand poetry this complicated, and the theme of the poem—to remember that we will all turn to leafmeal one day—isn't something young children are ready for. The *real* audience for this poem is people "grown older" whom Hopkins wants to remind of what is most important to him, namely, the certainty of death. As for Margaret, she is not a real audience but an **audience in the text**—a fictional character created by the author to aid him in making his rhetorical point.

audience in the text
a fictional character created
by the author to make a
rhetorical point

EXAMPLE 3–5   The Audience in the Text

Spring and Fall: To a Young Child

Margaret, are you grieving
Over Goldengrove unleaving?
Leaves, like the things of man, you
With your fresh thoughts care for, can you?
Ah! As the heart grows older
It will come to such sights colder
By and by, nor spare a sigh
Though worlds of wanwood leafmeal lie;
And yet you will weep and know why.
Now no matter, child, the name:
Sorrow's springs are the same.
Nor mouth had, no nor mind, expressed
What heart heard of, ghost guessed:
It is the blight man was born for,
It is Margaret you mourn for.

Gerard Manley Hopkins

Think of it this way: Hopkins creates in his poem a sort of miniature drama, complete with characters (Margaret and the speaker) and action (the leaves falling, the tears, the speaker's sermon). As real readers, we eavesdrop on and observe this created drama so that the author can achieve his rhetorical purpose—to remind us about our ultimate end.

What does all this have to do with technical writing? All writing, believe it or not, can be seen as a sort of verbal drama like the one in "Spring and Fall." It has fictional "speakers" just as this poem has a speaker; it has fictional "listeners" just as this poem has its Margaret; and it has verbal action. Sometimes the fictional listeners—the "audiences in the text," if you will—are as overt as young Margaret. For example, the "letters to the editor" in your local newspaper are really directed not to the editor but to readers in the community. The passage in Example 3-6 is from such a letter urging the State of Minnesota to continue its emissions testing program. The audiences in this text include people in the community as well as members of the legislature. Writing to the editor as the audience in the text is just a device for creating more drama and getting heard by other audiences.

So far we have focused on dramatic fictional characters within the text in order to explain what we mean by an "audience in the text." But every document is addressed to a fictional reader that is present, or implied, within the text. Occasionally, as our examples indicate, that character is indeed an overt presence in the text. But more often—especially in technical writing—the reader in the text is only implied. Look again at Example 3-2, which deals with increasing soybean yield. The reader implied in that example is not a specific farmer, but a fictional farmer interested in new techniques for improving yield.

EXAMPLE 3–6   The Audience in the Text

Continue Emissions Testing

If I were diagnosed with high cholesterol, subsequently went on a low cholesterol diet, reduced my cholesterol to an acceptable level, and then went back to eating fatty foods, I'd be pretty stupid, right?

After eight years of emissions testing, the Environmental Protection Agency is expected to certify that the metropolitan area is now in compliance with federal air-quality standards for carbon monoxide.

The Minnesota Pollution Control Agency says the program has kept 400,000 tons of carbon monoxide out of the air.

Now Rep. Barb Haake, R-Mounds View, wants to do away with the emissions testing program, because it "is not effective."

Say it with me. "Stupid!"

—David P. Sours, Minneapolis

Source: *Minneapolis Star Tribune,* letter to the editor, p. A26, February 13, 1999. Reprinted with permission of *Minneapolis Star Tribune.*

A reader in the text is implied by the language and conventions employed by the writer and by background knowledge assumed by the writer. Just as every document contains an implied author, a narrator (or *persona*, as we call the implied author in the discussion of *ethos* in Chapter 4) who may or may not have a lot in common with the real author, so too documents contain implied readers (created by writers) that may or may not have much in common with the real audience.

Let's look at the two documents in Example 3–7 for another illustration of creating an audience or reader in the text (you will see these documents again in Chapter 4 when we discuss persuasion). The Apple Computer manual creates an implied reader who even has a name, "you." That "you" is a "concerned" type of person, the kind of employee who worries about the company's equipment as if it were his or her own. That "you" is also assumed to be a worker. And that "you" is an informal and approachable worker, the kind of person who likes to drink coffee at the desk and perhaps likes the socializing that goes with coffee drinking at work. In addition, that "you" has a sense of humor.

On the other hand, the "you" in the Motorola manual is constructed not as the informal worker of the Apple document, but as a more formal "you"—the kind of person to whom the writer assumes he or she must say

---

EXAMPLE 3–7   Two Examples of Creating an Audience or Reader

Apple Computer

Many people are concerned at first about making a mistake that will damage their computers or their work.

But you don't need to worry. If you set up your Macintosh according to the instructions, avoid bumping it while it's turned on, and don't spill coffee on the keyboard, you won't hurt your equipment.

Source: Apple Computer, Inc., *Getting Started with Your Macintosh*, "Welcome," p. ix.

Motorola

Thank you for selecting Motorola and welcome to cellular telephone service. Your Digital Personal Communicator™ represents the state-of-the-art in personal cellular telephones today. The listing below shows just a few of the exceptional features that the telephone contains . . .

To cover all of these features properly, we will take you through a logical step-by-step learning procedure that explains everything you need to know to operate your new telephone.

Source: Motorola, Inc., *Digital Personal Communicator User Manual*, "Introduction," p. 4.

"You're welcome" and "thank you" when giving directions. In addition, the "you" is constructed as "logical" and thus presumably unlikely to panic when faced with instructions on how to use an unfamiliar product.

Thus, writers create implied, fictional readers in their texts. They create fictional roles for these implied readers by means of language choices and conventions. And then authors implicitly invite their real readers to assume the roles implied for them. Sometimes the authors of computer manuals, for example, create a reader in the text who is uncomfortable with the machine or needs to be cajoled into using it. The implied reader is given all sorts of commands, warnings, and directives ("do this, don't do that"); the implied reader seems anxious and error-prone (there is a great deal of emphasis on errors and the reader's fears); and the machine seems to call all the shots. Other times the writers of computer manuals create a different kind of reader in the text—a responsible person patient enough to sit still for a tutorial, an adaptable and intelligent and unintimidated person who makes decisions and creates things.

CREATING READERS IN TECHNICAL DOCUMENTS

By now you should be able to tell that creating the kind of reader in your text that your real reader will respond to effectively is important. The question is how you can do that. To understand better how the created reader works in technical documents, compare the introductions of two scientific articles reproduced in Examples 3-8 and 3-9. These two introductions were written for the *same real audience*—researchers in evolutionary biology, most of them university biology professors and their graduate students. We know that because they were published in scholarly journals read by those kinds of readers.

But note that the two documents construct very *different audiences* within the text. Both of them imply audiences made up of professional biologists—that's clear from the scientific jargon ("natural selection," "Darwinian fitness," "adaptation," "constraint," and so on) used in both introductions without the explanation that would be required if those terms were offered to readers outside evolutionary biology. The articles also situate their readers as biologists by invoking some of the shared assumptions, beliefs, and common interests of the field—for example, that "it is assumed that evolution has occurred by natural selection," that "there is nothing particularly new in this logic," and that "we do not impose our biological biases."

But beyond that, very different implied audiences are created. The first implies an audience of scientists who are *only* scientists. By employing passive and impersonal constructions ("It is assumed," "are to be interpreted," "has been a growing attempt"), by treating adversaries with respect and impartiality, by offering standard qualifications, and by keeping the language neutral and dispassionate, the author implies that his readers are idealized,

EXAMPLE 3–8 Introductory Paragraphs from an Article
in Evolutionary Biology

The article creates an implied reader within the text who seems like a student of the implied author.

In recent years there has been a growing attempt to use mathematical methods borrowed from engineering and economics in interpreting the diversity of life. It is assumed that evolution has occurred by natural selection, and hence that complex structures and behaviors are to be interpreted in terms of the contribution they make to the survival and reproduction of their possessors—that is, to Darwinian fitness. There is nothing particularly new in this logic which is also the basis for functional anatomy, and indeed of much physiology and molecular biology. It was followed by Darwin himself in his study of climbing and insectivorous plants, of fertilization mechanisms and devices to ensure cross-pollination.

What is new is the use of mathematical techniques such as control theory, dynamic programming, and the theory of games to generate a priori hypotheses, and the application of the method to behaviors and life history strategies. This change in method has led to the criticism (e.g., by Lewontin) that the basic hypothesis of adaptation is untestable and therefore unscientific, and that the whole program of functional explanation through optimization has become a test of ingenuity rather than an inquiry into truth. Related to this is the criticism that there is no theoretical justification for any maximization principles in biology, and therefore that optimization is no substitute for an adequate genetic model.

My aim in this article is not to summarize the most important conclusions reached by optimization methods, but to discuss the methodology of the program and the criticisms that have been made of it. In doing so I have taken as my starting point two articles by Lewontin. I disagree with some of the views he expresses, but I believe that the development of evolutionary theory could benefit if workers in optimization paid serious attention to his criticisms.

I first outline the basic structure of optimization arguments, illustrating this with three examples. I then discuss the possibility that some variation may be selectively neutral, and some structures maladaptive. I summarize and comment on criticisms made by Lewontin. The most damaging is undoubtedly the difficulty of testing the hypotheses that are generated. The next section therefore discusses the methodology of testing; in this section I have relied heavily on the arguments of Curio. Finally, I discuss mathematical methods. The intention here is not to give the details of the mathematics, but to identify the kinds of problems that have been attacked and the assumptions that have been made in doing so.

Source: Reproduced with permission, from John Maynard Smith and the *Annual Review of Ecology and Systematics*. "Optimization Theory in Evolution," *Annual Review of Ecology and Systematics* 9 (1978), pp. 31–32.

EXAMPLE 3–9   Introductory Paragraphs from an Article
in Evolutionary Biology

The author creates an implied reader within the text who is an equal to the implied author.

The great central dome of St. Mark's Cathedral in Venice presents in its mosaic design a detailed iconography expressing the mainstays of Christian faith. Three circles of figures radiate out from a central image of Christ: angels, disciples, and virtues. Each circle is divided into quadrants, even though the dome itself is radially symmetrical in structure. Each quadrant meets one of the four spandrels in the arches below the dome. Spandrels—the tapering triangular spaces formed by the intersection of two rounded arches at right angles—are necessary architectural by-products of mounting a dome on rounded arches. Each spandrel contains a design admirably fitted into its tapering space. An evangelist sits in the upper part flanked by the heavenly cities. Below, a man representing one of the four Biblical rivers (Tigris, Euphrates, Indus and Nile) pours water from a pitcher into the narrowing space below his feet.

The design is so elaborate, harmonious, and purposeful that we are tempted to view it as the starting point of any analysis, as the cause in some sense of the surrounding architecture. But this would invert the proper path of analysis. The system begins with an architectural constraint: the necessary four spandrels and their tapering triangular form. They provide a space in which the mosaicists worked; they set the quadripartite symmetry of the dome above.

Such architectural constraints abound and we find them easy to understand because we do not impose our biological biases upon them. Every fan-vaulted ceiling must have a series of open spaces along the mid-line of the vault, where the sides of the fans intersect between the pillars. Since the spaces must exist, they are often used for ingenious ornamental effect. In King's College Chapel in Cambridge, for example, the spaces contain bosses alternately embellished with the Tudor rose and portcullis. In a sense, this design represents an "adaptation," but the architectural constraint is clearly primary. The spaces arise as a necessary by-product of fan vaulting; their appropriate use is a secondary effect. Anyone who tried to argue that the structure exists because the alternation of rose and portcullis makes so much sense in a Tudor chapel would be inviting the same ridicule that Voltaire heaped on Dr. Pangloss: "Things cannot be other than they are . . . Everything is made for best purpose. Our noses were made to carry spectacles, so we have spectacles." Yet evolutionary biologists, in their tendency to focus exclusively on immediate adaptation to local conditions, do tend to ignore architectural constraints and perform just such an inversion of explanation.

Source: S. J. Gould and R. C. Lewontin, "The Spandrels of San Marco and the Panglossian Paradigm: A Critique," *Proceedings of the Royal Society of London: Series B (Biological Sciences)* 205 (1979), pp. 581–98. Reprinted with permission of the Royal Society and authors.

almost stereotypical, models of their kind—objective, impersonal, fair-minded, careful, reserved, and reasonable pursuers of factual truths.

In addition, the first introduction creates scientists who are somehow subtly inferior to the scientist writing. It is as if the author-director first puts implied readers into white lab coats and then conducts them to a lecture hall where they are asked to sit and take notes from a master scientist in front of them. The decisive personal "I" that is used to forecast the essay's contents establishes the author as an authority and the implied audience as a relative novice in need of direction. Like a student, the implied reader is directed to take notes—but not to take issue—with what is to follow from this presentation from a confident teacher.

The second introduction, by contrast (see Example 3-9), although it too conceives of its implied readers as professional scientists, creates its readers as *more than* scientists. By beginning with St. Mark's Cathedral and King's College Chapel, Voltaire and Dr. Pangloss, Biblical tradition and art history, metaphors and allusions, the authors create readers who are not merely scientists or specialists; these scientists are *worldly, well educated, and cosmopolitan*. By using the active voice and avoiding neutral phraseology, the authors make their readers less stereotypically scientific. In addition, by using "we" and avoiding heavy, pedantic forecasting, the authors create readers who are equals, not subordinates—readers who are invited to take issue with (not simply take note of) what is presented. If the first introduction cues readers into the role of students taking notes, the second removes readers from school and addresses them as colleagues.

Both of these introductions are well written; both are effective. They illustrate that there is a range of possibilities in scientific and technical writing for how to construct your implied reader. Your job as a writer is, in a sense, to create a drama, to cast readers into certain roles. As the producer and director of this drama, you can't help but create roles and action, so it makes sense to think carefully about the roles you are creating. In Chapter 4, you will learn how to create effective narrators or effective implied speakers. Here we want you to consider what kind of role you are implying for your reader. What specific signals place implied audiences into certain roles? How will your real audiences respond to the implied audience that you create? Are you creating an implied reader who is an ally and trusted partner or are you unwittingly creating readers who are adversaries?

SUMMARY

- In this chapter you have learned how to think about audience—the real consumers of your communications. You have learned that members of an audience are real, immediate, and concrete presences, and you have learned ways of thinking productively about their needs, attitudes, and background knowledge. You have also seen that an audience can be constructed in your mind to help you compose your documents. Finally, you have learned that an audience can be a textual presence that you can create in your communications to achieve certain effects.

■ At times you may address an audience that is simple and homogenous, with the same traits and similar identities. This audience may be easy to imagine and understand. However, you will often encounter multiple audiences or several people both inside and outside your organization who will make use of your communication. Meeting all their needs is more challenging. Finally, you may be challenged by a single audience with multiple needs. For example, your audience may use your communication differently at different times.

■ To understand and address your audience effectively, you need to determine your audience's needs, attitudes, and knowledge. An audience may need to solve procedural, informational, or decision-making problems. However, your audience's organizational roles must also be part of your consideration in creating a communication. If the members of your audience are transmitters, you'll need to make it easy for them to direct your communication to the proper audience. If they are decision makers, they will use your information and arguments to make decisions and to do activities. If they are advisors, they will use your communication to offer advice to decision makers. Finally, if they are implementers, the information you convey will help them to take or direct specific action.

■ You will need to assess your audience's attitudes, assumptions, and prejudices toward your message and to anticipate, address, reinforce, or counter these attitudes within your communication. The motivation or level of interest of your audience can be enhanced by dramatization, visual presentations, and even humor. Your audience's attitudes toward you, their knowledge of and respect for you, will either encourage or discourage them from listening to you or reading your words. Finally, if you assume that your audience is not passive, but constantly engaged in intellectual negotiation as they encounter your words, you can better decide how to appeal to them.

■ Your audience's experience in school and within their organizational roles gives them specialized knowledge that you need to assess in designing your communication. This specialized knowledge makes your audience familiar with topics and terminology, concepts, issues, controversies, and techniques within a given field. The conditions under which your audience encounters your communications—when, where, and how your documents will be consumed and consulted—makes a difference in how you design them. Remember that in our sample working papers, Ming's thinking about a manual for a summer internship and Mandy's preparing for an oral presentation for a meeting on a training problem involved all these aspects. The Writing Strategies 3–1 worksheet in this chapter offers questions that you might get in the habit of using as you think about or construct your audience.

■ Finally, audiences are created in texts. They are characters created by authors to aid them in making a point. Real audiences observe a sort of miniature drama involving audiences implied in the text, and authors invite their audiences to assume these roles, as we saw in the two computer manuals and the two introductions to the scientific articles.

■ In this chapter we have tried to challenge you to think not only about real, concrete, and immediate audiences but also about audiences you might construct in your mind or in your text. These are difficult concepts, but as you read through the rest of the chapters in this book, you will see how these concepts enable you to design effective technical communication in the workplace.

ACTIVITIES AND EXERCISES

1. Look again at the document, "How Tiny Organisms Helped Clean Prince William Sound" in Appendix A, Case Documents 2. The document, produced by Exxon and intended for publication in newspapers and similar publications, emerged from the aftermath of the *Exxon Valdez* oil spill in 1989. After the spill, Exxon used the bioengineering technique described in the article to help clean up the mess and reported on its success in this news release. Read the article not as a student but as a member of the general public. In that role, consider the following:

   ■ What is your reaction to the first paragraph? To the first section? Do you react to the first sentence cynically? Favorably? Neutrally? Do you seem open-minded and interested? Does the article interest you or not? Note: there are no incorrect answers to these questions.

   ■ What do you think is the purpose of this article? Does your sense of its purpose change at all as you read?

   ■ How does your reaction to the first section affect your reading of the rest of the document? Do your attitudes remain consistent throughout the article, or do they change? What specific words and sentences arouse your reactions?

   ■ Do you lose interest at any point? Do you find yourself skipping or skimming certain parts? Did you stop reading altogether before finishing? Did you read the article in the order in which it was written, or did you preview it first or skip around at all?

   ■ What was your general reaction after reading the article?

   ■ What did the authors assume about their readers of the article? How can you tell? How accurate are their assumptions?

   Compare your answers to these questions to those of your classmates.

2. Read the document "Bioremediation for Shoreline Cleanup Following the 1989 Alaskan Oil Spill" that appears in Appendix A, Case Documents 2. How would you describe the audiences of this document? How do you know? What devices do the authors use to engage or create their audiences?

   Write your findings in a brief report to your instructor.

3. Imagine that you are Dr. Karen Wartala, Dean of the College of Agriculture at the University of the Midwest. On August 2 you receive the letter from the Coalition of Citizens Concerned with Animal Rights that is reprinted in Appendix A, Case Documents 3. Remembering to do so not as a student but as Dean Wartala, read the letter as she would. Afterwards, answer these questions (there are no incorrect answers):

   ■ Did you react to the letterhead? Did the fact that the letter is coming from the Coalition of Citizens Concerned with Animal Rights make you put up your guard, or did you still read the letter "neutrally"? Would it make a difference if Dean Wartala herself were a member of the Coalition?

   ■ Did you react to the salutation of the letter—"Dear M. Wartala"? Would a salutation like "Dear Dean Wartala" or "Dear Dr. Wartala" or "Dear Ms. Wartala" have elicited a different response?

   ■ How did you react to the first sentence of the letter? To the first paragraph? To the first three bulleted accusations? Is your response a neutral processing of the information, or are you reacting emotionally to it?

   ■ How do your reactions to the first half of the letter shape your responses to the rest of it?

   ■ Are you likely to call the author of the letter at the numbers listed? Or will you respond by letter? Or ignore this communication?

   ■ How do you react to the fact that copies of the letter were sent to the news media before you have had a chance to respond? Do you resent that Mr. Younger wrote the letter (or copied it to others) before you have had a chance to respond to the accusations?

   Compare your answers to these questions with those of your classmates.

4. Fill in the answers to the questions in the Writing Strategies 3–1 worksheet for a document (or two or three) that you wrote recently for a class, a job or internship, or as part of your work in a social or extracurricular group. Or fill in the answers for a document that you plan to write for a class, for your job, or for another group to which you belong. Be prepared to turn your work in to your instructor.

5. By yourself or along with your classmates, examine carefully the documents reproduced in Case Documents 4 in Appendix A and consider these questions: Who is the implied audience of each electronic document? What textual features create this implied reader? How effective is the writer's choice of this implied reader?

WORKS CITED

Porter, James. "Preface." *Audience and Rhetoric*. Englewood Cliffs, NJ: Prentice Hall, 1992.

Young, Richard; Alton Becker; and Kenneth Pike. *Rhetoric: Discovery and Change*. New York: Harcourt, Brace, and World, 1970.

Persuasion in Technical Communication

Introduction ▌ A Continuing Case to Demonstrate Persuasive Strategies ▌ Ethos: Creating a Reliable, Trustworthy Persona ▌ Pathos: Appealing to Your Audience's Values ▌ Logos: Persuading with "Good Reasons" ▌ Supporting Good Reasons ▌ Which Good Reasons to Use? ▌ Special Case: Arguing Electronicallly

Understanding practical rhetoric as a matter of *conduct* rather than production, as a matter of arguing in a prudent way toward the good of the community rather than constructing texts, should provide . . . a locus for questioning, for criticism, for distinguishing good practice from bad. That locus is not the individual or any particular set of private interests but the human community that is created through conduct . . .

Carolyn R. Miller, "What's Practical about Technical Writing?" in *Technical Writing: Theory and Practice*. Bertie E. Fearing and W. Keats Sparrow, eds. (New York: Modern Language Association, 1989), pp. 14–24.

Although you may not have thought of technical communication as persuasive in nature, it quite often is. You persuade your audience that your solution to a problem will be effective; you persuade your readers that bioremediation is the best technique for cleaning up certain oil spills; you persuade university leaders to question the practice of trapping and killing birds in an experimental crop field; you persuade health care workers to take precautions against HIV infection; and you persuade a future employer that you can fulfill job expectations.

Some people believe that considering technical communication as persuasion is inappropriate. They think of technical communication as being objective and true, and they regard persuasion as a matter of opinion and emotion; they believe that scientific facts in technical documents somehow speak for themselves. Most scientists, engineers, and technical communicators in recent years have come to recognize that technical communicators argue rather than simply describe or demonstrate. And they realize that technical communicators try to win adherence for their claims by crafting compelling arguments. Even when writers or speakers are describing a product or a technique, summarizing events, or giving instructions, they are trying to persuade their audience that their version of things is correct.

Technical communicators must persuade their audiences that the solutions they recommend are effective, legal, ethical, financially feasible, and safe. Only if these audiences are persuaded will they use the technical document to help make decisions and take action. The sciences and applied sciences are very human disciplines, dedicated to developing arguments that create knowledge and solve problems in a social context. And technical work is a very human enterprise—emotional as well as rational, messy and unpredictable, as competitive in nature as it is cooperative. Scientists and engineers seek the truth, of course. But those "truths" are established through informal argument and reasoning more often than through formal, syllogistic means. In short, *technical communication is less an impersonal and rational demonstration than it is a set of arguments laid before a jury of readers, listeners, or viewers.*

Given that situation, you will have to think of yourself as persuading your audiences. You will have to develop effective arguments that will work in technical and industrial settings. Because **rhetoric,** as Aristotle noted, is "the faculty of observing in any given case the available means of persuasion," his account of persuasion is a good place to start. For Aristotle the available means of persuasion were what he called:

rhetoric
the faculty of observing in any given case the available means of persuasion (as defined by Aristotle)

- *Ethos* (the trustworthiness, credibility, and reliability of the speaker).
- *Pathos* (the appeal to the audience's most basic, most deeply held values and beliefs).
- *Logos* (the appeal of the evidence and the reasoning process).

In reality, it is sometimes difficult to separate ethos, pathos, and logos in particular communications, to tell where one kind of appeal begins and the other ends. But to discuss them, we separate them here. Learning these three means of persuasion allows you to develop successful communications that are overtly argumentative—things like résumés, proposals, sales presentations, and recommendation reports—as well as documents that are less overtly persuasive, such as informational reports and instructions.

A Continuing Case to Demonstrate Persuasive Strategies	To make the concepts in this chapter more concrete, imagine that together we are building an argument on a specific situation which you probably will be able to identify. Imagine that you are an engineering student at a state university and a member of your local student chapter of the American Society of Civil Engineers (ASCE), a group of prospective civil engineers.

Recently you and the other members of the chapter have read and heard a lot about the growing internationalism in engineering and the increasing internationalism of our economy. The articles, speakers, and teachers that you've encountered have been emphasizing the advantages of multilingualism and the benefits to engineers of knowing about other cultures. Just last week you heard a respected engineer visiting the campus bemoan the fact that civil engineers interested in roads spend so little time studying the techniques of Italian engineers, even though Italian highways are beautifully engineered. The speaker concluded that the single best way for engineering students to distinguish themselves from other new engineers would be to gain fluency in another language. You and some of the other members of ASCE are impressed—and a little frustrated. You know all too well how difficult it is to study foreign languages within the constraints of an already demanding engineering curriculum. Faced with so many requirements already, how can you even think about taking courses in another language?

You and your friends wonder whether space in your engineering curriculum might be carved out somehow to permit students at your university to take more courses in other languages. Indeed, you wonder whether it wouldn't be wise for your university to *require* engineering students to show competency in a second language (at present only liberal arts students on your campus have such a requirement). So you decide to ask your campus administrators to explore the possibility of encouraging or requiring engineering students to study a second language. You're not really recommending any change in the curriculum right now; all you want is some serious exploration of the possibility. After all, you aren't even sure if it's a good idea—if it's feasible to require a second language or worth the time and trouble to learn one, especially if it's at the expense of some other aspect of your education. But you do think the issue has merit and you want a serious study of the issue.

In short, you need to write a proposal that engineering administrators will use to decide whether a full-fledged feasibility study is worth the bother. Your goal is to ensure that a serious study is authorized. You don't want to be brushed off with a polite "thanks for the suggestion; we'll think about it." And you don't want a superficial study that can be easily ignored.

Now, as we proceed through this chapter, let's try to develop *ethos*, *pathos*, and *logos* for an argument in this case. We also clarify concepts through reference to the Case Documents in Appendix A.

Ethos: **Creating a Reliable, Trustworthy Persona**

You have probably noticed that in the course of listening to someone on the radio—sports announcer, disk jockey, or news commentator—you often get a sense of the kind of person who is speaking. Even if you have never met the speaker personally, you still have a sense of his or her character and personality. How you respond to the message is affected to some extent by how you respond to the person who delivers it. If you appreciate the person, you tend to believe and appreciate what that person has to say. Sometimes we refer to this sense of personality with the word *image*; in technical communication, the word is **ethos**.

ethos
the trustworthiness, credibility, or reliability of the communicator

Ethos refers to the persuasive value associated with the "image" of the "person" that is created in the text of a document or presented through a speech. Sometimes people use the term **persona** to distinguish this person from the real author. You have already encountered a great number of created personas in your reading.

persona
the created narrator within a communication; relates to the ethos of the communicator

As you learned in the discussion of creating an audience in Chapter 3, documents are delivered to their audiences by created characters or personas, who may not have much in common with the real authors. For instance, what kind of person seems to narrate Examples 4-1 and 4-2? What kind of "image" does the "author" create? These examples are the original and a revised version of one of Case Documents 3 reprinted in Appendix A.

Both of these letters contain much the same information, but their *ethos*—their ethical appeal—is very different. One of the writers comes across as a concerned but unconfrontational person; the Ed Younger in Example 4-1 seems eager to "discuss this matter" and to "alert" the reader to a problem that both he and his reader might deal with together. Respectful and concerned, this "first" Ed Younger also creates the impression that he is a careful and reasonable person. The cool accounting of factual observations and the bulleted lists suggest that this persona lets facts speak over feelings. Not that the first Ed Younger lacks feelings; his persona has strong feelings indeed about animal rights and the abuses of animals that his organization has observed. But he prefers understatement—to let the facts speak for him in creating his argument—and prefers to create a self that seems relatively objective, unbiased, and respectful.

On the other hand, the "second" Ed Younger, in Example 4-2, seems more confrontational than cooperative. Note how he constructs for himself a position of moral superiority by addressing Dr. Wartala and her colleagues as "you" and himself and his organization as "we." He uses morally charged language to suggest that his position is passionately felt. The Ed Younger who writes the last sentence even seems rather threatening. The language he chooses seems to say something about his personality; notice how he

---

EXAMPLE 4–1   The Original Letter to Dean Wartala

August 2, 1999

Dr. Karen Wartala
Dean, College of Agriculture
University of the Midwest

Dear Dean Wartala:

   The Coalition of Citizens Concerned with Animal Rights (CCCAR) was alerted by a concerned citizen two weeks ago that the University is trapping and killing large numbers of birds in the experimental crop fields on its campus. In our surveillance and investigation we have learned that:

- Birds are lured to the traps by bait placed in them, and the number of birds in these large traps has been observed to be at least 50 at times.
- The reported method of killing the birds was suffocation, by placing them, 400–500 at a time, in a bag.
- Approximately 500 birds are killed every day on this campus; 10,000 were killed in a three-month period.

We object to this for the following reasons:

- It is a waste of animal life.
- It is ineffective. Trapping and killing birds will not permanently reduce the bird population in the area. Even *temporarily* it will have no more than a minimal effect.
- The trapping and suffocation of these birds is cruel in the extreme.

   For these reasons we insist that the University immediately stop the killing of these birds. If you wish to discuss this matter, please contact me at 555-5537 or 555-5856.

Sincerely,

Ed Younger

Ed Younger
Vice President, CCCAR

---

refers to his reader and how he manages somehow to come off as an overconfident, less reliable, and more biased investigator.

    You may feel (and for good reason) that the second Ed Younger is a less effective persona than the first one, that the first Ed Younger seems more credible, more persuasive than the second. However, our point is not that one persona is more effective than the other in this instance, but that such a thing as a persona exists. Although their content is essentially the same, both letters seem to be the product of specific personalities with particular characteristics. Those personalities are created by the language and organizational choices the writer makes. And the nature of those personalities is responsible in a significant way for the persuasiveness of these documents.

    People react to the persona created in a message as well as to the contents of the message. Your job as a writer or speaker, because you are

---

EXAMPLE 4–2   Revised Version of Letter in Example 4–1

August 2, 1999

Karen Wartala
College of Agriculture
University of the Midwest

Dear Ms. Wartala:

   The Coalition of Citizens Concerned with Animal Rights (CCCAR) learned recently that your University is trapping, torturing, and killing large numbers of birds in the experimental crop fields on your campus. These outrageous actions must cease immediately.

   In the course of our investigation of your organization we have learned a number of shocking facts. Birds are being lured into traps when your employees place bait into them; we have observed that the number of birds in these large traps has often exceeded 50 at a time. In addition, your employees kill the birds by suffocating them; the employees murder the birds by placing them, 400–500 at a time, in a bag, sealing the bag, and watching them die a slow, painful death. Finally, your employees murder approximately 500 birds every day on your campus; we know that over 10,000 birds were slaughtered in a three-month period.

   We object to your reign of terror for several reasons. You are wasting animal life that we consider to be precious indeed. In addition, your actions are ineffective. Trapping and murdering birds will not permanently reduce the bird population in the area—that's a simple fact. Even *temporarily* the killings will have no more than a minimal effect—birds always quickly repopulate decimated areas. Finally, the trapping and suffocation of these birds is cruel in the extreme. While your employees trap the birds and joke about it, the birds flap about, desperately trying to get some air. They die slowly and painfully.

   For these reasons we insist that you immediately stop the killing of these birds. Otherwise we will be forced to take direct action.

Sincerely,

Ed Younger

Ed Younger
Vice President, CCCAR

---

aware that people will be responding to *you* as well as to your message, is to *create the kind of persona that will be most persuasive in a given case*. In most circumstances, that means creating a reliable, trustworthy, and fair persona that an audience is likely to find credible.

   Here are some of the principles for creating an effective persona in technical communication:

- Use an appropriate voice.
- Establish your credibility and trustworthiness.
- Adopt an appropriate attitude toward persuasion.
- Attend to the ethos of your company or organization.

CHOOSING AN APPROPRIATE VOICE

As you know from the previous discussion and from your own experience, documents can seem stern, friendly, condescending, gracious, or intimidating largely because the persona of those documents seems stern, friendly, condescending, gracious, or intimidating. Thus, you will want to make sure that the **voice** that comes across in your documents is one that you intend to come across.

For example, take a look at the four versions of the first paragraphs in an article on botany and ecology (Examples 4-3, 4-4, 4-5, and 4-6). The paragraphs discuss the flowering patterns of plants that are pollinated by ants, birds, and bees, not by the wind. The author is introducing his study of how flowering patterns attract or fail to attract the ants, birds, and bees—and how all that affects a plant.

First, note carefully the differences in the voice of each version. Which of these documents might be appropriate for a publication in a professional journal? Which might be appropriate for a classroom presentation or other oral presentation? Which might be best as a popular article discussing the research? What differences in each version create different

voice
the nature of the persona; can be stern, friendly, condescending, formal, gracious, etc.

EXAMPLE 4–3   Version One of an Article on Botany and Ecology

The flowering pattern of many animal-pollinated plants may be divided into three phases: a) an initial phase in which a low but increasing number of flowers is produced per day; b) a short peak phase in which most of the total flower production occurs; and c) a final phase in which a low and decreasing number of flowers is produced per day. In recent years studies by a number of evolutionary ecologists have focused much attention on the effects of this mass-flowering pattern on pollinator attraction and movement (e.g., Janzen, 1978, 1981; Gentry, 1984; Stiles, 1985; Augsperger, 1986). In these studies it is demonstrated that many species of floral visitors preferentially forage on individual plants that produce large numbers of flowers per day. Presumably these individuals are both conspicuous and highly rewarding to pollinators (Ravens, 1982; Stiles, 1983; Carpenter, 1986). In contrast, the low levels of daily flower production that characterize the initial and final phases of the flowering pattern may be insufficient for pollinators to be attracted from alternative floral resources. Consequently, this mass-flowering pattern poses a dilemma.

This study seeks to answer two questions: During what phase of the flowering pattern does outcrossing occur, and what factors promote outcrossing? Specifically, this study documents the flowering pattern of *Catalpa speciosa* (Warder ex Barney) Engelm (Bignoniaceae) and examines the effect of different floral abundances on the proportion of flowers that are self- and cross-pollinated on an individual plant.

Source: Adapted from A. G. Stephenson, "When Does Outcrossing Occur in a Mass-Flowering Plant?" *Evolution* 36 (1982), pp. 762–67. Reprinted with permission.

voices? For example, what are the differences in word choice, punctuation, and sentencing that distinguish these versions?

As these four examples indicate, there is really no single "correct" voice that you can adopt for all documents. Although you are likely to develop the characteristic voice you use most often, one that people will recognize and become accustomed to, you should realize that effective writers change voices when the situation calls for it. Look at Example 4–3 and Example 4–5. Both assume the "insider's" voice of the professional scientist, but the personality of one scientist seems somehow different from the other.

EXAMPLE 4–4   Version Two of an Article on Botany and Ecology

Many plants that are pollinated by animals—examples would be anything pollinated by ants or the proverbial birds and bees—develop their flowers in three fairly discrete phases. First, there's an initial phase; during this period the plant produces a small but increasing number of flowers each day. Next comes a short peak phase in which most of the total number of flowers are produced. And lastly comes a final phase when the plant produces a small and decreasing number of flowers each day. Recently many evolutionary ecologists, such as Dan Janzen, Al Gentry, Gary Stiles, and Carol Augsperger (right down the hall), have been concentrating their efforts on studying the effects of this mass-flowering pattern—on how pollinators like bees and ants are attracted to plants and how and why they move from plant to plant. These people are showing that many kinds of floral visitors (for example, the ants) prefer to eat individual plants that produce really large numbers of flowers per day, presumably because these individual plants are easy to spot by the pollinators (and are very tasty to them) and because it takes less effort for the pollinators to do all their pollinating when flowers are so close together.

In contrast, when daily flower production is low, during the first and last phases of the flowering pattern, the plant through its flowers might not be able to attract enough pollinators from other flowers. Consequently, this mass-flowering pattern poses a dilemma: When flowers are few and far between, the bees or ants fly or walk around and visit many different trees, since flowers are scarce; that's good in that crosspollination occurs (and crosspollination makes for a more vigorous next generation), but it's bad in that the birds and ants spend a lot of their time and energy visiting various trees and therefore don't make as many stops as usual. On the other hand, when flowers are many, the pollinators can save their energy and work more efficiently on one plant; that's certainly more efficient, but it cuts down on the crosspollination because the pollinators are sticking to one location. So what's a plant to do?

In our study, Terry Haas and I tried to answer two questions: During what phase of the flowering pattern does outcrossing occur? And what factors promote outcrossing? Specifically, we've examined the effect of different floral abundances on the proportion of flowers that are self- and cross-pollinated on an individual plant. What's more, we've documented the flowering pattern of the catalpa tree (*Catalpa speciosa*, Bignoniaceae).

EXAMPLE 4–5   Version Three of an Article on Botany and Ecology

A great many animal-pollinated plants develop their flowers in three phases:

   a. during an initial phase, a low but increasing number of flowers is produced per day;

   b. during a short peak phase, most of the total flower production occurs; and

   c. during a final phase, a low and decreasing number of flowers is produced.

Recently a number of evolutionary ecologists have been studying how this mass-flowering pattern affects the activities of various pollinators (e.g., Janzen, 1978, 1981; Gentry, 1984; Stiles, 1985; Augsperger, 1986). According to their studies, many species of floral visitors prefer to forage on individual plants that produce large numbers of flowers per day, presumably because these individuals are both conspicuous and highly rewarding to pollinators (Ravens, 1982; Stiles, 1983; Carpenter, 1986). In contrast, when daily flower production is low (during the initial and final phases of the flowering pattern), pollinators may not be sufficiently attracted from alternative floral resources. Consequently, this mass-flowering pattern poses a dilemma.

    We sought to answer two questions: During what phase of the flowering pattern does outcrossing occur? And what factors promote outcrossing? Specifically, this study documents the flowering pattern of *Catalpa speciosa* (Warder ex Barney) Engelm (Bignoniaceae); we examined the effect of different floral abundances on the proportion of flowers that are self- and cross-pollinated on an individual plant.

    Just as you have been able to develop many oral voices—stern, friendly, excited—to coexist happily with friends, rivals, co-workers, and loved ones, so too you should be able to modulate your voice in prose, depending on some of the following factors:

- Your relationship with your audience. Your voice can be less formal when you know someone well or more formal when you are communicating with a relative stranger or with someone elsewhere in a hierarchy.
- Your audience's personality. Different people respond sympathetically to different voices.
- Your message. Some messages are more routine that others; some messages contain more welcome news than others.
- Your aim. Sometimes you will desire an outcome that implies a certain voice.
- Your genre and medium—the kind of document and the medium in which it will appear—call for a certain characteristic voice: Consider the formal voice of the front page of a newspaper, for instance, and compare it with the tone of the opinion, sports, or comics sections inside; or compare the tone of a published academic essay with the voice you hear in a letter of application.

EXAMPLE 4–6   Version Four of an Article on Botany and Ecology

Ever wonder why the birds and bees seem to be attracted to some flowering plants and not to others? Well, some university ecologists have been wondering, too—and trying to find out.

Ecologists know that a great many plants that are pollinated by animals (not by the wind) develop their flowers in three phases. First, a plant will produce a small but increasing number of flowers each day. Next comes a brief (but glorious-to-behold) peak phase during which the plant produces most of its flowers. Finally, the plant will produce a small and decreasing number of flowers per day. In recent years many ecologists—most notably Daniel Janzen of the University of Pennsylvania, Albert Gentry of the Missouri Botanical Gardens in St. Louis, Gary Stiles of the University of Costa Rica, and Carol Augsperger of the University of Illinois—have been studying the effects of this mass-flowering pattern on pollinators; they have been considering how certain flowering patterns attract birds and bees (and even ants) and influence their movements. They and others have established that many species of floral visitors prefer to forage on individual plants that produce large numbers of flowers per day, presumably because these individuals are both conspicuous and highly rewarding to pollinators. To put it another way, birds and bees are attracted to large displays of flowers because such displays are easier to spot and because once the displays are located, the birds and bees can easily feed there and not have to search long and hard for more flowers. In contrast, when a plant produces just a few flowers per day, during the first and final phases of the flowering cycle, few pollinators may visit the plant; the birds and bees may be attracted to alternative flowers or fail to notice the few flowers produced.

This mass-flowering pattern poses a dilemma to birds, bees, and ants—and to the ecologists who study them. On the one hand, when flowers are relatively scarce, the bees and ants fly or walk to many different plants. From an ecological perspective that can be beneficial, because plants often benefit from cross-pollination: cross-pollination tends to make for a hardier next generation, while self-pollinated plants tend to be somewhat less vigorous. On the other hand, when flowers are out in full, the bees and ants tend to stick to one location; instead of visiting many different plants, they concentrate their efforts. From an ecological perspective that too can be beneficial: though such behavior leads to self-pollination and hence to relatively less vigorous offspring, it is highly efficient; many more flowers can be pollinated when the pollinators don't have to spend their time and energy locating new flowers. So what is a plant to do?

University professor Andrew Stephenson and his colleague Terry Haas are studying how plants handle this dilemma. In particular, they want to know the answers to two questions. First, they want to determine at what phase of the flowering pattern "outcrossing" occurs. (Outcrossing is the term used to denote when a plant "mates" with a neighbor as opposed to itself; most plants can either cross-pollinate or self-pollinate.) Second, they wonder what factors promote outcrossing? To find answers to these questions, they have been spending several summers looking carefully at the flowering patterns of the catalpa tree; they have been examining the effects that different floral abundances have on the proportion of flowers that are self- and cross-pollinated on an individual plant.

As a final example of the importance of voice, imagine for a moment our continuing case—the short proposal that you might write to the administration of your university suggesting a look at the feasibility of foreign language instruction for engineering students. Imagine the kind of voice that you would want to adopt for such a proposal. You'd probably want to sound rather formal and sober and respectful—but emphatic as well. Imagine, by way of contrast, how a presentation to fellow students on the same issue might go. There it might be best to be informal—perhaps to show a sense of humor or irreverence. The voice you choose in a situation will shape your audience's perception of your persona and ultimately affect your credibility and the success of your document or presentation.

CREATING TRUST AND CREDIBILITY

credibility
associated with ethos, the creation of a persona who appears knowledgeable and trustworthy

Let's concentrate for a moment on the problem of ethos in the proposal that we are creating on foreign language instruction for engineering students. Even more important than voice in this instance is the issue of **credibility**: How can you get the experienced educators that you are addressing to take you seriously? The answer lies in developing your facility with ethos. Ethos is always closely associated with *credibility*—with the creation of a persona that seems knowledgeable and trustworthy. Technical communicators need to establish credibility, and we suggest three ways in which you can do that:

1. Show that you have done your homework.
2. Establish your special expertise.
3. Exhibit a sense of fair play.

Show That You Have Done Your Homework

People appreciate arguers who come to a case after careful study, and they distrust those who jump into an argument before thinking things through. To give the audience the sense that you have done your homework, you can depend on two strategies:

1. Cite the work of experts, either informally in the course of your argument or in the form of Notes, References, Bibliography, and Works Cited. By referring informally to published scholarship on your subject in the text of your document ("Professor Marcia Tawney of Southwestern State University, who has studied foreign language instruction among engineering students since 1986, has noted that . . .") or through more formal means ("see Tawney, 1986, 1988"), you can indicate to your audience that your persona deserves attention. You might conduct and refer to interviews with chairs of other engineering departments at your school or with the registrar or provost of your institution to establish your knowledge of curricular and budget constraints.

2. Indicate that your ideas are the product of a systematic study or methodology. By indicating that you have a careful study ("After surveying the 40 top engineering schools identified by Dean Antonin Cellucci, we concluded that a number of fine universities are moving to find ways to permit their prospective engineers to study foreign languages"), or that you have performed some sort of experiment ("Our tests show . . ."), or that you have completed considerable library work ("Our search of the published literature on this matter indicates that . . ."), you establish the credibility of your persona.

Establish Your Special Expertise

Sometimes your credentials are established by virtue of the homework that you have done. Other times you can mention expertise that you have established by virtue of previous experiences. Sometimes such expertise is assumed as a matter of course. If you are given an assignment at work, it is usually assumed that you have the expertise necessary to do it, and so it would be pretentious to mention your credentials. At other times, establishing your credentials (the word **credentials** is associated with *credence* and *credibility*) is essential to your persuasive efforts. Sponsors of many proposals, for instance, typically require proposers to include résumés and information about previous work as a routine part of the proposal. If you have no special background on a given topic, that need not be a fatal flaw in your argument, especially if you follow some of the other advice in this chapter. But if you do have special expertise, let your audience know about it, if only indirectly.

credentials
experiences that give a communicator expertise

Create a Sense of Fair Play

Personas are most persuasive if audiences perceive them as fair-minded. Give your audience assurance that you are fair-minded in several ways:

- By attending responsibly to alternative ideas. For example, always summarize an opponent's argument in a way that the opponent could agree with.
- By showing respect for other points of view.
- By indicating that your personal biases are under control.

The last point doesn't mean that you should never reveal that you have a personal stake in the question at hand. Indeed, if you do *not* indicate that you are personally involved or engaged somehow, your audience may accuse you implicitly of covering up your personal involvement, feel that you are too remote from a situation to be trusted, or suspect that you somehow stand to gain personally if your position is adopted.

There is nothing wrong with being personally implicated in an argument, as long as you indicate that your involvement is not coloring your judgment improperly. For example, in our hypothetical proposal you might want to confess that your concern about foreign language instruction for engineers derives from self-interest: You want to get an edge on other new engineers who are with you in the job market. But you would also add that your concern is not purely self-interested; there are other good reasons for adopting your proposal.

Before reading on, you might take a moment to look again at Examples 4-1 and 4-2 as well as the four versions of the botany article in Examples 4-3 through 4-6. Look for the ways in which the writers of those documents establish—or fail to establish—their expertise and their fairness.

ASSESSING AND CHANGING YOUR ATTITUDES TOWARD ARGUMENT

The preceding discussion may remind you that a sense of fairness is often missing in our approaches to argument. In our culture many people grow up thinking that argument and persuasion mean winning and losing in all-out verbal combat. In courtrooms we see prosecutors and defendants arguing for one kind of clear-cut victory or another—guilty or not guilty—in front of a judge and jury. In our legislatures we see political parties arguing for victory on some piece of legislation. During election campaigns we see candidates engage in debates and then be declared the "winners" or "losers." Sometimes an adversarial approach to an issue makes good sense.

But there is also another way of thinking about argument that is often equally or more productive. The 20th century rhetorician Kenneth Burke, having observed the consequences of seeking all-out victory in rhetorical exchanges—consequences that have played out ultimately on battlefields—has suggested another theory of argument. He proposes that people seek not victory, advantage, manipulation, or coercion when they write. Instead, Burke proposes that writers and speakers conceive of themselves as cooperators, not competitors. To do so, writers and speakers must abandon the role of aggressive adversaries and assume the role of mutually supportive partners in the process of building bridges between people.

agonistic
descriptions of competition
rather than cooperation

Burke would have us conceive of language and rhetoric not in **agonistic** terms that suggest battle and strife, but in terms of cooperation. Burke says an effective arguer is "one voice in a dialogue," not a person in a monologue. Try thinking of argument not as a matter of beating an opponent but as a productive exchange, part of a continuing conversation designed to generate new insights.

As a general rule, construct a persona in your persuasive technical documents that your audience will see as a cooperator instead of a competitor. Show that you care about both your ideas and your audience.

Generally, audiences will appreciate it if you cast yourself as a person whose arguments are put forth in the interest of finding the *best* way, not "my way."

Here are a few suggestions for how to create a persona with whom your audience will want to cooperate:

■ Maintain an attitude of courtship, not coercion. It's important to articulate your position vigorously, but you don't want to "defend" your "position" so vigorously that your "opponent" is silenced. Try to argue in such a way as to seek dialogue and consensus.

For example, the original letter to Dean Wartala (Example 4-1) could have included the following statement to seek consensus:

> The Coalition of Citizens Concerned with Animal Rights wishes to propose a solution that would protect both the birds within the vicinity and the seeds and crops of the experimental fields.

■ Show that you understand and genuinely respect your audience's position, even if you think the position is ultimately wrong. In other words, argue against an opponent's position, not against the opponent. And often it means representing your opponent's position in terms that your opponent would accept. Look for common ground that you already share with your audience. See yourself as a mediator. Consider that neither you nor the other person has arrived at a best solution, and carry on in the hope that dialogue will lead to an even better course of action than the one you now recommend.

For example, the original letter to Dean Wartala (Example 4-1) might have contained a statement such as the following:

> The Coalition of Citizens Concerned with Animal Rights believes that the University crop experiments are important to the state. However, equally important to the state environment are the many birds that flourish here. It is in the mutual concern for state environmental welfare that we propose . . .

■ Expect and assume the best of your audience, and deliver your own best as well. Avoid threats, even if your opponent uses them. When someone else resorts to reprehensible tactics, your persona will be more appreciated if you respond by taking the high road. Avoid sarcasm and remain respectful.

For example, the original letter to Dean Wartala (Example 4-1) might have contained the following:

> Rather than pitting the welfare of the birds against the interests of the University and the experimental fields, the Coalition of Citizens Concerned with Animal Rights suggests a means of controlling, rather than killing, the birds . . .

- Cultivate a sense of humor and playfulness. Informal humor, used wisely, is a legitimate tool in technical communication. Like any other technique, however, humor should not be used thoughtlessly, and you will need to take culture, race, and gender into consideration when you make use of it. Humor does not mean necessarily telling a joke or making light of someone who is truly upset about something; it doesn't mean making fun of an embarrassed victim. Still, nothing creates a sense of goodwill quite so much as good humor, and humor can be especially welcome when the stakes are high, the sides have been chosen, and tempers are flaring.

Perhaps a statement such as the following in the original letter to Dean Wartala (Example 4-1) would have engaged the reader's interest and lightened the mood:

> It seems appropriate to propose this conversation to the largest university in the state most known for its populist spirit. Let's talk! We'll even bring the coffee.

To see this advice at work, consider once more your proposal to have administrators at your school authorize a study of the foreign language issue. In this instance you might work hard to show why foreign languages have traditionally not been associated with engineering education. You might indicate that you know there are good reasons why such instruction is not currently possible (it could get expensive), and acknowledge that you are not the first person to think of such an idea. You might name what everyone agrees to be the aims of engineering education at your institution and indicate your agreement with those aims (foreign language instruction, after all, is probably in keeping with such aims).

You might indicate that you appreciate the time and effort that would be required for a study like the one you are recommending. You might put forward your proposal for a formal and serious study of the issue as an effort toward negotiating something mutually satisfactory over the long run. You might in some way show a sense of humor that indicates that you consider yourself an equal, and that you are a pleasant colleague with whom to work.

ATTENDING TO THE ETHOS OF YOUR COMPANY OR ORGANIZATION

Because writers often speak not just for their creators but for the organizations that sponsor them, you will need to attend to your corporation's ethos as well as your own. Apple Computer, for example, has a **corporate ethos** that is very different from IBM's, and the people who work in those corporations must often adopt a voice that is in keeping with their surroundings. Examine Example 4–7, which you saw in Chapter 3. It contains selections from an Apple Computer manual and a Motorola instruction guide which reflect each corporation's distinctive ethos. As a member of a specific enterprise, you should consider your company's collective ethos when you are developing your own.

corporate ethos
the way corporate values and beliefs are manifested in the persona the corporation adopts in its communications with people inside and outside the organization

---

EXAMPLE 4–7   Two Examples of Corporate Ethos

Look over these examples of corporate writing and write a sentence or two to describe how they differ in terms of their ethos, the image they seek to project, and the way they see their relationship to the user of the product they produce. Note as well that ethos accounts for differences in what the introductory sections of the manuals are called.

Apple Computer

Many people are concerned at first about making a mistake that will damage their computers or their work.

But you don't need to worry. If you set up your Macintosh according to the instructions, avoid bumping it while it's turned on, and don't spill coffee on the keyboard, you won't hurt your equipment.

Source: Apple Computer, Inc., *Getting Started with Your Macintosh*, "Welcome," p. ix.

Motorola

Thank you for selecting Motorola and welcome to cellular telephone service. Your Digital Personal Communicator™ represents the state-of-the-art in personal cellular telephones today. The listing below shows just a few of the exceptional features that the telephone contains . . .

To cover all of these features properly, we will take you through a logical step-by-step learning procedure that explains everything you need to know to operate your new telephone.

Source: Motorola, Inc., *Digital Personal Communicator User Manual*, "Introduction," p. 4.

You can discover your corporation's ethos by looking at some representative documents and by doing a bit of analysis on your own. Here are some good places to go for insight into your company's ethos:

- The organization's style manual, which might specify such things as the use of second or third person in all documentation, or might encourage or forbid the use of contractions depending on how formal it wants to appear.
- Senior employees who oversee your first writing efforts.
- The changes that professional technical editors make in your documents when preparing them for publication.

Considering your company's ethos doesn't mean you have to give up your own. There is space for you to create your own persona within the context of your organization. You will have plenty of freedom to develop your own ethos, especially in documents intended for audiences within the organization. Even in documents that leave the organization, you can exert considerable influence over the ethos of the final product, especially if your choices are well considered.

Pathos: Appealing to Your Audience's Values

pathos
the appeal to the audience's most basic, deeply held values, attitudes, and beliefs

You already know from the previous chapter that you must take your audience's attitudes into consideration when you select and organize information for your communications. The technique of **pathos** fits right in with that approach. Pathos is the technique of appealing to your audience's most basic, heartfelt attitudes and values.

Sometimes *pathos* is defined simply as an "emotional appeal," as an overt appeal to an audience's basest instincts. For example, critics of mass advertising, particularly critics of television ads, contend that advertisers depend on purely emotional arguments when they tie automobiles to a drive for success, when a certain beer is depicted as part of a happy social life, or when insurance ads are associated with our fear of death and loss of security. At first glance, therefore, pathos seems an unlikely approach to use with the logic of technology and science. Yet there is nothing irrational in acknowledging human emotions in our arguments. For example, it would be appropriate to argue that "Our city should attend to its solid waste problems because we want our children to live in safe, healthy, and attractive communities," even if there is an emotional aspect to that argument.

Instead of thinking of pathos as an irrational appeal, think of it as an appeal to the most basic values of your audience. Suppose that a civil engineer writes, "If we make the changes recommended in this report, the intersection at Wayside Avenue and Salem Road will be much safer; injuries from traffic accidents will almost certainly decrease." This statement is couched in logical and rather matter-of-fact terms, but this argument also carries emotional overtones, "pathetic" overtones that make it much more powerful than the following argument: "If we make the changes recommended in this report, traffic at the intersection of Wayside Avenue and Salem Road will move much more efficiently." That's because the appeal to safety attends more to our basic human values. Example 4–8 illustrates some instances of pathos in workplace documents.

EXAMPLE 4–8 Examples of the Use of Pathos

Directions: Look over these passages taken from the Case Documents in Appendix A. What emotions or values are being appealed to in these examples? If you had to arrange these examples from most use of pathos to least use, what would be their order?

Bird Control

Birds are trapped in large cages baited with bread. The trapped birds are supplied with water and food until they are removed from the cages. There are eight traps located in the plot area. Approximately 150–200 birds in total will be trapped daily on the average. The number of birds per trap will vary a great deal. On unusual days as many as 500 birds may be trapped. The average number of birds trapped per year is usually about 5,000 with a range from year to year of 3,000 to 10,000. Bird populations are never permanently reduced as a result of the program. Our bird control program attempts to reduce bird populations in the plot areas only during the growing season. All songbirds are released. Only pest birds like blackbirds, starlings, and English sparrows are destroyed.

Exxon Valdez

After the spill, the U.S. Environmental Protection Agency (EPA) approached Exxon to discuss ways to clean up the oil. A joint effort between EPA, Exxon, and the Alaska Department of Environmental Conservation was organized to test the effectiveness and safety of bioremediation—a process in which nitrogen and phosphorus fertilizers are fed to naturally occurring, oil-hungry microbes, allowing them to multiply and consume oil faster.

Tuberculosis

Tuberculosis (TB), caused by *Mycobacterium tuberculosis*, is a contagious, airborne disease of the lungs. Sometimes TB affects other parts of the body. It is a major cause of death in many parts of the world today and was the leading cause of death in the early years of this century in the United States. As recently as 1950, Minnesota reported 5,000 new cases. TB has been called a "social disease with medical aspects" because it can be found wherever poverty, poor nutrition, homelessness, and substance abuse occur.

Year 2000 Web Sites

We advocate planning and preparedness in the event Year 2000 (Y2K) disruptions occur. The Y2K problem is not a mental exercise—it's a social challenge. Whether hype or disaster, we can and should take steps to protect our neighbors, friends, families, and ourselves, regardless. Time will determine who was right and who was wrong. Time is running out! There are only 521 days, 9 hours, 47 minutes left until the year 2000!

As a technical communicator you should use pathos with care. Explicitly emotional arguments are distrusted in technical discourse. But pathos can be used effectively. For example, if you are tying to convince the group of college administrators to authorize a study of the feasibility of foreign language instruction for engineers, you'll be on firmer ground if you depend mostly on logical appeals as described in the next section. However, you could mention that knowing a foreign language can help individuals in a multicultural world to understand and appreciate one another. By indicating (perhaps in an introduction or conclusion or some other emphatic point) that you feel strongly about something or that something deserves strong and close attention, you may convince your audience to give special weight to your argument.

Moreover, you need to be aware of pathos in others' arguments. Many of our current arguments about applied technology center on pathos. For example, as a society we often seem paralyzed by our need to agree on what to do with nuclear waste, with our old-growth forests, with our rangelands, with our prison systems, and with our cities' infrastructures. Audiences often resist "logical" arguments because of their deeply held beliefs. In technical discourse, then, especially if it has a public face, we often use pathos in our own arguments to counteract prior appeals that are working in the minds of our audiences.

Logos: **Persuading with "Good Reasons"**

In convincing people to take a certain action or to accept some assertion as true or likely, writers and speakers typically provide what rhetoricians call "good reasons" for their assertions. Good reasons are the grounds for beliefs; they are the reasons why we take the positions we take. You might think of good reasons as a series of clauses that begin with the word *because*; you believe certain things *because* there are various good reasons to support your position. Argument, then, may be seen as a process of devising **because-clauses** your audiences will find convincing, perhaps even compelling. A new highway should be put through downtown *because* it will improve safety, *because* it will improve traffic flow, and *because* it will improve aesthetics. A certain paint should be used on automobiles for good reasons—*because* it resists chipping and *because* it is inexpensive and easy to use. You support foreign language instruction for engineering students *because* other schools are doing it, *because* it might make students into more flexible, informed, and culturally aware engineers, and *because* it might give students an edge in the job market.

because-clauses
the good reasons that support arguments or proposed ideal situations

Finding good reasons is a necessary challenge for writers and speakers. Sometimes good reasons come from thinking about pathos. By pondering your audience's most fundamental values and beliefs, you can usually find some powerful because-statements. Most often, however, good reasons derive from mulling things over reasonably as well as emotionally; that is, good reasons derive from *logos.* **Logos** refers to the logic of what you communicate; in fact, *logos* is the root of our modern word "logic." Good reasons are thus commonly associated with logical appeals.

In the rest of this section, you will find a set of questions that you can use to find good reasons for your arguments. These questions will help you develop effective arguments, and they will equip you to communicate more effectively in various circumstances, even when you must think fast at a meeting or speak extemporaneously. But do not expect every question to be productive in every case. Sometimes a certain question won't get you very far, and often the questions will develop so many good reasons and strategies that you will not be able to use them all. Ultimately, you will have to select the ones most likely to work in a given case.

logos
the appeal to the evidence or the reasoning process; the logic of what is communicated

CAN I ARGUE BY DEFINITION FROM "THE NATURE OF THE THING"?

Probably the most powerful kind of good reason is an argument from definition. If you can get your readers or listeners to think of a thing in certain terms, to see it in the way you do, you are likely to convince them to believe what you want them to believe. You need not think here in terms of formal definitions (such as those we will discuss in Chapter 11). Rather, think in terms of an informal definition, or what is called a **categorical proposition.**

categorical propositions
informal definitions that consist of a subject (word to be defined), a linking verb, and a predicate (something about the word to be defined)

A categorical proposition sounds like something complicated, but it really amounts to the kind of simple definition statement—X is Y, something is something else—that we make all the time. To put it more formally, a categorical proposition consists of a subject (the word to be defined), a linking verb (such as "is" or "was"), and a predicate—something to be said about the word being defined. Here are some examples:

- Science and engineering curricula are usually rather demanding.
- A recycling program is something that this campus badly needs.
- Ms. Protevi was an ideal boss.
- Nursing is a more interesting profession than most people imagine.
- Certain climactic changes now in evidence are truly disturbing.

In each case it is easy to identify the elements of the categorical proposition. For example, in the first sentence above, "science and engineering curricula" is the *subject*, the *linking verb* is "are," and the *predicate* is "usually rather demanding."

We all use categorical propositions routinely. This statement, "new farming techniques are revolutionizing agriculture in the developing world," is really a version of the categorical proposition "new farming techniques are developments that are revolutionizing agriculture in the developing world." Likewise, the statement, "That substance contains toxins," can easily be transformed into categorical proposition form: "That is a substance that contains toxins."

Writers and speakers can develop categorical propositions to create because-clauses. If we go back to our study of foreign language instruction for engineers, we see that one way to persuade people to consider the idea is to define it in positive terms: "Foreign language study is likely to make engineers more flexible and resourceful employees in the 21st century." You have *defined* foreign language study as something very attractive to your audience, and you have developed a good reason—a because-clause—that you can use in your proposal: "You should study the feasibility of foreign language study for engineers *because* it is a course of study likely to make engineers more flexible and resourceful employees in the 21st century." One of the best characteristics about arguments from the nature of the thing is that you can develop any number of definitions in support of your arguments:

- Foreign language instruction can be defined as a means of ensuring that engineers can compete better in international markets.
- Foreign language can be defined as a way of helping engineers become more comfortable with the other cultures that they are likely to encounter in their work.

Of course, you have to do more than state a definition or categorical proposition. You must also present evidence that your definition is *accurate*. That is especially true when your audience either works from a different definition than you do or is unsure of the concept being defined. An example of the first instance—working from another definition—would be the person for whom foreign language study might be defined as "an unnecessary frill for an engineer," or "more painful than a root canal," or "an awfully expensive part of a curriculum," or "something that liberal arts students need, but not engineers." As a consequence, you should think of your categorical proposition as a substitute for the definition that your audience may bring to your proposal. You are offering a different predicate to replace the one your audience now has.

In the other case—an audience unsure of the concept being defined—you are providing the definition for your audience, not changing their views of the thing defined. For instance, if we are explaining a term like "categorical proposition" to someone like you, we'll be providing a complete, new definition rather than changing the one you already have. Or if a forestry expert wants to recommend to local government leaders the hormone-darting method of controlling a deer population over the trap-and-transfer method, that forester will simply need to provide complete definitions of the two methods for readers who have never heard of either.

Once you get a good definition to support your argument, you must defend it before moving on to your next good reason. Defending it will probably involve using some of the other strategies—citing examples and statistics, offering testimony, providing comparisons, and so forth. It may take you some time to defend your definition, but investing time in such a defense is worthwhile because argument from definition is one of the most powerful ways of winning support.

If you can get your audience to accept your definitions of things, you've won more than half the battle. That's why the most heated issues in our culture—abortion, affirmative action, gay rights, pornography, women's rights, gun control, the death penalty, among others—tend to turn on matters of definition. Is abortion a crime or a medical technique? Is pornographic speech protected by the First Amendment, or is it a violation of women's rights? Is the death penalty "just" or "cruel and unusual punishment"? People usually don't even care about the consequences: If a definition of something seems airtight, it often overwhelms questions of practicality. In other words, if you can prove that foreign language instruction is something that will make you closer to the ideal engineer, or show that foreign language instruction is in keeping with the engineering program's own definition of its mission, then you will have achieved a large part of your persuasive goal—even if some of the consequences of your proposal are rather negative (e.g., it might be costly or inconvenient, or require other difficult curricular adjustments).

CAN I ARGUE FROM CONSEQUENCE?

A second powerful source of good reasons—and one far less complicated than definition—comes from considering the possible *consequences* (good and bad) of your position.

If we return to our example of prospective foreign language instruction for engineers, we can see how arguing from consequences might work. You could, for instance, make your audience think positively about the prospect of such a curricular change if you established that good things would follow from the idea:

- Engineers with a second language could make their companies more profitable because they might open up new markets.
- Engineers with a second language would be more likely to profit from (and help import) engineering techniques developed in other nations.
- Engineers who speak more than one language are more sensitive to the needs of other cultures as well as to American subcultures, and therefore are likely to develop better products and services.

You could probably imagine other good consequences, but don't forget to think as well about what bad things would be eliminated by your idea:

- Engineers with proficiency in a second language would eliminate the insularity associated with American engineering.
- Engineers with a second language would help to eliminate errors in documents that are not translated correctly by others.

After considering the consequences, you'd be left with one or more good reasons, or because-clauses, to support your arguments: Administrators on your campus should study the feasibility of requiring foreign language instruction because it would make for more flexible, resourceful, and productive engineers and/or because it would permit engineering firms to compete in international markets, and so forth. Here again, you also have to show that these consequences would indeed follow from the idea or course of action that you are arguing for by citing statistics, providing examples, or offering the other kinds of supporting evidence that will be detailed in the next section. But by thinking of consequences, you can indeed think of good reasons that will provide the spine of your arguments.

Thinking about consequences will also encourage you to consider the feasibility of any proposals or recommendations you might be offering. A good idea has to be a practical idea, and you need to indicate how your ideas are feasible if they are not obviously so. Anyone proposing to make foreign language possible in engineering education, for example, has to deal with the practicalities: curriculum changes, accrediting agencies, budget, and institutional support. Thinking about the consequences of your ideas will bring questions like these readily to mind.

Arguments from consequence are especially common in proposals and recommendations. In the pragmatic, results-oriented, bottom-line culture of industry, results matter most; and most arguments will turn on arguments from consequence, particularly if moral and ethical issues are not seen as especially important to a given case. As you will learn in Chapter 15, proposals typically derive from problems that need to be solved—and the consequences that proposal writers cite typically include evidence that the problem will indeed be solved. The same is true in recommendations: When you are recommending a certain course of action, it is often to eliminate a problem or to answer a question—for example, which computer software should the organization purchase, or what reforms will eliminate excessive costs. In these cases, it makes sense to describe the consequences of the actions that you are recommending.

CAN I ARGUE FROM VALUE?

A special kind of argument from definition that often implies consequences is the argument from value. These arguments require you to support your thesis with one or more because-clauses that include an **evaluative component**. Essentially, your argument is that your position is the best one. Arguments of value sound like these:

evaluative components
a part of an argument
from value

- To produce really good engineers, you need to study the feasibility of foreign language instruction.
- You should give strong consideration to the treatment I'm suggesting because it is the best, the safest, and most efficient one available.

- The telephone system you are currently employing is inadequate, outdated, and cumbersome.
- Your harvest yield will turn out better if you follow certain planting principles.

Such statements generate powerful because-clauses like these:

- You should authorize this study because foreign language instruction can make people into good engineers.
- Take up this program of physical therapy because it is truly the best one for you.
- Buy our telephone system because it is the best one for your office.

Evaluative arguments usually proceed from the presentation of **criteria**, and these criteria come from the definitions of good and bad, inadequate and adequate, that prevail in a given case. A "really good chemist," for example, meets certain criteria or specifications; so does an effective treatment program, an outstanding telephone system, or a good harvest. In arguing by means of evaluation, therefore, you will need to define the criteria that hold for the item under discussion and then demonstrate patiently that the criteria are met. Sometimes the criteria—what makes something good or bad—are relatively obvious; a piece of machinery, for instance, should be cost effective, safe, easy enough to operate or maintain, and capable of performing its intended task. In making a case for it, you might need just one sentence to name the criteria, and you will probably spend most of your time applying the criteria to the piece of machinery to show that it has those characteristics. But in other situations the criteria aren't so clear. What makes a good scientist or a poor telephone system, for instance? In these cases you might need to spend some time defending your choice of criteria and the order in which you rank them before applying them to the topic under evaluation.

> **criteria**
> used in arguments from value

Evaluative arguments often are associated with *comparisons*. If you need to decide which piece of equipment to buy, which treatment to follow, or which employee to hire, you usually need to offer some choices. Otherwise there would be no reason to argue about the matter. Once the choices are clear, all you have to do is apply criteria to the available choices, weigh the results, and declare a winner. Offering choices can keep an argument from becoming too abstract.

CAN I COMPARE OR CONTRAST?

Evaluative arguments often generate comparisons. But even if they don't, you may need to think in **comparative terms**. Thinking in comparative terms in the course of developing an argument means asking what things are like or unlike the topic you're discussing. Aristotle called this habit of mind *argument by similitude*.

> **comparative terms**
> expressions of what things are like or unlike a topic in a communication; argument by similitude

Thinking in comparative terms can help you generate because-clauses like these:

- You should study the feasibility of foreign language instruction because our competitor schools are already offering it.
- You should offer that internship to me because it's the kind of thing I'm suited to (because of the experience I gained last summer) and because I have qualifications similar to the people that you've hired as interns in the past.

In turn, contrasts (which are really just "negative comparisons") can be equally effective in an argument. Consider statements such as these:

- You should study the feasibility of foreign language instruction for engineers because no one else seems to be offering it these days; it would give us a competitive edge.
- You should hire me for the internship because my qualifications are unlike (and superior to) those of the many people that you usually consider.

If you anticipate that your audience will be confused by or unfamiliar with your topic, think of comparisons that will make the topic more easily understood. Comparisons are an effective way of building common ground and effective premises.

Particular kinds of comparisons are metaphors, similes, and analogies. **Metaphors** and **similes** are figures of speech that compare an unfamiliar item with something more familiar in order to bring new insight to the topic under discussion. Similes are distinguished from metaphors because similes use "like" or "as" to make the comparison. Here are some metaphors used in the Case Documents included in Appendix A:

metaphor
a figure of speech that compares an unfamiliar item with something more familiar in order to bring new insight to the topic of a communication

simile
a figure of speech that uses "like" or "as" to compare an unfamiliar item with something more familiar

- Bioremediation refers to a process by which microorganisms "eat" oil.
- The Year 2000 problem is a blight.

Metaphors and similes are especially effective when you are explaining a concept that your audience might find obscure or hard to understand. Scientists especially are comfortable offering metaphors and similes to explain their work:

- "Well, the nectar that flowers produce can serve as a sort of intoxicant to the ants that consume it," explains an ecologist. "When the ants consume the nectar on a catalpa tree, they tend to wander around disoriented afterward. It's like they are drunk."
- An entomologist studying how trees in New England defend themselves against the gypsy moth might say, "The trees seem to talk to each other, in a sense. When one tree is infested with gypsy moths, it gives off a substance that seems to communicate with other trees—that seems to tell them to get their defenses up, to produce a toxin that the moths don't like too well."

The benefits of metaphors and similes are apparent to all who use and encounter them, and using them can be especially arresting and emphatic. But like other rhetorical techniques they should be used with care.

An **analogy** is a comparison that is extended for explanatory or persuasive purposes. Analogies are typically developed over several sentences or a paragraph. Here is an analogy provided by Richard Brennan in *The Dictionary of Scientific Literacy* to explain certain key terms in cell biology:

analogy
a figure of speech that extends a comparison for explanatory or persuasive purposes

> The relationship of the various elements of a cell may be thought of as follows: The nucleus of a cell is a library containing life's instructions. The chromosomes would be the bookshelves inside the library, the DNA would be individual books on each shelf, genes would be the chapters in each book, and the nucleotide bases making up the strands of DNA would be the words on the pages of the individual books. (p. 121)

Analogies such as Brennan's are especially valuable, but they work best with willing listeners or readers. They are less effective in convincing unwilling readers or listeners in more argumentative situations. Such an audience is likely to want other sorts of reasons, perhaps ones of value and consequence.

CAN I CITE AUTHORITIES?

Sometimes you can buttress an argument by citing other people, highly respected by your audience, who hold to the same position that you are promoting. Arguments from authority ("believe this *because* so-and-so believes it") don't make for the strongest because-clauses. Few people are persuaded to believe something on the basis of authority alone, but arguments from authority do work well as supporting arguments—as one because-clause among many. Not only can references to authorities give your audience another good reason to believe what you ask of them, but they can build your credibility as well.

CAN I COUNTER OBJECTIONS TO MY POSITION?

Another good way to find convincing because-clauses is to think about possible objections to your position. If you can imagine how your audience might counter your argument, you can include precisely those points that will address your audience's particular needs and objections and win your audience's assent. The strategy of mentioning counter arguments first and refuting them before they occur is called the **inoculation theory**.

inoculation theory
a persuasive strategy of mentioning counterarguments and refuting them before they occur to an audience

Inoculation theory suggests that you inoculate your audience with the opposing view; then when they hear it, they aren't overly worried because they already have your response in mind. Salespeople often use this technique effectively. They sometimes anticipate another company's argument and then supply a counterargument. For example, a car salesperson might say, "You'll be able to buy a sports utility vehicle for considerably less money from my competitor down the street, but when you hear the price, be sure to ask the salesperson about the company's plan to discontinue that line next year because it's not selling. You'll never be able to get parts if something goes wrong." In the case of our foreign language requirement example, you would have to think:

- Why would anyone oppose foreign language instruction for engineers?
- What practical objections might someone have?
- What philosophical objections might someone have?
- What keeps engineering educators from pursuing foreign language instruction right now?

By asking such questions as these in your arguments, you will develop effective because-clauses.

WHAT ARE THE SPECIAL LINES OF ARGUMENT THAT PEOPLE IN MY FIELD APPRECIATE?

So far the questions we suggested have been those that would pay off in any situation, whether the argument is carried out in your discipline or not, whether the argument is in an area of technical discourse or in the arena of personal or public affairs. But before you give up the quest for good reasons, you need to consider the specific lines of argument that people in your company or field especially respect. In fact, what you learn in those specialized courses in your major is both technical information and techniques as well as the specific lines of argument and special appeals people in your chosen field find compelling.

Specific lines of argument differ from company to company and discipline to discipline. Scientists in certain research laboratories, for example, describe their methods by means of general formulas and methodological rules that only members of their particular research community can appreciate. Transportation engineers rely so commonly on ridership surveys in their transit plans that a transit plan without a ridership survey would probably be found unpersuasive. Civil engineers looking at ways to improve highway safety rely on special topics such as sight distance, back plates, traffic signal phasing, channelization, total access control, the 85th percentile speed role, and other specialized concepts that civil engineers who design highways understand and appreciate.

When you join a discipline or a specific company, you learn to speak and argue like a member of that discipline or company; you learn to incorporate into your arguments the special lines of argument that distinguish your community from others. Use these special lines of argument along with the more common ones included in this chapter.

Once you have a proposition to defend with because-clauses, supported by good reasons, you must also verify those good reasons with evidence. Evidence consists of hard data, examples, narratives, episodes, or statistics relevant to the good reasons you are putting forward. To be effective, you will have to put forward not only propositions and good reasons but also evidence that those good reasons are true. Your evidence will consist of examples, accounts of past experiences, comparisons, statistics, calculations, quotations, or any other kinds of data that an audience will find relevant and compelling.

If your audience is likely to find one of your because-clauses hard to believe, then you should be aggressive about offering support: Present detailed evidence in a patient and painstaking way. As the one presenting an argument, you have a responsibility not just to *state* a case but to *make* a case with evidence. And so you must stick with a particular point until your audience is forced to agree with your conclusions. Arguments that are unsuccessful tend to fail not because of a shortage of good reasons, but because the audience doesn't agree that you have furnished enough evidence to back up the truth of the good reason being presented. On the other hand, if an assertion isn't especially controversial, don't belabor it.

FAILURES IN PROVIDING EVIDENCE

If you don't provide satisfactory evidence for a because-clause, an audience may feel that there has been a failure in your reasoning process. In your previous courses in writing and speaking, you may have learned about various "fallacies" associated with faulty arguments—faulty definitions, faulty analogies, hasty generalizations, faulty causal arguments (the so-called "*post hoc* fallacy"), *ad hominem* arguments (i.e., name calling), red herrings, and so forth. Strictly speaking, there is nothing "false" about these so-called logical fallacies. The fallacies most often refer to failures to provide enough convincing evidence to convince your audience. You can avoid such accusations if the evidence you cite is both *relevant* and *sufficient*.

Relevance refers to the *appropriateness* of the evidence to the case at hand. In science and industry, some kinds of evidence are seen as more relevant than others. Personal experiences or anecdotes are often viewed as of limited relevance, and experimental procedures and controlled observations have far more credibility. Compare someone who defends the use of a particular piece of computer software because "I like it and have used it with success" with someone who argues that "according to a journal article published last month, 84 percent of the users of the software were satisfied or very satisfied with it." On the other hand, in personal and more intimate circumstances, personal experience often is considered more relevant than other kinds of data. You might be convinced to use a particular

Supporting Good Reasons

relevance
the appropriateness of the evidence to the case at hand in a persuasive communication

over-the-counter remedy not because of the scientific studies cited in a brochure but because your roommate used the product with success.

Sufficiency refers to the *amount* of evidence cited. Sometimes a single piece of evidence will carry the day if it is especially compelling. Generally, however, people expect more than one piece of evidence if they are to be convinced of something. For example, you would probably need more than one anecdote about one engineer who succeeded after learning a foreign language if you are to persuade engineering administrators to require all students to study a foreign language.

sufficiency
judgment about the amount of evidence presented in a persuasive communication

If you anticipate that your audience might not accept your evidence, think carefully about the argument you are presenting. If you cannot cite adequate evidence for your assertions, perhaps those assertions should be modified or qualified in some way. If you remain convinced of your assertions, then think about coming up with additional evidence through further research in the lab or the library. Or, if you anticipate your audience will think you have overlooked important information, try to reassure them you have not.

Another strategy might be to acknowledge explicitly the limitations of your evidence. Acknowledging limitations doesn't shrink the limitations, but it does build your credibility as a communicator and convinces your audience that alternatives have indeed been explored fully and responsibly.

Which Good Reasons to Use?

We have now seen that using the questions described in the last sections can generate a series of because-clauses that you might use to achieve your aims. If used conscientiously, these questions are likely to generate far more information and far more premises or because-clauses than you can possibly use. The challenge then becomes how you can decide *which* points are likely to be most persuasive. In choosing these good reasons to use in your arguments, consider your audience's attitudes and values, particularly the values sanctioned by your community.

THINK OF YOUR AUDIENCE'S VALUES, NOT YOUR OWN

When they communicate, people tend to present the lines of thought that have led them to believe as they do. People have a good deal in common, and it is natural to think that the evidence and patterns of thought that have guided *your* thinking to a certain point will also guide others to the same conclusions.

But people are also different, and what convinces you may not always convince others. When you decide which because-clauses to present to others, try not so much to recapitulate your own thinking process as to influence the thinking of others. Ask yourself not only why you think as you do, but what you need to do to convince others to see things your way; think about what you can say or write that will be convincing to others. Pick the because-clauses that will seem compelling to your audience.

To test whether your arguments are likely to appeal to your audience, consult your audience during the composition of your communication or ask someone to take on the role of your audience and review your argument when you have completed a draft. Before you commit your arguments to a final draft, do something to get feedback from your audience or a representative of that audience. If you do that, you are almost sure to find the places in your argument that "leak."

THINK OF THE VALUES SANCTIONED BY YOUR COMMUNITY

It might be easy to pick arguments compelling to your audience if you belong to the same community. If you work for the same organization as your audience or if you belong to the same discipline (e.g., nursing, environmental engineering), you have probably learned to value many of the same things as your audience. As you decide what because-clauses to present, therefore, think of the ones sanctioned by your institution or discipline. For example, an argument that "foreign language study may need encouragement because it will make for more broad-minded engineers" might work within a university setting because broadmindedness is a value prized by the university community. However, an argument based on personal development might be somewhat less successful at work, where personal development is seen as less important than corporate development.

Special Case: Arguing Electronically

Before you leave this chapter, consider arguing in a rhetorical space that is already important and that will only grow more important over time: electronic space. One of the most important rhetorical locations for forging arguments in electronic space is electronic mail, better known as **e-mail**. In 1998 the American Management Association surveyed members attending its annual conference and found that although people preferred face-to-face communication, they used e-mail more than any other form of communication, including the telephone (Greenberg). In his "12 Rules for Succeeding in the Digital Age," Bill Gates argues that for companies to succeed managers must "insist that communication flow through e-mail" (Gates). E-mail is how much work gets done in the workplace.

In light of the increased use of electronic media in the workplace, the question to ask yourself as you move online is "How does an electronic medium—e-mail or the World Wide Web—affect arguments I make? Case Documents 4, Appendix A demonstrate a few ways that arguments can appear in Web sites, but here we want to focus on e-mail.

In broad terms, the same principles hold for arguing effectively no matter what the medium. The advice in this chapter should serve you very well

e-mail
electronic messages sent by computer over the Internet or over a company's or organization's internal electronic communication system

no matter what kind of document you are composing. However, there are special conditions worth considering when you work with e-mail.

You probably already sense from your experience that most e-mail is informal in tone. E-mail does tend to be more "oral" than "written" in its feel and texture. This informality is not a problem when you use e-mail for your personal communications, but it can be a problem if you let informality get out of hand in the workplace.

You will need to consider how your e-mail will be used as well as your potential audience if you hope to craft effective messages in this electronic space. Currently, the major uses for e-mail in the workplace are the following:

- To support collaboration by allowing writers to circulate drafts of documents.
- To communicate urgent information.
- To help companies get more out of their investments in communication infrastructure.
- To facilitate interaction up and down the company hierarchy.

Even though e-mail invites informality, you cannot afford to be too informal or careless if you want to be effective in this medium.

A FEW GUIDELINES FOR PERSUASIVE ELECTRONIC DISCOURSE

When you write e-mail, remember that the e-mail you send at work does not belong to you. In the work setting, your employer owns the system and none of your messages are private. Moreover, e-mail is generally stored somewhere on the system *forever*, even if you delete a particular message. What you say can come back to bite you, if you are not careful about your communications.

When you are using e-mail to collaborate with others on writing projects, your informal tone will stand you in good stead. People will have a sense of who you are, and you can help maintain the relationship dimension of your task. Remember, to keep your messages short. No one wants to read messages so long and complicated that they always have to be printed out. And, just because your e-mail is informal is no excuse for being sloppy with the drafts of documents you exchange. In addition, remember to be tactful when you criticize another's work because e-mail can easily be forwarded.

When you use e-mail to communicate urgent information, make sure that you present your information accurately and clearly. You don't want to have to send out clarifying follow-up messages. Make sure your urgent e-mail is spelled correctly and proofread. If your e-mail application doesn't support spell checking, compose your message first in a word processing program, check the spelling, proofread the document, and then copy it onto your e-mail sender. Mostly, urgent e-mails are sent to others within the company, but your message might well end up being sent to a client or vender outside the company. Plan carefully so that the information with your name on it reflects well upon you and the organization you represent.

Many companies are relying heavily on e-mail and Web sites to carry out their routine business. That means increasingly business forms and data are online. Many of these forms are read by a wide variety of people in human resources and management. Take care to maintain your ethos by responding carefully to documents and requests for information you receive online.

Finally, many companies encourage their employees to send both good and bad news up the management line. If that is the case at your organization, then craft your messages carefully. Consider tone; you don't want to appear to be a disgruntled employee if you are pointing out some problems to upper-level administration. Rather, you want to appear as a concerned team member. E-mail makes it possible for you to contact the president or CEO directly in ways that have never before been possible, especially in large organizations. Bill Gates, president of Microsoft, has said that he reads all employee e-mail sent to him. Here as elsewhere, don't be too informal or long-winded, and send a copy to your immediate supervisor so he or she isn't kept out of the loop.

The strength of this medium is the speed with which it travels, so take a few moments to make sure you really want others to read what you have to say and that you have said it well. At work, resist the temptation to put things online or in e-mail too quickly. Think carefully about your readers and the rhetorical situation, and craft your electronic communications with care.

SUMMARY

- In this chapter you have learned to think about technical communication less as an impersonal and rational demonstration of "the facts" and more as a set of arguments laid before a jury of readers and listeners. We have given you ways of thinking about arguments that may be somewhat different than the ways you thought about them before—especially by encouraging you to think of your reader as a partner instead of an adversary. We have encouraged you to construct arguments that are particularly responsive to your audience's views and values; we have encouraged you not simply to state your case, but to make your case by means of an appropriate use of *ethos*, *pathos*, and *logos*.

- To help you achieve these goals, we discussed ethos as creating a reliable, trustworthy persona. To create the kind of persona that will be most persuasive in a given case, you need to choose an appropriate voice based on your relationship with your audience, your audience's personality, your message, your aim, and your genre. You also must create trust and credibility, and you need to show that you have done your homework by citing your sources and indicating how you have conducted your study or gathered your information. Establishing your special expertise and creating a sense of fair play also contribute to a reliable and trustworthy persona. Finally, engaging in a cooperative dialogue with your audience and attending to the ethos of your company or organization are essential to creating persuasive technical communications.

■ Although at one time it was unusual to discuss pathos in technical communication, we now see that appealing to the values, attitudes, and beliefs that your audience holds most deeply can be effective.

■ Logos, the third factor we discussed in persuasion, is largely a matter of finding and selecting "good reasons" for your arguments. In this chapter, you learned to create because-clauses that reflect compelling reasons for your audience to accept your solutions and recommendations. To help you find good reasons for your argument, we provided you with a set of questions:

—Can I argue by definition?
—Can I argue from consequence?
—Can I argue from value?
—Can I compare and contrast?
—Can I cite authorities?
—Can I counter objections to my position?
—What are the special lines of argument that people appreciate in my field?

■ To support your good reasons, you learned that verification by relevant and sufficient evidence is essential. And you saw that to be persuasive you need to think of your audience's values and those sanctioned by your community.

■ Finally, you learned some considerations to keep in mind in the use of e-mail. Consider the purpose of the message you are sending and its potential readers. Apply what you have learned about creating effective arguments in print and oral communication to this medium as well.

We hope that the chapter will make you not only more successful in constructing your own arguments but also a better, more discriminating consumer of the arguments that others present to you.

ACTIVITIES AND EXERCISES

1. Individually or in groups, try to develop as many good reasons as you can for supporting a position on your campus or in your major. Some examples would be a reform in the curriculum or an improvement in the facilities and processes at your school, but you should pick something current and topical.

(Perhaps you could discuss a technical or scientific controversy current on your campus.) Once you have devised a series of good reasons for your proposition, try to develop a set of good reasons for *not* taking that position.

2. Examine any three documents in Appendix A of this book. First, determine whether the document could be considered an argument. If it can be, discuss the writer's *ethos* in the argument: How does the persona present himself or herself? What things make the persona seem credible or not? Then consider the "good reasons" that are given in support of the thesis: What kinds of good reasons are presented? Could any of them be considered as appeals to *pathos*? Does the author provide sufficient support for the propositions offered? In sum, how successful is the argument?

3. Reexamine Examples 4-3 through 4-6. Working with a partner, make a complete list of the differences between the documents. Pay attention to audience and purpose, but focus on all aspects of voice discussed in this chapter. Be prepared to present your findings to the rest of the class.

4. Using the position you selected in Activity 1 above, write a letter to your campus newspaper, your department head, your major advisor, or the most appropriate person to argue your case. If you completed Activity 1 in a group, complete the letter individually and then share results.

5. Throughout this chapter we offered as an extended example the argument for a foreign language requirement for engineering students. Your goal was to write a proposal that engineering administrators will use to decide whether a full-fledged feasibility study would be worth the time and expense.

    Although you'll learn more about the format of proposals in Chapter 15, now write an initial draft of that proposal (your instructor may ask you to revise the proposal after you study proposals formally). You may use and add to any of the reasons or strategies suggested in our extended example.

6. Examine the way certain public interest groups conduct advocacy efforts on Web sites related to various issues. For instance, Yahoo (www.yahoo.com), under its "Society and Culture" heading, has a subheading called "Issues and Causes" that includes public discourse on science and technology issues such as animal experimentation, cloning, nutrition, environmental health, the human genome project, ozone depletion, pollution, agricultural sustainability, and pest management: see <http://www.yahoo.com/Society_and_Culture/Issues_and_Causes/>. Look closely at some sites that interest you and note how they are designed. How well do the sites observe the suggestions for conducting arguments on the Web that you read about in this chapter?

7. Go to the library or a good bookstore and get a current copy of *Science*, which is a publication of the American Association for the Advancement of Science. *Science* has a number of sections: News, Research, Compass, Reports, and a special section. In each section, look for examples of the ways arguments are constructed using appeals to *ethos*, *logos*, and *pathos*. How do the sections vary in their use of appeals? What reasons can you deduce for the variety of styles, tones, and arguments used in these sections?

WORKS CITED   Brennan, Richard. *The Dictionary of Scientific Literacy*. New York: John Wiley, 1992.

Burke, Kenneth. "Rhetoric—Old and New." *Journal of General Education* 5 (1950-51), pp. 203-9.

Gates, William H. III. *Business @ The Speed of Thought: Using a Digital Nervous System*. NewYork: Warner Books, 1999.

Greenberg, Eric Rolfe. "E-mail Usurps the Phone as Communication Tool." *HR Focus* 75, no. 5, p. 2.

Miller, Carolyn R. "What's Practical about Technical Writing?" In *Technical Writing: Theory and Practice*. Bertie E. Fearing and W. Keats Sparrow, eds. New York: Modern Language Association, 1989, pp. 14-24.

PART

2

Acquiring the Tools of
Technical Communication

Collaboration

Rarely do even Big Ideas emerge any longer from the solitary labors of genius. Modern science and technology is too complicated for one brain. It requires groups of astronomers, physicists, and computer programmers to discover new dimensions of the universe: teams of microbiologists, oncologists, and chemists to unravel the mysteries of cancer. With ever more frequency, Nobel prizes are awarded to collections of people. Scientific papers are authored by small platoons of researchers.

Robert Reich, *Tales of a New America* (New York: Time Books, 1987).

I feel it is important to learn how to work with other people in whatever you do because it is essential when you are out in the working world.

College senior, Technical Writing class.

Introduction   As we mentioned in earlier chapters, professionals frequently collaborate or work in groups or teams to produce documents in the workplace. In 1982, for example, Faigley and Miller's survey of 200 college-educated individuals in a range of occupational settings indicated that 73.5 percent contributed to collaborative writing efforts. A broad range of collaborative writing goes on in the workplace, including multiperson authorship of reports, articles, books, briefs, memoranda, and proposals.

Professionals collaborate because groups often produce better solutions than individuals. Although people collaborate regularly to create proposals, plans, reports, and instructions, they often talk of the difficulties surrounding the collaborative experience. These difficulties may arise because most writing teams begin by attacking tasks without first completing some necessary preparatory steps. Teams function best when they first build the interpersonal environment necessary for their creative problem solving, consider possible approaches toward collaborating, decide on the ways they might handle conflict, access computer technologies to facilitate their work, and identify the strengths of their diverse members. The teams that succeed are those that come to an understanding about their collaborative relationships and the process they will follow to complete their tasks.

This chapter helps you prepare for the range of collaborative writing activities you will encounter in your classes and in the workplace. Deciding how and when to collaborate is an important part of any writing plan. However, you may have little or a lot of control over your collaborative writing assignments. At times you will be assigned to a team with the schedule and process mapped out for you by a supervisor; at other times you might have the freedom to design effective collaborative processes within your writing team. And, perhaps in a future managerial position, you may be assigning and directing the collaborative efforts of others. Whatever your situation, this chapter provides you with a common language for initiating, executing, and presenting a collaborative project. To begin, let's define what it means to collaborate on writing tasks.

collaborative writing
team or group writing
on a project

Collaborative writing is often defined according to how people work together on a project. In a survey of 1,400 professionals across the United States, seven basic forms of collaborative writing emerged:

1. Team plans and outlines. Each member drafts a part. Team compiles the parts and revises the whole.
2. Team plans and outlines. One member writes the entire draft. Team revises.
3. One member plans and writes the draft. Team revises.
4. One person plans and writes draft. This draft is submitted to one or more persons who revise the draft without consulting the writers of the first draft.

5. Team plans and writes draft. This draft is submitted to one or more persons who revise the draft without consulting the writers of the first draft.

6. One member assigns writing tasks. Each member carries out individual tasks. One member compiles the parts and revises the whole.

7. One person dictates. Another person transcribes and revises (Ede and Lunsford, 63–64).

Certainly, there are additional variations to these seven forms.

Because collaboration is embedded in the structure of many organizations, writers know they are working with others, but they rarely can explain the exact process they follow when collaborating. Similarly, you have collaborated in the past, but you probably remember more about the people involved and the working relationships than you do about the exact processes you followed.

At the heart of collaborative writing is the fact that you produce something with the direct or indirect help of others, that what you produce takes on a new form based on this input and interaction, and that your identity as an individual writer is likely to take second billing to the needs of the team. However, as you work together to solve a problem, create something, or discover something, the goal is to ensure that your collective work results in a better response, system, or discovery than is possible through individual efforts. Sometimes, complex challenges—reducing environmental degradation, ameliorating the effects of epidemic disease, addressing the need for more efficient energy production—demand collaboration, and as the needs of society and technical communication become more complex, we witness even greater challenges.

In order to understand challenges, generate solutions, and function as a team, groups need to develop on both the relationship and task levels. When researchers from Bell Communications Research interviewed 50 pairs of collaborators from social psychology, management, and computer science, they found that successful collaborations followed three stages: **initiation, execution,** and **public presentation.** What follows describes a process for initiating your collaborative project, executing it, and presenting it to your audience. We'll look at the relationship level and the task level during each of these stages, and we'll discuss some common structures of collaborative writing groups after our description of the initiation stage.

initiation, execution, and public presentation the three stages of successful collaboration, each of which has a relationship and task level

---

At each stage of the collaborative process, and especially at the initiation stage, your writing team needs to develop at both the **relationship level** and the **task level.** Look at the collaboration model in Example 5-1. The relationship level comes first; it's on top. If a collaborative team neglects relationships, task-level work will not be as effective, or it might fail.

Initiating a Collaborative Writing Project

relationship level
the interpersonal activities through which collaborative teams get to know each other, share assumptions, maintain trust, resolve conflict, and acknowledge each other's work

task level
the task completion activities in which collaborative teams clarify goals, generate ideas, establish procedures, draft, revise, edit, and present their work

---

EXAMPLE 5–1   Collaboration Model

	Initiation Stage	Execution Stage	Public Presentation Stage
Relationship Level	■ Find team members ■ Get to know each other	■ Maintain trust ■ Establish ways to resolve conflict	■ Establish final division of credit ■ Acknowledge and appreciate each person's work
	■ Share background assumptions	■ Clarify work styles	
Task Level	■ Clarify goal(s) ■ Choose approach and schedule meetings ■ Generate ideas and plan	■ Share information ■ Establish procedures to assess progress ■ Do the work (plan, draft, revise, edit)	■ Reach closure on your work ■ Package your work for public presentation

Source: The three stages and two levels in this table come from the research of Robert E. Kraut, Jolene Galegher, and Carmen Egido. "Relationships and Tasks in Scientific Collaboration," *Journal of Human-Computer Interaction* 3, no. 1 (1988), pp. 31–58, specifically p. 34, and are used with permission. The majority of these points have been modified by Ann Hill Duin.

RELATIONSHIP LEVEL

When you initiate a collaborative project, you first need to find partners and share background assumptions (refer back to Example 5-1 as you read the rest of this chapter). The most fundamental requirement is for group members to make contact with each other, to become acquainted, and to identify common interests and mutual benefits that might come from working together.

Carl Larson and Frank LaFasto, from interviews with numerous teams ranging from presidential commissions to championship football teams, noted that all too often people are chosen as collaborative team members for the wrong reasons or that people choose to work with other people for the wrong reasons: "Laura should be on the team because she's interested in the topic." Or, "Al's feelings would be hurt if he were left off." Or, "Jeremy should be included because he reports to Donna." Instead, what should be paramount is selecting or choosing people who are best equipped to achieve the team's objective. You should choose people based on the personal competencies required to achieve excellence while working with others on the project and on the technical skills and abilities necessary to meet your challenge.

Personal competencies include the qualities, skills, and abilities necessary for your team to function and write together. Part of creating a successful collaborative team is deciding what type of personal qualities and interpersonal skills you want members to have. For example, the team leader of the group that developed McDonald's Chicken McNuggets, Bud Sweeney, stated that he specifically looks for people who are "intelligent, creative, tenacious, and a bit of a maverick—willing to buck the system and not be bound by traditional thinking and reporting relationships" (Larson and LaFasto, 63–64).

Glenn Parker notes that collaborative team members should have the ability to:

- Participate in open discussions which include consensus building, ambiguity, and disagreement.
- Listen carefully and employ listening techniques such as questioning, paraphrasing, and summarizing others' ideas.
- Assume a leadership role when necessary because effective team members often change roles.
- Share all information freely with other members of the team as well as with members of contingency teams.
- Build relationships with others outside the team, including important resources in other parts of the organization.
- Appreciate different learning and working styles.

personal competencies the qualities, skills, and abilities necessary in each collaborative team member for the team to function well

Although needed personal competencies vary from team to team, it's extremely important to take the time to identify and promote interpersonal skills or the relationship level of your team.

Technical knowledge enables a collaborative team member to solve a problem, create something, or execute a specific task. Linda Alvarado, CEO of Alvarado Construction, notes a need for a balance between "nuts and bolts" members and creative and conceptual members. In her construction business, she organizes collaborative teams with members who can see the big picture and know when major changes are necessary as well as members who are very detail oriented (Larson and LaFasto, 63). To create a successful team, decide if you need people who know specifics about a topic, people who are creative and conceptual, or a blend. If you need people whose skills are radically different, you need to integrate them carefully into the team at the initiation stage.

technical knowledge the technical competencies necessary in each collaborative team member for the team to solve a problem, create something, or execute a specific task

Because so much of success in collaborative teams involves intangibles like attitude and energy, remember to take time at the initiation stage to find the right collaborators (if you are choosing your own collaborative team) and to become acquainted and to identify your common interests and needs (regardless of who initiates the collaborative team). Going out to eat together or playing volleyball, for instance, may help you and your team members get to know each other, bring your assumptions about the project out in the open, and share your concerns and your excitement.

TASK LEVEL

Only after you have talked informally at the relationship level should you proceed to the task level. At this step, you can begin to generate ideas and plans (look again at Example 5–1). If team members are located near each other, plan numerous face-to-face meetings; if not, use e-mail, fax machines, voice mail, and conference calls to link yourselves (note possible ways to use technology in "Employing Collaborative Technologies" later in this chapter).

At the task level, you should first clarify your goal by asking exactly what it is that your team wants to accomplish. Successful teams have a clear understanding of their goal, a sense of mission. Consider this example concerning clarifying a goal:

> The Apollo Moon Project provided a superb demonstration of the power of a clear and compelling vision. By committing themselves to "placing a [person] on the moon by the end of the 1960's," the leaders of the project took a stand. The clarity and conviction they generated touched people at all levels of the enterprise. One can imagine how much less spectacular the results might have been if they had adopted an alternative mission statement such as "to be leaders in space exploration." (Kiefer and Senge, 112)

In addition to clarifying your team's goal, you need to identify the audiences who will read the document your team produces and their specific reading and information needs. Will the audience expect the document to be structured in a certain way? What information will they need most immediately? What level of detail and explanation is most appropriate for readers? Finally, a collaborative team needs to visualize a final product. Will you develop a proposal that gets funded? Will you design a process that your company will follow? Will you write a position report that clarifies a current controversy? Will you create a Web site that introduces and organizes some complex information? Using techniques described in Chapter 2, collaborative teams will find the answers to these questions useful in formulating a short written statement of purpose and goals for a document. Such a brief statement will help ensure that all members of the team agree on what that product will look like and can participate in the production of it.

THE STRUCTURE OF COLLABORATIVE WRITING TEAMS

Once your collaborative team has clarified and visualized its goals and done some preliminary work to visualize the key parts of the written document it needs to produce, it can begin to make other decisions about how to approach the collaborative writing task. Collaborative writing is rarely an assembly-line process. Perhaps a division head or manager can jump into a group and say, "You do this, she'll do that, he'll do this, and I'll do that." And if work was always routine and predictable, this might be a suitable model. But because solving problems is rarely so simple and because people's strengths vary, you will want to consider a variety of models or approaches for structuring your collaborative writing team. We describe five models in this chapter. Perhaps your supervisor will suggest one of them. If you have the freedom to structure your own group, you need to read about each model and talk about them within your group. Your group will need to decide which will work best with your group's chemistry, commitment, strengths, weaknesses, and challenge.

Divide-and-Conquer Model

The **divide-and-conquer model** is the simplest one. In this case each team member takes a segment of the project and writes, revises, and edits only that part. This model assumes that every member of the team is alike, so that each person could replace another at any time. Although industry often resorts to this model to expedite the collaborative process, most writing challenges rarely allow for this simple approach. However, in cases where your problem is relatively simple and your timeline extremely short, you can divide the project or assignment into subtasks, decide on a group leader and have that person assign jobs, or have members choose jobs themselves.

divide-and-conquer model
a collaborative model in which each team member writes, revises, and edits one part of the project; assumes each team member is alike and replaceable

Specialization Model

The **specialization model** operates on the principle that each member gets a task that relates to his or her special talent or expertise. For example, one member may work best as a writer or editor; another may work best as a graphics specialist or a project manager. This approach doesn't allow you to exchange tasks; it assumes that the group's strength is based on its collective expertise. You should follow this approach when the project requires the expertise that certain members can offer. Whether or not you choose to follow this model, one of the first things you should do when establishing a collaborative team is to identify each person's special talents.

specialization model
a collaborative model in which each team member is assigned a task according to his or her special expertise

Dialogic Model

The **dialogic model** is marked by flexible roles that may shift as a project progresses. Working in this model, a member might first be a project manager and later take on the role of editor or graphic artist. As the project proceeds, needs and roles change. For example, after producing first drafts, the writers can serve as editors for their own and each other's words. This model best accommodates possible gender or cultural differences among team members. Because the collaborative process is not rigidly structured, a hierarchy or power structure seldom emerges. And, as team members assume new roles as the project proceeds, a high level of

dialogic model
a collaborative model in which team members' responsibilites shift during the project; requires a high level of interaction

interaction is required. When you use this model everyone will need to be heard from frequently, and voices that might receive less attention or be less welcome in other situations are essential here. As you develop a flow-chart of how your project might proceed, consider how your roles could change and how you might take on the strengths of others.

Sequential Model

sequential model
a collaborative model in which one team member writes a draft and passes it to another for comment or sign-off, who in turn passes the draft to the next team member, and so on

In the **sequential model,** one person writes a draft and passes it to an-other who comments and/or revises it and returns or forwards it to the next person. This next person again comments and/or revises and returns or forwards the document to the next person in the sequence. In many in-dustries a required series of "sign-offs" or signatures of approval results in a sequential model of collaboration. This review process allows a corporation to transmit its values and corporate culture quite effectively. The greater the impact of the document, the more likely it is that the sequential review will be extensive and involve drafts that are passed back and forth between people and corporate levels. For example, when Susan Kleimann studied how reports were created at the United States General Accounting Office, she found an elaborate review process (see Example 5-2). If you are com-pleting a task for an outside client or corporation, you may need to have your client(s) sign off at specific stages of your project. Even if you do not need these approvals, you may still want to follow this model if you are on a tight schedule.

Synthesis Model

synthesis model
a collaborative model in which team members are brought together because of their different views or perspectives

Often team members are brought together because of their different views or perspectives toward a controversial issue or problem. Such teams might follow the **synthesis model.** For example, if a company needs to solve a problem and three subdivisions have three distinct solutions, it's best to bring a subset of people from each division together to collaborate. In this case the members must synthesize their different perspectives, melding them into a recommendation acceptable to all. In the synthesis model, members belong to the team because of their different views or solutions. They are likely to work on those aspects of the project that represent their individual perspectives; however, through the process of collaborating, they will begin to resolve some, though probably not all, of their differ-ences. It's also natural that new areas of difference will emerge. In less con-troversial situations, you might follow this model if each team member sup-ports a particular product and your challenge is to recommend only one product in a collective report.

Be sure to study each of these models and identify which one you intend to follow. At each stage of your collaboration, evaluate how the model is working, make refinements to the model, or switch to a different one. Or, if you are assigned to follow one model by a supervisor, be aware of the bene-fits and challenges of that model. Finally, if you are supervising a collaborative

EXAMPLE 5–2   The Approval Process at the United States General Accounting Office

```
                          ┌──────────────┐
                          │  Work Team   │
                          └──────────────┘
                                 │
                          ┌──────────────┐
                          │  Work Team's │
                          │Assistant Director│
                          └──────────────┘
                                 │
                          ┌──────────────┐
                          │  Work Team's │
                          │   Director   │
                          └──────────────┘
                                 │
                          ┌──────────────┐
                          │ Report Review│
                          └──────────────┘
```

| Head of Report Review | Director of Reporting | Report Reviewer | Technical Reviewer | Editor |

```
                          ┌──────────────┐
                          │   Head of    │
                          │ Report Review│
                          └──────────────┘
```

| Work Team's Director | Work Team | Work Team's Assistant Director |

```
                          ┌──────────────┐
                          │  Work Team's │
                          │Assistant Director│
                          └──────────────┘
                                 │
                          ┌──────────────┐
                          │  Work Team's │
                          │   Director   │
                          └──────────────┘
```

| Head of Report Review | Director of Reporting | Report Reviewer | Technical Reviewer | Editor |

```
                          ┌──────────────┐
                          │Head of Division│
                          └──────────────┘
                                 │
                          ┌──────────────┐
                          │Final Processing│
                          └──────────────┘
```

Source: Susan D. Kleimann. "The Complexity of Workplace Review." *Technical Communication* 38, no. 4 (1991), pp. 520–26, especially p. 523. Reprinted with permission from *Technical Communication*, published by the Society for Technical Communication, Arlington, VA.

writing team, choose the model that will best help the team meet your organization's goals and ensure that the collaborative process will bring satisfaction, not frustration, to that team.

Executing a Collaborative Writing Project

In 1985 a 20-member team attempted to climb Mount Everest and identified their goal as getting one member to the top. Unfortunately, the team overused its best climbers during the first part of the climb and didn't share responsibilities evenly. They got to 28,200 feet, only 800 feet from the top. Team member George McLeod stated that "I think we should have sat down as a group, which we seldom did, and discussed how we were going to make this happen . . . We didn't take the summit" (as reported in Larson and LaFasto). In contrast, the first women's team to climb Mount Kongur in China identified a different goal and structure. Their goal was based more on consensus and was not to get one person on top, but to get as many people as high on the mountain as they could. They were successful in that goal.

What we learn from these experiences is that at the execution stage, a collaborative team needs to continue to develop at both the relationship and task levels. Team members need to sit down with each other, clarify their work styles, and reach consensus about how they plan to proceed. Part of this process means coordinating how to share information as you plan, draft, revise, and edit your collaborative text.

RELATIONSHIP LEVEL

As you execute your project, you need to maintain trust. In other words, you need to continue to attend to the relationship level. One way to maintain trust is to talk openly about the five areas in which conflict might occur:

1. Struggles for leadership: efforts to control, dominate, exert power over, or lead the team.
2. Unequal workloads: unequal contributions from members.
3. Personality differences: differences due to interpersonal likes and dislikes.
4. Procedural matters: organizational, procedural, or mechanical problems.
5. Ideational matters: problems relating to ideas, goals, and values associated with the team's task (Wall and Nolan).

One study of numerous teams found that when a conflict centered on the task level, 83 percent of the teams chose to discuss the conflict and work out a procedure for resolving it. However, when the conflict focused more on the relationship level (struggles for leadership, unequal workloads, personality differences), only 17 percent of the teams chose to resolve the conflict while the others chose to avoid or ignore the problem (Wall and Nolan 188). It's not surprising that teams find discussing relationship conflicts difficult; most people don't like interpersonal conflict. Yet, if teams ignore problems and the inevitable conflict that arises, the results can mean disaster for the project. Before conflict arises, set up a process for handling it; then use the process at the first sign of trouble. To ignore disputes is not a good strategy; to establish and use a procedure to resolve them is the best strategy.

Managing Conflict in Collaborative Relationships

Three specific procedures can be used to manage conflict, depending on the source of the disagreement. Notice that we talk about managing rather than resolving or eliminating conflict. As long as you are working with human beings, you have to expect conflict. It's not always a bad thing, and it can almost always be managed in such a way that your project and the interaction of group members don't suffer. How you manage conflict will depend on whether the source of conflict is a misunderstanding of the problem or solution—**pseudo-conflict**, individual disagreement over which course of action to pursue in a project—**simple conflict**, or defense of a member's ego—**ego-conflict**.

pseudo-conflict
a misunderstanding of the problem or solution

simple conflict
disagreement over which course of action to pursue during a project

ego-conflict
disagreement resulting from the defense of a collaborative team member's ego

Example 5–3 presents these three sources of conflict and suggestions for managing them. For example, remember the case of the engineering students who wanted to include foreign languages in the curriculum? Let's say that you now decide to write the proposal that engineering students should take foreign language courses, but the task is too large for you alone. You ask other members of your local chapter of the American Society of Civil Engineers to form a writing team. Perhaps one member of the writing team, Sandy, thinks that the proposal should include a recommendation that engineers take more history courses, because history courses teach engineering students more about other cultures than foreign language courses. Sandy's point is well taken, and she feels very strongly about it, but you might suggest an alternative to putting too much into one proposal. You could assign the history proposal to another group (see suggestion 6 under Simple Conflict in Example 5-3), delay this recommendation until you see how your foreign language proposal flies (suggestion 8), or postpone vetoing Sandy's proposal until your group can do some research on the matter of history courses (suggestion 9).

If Sandy feels so strongly that her suggestion is better than yours, you might encourage her to describe the problem as she sees it and how her solution would work, rather than labeling one suggestion good and the other poor (see suggestion 5 under Ego-Conflict in Example 5-3). If Sandy

---

EXAMPLE 5–3   Suggestions for Resolving Conflict

Pseudo-Conflict	Simple Conflict	Ego-Conflict
	Source of Conflict	
Misunderstanding individuals' perceptions of the problem	Individual disagreement over course of action to pursue	Defense of ego; individual believes he or she is being attacked personally
	Suggestions for Managing Conflict	
1. Ask for clarification of perceptions **2.** Establish a supportive rather than defensive climate **3.** Employ active listening: 　　Stop 　　Look 　　Listen 　　Question 　　Paraphrase content 　　Paraphrase feelings	**1.** Listen and clarify perceptions **2.** Make sure issues are clear to all group members **3.** Use a problem-solving approach to manage differences **4.** Keep discussion focused on issues **5.** Use facts, not opinions, for evidence **6.** Look for alternatives or compromise positions **7.** Make conflict a group concern **8.** Determine which conflicts are most important to resolve **9.** If possible, postpone decision while additional research is conducted; delay helps relieve tensions	**1.** Let members express concerns but do not permit personal attacks **2.** Employ active listening **3.** Call for cooling-off period **4.** Keep discussion focused on issues (simple conflict) **5.** Encourage parties to be descriptive, not evaluative or judgmental **6.** Use problem-solving approach to manage differences of opinion **7.** Speak slowly and calmly **8.** Develop rules or procedures that create a relationship allowing for the personality difference

Source: From *Communicating in Small Groups: Principles and Practices,* 3rd ed. Copyright 1990, 1986, 1982 Scott, Foresman and Company. Reprinted by permission of Addison-Wesley Educational Publishers Inc.

misunderstands the central issue, that engineering students need to understand *contemporary* cultural differences—for example, modern Italian highways—you can try to clarify your perceptions and have Sandy do the same (suggestion 1 under Pseudo-Conflict in Example 5-3).

Another possible procedure for resolving disputes is to appoint one member to regularly ask each member to talk about any difficulties he or she may be having. This person then brings these issues up at the next team meeting. Another possibility is to ask a person outside of your team to talk to each member about difficulties and then share these with your team. A third approach is to have each person write about difficulties, pass this information around during a team meeting, and let each person write constructive responses to the concerns. After that process is completed, then discuss these responses as a team.

As you share background information concerning a conflict, you may discover that the conflict stems from your different work styles or learning styles. Establishing a procedure to discuss rather than to avoid conflict will mean taking time to listen and explore the conflict, considering its complexity and the situation that surrounds it. However, such a procedure will benefit your team and the collective decisions you make.

TASK LEVEL

At the task level, you need to share information, prompt and support each other, and complete the collaborative writing task. A number of variables influence collaborators as they execute a writing task:

Control   Talk about the degree of control each member has over the text. How does each member's work contribute to the whole? Let members share how they see their work contributing to the project.

Credit   Begin to discuss how credit will be given for the work. If you are collaborating at school, ask your instructor this question. Will he or she evaluate each of you individually? Will he or she evaluate only the final team product? If you are collaborating at work, ask who will get credit for the assignment and who is responsible.

Response   Decide how you will continue to meet as a team and respond to each other's work. How will you respond to the modifications made by others?

Flexibility   Decide how flexible you can be with an established format (e.g., how closely you should follow a proposal or report format). Likewise, how flexible should you be with your approach to this collaborative project? If you have been following a specialization model, should you now follow a more dialogic model?

Constraints   Share the very real constraints facing you. What's your deadline? Who's your primary audience? On what criteria will this project be evaluated? (Ede and Lunsford)

At the task level, one form of conflict noted earlier can work to your advantage: *pseudo-conflict*. Pseudo-conflict is similar to what some call **substantive conflict**. It describes considering alternatives and voicing explicit disagreements about both content and rhetorical elements such as purpose, audience, organization, support, and design (Burnett, 534). Conflict of this sort typically enhances decision making. In this situation, you prompt members to clarify their perceptions, their positions, and their texts. Here you employ active listening; you stop, look, listen, question, and paraphrase. Writing Strategies 5–1 presents a list of prompts intended to help collaborative

substantive conflict disagreement about content, purpose, audience, organization, support, and design; enhances decision making

---

WRITING STRATEGIES 5–1   Prompts to Expand Ideas, Consider
Alternatives, and Voice Disagreements

Content

What information might we add or delete?
What will our audience expect? need?

Purpose and Key Points

I can't quite see why you've decided to _____ . Could you explain why?
What do we want the audience to do or think?
What is the point we want to get across in this section?

Audience

What problems (conflicts, inconsistencies, gaps) might our audience see?
How will our audience react to this (content, purpose, organization, design)?

Conventions of Organization, Development, and Support

How will we organize (develop, explain) this?
What support (or evidence) could we use?
What examples could we use?

Conventions of Design

Why do you like _____ better than _____ as a way to present
this information?
How will this design reinforce our point?

Synthesis of Plans about Content and Rhetorical Elements

Why do you think _____ is a good way to explain this point to our audience?
Will changing the example clarify or cloud the point we're making here?

Source: Rebecca Burnett. "Substantive Conflict in a Cooperative Context: A Way to Improve the Collaborative
Planning of Workplace Documents." *Technical Communication* 38, no. 4 (1991), pp. 532–39, especially p. 536.
Reprinted with permission from *Technical Communication*, published by the Society for Technical Communication,
Arlington, VA.

---

writing teams consider issues yet defer consensus. Use these questions to
expand your ideas, consider alternative ideas, and make your disagreements
explicit. These questions will move your team toward a better collaborative
document within a cooperative climate.

Preparing Your Collective Work for Public Presentation

The last stage of the collaborative process involves preparing your document for public presentation. At this stage, a collaborative team needs to establish final division of credit for its collective work and needs to package
its document for public presentation to an audience.

RELATIONSHIP LEVEL

Before your collaborative team presents its document to an audience, you should have resolved any disagreements you may have had. Now is not the time to rehash conflicts because your project must assume a unified public face. Even though you might be hurrying to complete the project, you need to take special care not to inadvertently forget to acknowledge each person's contribution because at this stage members often feel as if they're the only ones working or that they're working the hardest. Remember to acknowledge each person fully and to respond positively to each other's work.

TASK LEVEL

At this final stage, collaborative teams obviously need to reach closure on their written work—after all is said and done, the entire team is responsible collectively, for completing the assigned project. You might want to revisit the decisions you made during the execution stage with regard to control over the document. You will need to decide if one or two members should be in charge of the final packaging of the document, or if all of you will meet as an entire team. The team will need to consider what constraints might bog it down at this stage (for example, the computer programs you are using, your audience's expectations for the final packaging of the document). You also need to establish guidelines covering how flexible you will be with each other as well as with the established format of the document.

Last, remember that a document created collaboratively, like any other piece of writing, can always be revisited and revised up to the final deadline. For example, a team member might come up with a useful way to color code your document. Or, a team member might suggest, just at the end of your collaborative process, that the software your company is producing for university classrooms should be called "courseware," and you need to search your document for this change. A document is never finished, only abandoned. But you must reach consensus at this point; you must reach closure as you present your document to the public.

In today's world and workplace, **collaborative technologies** are available to support your team's work. These tools are designed to support your work at both the relationship level and the task level. Although some schools cannot afford this kind of technology now, many businesses and companies can. These discussions of collaborative technologies, therefore, will help you when you enter the workplace and join an organization that uses such equipment.

Collaborative technologies are used not only to help you produce a complete and polished document; they can also help you communicate

Employing Collaborative Technologies

collaborative technologies computer tools and software that support collaboration on relationship and task levels with team members, maintain records, receive feedback, and structure the final document. Whatever system you choose or have access to, remember to structure its use to support conversations, sketches, arguments, and discoveries; it should not merely help you write, edit, and design your final text. All members of a team should have access to any software or hardware systems that you decide to use, and collaborative technologies should be directly compatible with company standards and systems.

computer-supported team
a writing group that uses computer technology to support the collaborative process

Once you decide to use computer technology to support your collaborative process, you are, in effect, a **computer-supported team.** Four factors should determine the technologies you use:

- Your team's goal.
- Whether or not your members are physically separated.
- The deadlines you must meet.
- The type of computer systems available and their compatibility.

What follows are nine approaches to consider as a computer-supported team (for a more complete discussion, see Johansen). If you do not have access to this technology, you can sometimes conduct the activity with pen and paper, flip charts, chalkboards, and other tools. Although all of these technologies could be used at any stage of your collaborative process, most are best suited to a particular stage. In the following sections we describe how each might be used at a particular stage.

TECHNOLOGIES BEST SUITED FOR THE INITIATION STAGE

The technologies we discuss below are most useful as you discuss your audience and purpose and plan your writing process—and, of course, as you get to know the members of your team.

Face-to-Face Meeting Facilitator

Throughout your project, but especially at the initiation stage, you need ample face-to-face meetings to get to know each other, identify and clarify your goal, and generate plans. At this point you need frequent, open, and often lively discussions as you explore your options.

face-to-face meeting facilitator
a computer technology that enables group members to participate in meetings and record meeting notes

One way computer technology can support your collaborative process is through a **face-to-face meeting facilitator.** In this scenario, you ask a person from outside your team to facilitate by recording summary phrases from your meeting on a computer. If this computer is connected to an overhead projection system, the facilitator can project these statements on a screen for the team to see. Essentially these notes are electronic versions of notes you could put on a flip chart or chalkboard. However, by following this process, each group member is free to participate in the discussion rather than take notes. In addition, after the discussion each person has the same recorded version of the meeting to use when generating his or her part of the collaborative document.

E-mail and Computer Conferencing Software

If your team members do not work in proximity, if your work schedules do not permit you to meet face-to-face, or if some members prefer to work at night and others during the day, you need a tool that allows you to collaborate without having to meet face-to-face. Various forms of e-mail and computer conferencing software provide such a service.

Perhaps the most widely distributed and most useful tool for collaboration is *e-mail,* which we discussed in Chapter 4. E-mail is increasingly widespread in the workplace and allows collaborative groups whose members are not in the same location to exchange information and files quickly. E-mail programs such as Eudora, for example, allow fully formatted text and graphics files to be attached to simple e-mail messages and distributed in speedy fashion to each member of a collaborative team. Similarly, such programs allow writers to set up team distribution lists and discussion groups that keep team members in touch and up to date on exchanges within the group.

E-mail exchanges work best for teams when they are exchanging information that can be communicated asynchronously rather than in real time. Such exchanges allow team members to take time to consider their responses and polish their contributions before exchanging information.

Although it can be cumbersome because of the short delays that necessarily characterize the exchange of *asynchronous* messages, e-mail can be useful for substantive discussions. E-mail eliminates the interruptions that mark face-to-face discussions and, thus, allows all team members to have their full say on matters of importance.

Although e-mail systems provide one means of electronic communication, **computer conferencing software** provides another, sometimes allowing members of a collaborative team to communicate in real time (synchronously) within an electronic forum. For example, synchronous chat software such as the Daedalus Integrated Writing Environment, designed for local area networks, or WebChat, designed for use on the World Wide Web, allows team members to discuss and exchange written materials in real time from various locations. Some conferencing software, such as the Daedalus Integrated Writing Environment, allows for threaded discussions, discussions that are separated according to particular themes, or "threads," within a larger group exchange. Threaded discussions facilitate subgroup work and communication on specific tasks and allow for the archiving of particular discussions so that records of group decision making can be preserved.

computer conferencing software
a computer technology that enables group members to communicate within an electronic forum

Many campuses today have some form of computer conferencing software available. By using such a system, you can send messages to team members; you can use the network as a place to store and share information; you can argue and reach consensus on content, tone, style, and direction of a document; and you can request feedback from people outside your team as well as from your instructor.

Desktop Videoconferencing

desktop
videoconferencing
a computer technology that
enables group members to
collaborate over distances
with simultaneous audio
and video transmission

Teams that collaborate via computer conferencing systems often miss the nonverbal dimension to communication that is present in face-to-face meetings. If you need to collaborate across distances, you might use a **desktop videoconferencing** system that provides simultaneous audio and video. The simultaneous audio and video connections allow you to generate goals and plan ideas as well as get to know each other faster than with traditional computer conferencing systems. Videoconferencing packages such as Microsoft NetMeeting, for example, allow members with compatible equipment to view a common set of graphics, a piece of equipment, or a manufactured part as well as to engage in a discussion. However, this system, like many others, allows simultaneous or synchronous exchanges only when all individuals have compatible software and hardware that include storage devices (such as CD-ROMs or a laser disk) capable of handling large amounts of data. Thus, although collaborators can "meet" while at different locations, all members must be in front of their computers at the same time, and their computers must be equipped with the requisite software and hardware connections.

Desktop videoconferencing systems, although ever more common in the workplace, are slightly less available on college campuses. However, look around your media resources center, which is likely to have some system that supports videoconferencing. If you need to get to know each other quickly and make decisions across distances, these systems are valuable. It's worth the time and effort expended to track down a system you can use. For example, if you are developing a document for a corporate client, the corporation might have such a system available.

TECHNOLOGIES BEST SUITED FOR THE EXECUTION STAGE

The technologies we discuss in this section are most useful as you coordinate your research and drafting.

Project Management Software

project management
software
a computer technology that
enables group members
to keep track of task
assignments and schedules

Most collaborative teams have obvious and often pressing needs for planning and coordinating their work. Specialized software exists to help you plan what needs to be done, track your progress in reaching your goals, and coordinate the activities of individual team members.

At the execution stage/task level, your team needs to focus on generating the content of your work; however, you also need to keep track of what has been accomplished, what needs to be accomplished, and who is in charge of what. **Project management software** can help. While your team focuses on generating answers to the problems facing you, the project management software keeps a basic record of tasks to be done, task assignments, subtask breakdowns, and schedules. At least weekly, team members review their progress with this system.

Professionals who use such systems or who have been required to use them often complain that the systems fail because they make tasks and deadlines so explicit. Commitments made in the workplace are often ambiguous, and minor deviations from a set schedule are often overlooked. Thus, systems that make explicit any minor deviation, perhaps even flagging the person who first upset the schedule, sometimes create unnecessary resentments.

A decision to use this technology must be made after considering team members' work styles. As with any collaborative technology, *all* team members have to use the technology for it to be useful. Because team members' lifestyles and work styles inevitably vary, structure some flexibility into any project management software you might use.

Group Writing Software

When you meet face-to-face, how do you write as a group? Perhaps you share printouts and scribble on them, offering ideas in the margins for how to rephrase or rearrange the text. Perhaps you all squeeze around one computer and designate one person to enter and edit text as the rest of you rattle away. **Group writing software** allows team members to make document revisions without meeting face-to-face, and such systems record and remember who made what changes. Thus, team members can revise without wiping out the original file, and members can compare alternative drafts and suggestions. Often, a full-featured word processing package, such as Microsoft Word 98, includes group writing features for making written or oral comments on a common draft, tracking the changes made to written documents by individual members, or focusing on specific sections of a shared document.

group writing software
a computer technology that enables group members to make document revisions without meeting face-to-face and to record who made what changes

Group writing software works especially well if your team physically cannot meet together. As with project management software, this software brings work style differences to the forefront. But the ability to see and record exactly how each member plans, drafts, and revises a collaborative text can be extremely powerful. The key is to structure the system so that you don't get lost in the many drafts and revisions. One way to do this is to develop a system for naming drafts that identifies the document, the collaborator, and the stage in the drafting process. When we collaborated on this book, for example, we used a combination of chapter numbers, initials of authors, and dates to let us know which version of a chapter we were considering. Because we had agreed on a word processing program ahead of time, we did not have to indicate software in our naming process (see Example 5–4).

EXAMPLE 5–4   One System for Numbering Drafts

The system we used to revise this book identified the first draft of a chapter by its chapter number and its original author: **Ch. 5 CLS.**

When chapters were revised, they were renamed, and a date was added: **Ch. 5 3/5/99.** Because one person did all initial revisions, no initials were required to identify whose draft it was, but because chapters went through multiple drafts it was important to distinguish earlier drafts from the most current, by inserting the date.

When the first revision was completed, chapters went out to additional readers who made changes on the hard copy. Those changes were incorporated into the next version which was identified by the date the changes were completed: **Ch. 5 4/2/99.**

If a chapter needed several revisions, the manuscript was circulated and new versions were given subsequent dates. All versions were kept within a folder on the computer so that earlier versions could be reconsidered if necessary. When all revisions were completed, the final draft was renamed: **Chapter 5 Final.**

Copies of final drafts were sent at the end of the project to all writers involved so everyone could have a complete record of all revisions.

Screen-Sharing Software

screen-sharing software a computer technology that enables group members to link computers and simultaneously see their conversation and resulting changes in a text on screen

Collaborators can "meet" across distances and also share the same screen through **screen-sharing software.** For example, after one collaborator types, "I think we should change the beginning to start like this . . .," another responds, "Yes, let's do it," and both simultaneously see their conversation and resulting changes on screen.

Although such a system is a powerful tool for collaborators, currently no more than two computers can be linked this way. In addition, collaborators talk of the difficulty in sharing a screen with someone else: Who gets control of the screen first? What if you like a specific change and your collaborator immediately deletes it (when it hasn't yet been saved)? Such systems already exist; for example, Timbuktu and Aspects allow you to share screens during collaboration. For these systems to be used successfully, however, teams will need additional discussions at the relationship level: Who will be in control of these shared texts? How will members respond to changes other members make?

TECHNOLOGIES BEST SUITED FOR THE PUBLIC PRESENTATION STAGE

When you have finished your drafting tasks and are ready to present your project to your audience, computer technologies can also be helpful.

Presentation Support Software

The reality of the workplace is that you will make presentations, either to other members of your team or to outside clients. Numerous software programs can help you present your document, product, system, or solution publicly. **Presentation support software** often includes graphics programs and other desktop publishing software. In some cases, your team can print out its presentation and then make overheads for public display. If you have a computer projection system available for the presentation, the team can design a computerized presentation. The value of this computer technology is that the electronic version of the project you have created now can be used directly in your public presentation. Software packages such as Powerpoint, for example, also have group work features that allow individual members to insert comments in a presentation draft, send presentations to several team members at once for review, and modify presentations for team or audience members with visual handicaps.

During the initiation stage of your work, identify those members of the team skilled with software packages such as these. At the public presentation stage, some roles may need to change as those members more skilled in graphic arts may take charge of formatting and arranging your public presentation.

Group Memory Management

All teams need some form of group memory of the many electronic notes, plans, drafts, and revisions generated. Nothing is worse than vaguely remembering the "perfect" idea, example, or statistic, but not being able to recall it completely or remember who said it, when, and in what context. In addition, teams need systems that allow them to store and retrieve these documents by topic as well as by date. Hypercard is one widely available software product that can be used as a **group memory management**. As your team collaborates, you could develop a stack of hypercards that can be linked and cross-referenced easily. You could, in effect, make each note, plan, draft, and revision a card.

At the public presentation stage, you could access these cards and project only those needed for your presentation. Or, if you remember a specific idea talked about at one meeting, and you recorded it on a card, you could search all the cards and locate your idea, placing it in your final presentation.

Comprehensive Work Team Support

After skimming through these collaborative technologies, you possibly have highlighted two or three that look attractive—perhaps your team has read these sections of this book and is now ready to request access to the most useful technology. Your combination of these systems would result in a form of **comprehensive work team support.** In the best of computer worlds, you would be able to link a system especially helpful at the initiation stage with one at the execution stage, and link that in turn with one at the public presentation stage. Unfortunately, we're not there yet. However, by talking about these technologies as part of your collaborative process,

presentation support software
a computer technology that includes graphics programs and desktop publishing options that enhance collaborative group presentations

group memory management
a computer technology that enables group members to store and retrieve documents by topic or date

comprehensive work team support
a computer technology that enables group members to collaborate effectively through the initiation, execution, and presentation stages

---

EXAMPLE 5–5   User Approaches to Computer-Supported Teams

Initiation	Execution	Public Presentation
Face-to-face meeting facilitator	Project management software	Presentation support software
Computer conferencing software or e-mail	Group writing software	Group memory management
Desktop videoconferencing	Screen-sharing software	Comprehensive work team support

---

your group can structure its own comprehensive work team support system. By locating what's available on your campus to aid your collaboration, you can make informed choices about what computer tools to use at what stage.

Although the first vision of comprehensive team support was proposed by Douglas Engelbart in the 1960s, corporations are only now developing systems that integrate several collaborative technologies into one package. One integrated system should be more powerful than several stand-alone systems.

The bottom line is that all of these collaborative technologies raise new sorts of questions and feelings in collaborators. The point is to pay attention to these questions about styles of collaboration and how, when, and in what ways electronic support is appropriate. Always remember that when you link computers, when you share screens, when you project your prose, the computer is no longer personal, it's interpersonal. It becomes a medium for sharing information at the task level and the relationship level.

In Example 5-5, we list again for you these computer technologies according to their potential use at the initiation, execution, and public presentation stages of your collaborative project.

COLLABORATIVE TECHNOLOGIES AND THE OPEN EXCHANGE OF IDEAS

As you decide which collaborative technologies to employ, keep in mind that collaborating by means of interactive technologies involves changing the ways you have traditionally accessed and used information; it involves a change in the status of communication or of who owns information; and it can, of course, result in the possibility of conflict. However, collaborative technologies, when designed and used properly, can offer each member of your team an equal voice. Collaborative technologies can also ease your working with others at sites outside your own country.

Part of structuring these technologies—indeed of structuring your entire collaborative experience—should be to encourage an open exchange

---

EXAMPLE 5–6   Considerations for Choosing Technology
                 to Support Collaboration

Open Conversations

How will the system promote open conversations; that is, how will the technology enhance collaboration and interaction among yourselves, other students, your instructor, your co-workers, and the larger community?

Distribution of Power

How can you structure the technology so that each member has equal access to the system? Are there existing models of authority that will be affected by the technology? For example, if everyone has access to and uses the technologies, will a "boss" not emerge within your group and your team members speak more freely? Or will you need to appoint a leader to coordinate your efforts?

Accessibility

What system and technologies will all members be able to learn and use? How can you design a system that provides accessibility? Will this design allow all participants equal access to the technology?

---

of ideas. From diversity comes new ideas and solutions. When you collaborate, you are one among others. You no longer are the sole owner of an idea, question, process, or solution. You share your ideas and take on the ideas of others.

Earlier in this chapter we spoke of substantive or pseudo-conflict, and ways to defer consensus and explore additional options. But these options and additional ideas are never heard if one member stifles another. Open and lively debate over ideas enhances writing and collaboration. The key is to avoid unnecessary interpersonal conflict and to manage the conflicts that do occur. Collaborative technologies often allow team members to express their ideas without being silenced by traditional or cultural hierarchies. For example, research shows that in a face-to-face group setting, women are more often interrupted than men (Edelsky). On the other hand, collaborative technologies have the potential to allow all team members to contribute without the interruption that may come during face-to-face contact.

When you structure and use your technology, make sure that all members have access to it, that all members understand how to use it, that all members then actually use it, and that all members can call for changes in its use at any time. Never isolate members from each other through technology; use it to integrate and empower the entire team.

Therefore, when choosing and structuring a specific technology to facilitate your collaboration, remember to consider the issues given in Example 5-6.

Characteristics of Effective Teams and Assessing Team Members' Collaborative Efforts

Throughout this chapter you have read about how to approach collaboration at the relationship and task levels and possible technologies to employ at the initiation, execution, and public presentation stages of your work. But you may need some very practical tools for starting and carrying out the collaborative writing projects that you encounter.

Writing Strategies 5-2 provides a series of questions, based on the information in this chapter, that can help a collaborative team get started

---

WRITING STRATEGIES 5–2   Questions for Collaborative Team Members

Interpersonal Relationships
Rate from 1 (low) to 5 (high) your level of comfort/expertise with the following roles in a group or collaborative setting:

	Low				**High**
Leader	1	2	3	4	5
Follower	1	2	3	4	5
Compromiser	1	2	3	4	5
Conflict manager	1	2	3	4	5
Synthesizer	1	2	3	4	5
Encourager	1	2	3	4	5
Summarizer	1	2	3	4	5
Questioner	1	2	3	4	5
Enabler	1	2	3	4	5
Agenda setter/timeline manager	1	2	3	4	5
Progress assessor	1	2	3	4	5
Technology specialist	1	2	3	4	5
Researcher	1	2	3	4	5
Interviewer	1	2	3	4	5
Writer	1	2	3	4	5
Graphic contributor	1	2	3	4	5
Layout artist	1	2	3	4	5
Editor	1	2	3	4	5
Other _____	1	2	3	4	5
Other _____	1	2	3	4	5
Other _____	1	2	3	4	5

Writing Task
Respond to the following questions in a sentence or two.

- What is the primary purpose of this document? Secondary purposes?
- Who is the primary audience for this document? Secondary audience?
- What information needs to be included in this document? How best can it be organized to serve the audience's needs? Where can we get this information?
- How should the document be designed (e.g., use of graphics, layout, presentation, organization, media) to accomplish its purpose for the designated audience?
- What resources do we need to accomplish this writing task (e.g., information, technology, people, time, training/education)?

on a major writing project. Note that the questions ask the team to think about both interpersonal relationships (by asking members to assess their own strengths and weaknesses) and the writing task (by asking team members to refine their sense of the project's purpose, audiences, and format). Taking the time to have all the members of your team respond to these questions may alert you to differences of opinion on collaboration strategies and approaches, differing writing and collaboration styles, and differing understandings about the nature of the writing task itself. The questions should also help your team get to know one another and begin to find out about one another's strengths as collaborators and to build trust.

Taking the time to answer these questions will also help to establish a regular habit of communication among team members. The most important thing you can do to ensure the effectiveness of a collaborative team is to build this habit of regular communication through weekly meetings, electronic exchanges, progress memoranda, telephone calls, and exchanges of drafts in hard copy or digital form.

ASSESSING THE COLLABORATION SKILLS OF INDIVIDUAL MEMBERS

Some teams, especially those whose members are not experienced collaborators or who have members interested in developing their skills in collaboration, may find it useful to provide members with feedback on their skills as collaborators so that individuals can gauge some of their own effectiveness. Students who have just begun to work on collaborative writing teams, in particular, can find such feedback useful in determining how their collaborative efforts are perceived by other team members and comparing these perceptions to their own assessment. Depending on the focus of feedback instruments, these tools can provide information on interpersonal skills, teamwork skills, teamwork roles, and collaborative effectiveness. Many such instruments are filled out by individual team members and exchanged only with other members of the team. In some organizations, the evaluation instruments are shared with a supervisor who can provide team members with additional training in effective collaborative/teamwork strategies.

A number of different instruments are available for such purposes, and each has a slightly different focus. The peer evaluation form represented in Example 5-7, for example, focuses on the communication skills of listening, speaking, interpersonal relationships, and writing. In contrast, the peer feedback form represented in Example 5-8 focuses on a list of interpersonal skills identified as necessary for effective collaborative engagement. Finally, the peer feedback form represented in Example 5-9 focuses on the effectiveness of individuals as collaborators.

All collaborative teams are different, however, and each team must work within the local constraints of a corporation, business, or organization that has its own standards for collaborative effectiveness. In some

EXAMPLE 5–7   Peer Feedback on Collaborative Group Work

Name: _____   Date/Meeting _____

Using a scale of 1 (poor) to 5 (excellent), evaluate each team member on the following skills:

Listening:
Is the person attentive and responsive to what other people are saying?

Speaking:
Does this person clearly explain his or her points? Is she or he willing to engage in discussion? Is his or her discussion productive?

Interpersonal Skills:
Does this person help the group identify and address problems? Is he or she willing to compromise and seek and offer compromise situations? How well does the person deal with conflict, the recognition of conflict, and the management of conflict? Does this person communicate consistently and effectively with other team members?

Writing:
Is the person willing to take on his or her fair share of writing work? Are this person's writing skills adequate to the group's task?

	Poor				**Excellent**
Member _____					
Listening	1	2	3	4	5
Speaking	1	2	3	4	5
Interpersonal	1	2	3	4	5
Writing	1	2	3	4	5
Member _____					
Listening	1	2	3	4	5
Speaking	1	2	3	4	5
Interpersonal	1	2	3	4	5
Writing	1	2	3	4	5
Member _____					
Listening	1	2	3	4	5
Speaking	1	2	3	4	5
Interpersonal	1	2	3	4	5
Writing	1	2	3	4	5
Member _____					
Listening	1	2	3	4	5
Speaking	1	2	3	4	5
Interpersonal	1	2	3	4	5
Writing	1	2	3	4	5

EXAMPLE 5–8   Peer Feedback on Interpersonal Collaboration Skills

Name: _____ Date/Meeting _____

Using a scale of 1 (poor) to 5 (excellent), evaluate each team member on the following skills:

Member _____

	Poor				Excellent

Social

Knows names	1	2	3	4	5
Identifies commonalities	1	2	3	4	5
Communicates consistently/appropriately/regularly	1	2	3	4	5

Comments:

Organization

Takes on productive roles in groups (e.g., consensus builder, conflict manager, leader, follower, synthesizer)	1	2	3	4	5
Maintains cooperative responsibilities to group	1	2	3	4	5
Encourages participation by all	1	2	3	4	5
Summarizes (agenda/process)	1	2	3	4	5
Integrates ideas into single position	1	2	3	4	5
Checks group's progress	1	2	3	4	5
Sets agenda	1	2	3	4	5
Assigns and takes on tasks	1	2	3	4	5
Develops process (e.g. action items, conflict management, decision making)	1	2	3	4	5

Comments:

Communication and Discussion

Takes turns in talking	1	2	3	4	5
Maintains eye contact when appropriate	1	2	3	4	5
Avoids offensive behaviors	1	2	3	4	5
Uses effective listening habits	1	2	3	4	5
Shares feelings and offers ideas	1	2	3	4	5
Displays sensitivity to other's positions	1	2	3	4	5

Comments:

Inquiry

Asks for facts and reasons	1	2	3	4	5
Asks for help and clarification	1	2	3	4	5
Expresses support and acceptance of divergent ideas	1	2	3	4	5
Seeks accuracy	1	2	3	4	5
Seeks appropriate elaboration and examples	1	2	3	4	5
Questions appropriately	1	2	3	4	5
Criticizes ideas, not people	1	2	3	4	5
Explains and clarifies	1	2	3	4	5
Extends and expands answers	1	2	3	4	5

Comments:

(continued)

EXAMPLE 5–8 (Continued)

	Poor				Excellent
Conflict					
Is open about conflict	1	2	3	4	5
Understands others' positions	1	2	3	4	5
Checks for consensus	1	2	3	4	5
Seeks basis for lack of consensus	1	2	3	4	5
Expresses disagreement constructively	1	2	3	4	5
Seeks higher level syntheses	1	2	3	4	5
Manages conflict	1	2	3	4	5
Comments:					

Source: Adapted from Sharon J. Hamilton and Kris Bosworth, "Creating a Collaborative Community: Some Skills and Opening Moves," in *Collaborative Learning,* 2nd ed. (Indianapolis, IN: IUPUI Center for Teaching and Learning, 1997), pp. 34–41.

EXAMPLE 5–9   Peer Feedback on Collaboration Effectiveness

Name: _____   Date/Meeting _____
Check the description that best evaluates the group member's skills.

Member _____

Preparation

_____ Outstandingly well prepared for discussions and writing tasks. Had clearly given thorough attention to reading drafts/materials/information/research, carefully considered all proposals/questions/ideas, responded productively to issues, problems, tasks.

_____ Generally well prepared for discussions and writing tasks. Read drafts/materials/ information, considered proposals/questions/ideas, responded to issues, problems, tasks.

_____ Adequately prepared for most discussions and writing tasks.

_____ Some deficiencies in preparation for discussions and writing tasks. Had not read all drafts/materials/ information, had not considered many proposals/questions/ideas, responded infrequently to issues, problems, tasks.

_____ Marked deficiencies in preparation for discussions and writing tasks. Read few drafts/materials/ information, considered few proposals/questions/ideas, responded very infrequently to issues, problems, tasks.

Collaboration

_____ Outstanding collaborator. Listened well to others, exhibited skill in summarizing and synthesizing data/positions/directions/proposals that emerged in group discussions, was key player in compromises/conflict management/moving group forward/producing document, led group in fulfilling task and charge, contributed generously to group efforts. Communicated important information regularly and consistently with team members.

(continued)

EXAMPLE 5–9 (Continued)

Collaboration (Continued)

_____ Above average collaborator. Listened to others, exhibited skill in summarizing and synthesizing data/positions/directions/proposals that emerged in group discussions, contributed to compromises/conflict management/moving group forward/producing document, helped group in fulfilling task and charge, contributed consistently to group efforts. Communicated important information with team members.

_____ Adequate or average collaborator.

_____ Less effective as collaborator. Had some difficulty listening to others, exhibiting skill in summarizing and synthesizing data/positions/directions/proposals that emerged in group discussions, contributing to compromises/conflict management/moving group forward/producing document, helping group in fulfilling task and charge, contributing consistently to group efforts. Sometimes failed to communicate important information regularly and consistently with team members.

_____ Marked deficiencies as collaborator. Did not listen to others, exhibited little skill in summarizing and synthesizing data/positions/directions/proposals that emerged in group discussions, contributed little to compromises/conflict management/moving group forward/producing document, did not help group in fulfilling task and charge, contributed only inconsistently to group efforts. Did not communicate important information regularly and consistently with team members.

Source: Adapted from Kathy Egawa, "Collaborative Learning and Evaluation," in *Collaborative Learning,* 2nd ed. (Indianapolis, IN: IUPUI Center for Teaching and Learning, 1997), pp. 82–100.

cases, organizations will have their own methods and tools for assessing the effectiveness of teams. In other cases, individual writing teams can create their own feedback instruments that give more information more specifically geared to a particular writing task.

SUMMARY

- This chapter provides readers with a common language for initiating, executing, and presenting collaborative projects, and it presents technologies that can facilitate your collaboration. People collaborate to solve problems, create things, or discover new information. To collaborate effectively, people need to focus on both the relationship and task levels.
- At the beginning of a collaborative project, you need to find team members, get to know each other, and share background assumptions about working with others. You also need to clarify your team's goal, choose an approach and schedule meetings, and generate a plan about how to proceed. Part of this plan should

include accessing appropriate collaborative technologies to aid your work. You might pick from or be assigned one of several collaborative models, depending on your situation: the divide-and-conquer model; the specialization model; the dialogic model (marked by flexible roles); the sequential model (often characterized by a series of signatures of approval); and the synthesis model (formed with team members representing different views).

■ As your team executes its project, members need to maintain trust by establishing ways to resolve pseudo-, simple, and ego-conflicts. A team can clarify its work styles by talking about control, credit, response, flexibility, and constraints on the project. And a team can share information and assess its progress by using question prompts to help each other expand ideas, consider alternatives, and voice disagreements. Collaborative technologies that allow a team to share texts and perhaps even share screens especially aid team work at this stage.

■ At the end of a team collaboration project, members need to acknowledge and appreciate each other's work as well as package the whole team's work for presentation to a client. Any disputes during the process should now be resolved. Once again the team should identify those members with special talents to format and present its work, using appropriate technologies in effective ways to develop a presentation.

■ Collaborative work goes better if members participate in open discussions that include frequent and systematic communication about both relationship and task concerns. Remember to listen carefully to other members and to change models and roles throughout the project when the need arises. Take time to manage conflict effectively, identify different learning and working styles, and capitalize on diverse talents.

During your study of technical communication, you may frequently be asked to participate in collaborative writing activities. After you enter the workplace, you may be assigned to a collaborative writing team or, in the future, be asked to establish and supervise one. Keep in mind our suggestions for making this process an enjoyable and productive experience.

ACTIVITIES AND EXERCISES

1. Do the following individually, and then share this information with your team as part of building understanding and trust as you begin your project:

   a. Quickly write down the most recent writing project that you collaborated on either in college or at work—essay? speech? proposal? letter? instructions? memo? computer documentation?

   b. When you collaborated, what aspects of the project (and the writing) were you most concerned with? List the first three things you think of.

   c. How much control did you have over the final collaborative product? How much control did others have?

   d. How were you given credit for the work that you did?

   e. What process did you follow when responding to the work of others?

   f. What procedures did you follow for resolving conflict?

   g. Did each member of your team understand how flexible you could be with preestablished formats/specifications/style guides?

   h. When did your team talk about constraints facing the project—deadlines, specific client or instructor needs?

   i. What was the status of your project in the class? In the organization? Was the status different for different team members?

   j. What were your personal goals for being involved in the project? Simply to get the job done? To get promoted? To get an *A*?

   k. How did you approach your part of the work?

   l. Could you see any personal characteristics—gender, age, personal background, cultural heritage, and so on—that affected the collaborative process for you and for your team members?

2. As a group, list all of the technologies available for your collaborative work (remember to include the phone, fax machines, copy machines, and tape recorders as well as computer technologies). Define each and state whether you think it would be best used at the initiation, execution, and/or public presentation stages of your collaborative work. In a memo written to your classmates and instructor, discuss your findings.

3. Interview someone who is active in your present or future profession, asking that person to talk about how he or she:

   ■ Identifies possible collaborators.

   ■ Approaches the collaborative task.

   ■ Gets to know individuals' work styles.

   ■ Handles conflict.

   ■ Reaches closure on a project.

In a table or chart, display what you learned and how this matches your personal responses.

4. On August 2, 1999, the Coalition of Citizens Concerned with Animal Rights (CCCAR) sent a letter to Dr. Karen Wartala, Dean of the College of Agriculture, University of the Midwest, detailing their surveillance of and investigation into the trapping and killing of birds in the experimental crop fields on its campus (Appendix A, Case Documents 3). This letter started a long process in which university personnel met to determine what they should do. In turn, they met with CCCAR representatives and worked to clarify the issues and their agreements.

     Collaborate with a partner to draft a document that includes the issues from both sides and the agreements made as stated in the two letters dated August 13 (from the CCCAR to Dr. Wartala) and August 22 (from Dr. Wartala to the CCCAR). Assume that one of you represents the CCCAR and the other the University. The audience for this document will be people from the CCCAR and the University.

     As part of your collaborative work, remember to begin at the relationship level. That is, what are your personal stands on the issue of trapping birds?

What is the right thing to do—for the CCCAR—for the University—for both sides combined?

Then, work at the task level to draft your document.

5. Look at the Year 2000 case study at the end of this book (Appendix A, Case Documents 4). Identify a partner who will help you draft a one-paragraph summary of each of the sites represented. Make sure to talk about the purpose, audience, tone, information, and format of each site in your summary.

When you have completed this collaborative task, both you and your partner should answer the following questions separately and in writing.

a. When you collaborated, what aspects of the project (and the writing) were you most concerned with? List the first three things you think of.

b. How much control did you have/take over the final collaborative product? How much control did your teammate have/take?

c. What process did you follow when collaborating on this task?

d. What procedures did you follow for resolving conflict?

e. What were your personal goals for being involved in the project?

f. Could you see any personal characteristics—gender, age, personal background, cultural heritage, and so on—that affected the collaborative process for you and for your team members?

When you have completed this task, exchange your descriptions of the collaboration and compare yours with that of your partner.

6. This case study focuses on the executive letter of the 1986 annual report of the Auldouest Insurance Corporation (pseudonym), described by one employee as "the most revised piece we do all year." This executive letter would go to over half a million policyholders.

The seven collaborative writers were the following:

■ Newly elected president (48, vocational college graduate in computing, 26-year veteran)

■ Senior vice president (49, liberal arts college graduate, 18-year veteran)

■ Chief executive officer's secretary (52, high school graduate, 34-year veteran)

■ President's secretary (40, liberal arts college graduate, 19-year veteran)

■ Vice president of corporate communications (46, high school graduate with college coursework in public relations, 19-year veteran)

■ Supervisor (37, university graduate in broadcast journalism, 5-year veteran)

■ Writer (23, liberal arts college graduate summa cum laude in English/communications, 6 months at Auldouest)

In the 33-story Auldouest Building, the first 10 floors were for employees, the 33rd floor for top-ranking officials.

The collaborators began on October 13, 1986, and were expected to complete the letter in about five weeks, but the actual process required seven major drafts and the final letter was approved on December 29, 1986. This 77-day collaborative process exceeded its deadline by six weeks.

In a small group, read through the following collaborative process as well as data about the company (Example 5–10), perceived audiences for the document

(Example 5-11), and the actual letter (Example 5-12). Then develop a table or chart of this case study in terms of what the collaborators did at the initiation, execution, and public presentation stages (and at relationship and task levels) of their work. Analyze and evaluate their process in terms of:

■ Distribution of power.
■ Audience(s) for the letter.
■ Purpose(s) of the document.
■ Expectations for the document.
■ Collaborative model that this group followed.

Then, share your table or chart and your evaluation of this collaborative process with other groups.

The actual collaborative process went through three periods: a period of stability, a period of instability, and a resolution.

Period of Stability
■ Corporate Communications held two departmental brainstorming meetings. The president had only been on the job six months. The vice president encouraged his subordinates, including the supervisor and the writer, to present fresh approaches and ideas.
■ At their second meeting, collaborators learned that they were to ghostwrite the letter. They agreed that the letter should discuss a recovering Auldouest in the context of the troubled insurance industry.
■ The supervisor wrote an annual report outline that presented this image of Auldouest. It included the statement, "We are a forward looking company dealing successfully with difficult problems."
■ The supervisor and vice president met with the president and CEO, who suggested only a few additions (namely, that statistics/numbers were to be placed at the beginning of the letter).
■ The supervisor and vice president were elated that their first outline was approved. What had not surfaced was the new president's strong commitment to positive emphasis.
■ The supervisor delegated the writing of the first draft to the writer, but the writer had not attended the letter-planning meeting with top management.
■ After finishing her draft, the writer discussed it with the supervisor. The supervisor made suggestions, but didn't talk about her perceptions of the statement that the writer "must understand that surplus is a sign of corporate health, not corporate greed."

Period of Instability
■ The vice president reviewed the revised draft of the letter and rejected it, objecting to the phrase, "The public must understand." The supervisor challenged this decision. The vice president saw the public as the chief audience and felt the words were dictatorial (the vice president was not aware that the writer had written the letter and assumed that the supervisor had misunderstood his advice).

- Challenged by his subordinate, the vice president (along with the supervisor) consulted the senior vice president, who supported the vice president's judgment. In order to defend Auldouest's financial position, the senior vice president told the supervisor to include the number of recent bankruptcies of insurance companies and other information related to performance. The senior vice president also wanted the supervisor to discuss problems at the beginning of the letter.
- The supervisor communicated these needed changes to the writer, who was asked to revise.
- The supervisor and senior vice president reviewed the revised letter, and a clean draft was sent to the 33rd floor (executive offices).
- The executive secretaries incorporated several of their own editing changes, substituting words like "halt" for "stop."
- The CEO reviewed the draft and asked for more statistics.
- The letter went to the president who made suggestions that required a global change, that emphasized certain audiences and purposes, ignored others, and altered the company's account of itself in 1986. The president communicated this new concept for the letter to the senior vice president. They decided that the senior vice president, the supervisor, and the writer would each write his or her own new version of the letter.
- The supervisor and the writer submitted new letters to the senior vice president, who chose not to use any parts of these letters in the final letter.

Resolution
- The senior vice president wrote a "success story" draft that ignored the interests of domestic policyholders but met the chief intentions of the president and CEO.

EXAMPLE 5–10   Disclosure of Underwriting Losses in Auldouest Executive Letters

Year	Message
1980	As a result of inflation and a series of summer storms, our experience did not measure up to our expectations, and we suffered an operating loss for the year.
1981	The year's underwriting experience produced a loss.
1982	These elements, coupled with increased loss frequency and severity, produced record underwriting losses for the industry and for [The Auldouest] Mutual Insurance Company.
1983	Even with improving economic conditions, [Auldouest] Mutual, as well as the casualty/property insurance industry, was once again hit with unfavorable underwriting results, a condition that has prevailed since the early 80s.
1984	Actually, the industry paid out in claims and operational costs 18% more than the amount of insurance premiums received. Our (the industry and [Auldouest]'s) task is to identify the reasons for such adversity and, where possible, to develop solutions.
1985	The underwriting loss was $18, 896, 203.

EXAMPLE 5–11   A Comparison of Actual and Perceived External Audiences of the Executive Letter

Actual Audience	Perceived Audiences for Each Writer and/or Editor						
	President	Senior Vice President	Chief Executive Officer's Secretary	President's Secretary	Vice President	Supervisor	Writer
Family/friends of the top executives	×						
Other chief executive officers	|[a]			×		×	|[b]	|[b]
Other insurance staff	×	×	×		×	×	
Bankers					×		
Chamber of commerce	×						
Company representatives with clients		×					
Potential commercial policyholders	×	×			×		
Commercial policyholders	×				×	×	
Potential agents	×						
Agents	×	×				×	×	
Potential employees		×			×	×	
Potential domestic policyholders							
Domestic policyholders		×					×	
Media					×	×	
Public						×		

|[a] = most important; |[b] = most important external audience. The supervisor's most important audiences were the president, the CEO, and other CEOs. The writer's most important audiences were the president and the CEO.

Source: Geoffrey Cross, "A Bakhtinian Exploration of Factors Affecting the Collaborative Writing of an Executive Letter of an Annual Report," *Research in the Teaching of English* 24, no. 2 (1990), pp. 173–203. Copyright 1990 by the National Council of Teachers of English. Reprinted with permission.

EXAMPLE 5–12 Executive Message

We are pleased to present this 1986 annual report. It reflects not only the year's achievements, but also the positive effect from prior years' actions. A result of these measures is that the combined assets of all Auldouest Companies now exceed $400 million.

The assets of Auldouest Mutual grew to $324 million in 1986, an increase of more than $34 million. The company's surplus (equity) account increased to $88 million, which represents a 23.7 percent growth from the preceding year. The combined loss and expense ratio decreased from 1985's 109.6 to 103.2.

Auldouest Life improved its presence in the marketplace by adding Universal Life to its portfolio. Assets increased to over $56 million, while the surplus account grew to more than $14 million, which represents a 20.5 percent improvement from the previous year. The amount of life insurance in force increased from 1985's $470 million to $601 million.

AULDCO, Auldouest's nonstandard company, continues the trend of positive results, with assets increasing 73.7 percent to $15 million. Its surplus account now exceeds $8 million, which represents a 41 percent increase. AULDCO currently conducts business in Michigan and Wisconsin.

AULDLEASE, a wholly owned subsidiary of Auldouest, has as its primary product the offering of private vehicle leases to individuals and small groups. At the end of 1986, AULDLEASE had assets of $5.5 million with a year-end net worth of $2.5 million.

During 1986, we initiated many changes in our operations—changes we believe will have both immediate and long-term positive effects. To name just a few: installation of a new IBM computer to enhance automated processing; construction of a new claims training facility; inauguration of additional training classes for employees and agents. Thus, we anticipate greater efficiencies and an even greater degree of professionalism in providing protection and service to our policyholders.

As we look to the future, the insurance industry faces numerous challenges—some old and some new. One issue, the need for tort reform, is an external factor which has been present for several years. Although 39 states enacted some type of tort reform in 1986, the issue is far from resolved. As insurers, we are concerned that unpredictable court verdicts and the awards associated with those vehicles will continue negatively to influence insurance availability and affordability.

The Tax Reform Act of 1986 is another external factor that will affect the affordability of insurance products. It is estimated that taxes for the property-casualty industry will be increased in excess of $7 billion over the next five years. The life insurance industry will also be adversely affected.

While the future may present some obstacles, the Auldouest Companies are prepared. We are confident that through cost containment programs, increased operating efficiency, and enhanced professionalism of employees and agents, all Auldouest Companies will continue to grow and prosper.

Steven P. Norton
Chairman of the Board
and Chief Executive Officer

John H. L. Vogel
President and Chief
Operations Officer

Source: All figures and examples come from Geoffrey Cross, "A Bakhtinian Exploration of Factors Affecting the Collaborative Writing of an Executive Letter of an Annual Report," *Research in the Teaching of English* 24, no. 2 (1990), pp. 173–203. Copyright 1990 by the National Council of Teachers of English. Reprinted with permission. For a full description and analysis of this collaborative writing process and a summary of related research on group writing and conflict, see Geoffrey A. Cross, *Collaboration and Conflict: A Contextual Exploration of Group Writing and Positive Emphasis* (Cresskill, NJ: Hampton Press, 1994).

Beebe, Steven A., and John T. Masterson. *Communicating in Small Groups: Principles and Practices.* 3d. ed. New York: HarperCollins, 1990.

Burnett, Rebecca. "Substantive Conflict in a Cooperative Context: A Way to Improve the Collaborative Planning of Workplace Documents." *Journal of Technical Communication* 38, no.4 (1991), pp. 532–39.

Cross, Geoffrey. "A Bakhtinian Exploration of Factors Affecting the Collaborative Writing of an Executive Letter of an Annual Report." *Research in the Teaching of English* 24, no. 2 (1990), pp. 173–203.

Ede, Lisa, and Andrea Lunsford. *Singular Texts/Plural Authors: Perspectives on Collaborative Writing.* Carbondale: Southern Illinois University Press, 1990.

Edelsky, Carole. "Who's Got the Floor?" *Language in Society* 10, no. 3 (1981), pp. 383–421.

Egawa, Kathy. "Collaborative Learning and Evaluation." In *Collaborative Learning,* 2nd ed. Indianapolis, IN: IUPUI Center for Teaching and Learning, 1997, pp. 82–100.

Engelbart, Douglas. "A Conceptual Framework for the Augmentation of Man's Intellect." *Vistas in Information Handling.* P. W. Howerton and D. C. Weeks, eds. Washington, DC: Spartan Books, 1963, pp. 1–29.

Faigley, Leslie, and T. Miller. "What We Learn from Writing on the Job." *College English,* 44 (1982), pp. 557–69.

Hamilton, Sharon J., and Kris Bosworth. "Creating a Collaborative Community: Some Skills and Opening Moves." In *Collaborative Learning,* 2nd ed. Indianapolis, IN: IUPUI Center for Teaching and Learning, 1997.

Johansen, Robert. "Use Approaches to Computer-Supported Teams." In *Technological Support for Work Group Collaboration.* Margrethe H. Olson, ed. Hillsdale, NJ: Lawrence Erlbaum Associates, 1989, pp. 1–31.

Kiefer, C. F., and P. M. Senge. "Metanoic Organizations." In *Transforming Work.* J. D. Adams, ed. Alexandria, VA: Miles River Press, 1984, pp. 109–22.

Kleimann, Susan D. "The Complexity of Workplace Review." In *Technical Communication* 38, no. 4 (1991), pp. 520–26.

Kraut, Robert E., Jolene Galegher, and Carmen Egido. "Relationships and Tasks in Scientific Collaboration." *Journal of Human-Computer Interaction* 3, no. 1 (1988), pp. 31–58.

Larson, Carl E., and Frank M. J. LaFasto. *TeamWork: What Must Go Right/What Can Go Wrong.* Newbury Park, CA: Sage, 1989.

Paradis, James, David Dobrin, and Richard Miller. "Writing at Exxon, ITD: Notes on the Writing Environment of an R&D Organization." In *Writing in Nonacademic Settings.* Lee Odell and Dixie Goswami, eds. New York: Guilford Press, 1985, pp. 281–307.

Parker, Glenn M. *Team Players and Teamwork: The New Competitive Business Strategy.* San Francisco: Jossey-Bass, 1990.

Reich, Robert. *Tales of a New America.* New York: Times Books, 1987.

Wall, V. D., and L. L. Nolan. "Small Group Conflict: A Look at Equity, Satisfaction, and Styles of Management." *Small Group Behavior* 18 (1987), pp. 188–211.

Secondary Sources
of Information:
Libraries and Online

Introduction ▮ A Continuing Case to Demonstrate Gathering Information
▮ Conducting the Search ▮ Doing Research on the Internet
▮ Evaluating Your Materials ▮ Summarizing Information: Developing
Summaries and Abstracts ▮ Crediting Sources

Some of our students, perhaps more than we imagine, will actually contribute to the published, indexed literature. As authors of journal articles, technical reports, or conference papers, or as contributors to electronic databases of various sorts, they will need to make decisions about how to title, to abstract, and to list keywords for their work. Many more students will need to use indexes and computerized literature searches; they need to become efficient and intelligent users of the electronic literature.

> Donnelyn Curtis and Stephen A. Bernhardt, "Keywords, Titles, Abstracts, and
> Online Searches: Implications for Technical Writing,"
> *The Technical Writing Teacher* 18, no. 2 (Spring 1991), p. 143.

[T]he World Wide Web has become a ubiquitous electronic medium of communication. It is also becoming a Universal Electronic Campus, a global "academical village," where thoughts, ideas, and processes are shared, and where people learn and produce in bits, not atoms.

> David C. Leonard, "The Web, the Millennium, and the Digital Evolution of
> Distance Education," *Technical Communication Quarterly* 8, no. 1 (Winter 1999), p. 9.

Introduction

When you are charged with creating a print or online document or preparing an oral presentation, you often have to gather information about your subject as part of your planning. Doing research enables you to uncover what others have already learned about that subject—what solutions others have proposed, what tests they ran, what decisions they made, what arguments they posed, what "good reasons" they used, and what problems they encountered. If you find no answers to your questions after looking at the materials others have generated on your subject, then you may need to conduct primary research yourself. We'll look at ways of doing that kind of research in Chapter 7; here, we're going to focus on the first steps of learning about any topic, conducting careful and thorough secondary research.

When you gather information from existing sources, you are looking for already published information that can cast light on the questions that you are asking. Such research is commonly referred to as **secondary research** because you are investigating the research that others have already done.

secondary research locating and evaluating published or online research and scholarship to gather information for some task

Whether you search for information at the library or online, a successful investigation will have several important steps:

- A thorough search.
- An evaluation of the sources.
- An effective way of summarizing and organizing the information.
- Accurate citations of the sources used.

In this chapter we're going to look at how to perform these actions whether you are looking for information in a library or online.

A Continuing Case to Demonstrate Gathering Information

In this chapter and the one that follows, we refer to a continuing case to discuss gathering information. Let's assume that you work for Biomed Corporation, a company that produces computer software for the health care industry. You have been asked to do research on your customers' knowledge and attitudes about a new product, Med-Ease Software. This product is designed to computerize the patient-management system of hospitals. With Med-Ease, hospitals can admit, track, and release patients, and monitor their treatments, medications, billing, and insurance claims. This is a large product, and Biomed has installed it at City Hospital in a pilot testing program designed to verify the performance of the software. If the product works well for three months in City Hospital, it will be released to the market.

You have been asked to begin a **usability study** of Med-Ease Software. You are told that the first step will be to develop information about the users' responses to the new product: How easy is it to use? How well do they like it? Are there differences in attitude among different types of users? You know that before you contact Med-Ease users, you need to read the published studies of software use in other hospitals. Others may well have studied the issues that you are investigating. You predict that you can gain further insight into your own project by comparing what others have found to what you may find when you contact users. Careful secondary research is often done before you begin primary research as a way to avoid "reinventing the wheel." Other people's research can give you ideas and information you didn't have before, and it can often save you time by pointing out dead ends. To benefit from the work of other researchers, you will need to conduct both library and online searches for appropriate information.

usability study
an investigation of how well a product meets its users' needs; conducted on-site or in a laboratory

Conducting the Search

The first decision to make in beginning research on any topic is to decide whether to search library or online resources, or both. To some extent whether you will rely on materials in libraries or on the World Wide Web will depend on what you're looking for. If you need to use journals and scholarly books in print because your instructor or supervisor expects you to use refereed material and you have a research library handy, then you may end up getting all your resources from the library. Yet even in this case, knowing about online searching will be helpful because most major libraries have abandoned their traditional card catalog in favor of online catalogs that can be searched at the library itself or on the Internet from your desktop.

Sometimes, however, the information you are looking for is more readily obtained online than in the library. Because so many companies have Internet connections, it's clear that simply knowing how to navigate the library on your campus is not enough to prepare you for the workplace. Most major newspapers publish an online edition and these copies are archived, indexed, and can be searched. Most professional societies have Web sites, **discussion lists,** and **newsgroups** where you can query professionals in your field about your specific question. In addition, the Internet has enormous and continuously expanding reservoirs of information on almost any topic you can imagine, and the Internet as a whole is searchable using any of a variety of search engines. Whether you use the library in person or online, use Internet resources, or use both library and Internet sources will depend on your needs, what's available, and your skills as a researcher.

discussion lists
e-mail discussion groups on specific topics; they can be moderated, and users subscribe to them

newsgroups
specialized electronic bulletin boards to which anyone can post or respond to a message

DOING LIBRARY RESEARCH

In this section, we provide you with only an introduction to using the library. You might also ask your librarian or instructor to arrange a tour of your library. Some colleges and universities provide courses for new researchers on how to use library resources. However, any library, and especially those at large research universities, can be intimidating the first couple of times you use it. As a consequence, you should read this chapter over before you go to the library to begin doing research. You should also know that the **reference librarians** at your institution are there to help you conduct research accurately and efficiently.

When you are doing secondary research, the information you seek from libraries may be stored in a variety of media, including books, magazines, newspapers, printed journals, government documents, microfiche, audiotape, videotape, and on **CD-ROMs.** You can find most of the sources that you seek by following the library's guidelines for searching. Most university and college libraries have done away with their traditional card catalogs and now list their holdings, whatever their type, only online. In this way, libraries across the country can list each other's full resources so people can find what they need somewhere. Online catalogs are available at the library itself, and in most cases are also available through a library Web site from your desktop. Our focus here is on searching the online catalog.

Despite their size and the variety of their holdings, libraries are highly organized places. In this way they differ significantly from the Internet, which is vast, of uneven quality, and constantly changing. One of the first keys to a successful search is understanding the library's method of organizing its holdings. Most libraries use a standard system for organizing its materials called the Library of Congress system. This system generates a **call number** to identify a particular holding. Once you've identified the material you want, the call number will tell you where the material is held in the library, and you can go there and get it or ask for the material using the library's retrieval system.

BEGINNING THE SEARCH PROCESS

In order to begin our secondary research on Med-Ease, the computerized patient-management system used by City Hospital, we would first have to think about how to conduct our search of the library's holdings. Because we're just beginning this process, we may not yet have the names of authors or of any particular books, so we can begin examining the library's catalogs using **keyword search** techniques. Most libraries give you an initial screen that gives you some choices for searching beyond a library's own catalogs. Those choices might include indexes, electronic journals, electronic texts, reference sources, and news sources. We'll look at some of the

reference librarians
specially trained librarians who help researchers find needed information

CD-ROMs
compact disks that serve as storage devices for anything in digital form; multimedia, encyclopedias, census data and other demographic information may be stored this way

call numbers
coding devices used to identify library holdings; no two holdings will have the same call number, but related holdings may have related call numbers

keyword search
a way to examine Internet sites or library catalogs to find materials that contain selected key words

---

EXAMPLE 6–1   Some Libraries with Online Catalogs

Library of Congress WWW Home Page
  http://lcWeb.loc.gov
Stanford's WWW Virtual Library
  http://vlib.stanford.edu/overview.html
University of Minnesota Libraries
  http://www.lib.umn.edu
California Digital Library (CDL)
  http://www.cdlib.org/
The Center for Research Libraries
  http://wwwcrl.uchicago.edu/info/crlcat.htm

---

other choices later, but for now just be aware that most libraries provide you with tools to conduct searches beyond their own holdings.

On many campuses, you don't even have to go to the library itself to do this initial search. If your college or university is online, you can begin your search of library catalogs from the desktop. The Uniform Resource Locator (URL) for the library should be listed in the campus phone book, but, if not, a call to the reference desk should provide you with the information you need to begin. Example 6–1 gives URLs for some libraries whose catalogs are online so you can examine them and note what kinds of searching tools and sites they provide.

Although electronic catalog systems may differ, most are organized on the same principle as the traditional catalog: author, title, subject, keyword. Let's say that because you want to use secondary materials in your usability study of Med-Ease patient-management software, you decide to search first for materials on the subject. Using the library's computer terminal or working from your desktop, you type in the instructions to search for each of the key words you think might yield information.

In the case of the Med-Ease project, you have given your problem some thought, and you decide to look for materials on the following subjects: *computer, software,* and *usability.* Some electronic library search systems ask that you type something that looks like this— **s=computer**—to indicate that your subject is computers. Other library search systems only give you a space in which to type your title, author, subject, or keywords.

In this case, your subject search is both successful and not. You have turned up 5,000 entries on the subject of computers, 153 entries on software, and 0 on usability. You decide to try to search for the subject *management,* but it turns up 3,617 entries. Fortunately, you don't have to stop here. On most electronic card catalogs, subject or keyword searches are greatly enhanced because you can use a more thorough method of searching. Keyword searches allow you to combine subject names and narrow your search accordingly.

NARROWING THE SEARCH

In our example, let's assume that because you have access to an electronic card catalog system at your library, you try several combinations of subjects to see what you get. The first time you try *management software.* After you browse through the screens listing the titles of the sources you turned up, you decide that none are what you had in mind. Next you try *hospital software* but that fails to turn up any entries. Finally, you try *medical software* and find six entries:

1996	Directory of software vendors
1996	The HCP Directory of medical software
1995	Twin Cities industry cluster study
1992	Medical software reviews
1989	Software for health sciences <data file>
1986	The Code Blue Professor

Now you're getting nervous and frustrated because you haven't turned up anything that seems to suit your needs. Therefore, let's consider how to use Boolean searching techniques to refine your search. Boolean searches may not be listed on the library page, but there is likely to be a button that says "Advanced Search," and the advanced search is generally based on Boolean logic. The nice thing about learning these techniques is that you can use them with library catalogs and when you are searching more broadly on the Internet as well.

CONDUCTING BOOLEAN SEARCHES

search engines
searching tools that use automated processes to search Internet sites

Boolean searches
database searches that follow Boolean logic by making use of three key operators: AND, OR, and NOT

inclusive searches
searches that use the Boolean key operator AND and require that all included terms be found in a given work

Search engines for libraries and on the Internet perform what are known as **Boolean searches.** These searches are named after George Boole (1815–1864), a mathematician who invented a way of organizing questions called Boolean logic, which uses three key operators: AND, OR, and NOT. Knowing how to use these terms and a few other techniques will make the difference between a successful online search and a failure.

Each of the key operators refines and limits your search in a particular way. AND is used to construct **inclusive searches** in which you require that all included terms be found in a given work. OR is used to conduct **conditional searches** in which any included term is acceptable. NOT is used to conduct **exclusive searches** in which you do not want to search for a given term. Quotation marks are used when you want to keep words

---

EXAMPLE 6–2   Using Boolean Search Terms

Here are two examples of how Boolean terms were successful in narrowing searches. We used the search engine *http://www.hotbot.com* to conduct this search.

Example 1— Star Wars	Search Term	Number of Hits
	Star Wars	145,360
	"Star Wars"	128,320
	"Star Wars" AND "Episode 1"	5,440
	"Star Wars" AND "Episode 1" NOT "Rebel Alliance"	230
Example 2— Land Reform	Search Term	Number of Hits
	"Land OR Agrarian AND Reform"	1,240
	"Land OR Agrarian AND Reform NOT Mexico"	470

---

together (e.g., when you use an author's name "Cynthia L. Selfe" or use a phrase, e.g., "Give me liberty or give me death"). Without the quotation marks around a name requiring that the whole thing be searched for as a unit, for example, an Internet search would give you hits for every time the name "Cynthia" or "Selfe" appeared in any site, and depending on the search engine, it might even return sites that have "L." also.

Example 6-2 shows how powerful Boolean logic is for narrowing a search and finding what you want. Each time the search engine matches the terms you've given it, a *hit* is listed. Most of the time when you're searching, you don't find too little material; you find way too much. Refining your search will reduce the number of hits you have to consider and will turn up materials more closely matching what you want. As you see what materials you turn up, you can keep refining your search until you get exactly what you want, if it is part of the library's (or the Internet's) collection.

After you refine your search for information relevant to the Med-Ease project, you will see some titles on the results page that look as if they might be winners. The title of each hit includes an embedded link to the work that has your search terms in its title, author, or subject entries. The first title looks like a possibility so you choose it by clicking on the title, and the next page you receive gives you more information. Example 6-3 indicates what kind of information comes up on the screen at this point. It looks a lot like the actual cards in the old catalog, and so you are prepared for the information on it. Once you get this far, you can tell if this source will be useful to you or whether you need to keep looking.

conditional searches
searches that use the Boolean key operator OR and specify any included term is acceptable

exclusive searches
searches that use the Boolean key operator NOT and specify that certain words are to be excluded from the search

EXAMPLE 6–3   Sample Electronic Catalog Entry

Title:	Software for health sciences education <computer file>: an interactive resource/designed and developed by the Learning Resource Center, University of Michigan Medical Center
Edition:	2nd ed.
Published:	Ann Arbor, Michigan: The Center, c1989
Description:	1 computer disk; 3 1/2 in.
Subjects:	Library of Congress: Medical education—software—catalogs:
Contributors:	University of Michigan. Medical School. Learning Resource Center.
Notes:	Comprised of 4 HyperCard Stacks. System requirements: Macintosh; 1 MB RAM; two BOOK disk drives or a hard disk drive; Hypercard version 1.2.1
Location:	Bio-med Learning Resource Center Library
Call No.:	W18 CMM1989
Status:	Not checked out

TAKING ADVANTAGE OF SERENDIPITY

One tremendous advantage and pleasure of working in the actual library that you can't get working from your desktop is *serendipity*—being in the right place at the right time. Once you find some call numbers identifying books that may be useful to you, you can head for the stacks, or shelves, to gather these materials. Related materials are likely to be classified near the specific materials you have identified; therefore, when you are in the stacks and you have located your book, look around on the same shelf and above and below your book. Also, look through the bibliography that appears in one book you have found. It could lead you to other sources. Serendipity can often help you discover additional related materials not turned up in your search. The library is systematically organized, and the librarians may have identified related materials that escaped your search. Serendipity does not substitute for systematic examination of the card catalog, but it can play a helpful role in your research as you browse through related materials and discover something particularly useful.

CONDUCTING A PERIODICAL SEARCH

Although library book holdings can be vast, the amount of information available through periodicals is almost too large to imagine. Through electronic advances in information storage and retrieval, a research library can provide literally millions of sources of information through periodical searches. Once you begin to learn the ins and outs of periodical searching, you will rarely face the problem of too few sources. Instead, you are more likely to be seeking ways to eliminate sources before you are overwhelmed by information, analyses, opinions, and data.

As you search for information in periodicals, you will find a variety of useful publications. Generally, research journals, particularly those that report scientific and technical studies, will offer you the latest data and interpretations of these data available in your field. Researchers at university or industrial sites report on their studies—including the research question asked, the method used to collect data, and the findings—in such journals. For example, in your search for material on user attitudes about health care computer software products for the Med-Ease project, you might find results of similar user testing in computer science research journals as well as recommended methods for conducting user testing in psychology research journals.

On the other hand, trade journals, periodicals that address practitioners in a particular field, usually suggest how to improve workplace practices. Although not all libraries carry trade publications because some publish for only a few years before others take their place, these journals are designed to help practitioners interpret and apply the recommendations and results that have been reported in research journals. For example, in your Med-Ease project for Biomed Corporation and City Hospital, you might find that hospital association trade journals offer suggestions on how to use new software products and computer science trade journals offer recommendations on how to test the impact of such products.

Popular magazines both inform and entertain their readers and can give you a glimpse at how society in general might view your subject. For example, in your search for information for health care computer software programs, you might discover general social attitudes toward hospital care in popular magazines. Finally, newsletters produced by associations and corporations offer a wide variety of information. Usually newsletters update readers on events and people within a specific corporation or professional group. Perhaps you'll need to read back issues of the City Hospital newsletter to understand the climate of the hospital before you question Med-Ease users.

As with the keyword search, you can follow electronic channels to search for periodical information. Either way you will follow one of two avenues: first, if you begin your search knowing the particular article that you wish to find, you read the printed listing of your library's periodical holdings, find the call number, and proceed with your search. If you don't already have a title and journal in mind, then you must begin subject or keyword searching.

Indexes and Abstracts

When you logged on to the library's home page looking for books, you conducted keyword searches of the library's catalogs. On that first computer screen when you entered the electronic library, there were probably other choices than just searching the catalog as we noted earlier. The two most commonly used types of references for searching periodicals are **indexes,** which contain listings of citations organized by title, subject, and author, and

indexes
reference texts that include listings of citations

abstracts

brief summaries that stand alone in a document or represent a document in a list or index; provide a glimpse of the entire document to help the reader decide whether to read, skim, or skip this source

abstracts, which include, in addition to the bibliographic information provided by indexes, brief summaries of the periodical articles. Often both indexes and abstracts exist in physical form in the library and are also accessible online by choosing the "Search Index" option from the library's initial Web pages.

There are hundreds of different indexes and abstracts to consider. For example, if you are searching for articles about stock market trends, you could use the *Business Periodicals Index* or *The Wall Street Journal Index.* If you are searching for articles about alternative energy sources, you may find them in *Energy Research Abstracts.* Two of the most commonly used indexes cover the popular press: *Reader's Guide,* which indexes hundreds of magazines and journals, and *The New York Times Index,* which indexes that newspaper. Many specialized indexes exist as well. What you choose will depend on the subject you are researching and how technical you need to be. Example 6-4 lists some of the print and online specialized abstracts and indexes.

EXAMPLE 6–4   Selected Abstracts, Indexes, and Bibliographies of Indexes

Abstracts and Indexes

Traditional Periodical Indexes and Abstracts

Aerospace Database	Energy Research Abstracts
Animal Behavior Abstracts	Engineering Index
Applied Mechanics Review	Environment Abstracts
Applied Science and Technology Index	Index Medicus
Astronomy and Astrophysics Abstracts	Metals Abstracts
Biological Abstracts	Microbiology Abstracts
Ceramic Abstracts	Nuclear Science Abstracts
Chemical Abstracts	Nursing Studies Abstract
Chemical Engineering Abstracts	Oceanic Abstracts
Computer and Control Abstracts	Plant Breeding Abstract
Computer Literature Index	Pollution Abstracts
Ecology Abstracts	Science Citation Abstract
Education Abstracts	Social Science Citation Index
Electrical and Electronics Abstracts	Wildlife Review

Online Periodical Indexes and Abstracts

Agricola
AgEcon Search
Arts and Humanities Citation Index
Biological and Agricultural Index
Computer Data Base
Fish and Fisheries Worldwide
Food Service and Technology Index
Microcomputer Index
National Newspaper Index
Sociological Abstracts

When you search an index, you are generally looking at materials outside of the library's catalogs. Libraries subscribe to indexes and list them on their catalog or "research tools" pages. If you are searching online and know the journal you want to search or the index that lists it, sometimes you can go directly to that index from a search page on your Internet browser.

When you conduct a keyword search on an index, make use of Boolean logic just as you did searching the library's holdings. When the search is complete, you will end up with a results page just as you did earlier. And in this case as well, the hits that are identified have embedded links that let you move directly to the article you have chosen, if it is online. Example 6-5 shows the kind of information you will find in an index. Index entries provide authors, titles, and abstracts, and they direct you to the correct journal. Indexes frequently abbreviate the titles of journals, as Example 6-5 indicates. A key to these abbreviations is given somewhere at the front of the print document or at the beginning of the online index site, but sometimes you have to look for it.

EXAMPLE 6–5   *Sample of Psychological Abstracts*

Heading	Psychopharmacology 38374.
Authors	Abel, E. L.; Bilitzke, P. J. & Cotton, D. B.
Site of research	(C. S. Mott Ctr for Human Grown & Development, Detroit, MI)
Title	**Alarm substance induces convulsions in imipramine-treated rats.**
Journal	*Pharmacology, Biochemistry & Behavior*
Date of publication	1992 (Mar), Vol 41(3), 599–601.
Abstract	—Male rats were injected with imipramine (0-30 mg/kg) and tested in the forced-swim test in either fresh water or water soiled by other rats, which presumably contained an alarm substance. Imipramine did not affect the behavior of rats in fresh water. More than half of the Ss given the combination of imipramine (30 mg/kg) and stress from alarm substance had clonic convulsions. Adrenalectomy did not affect this relationship, indicating that corticosterone is not involved, although imipramine and stress cause increases in plasma corticosterone levels. The convulsive effect of combined stress and imipramine may be due to their common actions at central noradrenergic and serotonergic receptors.

USING OTHER REFERENCE MATERIALS

Although much of your research will grow from books and periodicals, there are a number of other potentially fruitful avenues to explore. Reference librarians will be able to tell you what sources are available at your location. Even if you think that you will be using only books or periodicals, it often is a great time saver to ask for the reference librarians' advice as you begin to search. Often, they will direct you to additional sources of the type we describe below. Many of these resources can be searched online, using the same techniques we have discussed earlier, and initial links are listed on the library's home page. If you don't want to go through a library to conduct your search, you can visit these sources directly by searching for them on the Internet.

Government Documents

Many libraries are registered repositories for U.S. government documents that cover topics from economic forecasts, to environmental studies, to declassified intelligence reports, to census data. The federal government publishes more than 25 million different documents yearly aimed at both specialized and general audiences. You can find reports from departments and agencies, including the Departments of Defense, Interior, and Labor, the U.S. Information Agency, Environmental Protection Agency, NASA, the Copyright Office, and the Central Intelligence Agency.

Libraries that serve as repositories of federal documents often employ a special reference librarian. If your library is such a repository, ask at the reference desk where to find the librarian in charge of the collection. Example 6–6

EXAMPLE 6–6   Selected Government Documents

Monthly Catalog of U.S. Government Publications
GPO Publications Reference File
Government Reports Announcements and Index
Census of Population and Housing
County and City Data Book
National Retail Trade
Census of Agriculture
Dept. of Defense Hazardous Materials Information
National Economic, Social, and Environmental Data Bank

Don't miss the Library of Congress <http://lcweb.loc.gov/rr/collects.html> which gives you access to government documents and the Library of Congress holdings which include about 115 million items in all formats and languages and about all subjects. From the Library of Congress page, you can get direct access to the National Agricultural Library <http://www.nal.usda.gov/> and the National Library of Medicine <http://www.nlm.nih.gov/>.

---

lists a few additional sources of information on government documents. Don't forget that you can search these holdings online by going to the Library of Congress Web site.

Many state government agencies also provide publications. If your state has a land-grant university, it also has an Extension Service—which publishes a wide variety of information on agriculture, rural sociology, food science and nutrition, gardening—and the Internet. Many of these resources are cataloged and can be searched online. These and other materials, often housed separately in the library, are designed for both lay and specialized audiences. Check with your librarian to see how these are indexed. State and most local governments also maintain Web sites and listings, and many of these sites are searchable.

Reference Works

Other sources of information from the reference library include specific disciplinary *dictionaries, encyclopedias,* and *handbooks.* There are both general and specialized versions of these reference materials (see Example 6-7). When you know very little about a topic, general reference works will help you get started. Later, when you have a clearer idea of your subject and its specialized vocabulary, you may wish to look at more specialized reference materials. Many libraries list their reference sources on their Web pages and provide search engines to examine them.

EXAMPLE 6–7   Selected Reference Works

CRC Handbook of Chemistry & Physics
CRC Standard Math Tables
Dictionary of Advanced Manufacturing
   Technology
Dictionary of Ceramic Science
   & Engineering
Engineering Acronyms & Abbreviations
Electronic Engineers Handbook
Encyclopedia of Chemical Technology
Encyclopedia of Fluid Mechanics
Encyclopedia of Materials Science &
   Engineering
Encyclopedia of Physical Science &
   Technology
Encyclopedia of Polymer Science &
   Engineering
Engineering Mathematics Handbook:
   Definitions, Theorems, Formulas, Tables
Handbook of Industrial Toxicology

Handbook of Mechanics, Materials, &
   Structures
IEEE Standard Dictionary of Electrical &
   Electronic Terms
McGraw-Hill Dictionary of Scientific &
   Technical Terms
McGraw-Hill Encyclopedia of Science &
   Technology
Machinery's Handbook
Metal's Handbook
Perry's Chemical Engineers' Handbook
Standard Handbook for Electrical
   Engineers
Standard Handbook of Engineering
   Calculations
Standard Handbook of Industrial
   Automation
Standard Handbook for Mechanical
   Engineers

EXAMPLE 6–8   A Few Online Dictionaries and Encyclopedias

Encarta
  http://encarta.msn.com/
Encyclopedia Britannica
  http://www.eb.com
Grolier
  http://grolier.com/
Biotech Dictionary
  http://www.lib.adfa.oz.au/web/listlib/dict.htm#engin
Hypertext Webster Interface
  http://c.gp.cs.cmu.edu:5103/prog/webster

Encyclopedias and handbooks are particularly useful in getting started because you will find a bibliography of other sources on the topic at the end of most entries. Most encyclopedias are online these days, but many charge a fee for using their entries. Example 6–8 lists a few of the most familiar encyclopedias online. Before you decide to subscribe to one of these services, check with your library to see if they are already paying the fees for access to these resources. You can save yourself time and money if the library has subscribed for you.

Doing Research on the Internet

As you can see from earlier sections in this chapter, you won't need to use the library at all to conduct productive searches on some topics. Surfing the net is a popular activity for people wanting to find out a wide range of information—the cheapest airline tickets, the weather report for the area where they will travel tomorrow, the latest research conducted on the disease they might have, or the age when their puppy should be spayed.

Internet
an enormous network of computer networks; the World Wide Web is the name for all computer files accessible through the Internet

The potential for individuals to have vast databases available at their homes and offices is being realized through the large web of computer networks known as the **Internet.** This system isn't merely a network of computers; it is a network of networks, each of which is composed of a group of linked computer systems in organizations such as universities, companies, nonprofit organizations, government agencies, or research labs. The World Wide Web debuted in 1991 and the Web and the Internet have grown, often exponentially. No one knows how big the Internet community really is, but in 1993, the best guess was that it included about 1.3 million computers, 8,000 computer networks, and linked about 10,000,000 users. And in 1996, when the Web was only five years old, there were already 40 million users. For a while the number of users was doubling every 50 days or so. At the beginning of the millennium, the number of users is doubling about every 100 days.

EXAMPLE 6–9   Some Internet Search Engines

AltaVista *<http://altavista.digital.com>*
HotBot *<http://www.hotbot.com>*
Infoseek *<http://www.infoseek.com>*
Lycos *<http://www.lycos.com>*
Excite *<http://www.excite.com>*

EXAMPLE 6–10   Some Online Resources

Some Newspapers Online	The New York Times *<http://www.nytimes.com/>* Minneapolis StarTribune *<http://www.startribune.com>* Washington Post *<http://www.washingtonpost.com/>* Seattle Times *<http://www.seattletimes.com/>* St. Paul Pioneer Press *<http://www.pioneerplanet.com/>* USA Today *<http://www.usatoday.com>*
Journals and Magazines Online	Time Magazine *<http://pathfinder.com/time>*
News Sites Online	MSNBC *<http://msnbc.com/>* CNN *<http://cnn.com/>* CBS *<http://cbs.com/>* NPR *<http://www.npr.org/news/>*

When you log on to the Internet, you do so through one of the Internet providers such as AOL (America Online). If you are at a university that provides Internet connections, you may connect to the Internet directly and then interact with it via a Web **browser** such as Internet Explorer or Netscape. Whatever browser you use, it is likely to have Internet Search as one of the options available on its home page. You can use any of the search engines offered by your browser or any other you've discovered to conduct an Internet search.

browsers
user interfaces that provide access to the World Wide Web

INVESTIGATING SOME INTERNET SEARCH ENGINES

Although theoretically each search engine could examine every Web site, the Web has grown so large that each search engine only visits a fraction of available sites. That's why you might want to try more than one. If you find one you especially like, **bookmark** it; that is, add it to a list of sites whose addresses are saved for you on your browser. Example 6-9 lists a few of the search engines available.

Online searching connects you to databases containing millions of sources of information. Example 6-10 indicates some online resources that

bookmarks
Web pages visited and tagged by entering their address into a special file on the browser so that one can return to these sites directly

may be useful as you begin. After trying some of these out, you can open your search to the entire Internet. Conduct your Internet search using the same techniques of searching we discussed earlier. Each search engine, however, has its own *syntax* or way in which words are put together to form phrases or sentences; therefore, you may need to look at each site's instructions on refining your search using its system.

The major advantage of searching online is that the databases are continually updated so that you will have access to the most current information available through research. A disadvantage is that without good searching skills a single query can generate more than 1 million hits, more sources than anyone could hope to evaluate.

FINDING SOMEONE TO TALK TO ONLINE

Sometimes just searching on the Internet will not give you the information you need. You may want to "speak" with knowledgeable people who can give you insights into the problem you are researching. The Internet is very useful for finding human resources too. If you know the persons you want to "talk" with, you can use Internet search engines designed for finding specific people. By entering a person's name, you can get access to their current e-mail address. Unfortunately, in trying to find out about something like our Med-Ease usability study, you may not know with whom you want to talk. Even so, the Internet provides us with many promising places to look for people who have the expertise that we need. We're going to look at two: discussion lists and Usenet groups.

Discussion Lists

Discussion lists are basically mailing lists, collections of e-mail addresses, set up to allow one message to be sent to all members of the list simultaneously. Discussion lists are "owned" by someone who takes responsibility for monitoring entries and maintaining the list on his or her **server,** or central computer. The server stores the messages from the list, automates subscribe and unsubscribe messages, and maintains the mailing list.

servers
special computers devoted to maintaining such things as networks

Imagine that you were unable to find any specific information about Med-Ease during your library search. You're not surprised by this because you know that this software is just being piloted, hence your need to do a usability study. What you did discover, though, were a number of articles mentioning MedSoft, a competing hospital software program that was released a few months ago. That sounds like a promising lead, and you'd like to follow it up by finding out who has been using this software, how they studied its effectiveness, and what the results of their study were. You might want to try subscribing to a variety of discussion lists and sending out queries, or requests for information.

There are thousands of discussion lists to consider, and your developing skill doing keyword searches will help you to find out which discussion

---

EXAMPLE 6–11   Discussion List Indexes

Liszt, the Mailing List Directory <*http://www.liszt.com*>
Tile.Net/Lists: The Reference to Internet Discussion and Information Lists
<*http://tile.net/lists/*>

Sample Discussion Lists

Ancient-dna-1	*Majordomo@coombs.anu.edu.au*
Bee-L (Bee Biology)	*listserv@cnsibm.albany.edu*
Consbio (Conservation Biology)	*listproc@u.washington.edu*

---

lists are discussing the topics of interest to you. The easiest way to find the right listservs is to search discussion list indexes. Example 6–11 provides you with two places to begin and an idea of the range of lists you can find. Be patient and thorough in your search. Liszt, for example, lists more than 90,000 different discussion lists.

Once you find the discussion lists you want, you will have to subscribe to them in order to ask your research questions and receive responses. When you identify a list name, you send them a "subscribe" message via e-mail. In the first line of your e-mail message, you put the list name e-mail address. In the body of the message you would type a "subscribe" message (see Example 6–12).

You will receive an e-mail notice that you have subscribed successfully to the group. Within 24 hours, you will begin getting postings from the discussion group.

EXAMPLE 6–12   Subscribing to a Discussion Group

To: Here you put e-mail addresses of the three main kinds of discussion groups (Listserv, Listproc, or Majordomo):

- listserv@
- listproc@
- majordomo@

In the body of the text type the following words:

subscribe (Name of the discussion group) and your first and last name

Here's what it looks like when one of the authors of this book subscribes to the technical writing listserv:

To: Techwr-L@listserv.okstate.edu
Subscribe TechWr-L Billie Wahlstrom

Once you have subscribed, it's time to query list members about your usability study. Remember what you've already learned about audience and purpose as you prepare your query. Briefly establish your ethos by letting people know who you are, and then ask them your questions, making it clear why you are asking and what you will do with the answers. If you are planning to quote the people who respond, you should let them know that too. Discussion lists are public, so you would be within your rights to reproduce conversations found there, but common sense and courtesy suggest strongly that you let people know that you are going to write a report to your supervisors about your findings.

Newsgroups

Newsgroups are the broad title for thousands of discussion and news sites. By the end of the 1990s, there were about 20,000 newsgroups online with about 10,000,000 users. Most Web browsers have a newsreading interface that allows you to read and to post or send responses to these groups. To find out if there is a newsgroup on the topic about which you are interested, turn again to an index and search out the right group. One of the most complete indexes can be found at Liszt Newsgroups: *<http://www.liszt.com/news/>*.

Unlike a discussion list, you don't have to subscribe to participate in a newsgroup. If you find a newsgroup dealing with the topic you are researching, read a few of the entries to get a sense of the local etiquette and once that's done tell people about your project and let them respond to any questions you might pose.

SAVING YOUR INTERNET RESEARCH FINDINGS

When you work with library materials, you often have a book in hand that has a call number. You can photocopy materials you need, including the title page with the author, title, and publisher on it, write the call number on the top, and put your materials in a file folder to return to another day. Internet findings are not so easily managed. In the first place, the Internet changes daily, and a useful site may not be there tomorrow. Therefore, it pays to cut and paste materials you find into folders on your desktop. Always bookmark your good sources because often you come to these sources in a roundabout way that you might be unable to retrace the next day. If you are going to print out some of the materials you found on the Internet, set your Web browser to print the URLs as a header so you have a record of where you found what. Make sure that you have taken good notes on where you found material so that you can return to it if necessary and can provide a full citation of it for your final project.

In this section we discuss evaluating the materials you have found, in particular, assessing the credibility of your sources. You're going to have to sort your data by facts, inferences, and opinions. Depending on where you found your information and what kind of information you have, you will have to approach the evaluation process slightly differently.

Evaluating Your Materials

EVALUATING LIBRARY SEARCH MATERIALS

If you have found what you're looking for in the library, a lot of the evaluation of the quality of materials has already been done for you by the library staff. Libraries have standards that determine what holdings to collect. Books and journals are from well-recognized presses which impose a rigorous peer review of the materials they publish. **Peer review** means that the materials in the library have been subjected to review by the author's peers, verifying the quality of the research and findings reported. Peer review means that articles in medical journals have been read by other physicians and scientists who, by accepting an article for print, testify to the qualifications of the researcher and the rigor with which research was conducted. Therefore, what you find in the library can generally be cited with some assurance.

peer review
the process whereby research and findings are validated by peers in the same discipline

What you must do to evaluate the materials you have found in the library is to search carefully to make sure you uncover any controversies surrounding the information, noting precisely what the objections are. You also have to make sure that the materials you collect are up to date, that they represent the latest findings on the subject.

EVALUATING ONLINE SEARCH MATERIALS

When you work with online materials, your evaluation process is more complex. A lot depends on what materials you have found. If you have found online articles, you will need to determine the credentials of the people you cite. Who is the author? What are his or her credentials for claiming expertise on this matter? How recent is this material? Was there a peer review process in place for the online journal in which the material appeared?

If you have relied on materials from online newspaper indexes, you will have to establish the credibility of the source. Because of their editorial policies and the quality of their reporters, we are more inclined to find what we read in *The New York Times* or *The Wall Street Journal,* whether in print or online, more credible than what is published in a tabloid.

If you have relied on discussion lists or newsgroups to get current information by professionals in the field, giving the name, title, and credentials of your source will be helpful in establishing his or her (and ultimately your) credibility. If you retrieved your information from Web sites, then you have to make a special effort to evaluate your sources. Unlike the library, where all sources are chosen by librarians trained in evaluating the quality of materials and the materials themselves have undergone peer review, the Internet is largely uncharted and unregulated territory. Many materials are credible, but anyone with a smattering of **HTML** knowledge and access to a server can put up a Web site, and no one is there to check the validity of the material on the site.

There are several things you can do to evaluate the quality of the material you find online. First, look at the **domain type** associated with the site. At the present time there are a limited number of domain types available, and they are assigned more or less by a set of rules that lets us know in a general way who has set up the Web site. You probably have noticed these components of **URLs,** or Web addresses:

- .com   (commercial organization)
- .edu   (educational institution)
- .org   (nonprofit organization)
- .gov   (governmental organization)
- .mil   (military institution)
- .net   (Internet service provider, e.g., AOL)

URLs that end with *.com* are commercial sites. Information on these, especially about products, has to be taken with a grain of salt because these are sites established to market or support products and services. URLs with *.edu* are educational institutions. Before the *.edu* in the Web address you will find an abbreviation of the educational institution itself. For example, the University of Minnesota's address is *umn.edu*, Michigan Technological University's address is *mtu.edu*, and Texas Tech's address is *ttu.edu*. With this information you can make some assessment of the validity of the material you find at the Web site.

Remember, however, that just because a URL has an *.edu* attached, not all material at that site is "official." Students and faculty often put up unofficial pages that contain information not submitted to peer review. Assessment of this kind of material will have to depend on the credentials of the person putting up the site and what kinds of references you find on the Web page. Official university sites carrying the university's logo or word mark are generally credible. Typical sites of this sort include access to the digital collections of a college or university (e.g., insects or soil types) or other informational resources (see Example 6–13).

Web sites with *.gov* are put up by the government. You can find local, state, and federal government Web sites with a wide variety of information, and these sites generally are credible. Web sites with *.org* are sponsored by nonprofit organizations. Most of the information at these sites is in the public interest and can be considered credible. Remember, though, that you need to take some time to look at what kind of nonprofit organization you're dealing with. Racist hate groups like the Ku Klux Klan have *.org* Web sites just as the Red Cross and environmental groups do. Countries

EXAMPLE 6–13   Material Found at a University Web Site: Nutritional Tools

You can visit this site at this address *www.agricola.umn.edu/NutritionTools*

UNIVERSITY OF MINNESOTA

kCal-culator®

Copyright 1996, Regents of the University of Minnesota
All rights reserved
Created by PBrady
Department of Rhetoric

BMR or basal metabolic rate is the energy requirement to maintain life. It is measured at rest, but not asleep in a thermo-neutral environment in the post-absorptive state. It can be measured directly or indirectly, or it can be estimated as we are doing here. We are using the equations of Harris & Benedict (1919) to estimate BMR.

To estimate your BMR, input:

Physical characteristics:	BMR		
Age in years		kcal/day for men	Enter Clear
Weight in pounds		kcal/day for women	
Height in inches			

BMR can also be estimated as $70 \, W^{0.75} =$ [ ] kcal/day, where W is body weight in kilograms.

Obviously, there is more to life than just resting in that temperature-neutral environment. You must also have energy from your diet to support your activities above basal. Once you have calculated the BMR above, you can enter the **minutes** you spend in your various other activities each day. We have divided these into five levels from *very light* to *very heavy*. We have included a few examples of each category to allow you to gauge where a given activity might fit. The result is only an estimate, but should give you an idea of your daily caloric needs.

Physical Activity, minutes/day					Total Energy Requirement	
Very Light	Light	Moderate	Heavy	Very Heavy		
[ ]	[ ]	[ ]	[ ]	[ ]	[ ] kcal/day for men	Enter
Examples include: *Reading Sitting Driving Eating*	**Examples include:** *Walking Sweeping Playing Piano Bicycling (easy)*	**Examples include:** *Fast walk Dancing Ping-Pong Skating*	**Examples include:** *Swimming Running Bicycle Race Basketball*	**Examples include:** *Boxing Rowing Mountain climbing*	[ ] kcal/day for women	Clear

Thank you for visiting,
webmaster@mail.agricola.umn.edu
URL = http://www.agricola.umn.edu/nutritiontools/kcalculator.htm

outside the United States identify themselves on the Web with a two-letter identifier at the end of their Web address (e.g., *.uk* represents the United Kingdom).

Ultimately, you will have to stand by the evaluations you make of Internet resources, so take some time to determine who and what you consider expert.

SORTING OUT FACTS, INFERENCES, AND OPINIONS

To make evaluations of the materials themselves, we are going to look at some ways to assess the quality of information, regardless of where you found it. As you read or listen to the information you gather from primary or secondary sources, you will need to assess your source's credibility or *ethos;* you need to determine whether your source uses *pathos* to appeal to your values and emotions; and you need to decide what "good reasons" or *logos* your sources offer in their own arguments. The lessons you learned in Chapter 4 should help you analyze the persuasive strategies that your sources use in convincing you to listen, read, and accept their arguments.

In this section we offer you additional help in evaluating the logos of your sources. Recall for a moment the *because-clauses* that we discussed earlier. A new highway should be put through downtown because of improved safety, better traffic flow, and better aesthetics. A certain paint should be used on automobiles for good reasons, because it resists chipping, for instance, and because it is inexpensive and easy to use. And, recall that these because-clauses support the ideal situations that you saw when we discussed communicating to solve problems. The *ideal situation* is to have foreign language study within the engineering curriculum, but at the moment it is not required or encouraged (*actual situation*), so you are proposing how foreign language might be included in the curriculum (purpose) because other schools are doing it, because that might make students into more flexible and informed and culturally aware engineers, and because it might give students an edge on the job market. Looking at ideal and actual situations enables you to solve problems for your own audience; solving these problems through persuasive because-clauses ensures that your audience will respond to you. You can use these same tools to evaluate the information you gather from your sources.

In particular, you need to determine whether your sources back up their good reasons—their because-clauses—with facts, inferences, or opinions. Of course, what appears to the source as fact might seem opinion to you and so on. Your evaluation will affect how you present the material to your own audience and how you interpret the importance, relevance, and value of the material. Your decisions about what information to include will be based on your determination of whether the information seems verifiable, logically inferred from sufficient evidence, or based on expert opinion. No information comes without context, interpretation, and bias—both your source's and your own.

Facts

If you gather and present information that can be verified or proven by personal observation or research in qualified sources, that information comes as close to **fact** as we can get. We add that caution because the context in which the fact is presented and even which fact is selected adds subjectivity or potential bias to any objectivity your source or you might attempt to achieve. For example, by using a standard unit of measurement, a source might say "the computer screen is 8 inches from top to bottom," but the choice to use inches instead of millimeters and the decision to convey that dimension instead of the height of the terminal itself reflect the source's personal choices, goals, perspectives, or values, and affect you and your own audience. Given this qualification, keep in mind that a fact is universally agreed upon or can be verified by anyone repeating the observation, experiment, or calculation. For those reasons, you may more readily accept as "fact" a source's description of an actual situation than of an ideal situation. And, you need to look at your source's because-clauses to see how many of them could be described as factual.

When reading or hearing what appears to be a fact, look carefully at every word. For example, if your source refers to an actual situation such as, "The University of Minnesota trapped and killed 10,000 birds in a three-month period," you will need to ask whether this number reflects those birds trapped in experimental agricultural fields only; whether the number reflects sparrows, blackbirds, grackles, starlings, cowbirds, or other types; whether the source refers to only the Twin Cities branch of the university or includes coordinate campuses; and what three-month period was involved. Generally, qualifying words, particularly of time and place, add to the credibility of any statement (e.g., "Ten thousand blackbirds and starlings were trapped and killed in the seven experimental fields, numbered 1001–1007, at the Twin Cities branch of the University of Minnesota between March 1, 1999, and June 2, 1999").

Finally, when you believe you have encountered a fact, look carefully at your source. The statement about the number of birds trapped in university fields might be more credible if offered by the person in charge of counting the number of birds rather than an unqualified observer. It would also be more credible if confirmed by more than one person and if published in a research report rather than stated in a phone call.

facts
information that can be verified or proven by personal observation or research

Inferences

Inferences are usually theories and hypotheses, statements backed up by some observations or experiments but not completely verifiable. Much of the information you gather from your primary and secondary sources, many of your source's good reasons and because-clauses, will probably be inferences, or those logical conclusions and statements of probability, supported by as many "facts" as the observer or experimenter can provide. For example, the statement, "Nevertheless, extensive laboratory studies suggested that enhanced bioremediation might be applicable to stranded oil on the beaches of PWS [Prince William Sound] and the GOA [Gulf of

inferences
theories, hypotheses, and statements backed up by some observations or experiments but not completely verifiable

Alaska]," reflects the source's inference that bioremediation would be effective. The laboratory studies might indeed predict that the conditions in Prince William Sound would be conducive to bioremediation. However, look again carefully at each word within the statement: What does the source mean by "extensive"? By "enhanced"? Who did the laboratory studies? How many studies were done? In what environment? Even though the inference may seem logical or reasonable, you will need to evaluate your source's background, experience, values, organizational affiliation, and so forth before presenting the inference to your own audience.

Opinions

opinions
personal attitudes

Finally, at times you might encounter what seems to be an **opinion,** or a personal attitude, in the information you gather and must interpret. That opinion might appear in your source's description of an ideal situation, or one of your source's because-clauses might seem an opinion to you. To say that trapping and killing birds in University of Minnesota experimental fields is "a waste of animal life" is an opinion. What is a "waste" to some might be a necessity to others. If the birds killed are destroying valuable research seeds and if they are used to feed injured raptors being cared for in the veterinary school, another person might have a different opinion about the university's actions. However, do not dismiss opinions entirely from the information you gather but instead evaluate the ethos of your source and the information upon which the source has based the opinion. A strongly stated opinion from an expert on the topic might carry more weight than an inference offered by a less qualified person.

Summarizing Information: Developing Summaries and Abstracts

summary
a way of reducing information from many sources or a single long source into a paragraph or two; sometimes called an *abstract*

Now you have gathered material, evaluated it, and using the strategies we discussed for organizing information in Chapter 2, you have thought about how to structure it. Here we're going to look at ways of summarizing or managing the data you've decided to use.

A useful tool to help you manage information is the **summary** or **abstract.** When you collect information from secondary or primary sources, you will often summarize or reduce the essence of that information to a paragraph or two of notes. When you decided what secondary research would be valuable for your project, you may have read the abstracts published at the beginning of research articles or within collections (see Example 6-5).

More often than not, when you do any kind of research, you will end up presenting the results of that research in a shortened form. After all, if your research generates pages and pages of notes and data, and dozens of books and articles, your written and oral reports serve, in part, as extended summaries of all of that information. Even so, within those reports you will need to provide shorter versions of the information you are presenting. You provide these shorter versions within carefully constructed abstracts or summaries. We cover summaries next, as a general technique for reducing your information to a manageable form, and then we cover abstracts, as formal sections of the documents you will read and write.

SUMMARIES

Summaries serve several functions in your research and communication. In order to manage the information you gather through primary and secondary research, there are times when you will record that information word by word. At other times you will record only the essence of that information in a summary. Also, when you write or speak, you often offer your audience short summarizing statements throughout a communication to help that audience remember and appreciate your arguments. You might conclude your argument with a summary of all your "good reasons" or main points. Moreover, your introduction to a document or speech often previews the structure of your communication by providing a kind of summary of your main points.

Let's begin by looking at some techniques for summarizing the information you encounter in your research or in summarizing the main points in your own documents. Example 6-14 on page 202 repeats the passage that you studied in Chapter 4 (Example 4-4) to demonstrate summarizing techniques.

When you write a summary of information you gather, first identify the main points and the important supporting evidence. You can either underline this information, record it on a note card, or enter it in a computer database that you create for your project. You cannot include all of the source's examples, but you might record the most revealing or universal example in your summary. You will encounter some words and phrases in the original that convey the meaning to such a great extent that you cannot change them within losing that meaning. Record those words and phrases as direct quotations and note the page number on which they occurred in the original source.

Your use of the information will also determine what you capture in your summary. For example, let's say that you are gathering information on cross-pollination of flowering plants by animals. You are particularly interested in what happens to cross-pollination when animals encounter plants in the peak or second flowering stage. The passage in Example 6-14 appears in one of the articles you read. We have placed in boldface the main points you would probably record for your summary.

You'll notice that the summary in Example 6-15 on page 203 begins with a short explanation of the three phases of flowering, contains the specific name for the phenomenon, and some essential information about the effects of this phenomenon on cross-pollination. Because you are concentrating

EXAMPLE 6–14   An Article on Botany and Ecology

Many **plants** that are **pollinated by animals**—examples would be anything pollinated by ants or the proverbial birds and bees—**develop their flowers in three fairly discrete phases.** First, there's an **initial phase;** during this period the plant produces a **small but increasing number of flowers each day.** Next comes a **short peak phase** in which most of the **total number of flowers are produced.** And finally comes a **final phase** when the plant produces a **small and decreasing number of flowers** each day. Recently many evolutionary ecologists, such as Dan Janzen, Al Gentry, Gary Stiles, and Carol Augsperger (right down the hall), have been concentrating their efforts on studying the effects of this **mass-flowering pattern**—on how pollinators like bees and ants are attracted to plants and how and why they move from plant to plant. These people are showing that many kinds of floral visitors (for example, the ants) **prefer to eat individual plants that produce really large numbers of flowers per day,** presumably because these individual plants are **easy to spot** by the pollinators (and are very tasty to them) and because it takes **less effort** for the pollinators to do all their pollinating when flowers are so close together. In contrast, when daily flower production is low, during the first and last phases of the flowering pattern, the plant through its flowers might not be able to attract enough pollinators from other flowers. Consequently, this mass-flowering pattern poses a dilemma: When flowers are few and far between, the bees or ants fly or walk around and visit many different trees, since flowers are scarce; that's good in that cross-pollination occurs (and cross-pollination makes for a more vigorous next generation), but it's bad in that the birds and ants spend a lot of their time and energy visiting various trees and therefore don't make as many stops as usual. On the other hand, when flowers are many, the pollinators can save their energy and work more efficiently on one plant; that's certainly more efficient, but **it cuts down on the cross-pollination because the pollinators are sticking to one location.** So what's a plant to do?

In our study, Terry Haas and I tried to answer two questions: During what phase of the flowering pattern does outcrossing occur? And what factors promote outcrossing? Specifically, we've examined the effect of different floral abundances on the proportion of flowers that are self- and cross-pollinated on an individual plant. What's more, we've documented the flowering pattern of the catalpa tree (*Catalpa speciosa,* Bignoniaceae).

on the second phase of the mass-flowering pattern, you record only one aspect of cross-pollination.

On the other hand, if you need to identify the research questions asked about this "mass-flowering pattern," the researchers asking the questions, and the dilemma that has caused some of these questions, you would have marked different parts of the same passage (see boldfaced words and phrases in Example 6-16).

EXAMPLE 6–15   Summary of Example 6–14

Plants pollinated by animals develop their flowers in three phases:

Phase one: a few flowers are produced.

Phrase two: almost all the total number of flowers are produced.

Phrase three: only a few flowers are produced.

This phenomenon is called a "mass-flowering pattern" (p. 3). Pollinators prefer to eat the plants in the second phase because they can find them easily and stay in one spot. However, less cross-pollination occurs because the pollinators are not moving around.

EXAMPLE 6–16   An Article on Botany and Ecology

Many plants that are pollinated by animals—examples would be anything pollinated by ants or the proverbial birds and bees—develop their flowers in three fairly discrete phases. First, there's an initial phase; during this period the plant produces a small but increasing number of flowers each day. Next comes a short peak phase in which most of the total number of flowers are produced. And finally comes a final phase when the plant produces a small and decreasing number of flowers each day. Recently many evolutionary ecologists, such as **Dan Janzen, Al Gentry, Gary Stiles, and Carol Augsperger** (right down the hall), have been concentrating their efforts on studying the effects of this mass-flowering pattern—on **how pollinators like bees and ants are attracted to plants and how and why they move from plant to plant.** These people are showing that many kinds of floral visitors (for example, the ants) prefer to eat individual plants that produce really large numbers of flowers per day, presumably because these individual plants are easy to spot by the pollinators (and are very tasty to them) and because it takes less effort for the pollinators to do all their pollinating when flowers are so close together. In contrast, when daily flower production is low, during the first and last phases of the flowering pattern, the plant through its flowers might not be able to attract enough pollinators from other flowers. Consequently, this mass-flowering pattern poses a dilemma: **When flowers are few and far between, the bees or ants fly or walk around and visit many different trees, since flowers are scarce; that's good in that cross-pollination occurs (and cross-pollination makes for a more vigorous next generation), but it's bad in that the birds and ants spend a lot of their time and energy visiting various trees and therefore don't make as many stops as usual.** On the other hand, **when flowers are many, the pollinators can save their energy and work more efficiently on one plant; that's certainly more efficient, but it cuts down on the cross-pollination because the pollinators are sticking to one location.** So what's a plant to do?

   In our study, Terry Haas and I tried to answer two questions: **During what phase of the flowering pattern does outcrossing occur? And what factors promote outcrossing?** Specifically, we've examined the **effect of different floral abundances on the proportion of flowers that are self- and cross-pollinated on an individual plant. What's more, we've documented the flowering pattern of the catalpa tree** (*Catalpa speciosa*, Bignoniaceae).

EXAMPLE 6–17   Summary of Example 6–16

Researchers Dan Janzen, Al Gentry, Gary Stiles, and Carol Augsperger have asked "how pollinators like bees and ants are attracted to plants and how and why they move from plant to plant" (p. 3). In asking these questions, a dilemma surfaced: if pollinators find lots of flowers clustered, as during the peak flowering phases of many plants, the pollinators stay in one spot, and less cross-pollination occurs. But if pollinators find few flowers clustered, they move around a lot but they don't make as many stops, and less self-pollination occurs.

In this article, Terry Haas and A. G. Stephenson have asked two questions: "During what phase of the flowering pattern does outcrossing occur? And what factors promote outcrossing?" (p. 4). They have researched: "the effect of different floral abundances on the proportion of flowers that are self- and cross-pollinated on an individual plant" on the catalpa tree (p. 4).

The summary of these researchers' questions and the dilemma that raised these questions appears in Example 6–17. You can use the same techniques, for example, when writing a summary of your own words and ideas for an introduction or conclusion to a long report.

ABSTRACTS

An abstract often stands alone or represents a document in a list or index and helps a reader decide whether to read, skim, or skip the document or to pass it on to others. Therefore, an abstract must be understood by audiences who do not have the document at hand. Imagine yourself in a library searching for articles on a research topic and finding an abstract in an index or electronic database. As a reader and researcher, you will want that abstract to describe the full article clearly enough so that you can decide whether it will be useful to find and read it. If the abstract is effective, its writer has achieved that stand-alone quality. Moreover, you will be required to write abstracts for the formal reports you produce on the job. Your readers will use your abstract to make the same kinds of decisions you would in your research: to read, skim, skip, or pass it on.

Abstracts provide readers with a glimpse of an entire document, such as a report, in shortened form. In general, abstracts follow the style and the order of the document. You will encounter several types of abstracts in your reading and will be asked to write a variety of abstracts on the job. We review the two common types of abstracts in this section.

EXAMPLE 6–18 A Descriptive Abstract

This 50-page analytical report summarizes the recent changes in the waste management system that services the city of Roseville, the closest northern suburb to St. Paul, Minnesota. The changes covered include the new sewer system installed in the twenty-block area bounded by Larpenteur Avenue, Lexington Parkway, Snelling Avenue, and County Road B2. The report includes a complete listing of independent contractors licensed for home trash collection in the suburb and the recycling contractor approved by the Roseville council. The report first reviews the problems the city experienced with the old system, including clogged or limited drainage and overloaded sewer lines. The report then narrates the installation process of the new system between April 1998 and September 1999. Finally, the report describes the expected financial and safety advantages of the new system.

Descriptive Abstracts

Descriptive abstracts describe the contents, including each major part, or facet, of a document. They describe the purpose, scope, and organization of the document but do not offer the findings or conclusions about the topic the document discusses. A descriptive abstract of the document on the Roseville waste management system is shown in Example 6-18.

Notice the use of transitions and verbs to describe the document itself.

descriptive abstracts long abstracts that describe the entire contents of a document

Informative Abstracts

Informative abstracts provide the audience with the purpose and organization of a document and also specific findings, evaluations, and conclusions about the topic covered in the document. They describe the topic, methods used to research that topic, experiments set up to test some aspect of the topic, the findings of that research or those experiments, and conclusions based on research or experiments. Although the audience may elect to read the entire document, informative abstracts can also serve as substitutes for the document based on the information within this kind of abstract. The audience knows enough about the specifics in the document and the significance of the findings to learn something new about the topic. An informative abstract of the document on the Roseville waste management system is shown in Example 6-19.

informative abstracts abstracts that provide an audience with the purpose and organization of a document and specific conclusions or findings covered in it

Notice that the verbs within the abstract ("proved to be," "found to be," "tested," "confirmed," and so on) reflect action and results pertaining to the waste management system.

Writing Strategies 6-1 on page 207 will give you suggestions for writing more effective abstracts.

EXAMPLE 6–19   An Informative Abstract

The city of Roseville, the closest northern suburb to St. Paul, Minnesota, had experienced problems with its waste management system after a period of growth and new housing development between 1960 and 1990. The city council reviewed the possibility of installing a new waste management system in November and December of 1996, which was approved by the voters in January 1997 and construction completed between April 1998 and September 1999. The new waste management system includes a new sewer system, type 678B7, installed in the twenty-block area bounded by Larpenteur Avenue on the south, Lexington Parkway on the east, Snelling Avenue on the west, and County Road B2 on the north. The effectiveness of this sewer system was tested by the engineering firm of Walzer and Gross in late September 1999 and found to be 95% effective in transporting household and business waste. The firm concludes that the previous problems with clogged or limited drainage and overloaded sewer lines could be further relieved by installing new sewer lines north of County Road B2 in the summer of 2000. In August 1999 the Roseville council contracted with five independent contractors to remove trash from households and businesses and published this list, with price options, for all Roseville residents in local newspapers. Recyclable waste is picked up curbside in containers provided by residents twice a month and includes paper, glass, aluminum, and plastic. The city has raised property taxes by 2% for 1998 and will raise an additional 2% for 1995 to cover the cost of the system, as approved by voters in January 1997, and has remained within budget. As confirmed by Walzer and Gross, water quality in Roseville has proved to be 10% higher than other nearby suburbs since the installation of the system.

Crediting Sources

Another aspect of managing the information you collect through research is identifying the source of the information clearly and accurately. The whole point of doing research is to obtain and analyze the knowledge of others to help you inform your readers and solve a problem. You must remember, however, that the knowledge you gain from your sources is not coming from you but from them; therefore, **citing sources** is essential in your work. Failure to do so is **plagiarism.** Plagiarism in school results in failure in a course and in disciplinary action; plagiarism on the job leads to a severe reprimand or dismissal. You always need to be accurate and honest in giving other people credit for their ideas and work.

On the job, however, you are often asked to collaborate with others to produce a document or you might use existing internal company reports and manuals (called boilerplate) to produce a new company document. It may be difficult to assign credit to each of your collaborators or to these internal company sources. Be conscientious, and when in doubt, ask your

citing sources
identifying clearly and accurately the sources of information

plagiarism
failure to cite sources; can lead to dismissal at work, failure at school, and a lawsuit

---

WRITING STRATEGIES 6–1 *Suggestions for Writing Abstracts*

The following guidelines should help you write more effective abstracts.

1. If possible, write your abstracts after you have written the full document. This practice provides two benefits. First, it helps you avoid discussing a topic in the abstract that has not actually been included in the document. Second, it enables you to construct an abstract that follows the structure of the document.

2. For a first draft of your abstract, scan the document and pull out passages that best characterize the overall document, the major conclusion of the document, and each section of the document. If you have written your document on a computer, this can be done quite easily.

3. If you are writing a long abstract, choose the most significant details and include them in the abstract in a way that will make them understandable. Although it is easy to pull out the key details of a document, it is far more difficult to make those details meaningful outside the context of the full document. Remember, in general, abstracts should be able to stand alone.

4. Once you have pulled out key passages from the full document, put them together, and read them through from beginning to end, to be sure each passage moves smoothly to the next. Add transitions. Edit the abstract so that it reads as a cohesive statement, not as a series of disconnected sentences. Make sure you have chosen the most appropriate verbs for the type of abstract you are writing. Also, be very sure that you delete any words or statements referring to points you covered in the document but don't need to mention in the abstract.

5. The best test to see if you've written a good abstract is to have someone read it who has not read your full document. If the abstract reads as a cohesive whole to that person, then you are on the right track.

---

supervisor. The guidelines we offer in this next section will help you credit your primary sources and secondary library sources properly.

QUOTING FROM YOUR SOURCE

When you quote from a source, you are conveying to your audience what that source said or wrote *word for word*. Generally, you should use a **quotation** if the words are particularly interesting, startling, or important to your topic; will catch the audience's attention; are difficult to paraphrase; or contain essential details or statistics. For example, let's say that you encountered the following source and passage:

quotations
word-for-word statements of what a source said or wrote

> Other parts of the world are watching North America's wolf reintroduction efforts. Scotland is pursuing public input for potential wolf reintroduction

> into its Highlands, where there is a low human population and about 300,000 red deer. A BBC poll reported that 75 percent of Scots favor wolf reintroduction. The polls are not nearly as favorable in Japan, but scientists are urging wolf reintroduction in the northern island of Hokkaido where wolves once ranged. Wild sika deer, serows and boars have overgrazed the island, causing serious soil erosion and a decline in the ability of the island's forests to regenerate.
>
> Source: Nancy Gibson, "The Two Sides of Wolf Restoration: Recolonization and Reintroduction," *International Wolf: A Publication of the International Wolf Center* 7, no. 2 (1997), p. 4.

If you find the passage essential to your document, you will need to quote directly much of the information. You might do so as follows and still use your own transitions to introduce the quotations:

> According to Nancy Gibson (1997), cofounder and current board member of the International Wolf Center, "a BBC poll reported that 75 percent of Scots favor wolf reintroduction" in the Scottish Highlands where "there is a low human population and about 300,000 red deer" (p. 4). Popular support is not nearly as high in Japan, where scientists argue for the reintroduction of wolves on the island of Hokkaido because "wild sika deer, serows and boars have overgrazed the island, causing serious soil erosion and a decline in ability of the island's forests to regenerate" (p. 4).

Of course, whenever you use a writer's exact words, you must give him or her credit in one of two ways: by mentioning the source's name or some other identifying label in your own text or by including the source's name or label parenthetically. The exact date of the publication and the page number where the quotation is found in the original must also be included. The passage above uses APA style and therefore relies on the name of the source and publication date. (The upcoming section "Academic Documentation Styles" provides information about citation styles.) Because it is clear that all quotations come from the same source, only the page number needs to be included parenthetically.

Whenever possible, work the quotation in smoothly with your own words and introduce the quotation in some way ("Researchers Jansen and Smith"; "According to a BBC poll"). Use ellipses if you leave out any words ("Wild deer . . . have overgrazed the island . . ."), and use brackets if you insert your own words into the quotation ("Wild sika deer, serows [Asiatic goat antelopes] and boars have overgrazed the island . . .").

PARAPHRASING YOUR SOURCE

paraphrases
restatements of a source that maintain the essence of the original

If your source's words themselves are not as important to your document as the ideas, or if the words from the original source would be difficult or unfamiliar to the reader, you might elect to **paraphrase** those words. When you paraphrase, you need to maintain the essence or main ideas of your source. A paraphrase is faithful to the structure of a source, listing most major and relevant minor points in their original order and with approximately the same emphasis. The following would be an acceptable paraphrase of the Gibson passage:

> North American wolf reintroduction attempts are being watched by people in other countries around the world. Scotland and Japan both are considering reintroducing the wolf, but to different public reactions. A BBC poll found that 75 percent of Scots favor reintroducing the wolf into the Highlands where there are few people and a large population of red deer. People in Japan are less enthusiastic about scientists' suggestion to reintroduce wolves on a northern island to reduce the population of wild sika deer, serows, and boars that have caused soil erosion (Gibson, 1977, p. 4).

Notice that when you paraphrase you still must indicate the source and the page number of the passage you are paraphrasing.

SUMMARIZING YOUR SOURCE

Finally, if you just need to convey the information, facts, or ideas, but you don't need the words or phrases themselves from your source, you can summarize them within your document. A summary, often much shorter than a quotation or paraphrase, captures the gist of a source or some portion of it, boiling it down to a few words or sentences. Don't forget to use a citation because your audience might want to go back to your original source for more information. The following passage summarizes the Gibson material on the reintroduction of wolves:

> North American efforts to reintroduce the wolf are being watched by other countries, such as Scotland and Japan, which are considering reintroducing wolves to reestablish ecological balance in areas where there is overgrazing, particularly by wild deer (Gibson, 1997, p. 4).

In general, your decision to quote, paraphrase, and summarize depends on your audience's interests and needs and the importance of the passage to your document.

The key to crediting your sources accurately is to identify which words and ideas are derived from your sources. You must cite your sources not only when you quote them directly but also when you paraphrase them or restate their information in your own words. You can give credit to your sources in a number of ways. Your company may supply you with a complete style guide, including preferred citation style. Whichever one you choose or are required to use in your class or in the workplace, you must remember to be thorough and consistent.

CREATING CITATION AND WORKS CITED SECTIONS

You are probably familiar with the practice of footnoting, in which references mentioned in the text are documented at the bottom of a page or at the end of a chapter. Generally speaking, most of the documents you create

EXAMPLE 6–20   In-Text Citation—APA Style

The key to creating a usable multimedia development lab is to gain the ability to digitize every audio, video, and mixed media input available. This can only be done with "a device that accepts a full range of data standards and that is not yet feasible" (Roth, 1991, p. 10). Some experts believe, however, that a universal standard—a common way to create and digitize all information—is close at hand (Grant, 1992; Person, 1993).

EXAMPLE 6–21   Excerpt from a Works Cited Page—APA Style

Works Cited

Grant, J. (1992). The coming revolution in digitization. *CD Journal, 10,* 157–159.
Person, B. (1993). Even the dog's bark can be digitized. *New Digital World, 2,* 120–121.
Roth, A. (1991). Digitization: New impact on the world. *New Digital World, 1,* 101–109.

will not have footnotes but will rely on brief in-text citations to identify references. At the end of a document, or in the case of a long document at the end of a chapter or part of it, complete references are given in a "Works Cited" section. Probably the best way for you to grasp this convention is to look at examples of the practice (see Examples 6-20 and 6-21). These examples follow the American Psychological Association or APA style guide (see the APA style also in Example 6-22).

Notice that Example 6-20 has citations for both a direct quotation (Roth, 1991) and for a paraphrase (Grant, 1992; Person, 1993). Notice as well the format for each reference is the same: the author's name is followed by a comma and then the date of publication, and this information is put in parentheses. This convention tells the reader that he or she can find complete information about the sources in the Works Cited section of the document, a portion of which is shown in Example 6-21.

FOLLOWING ACADEMIC DOCUMENTATION STYLES

Whenever you are documenting your sources in an academic setting, you should always verify the proper format for that setting. Different disciplines use different formats, and the key to being correct and accurate is to follow the format that is most appropriate. Example 6-22 lists a few examples from two of the most widely used formats: Chicago style, which refers to *The Chicago Manual of Style* (now in its 14th edition), is favored by the natural sciences; and the APA (American Psychological Association) style is

---

EXAMPLE 6–22 Citation Styles

Chicago Style

Book:
Schriver, Karen. *Dynamics in Document Design.* New York: Wiley Computer Publishing, 1997.

Edited Book:
Spilka, Rachel, ed. *Writing in the Workplace: New Research Perspectives.* Carbondale: Southern Illinois Univ. Press, 1993.

Essay in an Anthology:
Selzer, Jack. "Intertextuality and the Writing Process: An Overview." In *Writing in the Workplace: New Research Perspectives,* edited by Rachel Spilka, 171–180. Carbondale: Southern Illinois Univ. Press, 1993.

Journal Article (paginated by volume):
Winsor, Dorothy A. "Engineering Writing/Writing Engineering." *College Composition and Communication* 41 (1990): 58–70.

Article (author unidentified):
"A Business that Defies Recession." *Business Week,* 25 October 1982, 30–31.

Newspaper Article:
Whalen, Thomas. "Is Digital the Wave of the Future?" *Boston Beacon,* 10 October 1948, A2.

Government Publication:
U.S. Department of Interior. *Underwater Archeological Sites.* Washington, DC: Government Printing Office, 1987.

Electronic Sources:
World Wide Web Site:
McGann, Jerome J. "Dante Gabriel Rossetti: A Brief Biography." *The Complete Writings and Pictures of Dante Gabriel Rossetti: A Hypermedia Research Archive.* 19 March 1997. *http://jefferson.village.virginia.edu/rossetti/dgrbio.html* (23 March 1997).

E-mail Message:
Lay, Mary M. *mmlay@tc.umn.edu* "Professional Journal Guidelines." 4 April 1999. Personal e-mail (6 April 1999).

Discussion Group Message:
Wahlstrom, Billie J. *bwahlstr@mailbox.mail.umn.edu* "Query: Historiography of Pharmacy." 7 April 1999. *H-SCI-MED-TECH@H-NET.MSU.EDU* (12 April 1999).

APA Style

Book:
Yates, J. (1989). *Control through communication: The rise of system in American management.* Baltimore, MD: Johns Hopkins University Press.

Edited Book:
Spilka, R. (Ed.) (1993). *Writing in the workplace: New research perspectives.* Carbondale, IL: Southern Illinois University Press.

Essay in an Anthology:
Selzer, J. (1993). Intertextuality and the writing process: An overview. In R. Spilka (Ed.), *Writing in the workplace: New research perspectives* (pp. 171–180). Carbondale, IL: Southern Illinois UP.

(continued)

EXAMPLE 6–22 (Continued)

Journal Article (paginated by volume):
Winsor, D. A. (1990). Engineering writing/writing engineering. *College Composition and Communication, 41,* 58–70.
Article (author unidentified):
A business that defies recession. (1982 October 25). *Business Week,* pp. 30–31.
Newspaper Article:
Whalen, T. (1948, October 10). Is digital the wave of the future? *Boston Beacon,* p. A2.
Government Publication:
U.S. Department of Interior. (1987). *Underwater archeological sites.* Washington, DC: Government Printing Office.
Electronic Sources:
The APA treats electronic correspondence—discussion groups and newsgroup messages, as well as e-mail—as personal communications. Because such communications generally don't include data available to other researchers, don't include them in your reference list—they may only be cited in-text.
In-Text Citation:
Discussion groups and newsgroups:
Although the APA doesn't treat discussion groups or newsgroups differently from personal e-mail correspondence, you may want to include additional descriptive information—the name of the discussion group or newsgroup—in your citation.

> *Format:*   (First Initial. Second Initial. Last Name, electronic communication [name of discussion group or newsgroup, if applicable], Month Day, Year)
>
> *Example:*
>> Although most "flames" do not contain threats of physical violence, on at least one occasion a threatening note has been posted to an electronic mailing list respondent
>>
>> (K. W. Olsen, electronic communication [Cyber-L listserv], April 15, 1996).

World Wide Web or File Transfer Protocol (FTP) site
Bibliographic reference
> *Format:*   Author Last Name [if known], First Initial. Second Initial. (Date of Publication [if known]). Title of work: Underlined or italics, with sentence-style capitalization [Medium, e.g., World Wide Web page, CD-ROM, Online serial]. Available <specify pathname>.
>
> *Examples:*
>> Moulthrop, S. A. (1994). *Traveling in the breakdown lane: A principle of resistance for hypertext* [World Wide Web page]. Available <http://www.ubalt.edu/www/ygcla/sam/essays/pre_breakdown.html>.
>>
>> *National Health Interview Survey—Current health topics: 1991—Longitudinal study of aging* (Version 4) [Electronic data tape]. (1992). Hyattsville, MD: National Center for Health Statistics [Producer and Distributor].

In-Text Citation
> *Format:*   If you mention a work's primary author or authors in the text, then you simply need to append the work's publication date in parentheses after their name(s) as a citation.

EXAMPLE 6–22 (Continued)

Example:
>   Moulthrop (1994) investigated . . .

Although the APA hasn't settled on a single policy for citing specific information from electronic resources, they suggest that you employ a consistent, logical means for referencing your material. If you are referring to a specific piece of information within an electronic text, Web pages, for example which do not have page numbers, you should include a paragraph number in your parenthetical reference preceded by "para." for citation of a single paragraph or "paras." for a range of more than one paragraph.

Example:
>   Moulthrop argues that modern analysts of hypertext have uncritically ac-
>   cepted the notion of the liberatory nature of hypertext (1994, paras. 3–5).

If you do not mention a work's author(s) in-text, then you need to include the last name(s) in your parenthetical citation.

Example:
>   At least one study (Moulthrop, 1994, paras. 3–5) found that modern hy-
>   pertext analysts have been . . .

For additional information on citing Internet sources using APA style, visit <*http://www.uvm.edu/~ncrane/estyles/apa.html*>.

---

used in the social sciences. These styles are academic documentation styles; your company probably will provide you with its preferred documentation style in a style guide. This list of examples is extremely limited, and it won't answer all the questions you might have as you create your Works Cited section. It will provide you with a few of the most common formats, however, to get started.

For any citations that do not match these, you should consult the appropriate style guide. You will need to get used to noticing what is underlined (or italicized), what appears in capital letters, what is abbreviated, what is followed by a comma or period, and so on. Example 6-23 lists several different style guides.

CITING ONLINE SOURCES

There are many ways of citing online sources, and you will need to check with the particular style guide you use to see what's required. However, we can provide you with some general guidelines that will help you to know what kind of information to gather in your secondary research, so that you will have all the data necessary when you begin to create your Works Cited page.

EXAMPLE 6–23   A Selection of Style Guides

American Mathematical Society. *The AMS Author Handbook: General Instructions for Preparing Manuscripts.* Providence: AMS, 1994.

American National Standard for the Preparation of Scientific Papers for Written or Oral Presentation. New York: American National Standards Inst., 1979.

Bates, Robert L., Rex Buchanan, and Marla Adkins-Heljeson, eds. *Geowriting: A Guide to Writing, Editing, and Printing in Earth Science.* 5th ed. Alexandria, VA: American Geological Inst., 1992.

The Chicago Manual of Style. 14th ed. Chicago: University of Chicago Press, 1993.

Council of Biology Editors. *Scientific Style and Format: The CBE Manual for Authors, Editors, and Publishers.* 6th ed. New York: Cambridge UP, 1994.

Dodd, Janet S., ed. *The ACS Style Guide: A Manual for Authors and Editors.* Washington: American Chemical Soc., 1986.

The GPO Style Manual. 29th ed. Washington, DC: GPO, 1988.

MLA Handbook for Writers of Research Papers. 4th ed. New York: MLA, 1995.

Publication Manual of the American Psychological Association. 4th ed. Washington, DC: APA, 1994.

Many of these style guides maintain Web sites that can help you learn their citation style.

When you do research online, here is the basic information you will need about the materials you gather:

- Author's name
- Title of document
- Volume and issue number
- Date of publication
- Number of pages or paragraphs
- Date of access.

Also, depending on what Internet source you used, make sure you have the following information about the path you took to find your data:

FTP
File Transfer Protocol, defines how files are to be transferred from one computer to another; like World Wide Web servers, FTP servers are part of the Internet

MOO, MUD, IRC
three kinds of synchronous, or real-time conversations, among multiple users online

- If you are citing a Web site, give the web address (e.g., *http://www.agricola.umn.edu*).
- If you are citing an **FTP** site (File Transfer Protocol), give the full path (e.g., *ftp://www.agricola.umn.edu/Rhet8210/Library/Aristotle.htm*).
- If you are citing a **MOO (multi-user object-oriented environment), MUD (multi-user dimension), IRC (Internet Relay Chat)** or other synchronous communication site, identify speakers, type of communication (e.g., interview) and address (e.g., Web Chat address: *http://hal9000.rhetoric.umn.edu:6070/~InstructionalComputing*).
- For e-mail, discussion lists, and newsgroups include the subject line from the posting and the address of the discussion list or

newsgroup (e.g., *subject line*: P&A Posting for Classroom Manager in the Office of the Registrar; Address: umn-dd@tc.umn.edu).

CONSIDERING SOME INTERNET INTELLECTUAL PROPERTY ISSUES

Before we leave the Internet, we need to talk briefly about the digital dilemma it sometimes causes us with respect to intellectual property. **Intellectual property** is the term used to describe the creative work done by people that is protected by copyright (and other law). You have seen copyright notices by the copier in the library or at copier firms such as Kinko's when you have made copies of books and articles. They remind you that the material you are copying is protected by law, or they ask you to obtain permission to copy this material. On the Internet, the ease with which we can copy and paste words and images from a Web page into our own documents is sometimes problematic. We are inclined, perhaps, to "borrow" this passage from one Web site and that image from another, failing to give credit to the work of others. In addition to this kind of "borrowing" being a violation of copyright law, it's simply not right to fail to give credit to others for the work they've done. It's easy to provide correct citations for their work. If you fail to do that, and you claim the work of others as your own, then you are guilty of intellectual dishonesty as well as violating copyright law. Don't take a chance on your career. Always give credit where it's due.

intellectual property describes creative work (e.g., text, image, music, objects) protected by copyright, trademark, and patent laws

---

SUMMARY

- The suggestions and guidelines presented in this chapter are only introductions to the vast world of secondary research. Whenever you are beginning to learn new research techniques, whether primary or secondary, you should work with experienced researchers who can help you learn and refine your skills and guide you through the pitfalls and challenges of research. The first time you do it, library and Internet research can leave you overwhelmed and confused. Learning how to become proficient at both can take some time, but the payoff is potentially enormous.
- In doing secondary research, you are not generating new data about your specific situation as you are when you do primary research. Instead, you are seeking research already done by others and published in a variety of documents. Secondary research can be undertaken by examining library or Internet sources or a combination of them both.
- Although libraries are highly organized places, there is a staggering amount of information available in a variety of formats, including books, magazines, newspapers, printed journals, government documents, microfiche, audiotape, videotape, and compact disks.

As a consequence, you need good search techniques to find what you want in this mountain of information.

- Keyword searches can help you narrow the field of information you must examine, and Boolean search techniques can help you refine your search even further. You can use keyword searches to go through all library materials including books, periodicals, and government documents, as well as a variety of types of reference books, including dictionaries, encyclopedias, handbooks, and CD-ROMs. You can also use keyword searches on the Internet. Search engines allow you to examine the Internet for materials that meet your needs using the same techniques you used in searching library holdings.

- This chapter also offers some methods of managing your information. One of your first considerations is evaluating the material you encounter. Material gathered on the Internet needs to be examined first to see what you can discover about the credibility of the sources. Then all information needs to be examined in terms of whether it is presenting what appear to be facts, inferences, or opinions.

- A second consideration you face as you deal with information is summarizing it. Within many documents you will need to provide shorter versions of the information you are presenting. You provide those versions by carefully constructing summaries or abstracts. Typically, most abstracts appear in one of two forms: (1) the complete, descriptive abstract that describes the entire contents of a report in as few words as possible or (2) the informative abstract that provides an overview of the purpose, significance, scope, and important details of a report.

- Finally, you need to decide whether quoting, paraphrasing, or summarizing the information will be most useful to you and your audience. And you need to document your sources. A number of common style guides help you document information in an academic setting. Your company will probably provide you with its preferred style. If you have gathered information from the Internet, you will have to include additional information about the paths you took to find the information in your citations.

ACTIVITIES AND EXERCISES

1. Many product liability cases result from industrial and commercial accidents. Many of those cases involve faulty documentation of some sort. The following lists areas in which such accidents have occurred.

   Industrial accidents
   Manufacturing: steel, aluminum, auto, petroleum, aerospace, chemical
   Construction: commercial, residential, industrial
   Food service
   Agriculture

Forest products
Power utilities

Consumer product accidents

Automobiles
Motorcycles
Boats
Household appliances
Food
Medical care
Computers
Lawn, garden, and farm equipment
Pesticides and insecticides
Airplanes

Choose one of these areas (or another one not listed) and conduct secondary research on accidents that involved faulty documentation in that area:

a. Find the citations for five books that include information about accidents in the area you choose. List the bibliographic information for each entry (see the "Works Cited" section for the form these entries should take).

b. Search the popular press indexes and list the references to five articles about two or more accidents within the area you've chosen. In doing so, identify not only the entry but in what index or abstract you found the entry, and whether or not the sources that contain the entries are available in your library.

c. Search technical, scientific, business, or legal indexes/abstracts about one of the accidents in the area you've chosen. List five citations, the indexes or abstracts that you use, and the location and availability of the entries.

d. Search the Internet and find five additional citations for materials relating to accidents in the area you've chosen by using a keyword search. List the five citations and write a brief step-by-step description of the way you narrowed or refined your search. How many hits did you find on your first try? How do you reduce that number by refining the search?

2. Bioremediation, as explained in Appendix A, Case Documents 2, was the main treatment used by the Exxon Corporation to clean up the oil spill in Prince William Sound, Alaska. Assume that you are an engineer working for a competing oil company and have been asked by your supervisor to gather information in three areas:

■ The ultimate success of bioremediation in the *Exxon Valdez* spill area.

■ Other procedures and measures that Exxon used to supplement bioremediation.

■ Procedures and measures used in other oil spills since the *Exxon Valdez* spill.

Gather enough information to write a one- to two-page memo to your supervisor. You will need at least three sources (make sure that they are the very latest) for each of the three areas assigned. Be sure to refer to your sources in your memo and list them at the end of the memo.

3. Choose a topic from your major that interests you and complete the following:

a. A library search for the latest information on that topic. Ultimately, you want to find two or three relevant sources on the topic. Keep notes on your search procedures: where you started, where you went next, what indexes you used, how you chose key words, and so forth. Then, in an oral presentation to your classmates, describe your library search from beginning to end. You might also include advice, warnings, and helpful hints about library searches in general based on your experience.

b. An electronic search for the latest information on that topic. Follow the instructions and assignment as described in part *a*. Compare what you found in each search. What worked best?

4. Write a paragraph on the preservation of coastal habitats using information from the sources below. Choose parts of these sources to quote, paraphrase, or summarize as you see fit. Keep in mind whether the source is stating what appears to be a fact, an inference, or an opinion. Use your own thoughts to make connections between your sources and draw your own conclusions as needed.

Source 1

Burger, Joanna. "Factors Affecting Distribution of Gulls (Larus spp.) on Two New Jersey Coastal Bays." *Environmental Conservation* 14 (1987), 59-72. Taken from p. 59:

> Gulls of the genus *Larus* are important components of coastal and marine ecosystems, and hence can constitute a factor of substantial environmental significance. Yet their relative contribution to coastal avifaunas, and the basic factors affecting their distribution along coasts, are rarely studied. Because gulls are so common along coasts, and tend to frequent areas that are used extensively by people, they are usually considered to be undisturbed by the presence of human beings . . .
>
> Marine biologists who are interested in birds usually census marine birds over the oceans, or shorebirds along the coasts. Most censuses do not report on the numbers of all species found along coasts. Shorebirds and gulls are often omitted from these seabird censuses (Burger). However, data are available from the shelf waters off the US (Powers), Gulf of San José in Argentina (Jehl *et al.*), several bays in England (Prater), Mollendo in Peru (Hughes), and Cape Town in South Africa (McLachlan *et al.*). In those studies gulls and terns together accounted for between 12 percent (Weston, England) and 94 percent (Wash, England) of the birds. In general, gulls and terns comprised 30–60 percent of the inshore avifauna (Burger).

Source 2

Thorhaug, Anitra, and Beverly Miller. "Stemming the Loss of Coastal Wetland Habitats: Jamaica as a Model for Tropical Developing Countries?" *Environmental Conservation* 13 (1986), 72–85. Taken from p. 75:

> Rapid and accelerating development of coastal-based industry, infrastructure, and urban expansion is occurring throughout the developing world, of which many nations are tropical and have extensive coastlines. Owing *inter alia* to the World Conservation Strategy (ICUN-UNEP-WWF, 1980), increased attention is being

paid by developing nations to coastal-zone management for long-term conservation of critical coastal habitats (in the tropics especially mangroves, marshes, seagrasses, and coral reefs). Tropical nations are implementing new policies and regulations to protect coastal habitats and their associated fisheries from industrial and other "development," which erodes away the natural resource-bases if left unmanaged (Fig. 1).

To overcome the ill effects of rapid industrial, urban, and tourist expansion along the 420 miles (672 km) of coastline, with its accompanying transmigration of inland population towards coastal cities, increasing demophoric pressures, and over a million tourists per year, the island nation of Jamaica has provided a model for the concept of zero-loss policy of their coastal wetlands.

Source 3

Sumeriz, Philip. "Environmental Reconciliation of Opposing Views: The New Course." *Journal of Environmental Thought* 27 (1991), 47–90. Taken from p. 47:

> Surely, those with any sense will see the importance of preserving our coastal habitats. Each species contributes to the overall ecological balance of a habitat in numerous ways that we can only begin to fathom. Directly or indirectly, each species is delicately connected to all of the other species in a habitat. If we had known what we know now twenty years ago, maybe our cause would be less critical. But we must act with what we know now.
>
> What is needed is a consolidation of the many sides of the ecological debate. On one side are the industrialists, who are concerned more with the preservation of their products than with the niches of the seagulls and seagrasses. On the other side are the ecologists, who are concerned with allowing each species to play its role in nature undisturbed. On some of the other sides of the issue are those who would seek ways to reconcile the views of the industrialists and ecologists. This should be our course.

(**Source:** Exercise and third passage created by Ronald L. Stone and used with permission of the author. First and third passages reprinted with the kind permission of Elsevier Sequoia, Lausanne, Switzerland, publishers of *Environmental Conservation*.)

5. Write a limited, descriptive abstract and an informative abstract of the "AZT Step Fact Sheet" (Appendix A, Case Documents 1). In addition to your abstracts, explain in a brief memo the relative importance or unimportance of what you stated/left out.

6. Visit a few of the Web sites for the libraries listed in Example 6-1 or others you have found, such as your own school library. How do they list their holdings? How are their searches organized? What sources outside the library's own holdings can you reach from the library page? Write a short (1–2 page) memo to your instructor discussing the strengths and shortcomings of each of these online libraries.

WORKS CITED   Curtis, Donnelyn, and Stephen A. Bernhardt. "Keywords, Titles, Abstracts, and On-line Searches: Implications for Technical Writing." *The Technical Writing Teacher* 18, no. 2 (Spring 1991), pp. 142–61.

DeLoughrey, T. J. "Software Designed to Offer Internet Users Easy Access to Documents and Graphics." *The Chronicle of Higher Education* 39, no. 44 (1993), p. A23.

Gibson, Nancy. "The Two Sides of Wolf Restoration: Recolonization and Reintroduction." *International Wolf: A Publication of the International Wolf Center* 7, no. 2 (1997), pp. 3–4.

Krol, Ed. *The Whole Internet User's Guide and Catalog.* Sebastopol, CA: O'Reilly & Associates, 1992.

Laquey, T., and J. C. Ryer. *The Internet Companion: A Beginner's Guide to Global Networking.* Reading, MA: Addison-Wesley, 1993.

Leonard, David C. "The Web, the Millennium, and the Digital Evolution of Distance Education." *Technical Communication Quarterly* 8, no. 1 (Winter 1999), pp. 9–20.

Primary Sources of Information: Surveys, Interviews, Observations

Introduction ▌ The Interpretation of Information ▌ Primary Research Tools ▌ An Important Word about Human Subjects

What is research? For most, it is a search for knowledge, for understanding, for meaning—and, in some cases, it is a search for the ability to predict, and, thus, control.

> Mary Beth Debs, "Reflexive and Reflective Tensions: Considering Research Methods from Writing-Related Fields," in Rachel Spilka, ed., *Writing in the Workplace: New Research Perspectives.* (Carbondale: Southern Illinois University Press, 1993), pp. 238–52.

Introduction

primary research
research methods designed to help you create or discover new knowledge, such as interviews, questionnaires, observations, and experiments

In Chapter 6 we discussed secondary research—how to locate and use information from existing sources. When you conduct secondary research, you locate, evaluate, organize, and interpret the work of others for some purpose. In this chapter we cover **primary research,** the research methods designed to help you create new knowledge. This overarching goal may sound straightforward, but there is a great deal of complexity involved in exploring most research questions; moreover, depending on the context in which primary research is done, different standards exist for having confidence in the results that a researcher obtains.

The first question to ask yourself in the information-gathering process is when primary research is needed. We need primary research when a review of secondary or existing published sources reveals gaps or holes in what is known. Imagine that you are asked at work to write a position paper on which multimedia hardware and software the company should purchase. You don't know if you have to do primary research to find out this answer until you have examined the secondary research to see if someone else has already found this answer. You would have to ask your boss at least two questions before you could even begin your work:

1. How much money do you want to spend on this equipment?
2. What is it going to be used for? (Training? Public relations?)

Once you get answers to these questions, you can conduct secondary research to see if anyone has come up with an answer for your specific question. You might find in your secondary research that there is lots of information on the best high-end multimedia software and hardware, but almost nothing on more modest equipment setups in the price range your boss has identified. Or you might find that many people have written about using multimedia for training but there is very little information on using the equipment as your supervisor has envisioned. When you find such gaps in the information, you will have to create or discover new knowledge yourself.

A few more examples may help reveal some of the complexity involved in a primary research effort. The purpose of such a project, for instance, might be as simple and direct as a need to learn about a new process or product, which would require you to find published reports on the topic and to interview experts. But the interpretation of the data that you find might vary, especially if your research task involves applying the original findings—obtained under a very particular set of conditions that might significantly influence the outcomes of the research—to a different set of conditions: for example, a slightly different version of the product, different kinds of product users, products employed under different conditions, or a process used within a new context.

The complexity of a research project might also be magnified by many other factors involving the audience(s) who will use the results of the project, the methodologies (i.e., ways used to gather and interpret data) employed in the process of research, or the form that the final research project takes. If, for example, your company wants to develop an advanced multimedia series for its training and documentation efforts, you might be charged with researching the available multimedia hardware and software systems, discussing their relative merits, and recommending the best system for purchase. If you are charged with writing the final report of this project for decision makers in upper management, however, you might have to find, organize, and interpret this information for an audience who is unfamiliar with such systems and their technical specifications. And, because different departments within a company may need different capabilities in such multimedia equipment, you might also need to conduct a series of initial interviews to learn both about the nature of such needs and the current status of multimedia expertise in your company. Further, if your company is struggling with an existing multimedia system that is not performing up to expectations and needs to be replaced with the new system, you might want to question users—through survey questionnaires and interviews—about the shortcomings of the current system. Or you might want to ask users in various departments to engage in a particular task that helps to identify the current product's specific limitations.

All of these activities are different types of information development, which generally combine secondary and primary research. The information you find through secondary research or create through primary research often provides you with the "good reasons" that you need to persuade your audience, as we discussed in Chapter 4. The research will help you describe the actual situations and propose the ideal situations that help you solve problems for your audience, as we discussed in Chapter 2. Although the methods of developing information are many, this chapter helps you to become familiar with three of the most common primary research methods used in the workplace—*surveys, interviews,* and *observations.*

The Interpretation of Information

Regardless of your purpose or your method for developing information, you must realize that such activities are not neutral investigations of fact, particularly when you generate new knowledge through primary research. Your activities are much more likely to be interpretive actions based on your perceptions of fact and on the context within which you are observing and using the data. That is, the research you do is likely to lead to conclusions that are not absolutely verifiable facts, but instead are educated inferences based upon careful and disciplined interpretations of things you treat as facts. Whenever you conduct an experiment, write a survey question, or interview someone, you are communicating from a particular cultural, political, or organizational point of view. Likewise, the responses you get from others through surveys, interviews, or observations reflect particular points of view. Therefore, you should always try to recognize your inherent biases and try to uncover the biases of others with whom you develop information.

You can see this challenge if we go back to the examples at the beginning of this chapter. If you interview experts in your research on the products and manufacturers of multimedia systems, you will find that these experts have values and standards, developed over time and through experience. Two experts may differ greatly on their professional opinion of which multimedia system is best. One expert may value the longevity of the system and one may value low cost; one may value advanced applications for experts and one more basic applications accessible to beginners. Further, when doing interviews that will help you determine the various multimedia capabilities different divisions in your corporation need, you may want to seek some corroborative advice on your interpretation of the results: Does a division need a certain kind of system to do a *sufficient* job? A *better* job? An *outstanding* job? What are the implications—in terms of inadequate results or cost overruns—of choosing one system or another? Finally, in surveying a whole range of individuals involved in using the system under question, you might find that their responses vary greatly depending on their specific job, responsibility, or context within the company. For instance, a multimedia system may perform flawlessly for users who speak English as their first language, but the accompanying documentation may perplex users who speak English as a second or third language.

The data that a researcher gathers are important and understandable only when they are analyzed and interpreted carefully within the complex social contexts of a workplace environment. The best researchers, therefore, are those who approach their work and the humans who participate in their projects with both respect and care. Those researchers understand that it is their responsibility to listen, observe, and learn from the subjects with whom they work.

Primary Research Tools

The rest of this chapter focuses on three types of primary research tools: surveys, interviews, and observations. Each section introduces specific research strategies that will help you develop information. Remember that most of the research tasks you are assigned will require a combination of primary and secondary research.

Throughout this chapter we will refer to the continuing case we used in Chapter 6 to illustrate secondary research techniques. To refresh your memory, let's assume that you work for Biomed Corporation, a company that produces computer software for the health care industry. You have been asked to do research on your customers' knowledge and attitudes about a new software product, Med-Ease Software. This product is designed to computerize the patient-management system of hospitals. With Med-Ease, hospitals can admit, track, and release patients; monitor their treatments, their medications, their billing; and process their insurance claims. This is a complex product, and Biomed has installed it at City Hospital in a pilot testing program designed to verify the performance of the software. If the product works well for three months in City Hospital, it will be released to the market.

You have been asked to begin a **usability study** of Med-Ease Software. You are told that the first step will be to develop information about the users' responses to the new product: How easy is it to use? How well do they like it? Are there differences in attitude among different types of users?

usability study
an investigation of how well a product meets its users' needs; conducted on-site or in the laboratory

SURVEYS

If you need to find out what certain people know about a process, product, or concept, or how those people think or feel about those things, you may be able to develop this type of information through **survey questionnaires.**

A survey is not the only method you might choose to use, but certain constraints of your research task might suggest the use of surveys as the most effective tool for a particular purpose. To discover the answers to the questions you have asked in your Biomed-City Hospital task—for instance, How easy is Med-Ease Software to use? How well do users like it? Are there differences in attitude among different types of users?—you could employ several different research methods:

survey questionnaires
a primary research tool, often called a survey, that asks sources to complete a question-answer form; often conducted through the mail or over the telephone

- Interviewing selected users personally.
- Observing users at work on the system.
- Surveying users about their uses of the system.

All three of these methods could yield interesting information that would help you and others improve your company's products—and sections of this chapter explain all of these methods. But here we are going to assume that you want to get data from a population too large to interview individually.

Although you may have time to interview or observe a handful of users, you need to find a way to question a large number of users—enough users so that their responses will be representative of all users at City Hospital. The best way to do this is to develop and administer a printed questionnaire to survey the users. Example 7–1 provides an overview of this task.

Here's a note of caution before you go out and start gathering data. Before beginning any research involving human subjects, be sure to read about human subjects guidelines that we have included before the Summary to this chapter. If you are working at a university or college, you will need to follow its human subjects guidelines and may have to provide the human subjects committee at your institution with a list of your survey questions and get written approval to proceed. Your company may have similar research guidelines, and you will need to make yourself aware of them before you start working on this project. If you have to obtain written approval, don't forget to figure the time this will take into your planning process.

EXAMPLE 7–1   Survey Research for Biomed Corporation
Med-Ease Software Use at City Hospital

Research Questions

How easy is Med-Ease to use?
How well do users like it?
Are there differences in attitude among different types of users?

Research Goal

To ask these questions of a representative sample of the entire population of users at City Hospital

Research Method

Survey questionnaire

Choosing a Design

longitudinal survey
a primary research tool that allows the researcher to ask questions of the same or similar group over a period of time

cross-sectional survey
a primary research tool that allows the researcher to ask questions of members of a heterogeneous group about current phenomena

The first step in developing your survey is to decide its general design: will it be a **longitudinal survey** or a **cross-sectional survey**? A longitudinal survey allows you to ask questions of the same or similar group over a period of time. For example, you could survey Med-Ease users after they have used the product for one month, two months, and three months to get some idea of their attitudes toward the product over time. Or you could get some idea of the learning curve associated with the product by asking users to report how long it took them to perform the same series of tasks on each of the three administrations of the survey over a period of three months.

When you use a cross-sectional survey, you ask a set of questions only once because you want to get a current picture of the phenomena under study. Imagine a cross-sectional survey to be a snapshot of the target group you are studying. For the City Hospital survey, you decide that you need to conduct a cross-sectional survey because your company needs to learn about the current state of Med-Ease and its use (see Example 7–2).

Choosing a Sample

Once you decide the general design, you need to focus on the people you will survey. In some cases, you may be able to survey every single person that comprises the group you want to study. If you have enough time, money, and personnel to devote to this research project, you may be able to survey every single one of the 350 Med-Ease users at City Hospital. But, given realistic constraints on time, money, and personnel, such a move could also be impractical. If you were doing an even larger project—for example, a survey of all 6,900 Med-Ease users in New York City or the 23,000 Med-Ease users in the southeastern United States—surveying each user

---

EXAMPLE 7–2   Survey Design for Study of Med-Ease
                  Software Use at City Hospital

Design Options

   1. Longitudinal study of users from beginning to the end of their training
   2. Cross-sectional study of users near the end of their training

Choice

Cross-sectional study

Rationale

Because management wants to evaluate user attitudes and knowledge of the product and not the progress of their training, the one-time cross-sectional survey near the end of their experience was chosen.

---

might be even less practical. Clearly, for many projects, the most effective and efficient way to gather information is to survey a **representative sample** of a more **general population.**

   To make sure that your sample is representative in the most important way, you might want to identify the representative sample so that it contains the same proportion of employees from each department within City Hospital as does the general population. If you were trying to determine how Med-Ease was employed by various kinds of health care workers (nurses, physicians, emergency care professionals) in the general population, you might make sure your sample contained representative proportions of nurses, doctors, and emergency care professionals. If you wanted to ascertain how easy Med-Ease was to learn for employees of differing experience levels, you might structure your representative sample so that it contained the same proportion of experienced and novice employees as your general population. Some studies, of course, will require making a sample representative for several different factors. Collecting statistics about a representative sample in the ways we describe in this chapter often gives a researcher more confidence in generalizing findings to a larger population, although a good researcher always makes generalizations with great care and caution, recognizing that even slight variations in the contexts, conditions, or environments from which data are collected can produce great changes in findings.

   There may also be times when you want to survey only certain subgroups. If you are interested only in the uses that obstetrics workers make of Med-Ease, for example, you may wish to question a representative sampling of both staff, nurses, and physicians attached to the obstetrics department of City Hospital. If you are interested in determining how long various groups of nonmedical employees take to learn Med-Ease (with the aim of targeting the training that might be needed for some groups of employees),

representative sample
a selection of some of the people in the general population of those involved in the subject or topic being researched

general population
all of the people involved in the subject or topic being researched

you might choose representative sample groups from the insurance, billing, facilities management, and supplies departments. Focusing your research question in different ways, in other words, will mean that you need to focus survey questions in different ways and to involve different groups of people in answering a survey. Whether you can survey every single person in a subgroup depends on the time you have, the amount of money you can spend producing and distributing the survey, and the availability of everyone in the subgroup, among many other factors.

Although it's not the purpose of this book to explain the subtleties of probability sampling, you should be aware that you can achieve a representative sample by *randomly* sampling a relatively small group of respondents. For example, public opinion polls of presidential races are usually based upon a sample of 500 to 1,000 respondents, chosen randomly from a variety of regions across the country. In other words, gathering the opinions of only 1,000 randomly chosen citizens allows a pollster to predict—with a relatively high degree of confidence and accuracy—the opinions of the general population of potential voters across the nation! Only some research projects and some kinds of information and predictive tasks, however, are amenable to such random sampling techniques. You will want to consult one or more of the statistical research methods books listed at the end of this chapter or a specialist in research methods for additional information.

You can choose respondents randomly several ways. If you are conducting a survey of a relatively small population, the simplest way to get a random sample is to assign each potential respondent a number and then consult a random number table (which can be found in most statistics textbooks) or run a random number generation program on a computer. If, for example, the first random number is 57, then you choose person number 57 to be included in your sample. You continue this process until you have chosen the number of potential respondents you wish to have.

In the City Hospital project, for example, you might discover that although there are 1,550 employees, only 950 use Med-Ease in their work. Of those 950, 50 are managers who regularly receive reports generated by the system but who do not actually use the system to do their jobs. The remaining 900 employees are spread evenly across five different departments at the Hospital. You decide to choose 150 employees randomly from those five departments (see Example 7–3).

This does not mean, however, that you can necessarily count on 150 users to respond to your survey. In large surveys, often only 10 percent of those who receive a questionnaire actually fill it out and return it. It is difficult to generalize with confidence or accuracy when you get a small return rate, and you might need to encourage the people surveyed to respond by offering to send them the results of your survey or by stressing the importance of their feedback. You might also want to send a series of follow-up letters prompting employees to return their surveys—outlining

---

EXAMPLE 7–3   Survey Sample for Study of Med-Ease
                  Software Use at City Hospital

Total Population

900 nonmanagerial users evenly spread across five different departments

Sample

30 users from each department for a total of 150 potential respondents

Method Chosen

Random number generator to pick respondents from numbered list of users

---

the benefits that the hospital and employees will reap from this research project. After follow-up reminders and prompts in a closed population like City Hospital, you may still achieve only a 75 percent response rate. In any case, you must choose your sample knowing that only a percentage will respond.

Finally, you must consider how you will distribute your survey to your sample. If you send questionnaires through the mail, you must anticipate the costs involved in sending out a large number of surveys. Even if there are no mailing costs, you must consider the time, expense, and effort involved in copying and preparing a large number of surveys, coding and analyzing the response data you get in return, and interpreting the findings for the readers of your final research document.

Determining the Form of Questions and Answers

The questions you ask on a survey can take a variety of forms. You can allow your respondents to write anything they wish to write—this is called an **open question**—or you can write questions that force respondents to answer in specific ways—these are **closed questions**.

open questions
a survey questionnaire that allows your respondents to write anything they wish

closed questions
a survey questionnaire that forces respondents to answer in specific ways

Open Questions   The *advantages* of using open questions are that they are easier to write and, because you are not determining the answer, you may be able to get responses that accurately reflect the respondents' thinking. Ask open-ended questions when you want to learn about things that you cannot yet imagine. For example, in the City Hospital situation, if a researcher could not anticipate *all* of the potential problems with Med-Ease that respondents in the five departments could have encountered in the workplace, he or she might ask an open-ended question, like that in Example 7–4, at the end of a survey.

The answers to such a question may provide a researcher insights that closed questions would not have allowed.

EXAMPLE 7–4   Open-Ended Question

Example

Are there any other work-related problems that you would like to identify? Please explain in the space below.

The *disadvantages* of working with open questions are that:

1. It is time consuming to have to read the answers.
2. The validity of those answers is dependent on how articulate the respondents can be.

The latter point can be crucial. If your respondents use vague, unclear language, you may not be able to verify their meanings. Accordingly, if a small minority of the respondents are quite articulate, then you run the risk of weighing more heavily the opinions of that minority without a clear indication of how the majority feels about the topic. However, you can do a **content analysis** of the responses to all open-ended questions and then *code* these responses to find out how many people responded in certain ways.

content analysis
a particular research approach that allows a researcher to determine the relative frequency of themes or actions in a collection of texts by counting the number of times they occur

A content analysis is a particular research approach that allows you to determine the relative frequency of themes or actions in a collection of texts by counting the number of times they occur. Content analysis is often performed to determine what's going on in the mass media—for instance, the amount of violence in children's cartoons. To do this kind of analysis, you must develop a coding system that is appropriate for the data that you are gathering. For the open-ended question in Example 7–4, for instance, a researcher might develop a coding system that categorizes or codes problems according to their type: problems with documentation, problems with hardware, problems with software, problems with training. As Example 7–5 indicates, each kind of problem would then be assigned a preliminary code.

To check the validity and comprehensiveness of this preliminary coding system, the researcher would have to check it against all of the problems identified by the survey respondents: adding, deleting, or modifying the coding system until it accurately and fully accommodated all the responses that were provided. After the coding schema is finalized, it can be applied systematically to every problem identified by the respondents for the open-ended question on the survey. The codes can later be tabulated and analyzed for patterns or trends: What is the most frequent kind of problem? What kinds of problems did nurses mention most frequently? Doctors? New employees? Content analysis works only when the code is clearly spelled out, and two different people looking at the same example can get the same results.

EXAMPLE 7–5   Coding System for Content Analysis of Open-Ended Question on Med-Ease Software Use at City Hospital

Code	Problem Description	Examples
DP	Problems with documentation	Finding definitions of unfamiliar terms, understanding procedural directions, troubleshooting specific problems
HP	Problems with hardware	Using pressure pen, adjusting monitor height, using control adjustments
SP	Problems with software	Data entry, data reporting, spreadsheet work
TP	Problems with training	Timing of training sessions, competency of trainer, appropriateness of training

EXAMPLE 7–6   Closed Question

Example

Do you use the multitasking function when using Med-Ease?
Circle one of the choices below:

Yes   No

Closed Questions   Closed questions can take many forms. Some of the most commonly used are *yes/no questions, check-offs,* and *scaled questions.* Regardless of the type of question, it's important to tell the respondent how to reply—such as circle a response or mark all (or a defined number of) choices that apply. **Yes/no questions** allow the respondent to answer only yes or no to your question, as indicated in Example 7-6.

Yes/no questions are useful when you want to force respondents to make a choice, but they allow the researcher and the respondent a very limited level of response. Yes/no questions can frustrate users who do not feel that the options presented provide for an adequate level of freedom or complexity in response.

Check-off questions (see Example 7-7) present the respondent with a list of options from which to choose. You may ask a respondent to choose one or more items from a list by placing a check next to each choice. Check-off questions are useful when a researcher is sure of all of the possible choices respondents will have to choose from, but cannot anticipate their exact replies.

yes/no questions
a survey questionnaire that allows the respondent to answer only yes or no to a question

check-off questions
a survey questionnaire that presents a list of options from which the respondent must choose

---

EXAMPLE 7–7   Check-Off Questions

Example

Which of the following Med-Ease features have you used to do your work? Check all features you have used.

_____ Spread sheet

_____ Word processor

_____ Database manager

_____ Multitasking functions

_____ CD ROM

_____ Online help screens

EXAMPLE 7–8   Nominal Scaled Questions

Example

What is your job category?

_____ Nonmedical staff

_____ Medical staff

_____ Teaching faculty

_____ Administrator

_____ Clerical

_____ Other: _____

---

scaled questions
closed questions that ask respondents to place their answers in a list of categories or to rate them on a continuum of possibilities

nominal scale questions
closed questions that ask respondents to place their responses in one of a number of categories provided for them

demographic information
the vital statistics of the general population which can be easily collected with closed questions and nominal scales

ordinal scale questions
closed questions that ask respondents to place their answers in rank order

Scaled questions ask respondents either to place their answers in a list of categories or to rate their response on a continuum of possibilities. There are three common types of scaled questions:

1. *Nominal*   When you ask respondents to identify a category, you are using a nominal scale question. **Nominal scale questions** are best used when you want to identify your respondents in very specific ways. **Demographic information**, the vital statistics of general populations, can be easily collected with closed questions and nominal scales. Common nominal scale questions ask respondents to identify their sex, religious affiliation, level of education, employment and such. For our City Hospital survey, a nominal question might look like the one in Example 7–8.

2. *Ordinal*   When you ask respondents to place their answer in a rank order, you are using an **ordinal scale question.** Ordinal scales (Example 7–9) are useful in identifying when respondents feel strongly one way or another about a response. The questions in Example 7–9

EXAMPLE 7–9   Ordinal Scaled Questions

Example

At the beginning of the training period for using Med-Ease, did you consider yourself:
(circle one)

Highly motivated   Moderately motivated   Poorly motivated

Example

Identify the extent to which you agree or disagree with the following statement:
(circle one)

I am motivated to learn the features of Med-Ease.

Strongly agree   Agree   No opinion   Disagree   Strongly disagree

---

EXAMPLE 7–10   Interval Scale Questions

Example

What is your departmental traffic per month?
(check one)

_____ Under 10,000 patients/transactions

_____ 10,000 to 19,999 patients/transactions

_____ 20,000 to 29,999 patients/transactions

_____ 30,000 to 39,999 patients/transactions

_____ 40,000 to 49,999 patients/transactions

_____ 50,000 patients/transactions and above

---

show how you might use an ordinal scale to get information on a respondent's motivation for using Med-Ease.

3. ***Interval***   When you ask respondents to provide answers that are intrinsically numeric, like age or income, you are using **interval scale questions.** Using an interval scale question allows you to place responses in categories, when specific statistics are not necessary or when respondents are reluctant to offer specific information because of potential embarrassment or concern. Example 7–10 indicates how an interval scale question might be framed for the Med-Ease case.

interval scale questions closed questions that ask respondents to provide answers that are intrinsically numeric, such as age or income level

Another type of interval scale question (Example 7–11) asks respondents to count things that they do. Prompted by an interval scale question such as in Example 7–11, City Hospital employees who do not remember whether they save Med-Ease files three, four, or five times a day can still give researchers useful information.

EXAMPLE 7–11   Additional Interval Scale Questions

Example

How many times a day do you save the files that you are working on in the Med-Ease system? (check one)

_____ 0

_____ 1 to 5

_____ 6 to 10

_____ 11 to 15

_____ 16 to 20

_____ More than 20

What types of questions would be most useful for the City Hospital survey? The answer to this question depends on the specific focus and purpose of your research project. For the purposes of providing an example, go back to the specific questions with which we began this research section (Example 7–1). You want to gauge the attitudes of the users toward the new Med-Ease system, and you want to identify how different types of users like the system. To begin answering these questions, you might decide to use a combination of nominal and ordinal scale questions. The nominal scale questions will enable you to establish a **profile of the users** by asking them to identify factors including their age, years with the company, sex, education, and job function. The ordinal scale questions will give you insight into their attitudes. Finally, you might decide to add an open-ended question at the end of the survey to allow respondents an opportunity to bring up issues not addressed by the closed questions. Example 7–12 provides an overview of the question types you might choose to use for the Med-Ease project.

profile of users
includes the identify factors gained from nominal questions

Constructing Questions

Writing effective questions for surveys is often the most important and most difficult part of the entire enterprise. Your goal is to write *valid* and *reliable* questions.

valid questions
survey questions that accurately ask what the researcher intends, without any misunderstanding between the researcher and the respondent

Validity and Reliability   **Valid questions** accurately ask what you intend them to ask. In constructing such questions, the purpose is to bring the intent of the author and the interpretation of the reader into the closest possible congruence; of course, in practice this may not prove so easy. If you write a question that you believe asks one thing and your survey respondents believe the question asks something entirely different, you have not written a valid question. For example, if you asked respondents if the Med-Ease software made their job "easier," you might imagine that the software

EXAMPLE 7–12   Question Types for Study of Med-Ease
Software Use at City Hospital

Nominal questions for developing user profile, for example:
Please circle the correct answer.

   1.   You are:

   Male   Female

   18–25 years old   26–45 years old   46 years or older

   2.   Your highest level of education?

   High school graduate
   Associate's degree in college
   Bachelor's degree
   Master's degree
   Doctorate

   3.   Your job description is:

   Physician
   Nurse
   Paramedic
   Aide
   Nonmedical staff

Ordinal questions on attitudes, for example:
Please circle the response that mosts closely reflects your opinion about Med-Ease.

   4.   The menu system enables me to move to the functions that I need to do my job.

   Strongly agree   Agree   No opinion   Disagree   Strongly disagree

   5.   The help screens can be found quickly.

   Strongly agree   Agree   No opinion   Disagree   Strongly disagree

   6.   The help screens do not provide enough information to solve my problems.

   Strongly agree   Agree   No opinion   Disagree   Strongly disagree

Open-ended question:
Is there anything else we should know about the new Med-Ease system? Please write your response in the space below:

could enable these users to complete tasks more quickly. The respondents, however, might interpret "easier" to mean completing the tasks with fewer mistakes or with fewer bureaucratic steps.

   Reliable questions ask the same question every time, for every respondent—or as close to this ideal as humanly possible. If you write a question that appears to change its meaning over time, then it is not reliable.

reliable questions
survey questions that ask the same question of each interviewee without changing its meaning or form

For example, you would have asked unreliable questions if at the beginning of the survey you asked how the Med-Ease software enabled users to complete "tasks" and meant data entry but later in the survey "task" meant billing patients.

Although it is difficult to write perfectly valid, reliable questions, you can approach this goal by attending to the suggestions in Writing Strategies 7–1.

WRITING STRATEGIES 7–1   Five Strategies for Asking Valid and Reliable Questions

Writing Tip 1: Use specific language.

Use clear, concrete language and define terms. The more specific your language, the greater the chances that respondents will interpret a question as you hope they will.

Example of overly general language:
Is the multitasking feature *good*?

    Always   Sometimes   Never

Example of more specific language:
Does the multitasking feature enable you to complete more tasks in a work day than you could without multitasking?

    Always   Sometimes   Never

Writing Tip 2: Define technical terms.

Define any terms that might be unclear to your respondents.

Example of undefined term:
Does the multitasking feature enable you to complete more tasks in a work day than you could without multitasking?

    Always   Sometimes   Never

Example of defined term:
Do you prefer *multitasking,* a system that allows you to run more than one program at a time (e.g., make printouts while you continue to use the spreadsheet program), to the system you have used in the past?

    Always   Sometimes   Never

Writing Tip 3: Recognize and eliminate biases.

Recognize your biases and try to limit them. It does you no good to write a question that pushes respondents toward an answer that you may like them to give. Stating a question with the least possible amount of bias in the desired direction may require stating it in the negative.

Example of biased question:
Wouldn't you agree that our system, with its multitasking capabilities, is a step beyond the competitors'?

    Always   Sometimes   Never

(continued)

WRITING STRATEGIES 7–1 (Continued)

Example of less biased question:
Does the multitasking feature allow you to get more work done in less time?

Always   Sometimes   Never

Example of unbiased question stated in the negative:
Does multitasking impede your work?

Always   Sometimes   Never

Writing Tip 4: Limit questions to one idea.

Be sure to ask only one question at a time.

Example of question with dual focus:
Do the multitasking and multimedia features impede your work?

Always   Sometimes   Never

Example of question with a single focus:
Does the multimedia feature impede your work?

Always   Sometimes   Never

Writing Tip 5: Avoid untrue absolutes.

Be sure that if you ask your respondents to choose an absolute—yes/no, always/never, and so forth—that these absolutes are "true" to the given situation.

Example of question with untrue absolute:
Does the multitasking feature enable you to complete more tasks in a work day than you could without multitasking?

Always   Sometimes   Never

Example of question that eliminates untrue absolute:
The multitasking feature allows me to complete more tasks in a work day than I could before multitasking.

Strongly agree   Agree   Neutral   Disagree   Strongly Disagree

Presenting Questions

Once a researcher begins to write questions, he or she will also need to decide what form the questionnaire should take. The most common form is the printed, **self-administered questionnaire** sent to respondents through the mail. Another common form is the **telephone survey**. Using this technique, a caller asks the questions of the respondent and records the answers by marking an answer sheet. When you are developing a printed, self-administered questionnaire, you must pay attention to the format of the document. When administering the questionnaire by telephone, you must consider both your verbal presentation as interviewer and the impatience most respondents feel with telephone surveys in general.

self-administered questionnaire
printed survey sent through the mail

telephone survey
survey in which a caller asks questions over the telephone and records answers by marking an answer sheet

Self-administered Questionnaire Format Because researchers want respondents to be motivated enough to fill out any questionnaire that they design, they generally attempt to make such documents as brief as possible. At the same time, however, questionnaires need to be easy to read. Therefore, the format of a questionnaire can affect a respondent's willingness to answer questions. Will respondents be less likely to fill out the survey if they see that it extends for three, four, five, or more pages? Maybe. Will they be less likely to fill it out if it appears to be one page of small type and dense copy? Perhaps. Try to make the form both brief *and* readable.

Usually, researchers provide brief but clear *instructions* at the top of the page for respondents. Sometimes researchers may also may want to provide a brief *rationale,* an explanation of why the respondent is being asked to fill out the survey.

Finally, researchers must consider the order in which they present the questions on the survey sheet. Although there are no hard-and-fast rules on this matter, researchers may wish to present simpler questions first, followed by questions that may take more thought and reflection. The reasoning behind this pattern is simple: if respondents are hit immediately with challenging or difficult questions, they may be less inclined to complete the survey. But if a survey is easy to start, then the respondents build a kind of momentum that may carry them through to the end of the questionnaire.

For the City Hospital survey, then, you might decide to begin with a brief rationale, followed by simple instructions. Begin with the profile questions, followed by the attitude questions, and end with the open-ended question (see Example 7–13).

EXAMPLE 7–13   Self-Administered Survey Format
Survey Format for Study of Med-Ease
Software Use at City Hospital

Rationale

The Biomed Corporation needs to learn more about your attitudes toward the new Med-Ease system, so that it can determine whether or not to continue with the testing and adoption of this package. Please take a few moments to fill out this questionnaire.

Instructions

For each question, please circle the best answer.

Order of Questions

  1. Profile questions
  2. Attitude questions
  3. Open-ended question

Telephone Survey   An alternative to sending printed forms to respondents and hoping they send them back is for you to conduct a survey interview. The advantage of this method is that researchers can achieve a higher response rate. It's generally much harder to hang up on a person conducting a short telephone survey than it is to throw away, often unopened, a questionnaire. However, the recent increase of telemarketing schemes may make people impatient with telephone surveys that do not have a clearly defined purpose, that lack clear benefits for the respondent, or that require time after business hours. The primary disadvantage of telephone surveys is that they do not allow researchers to reach as many potential respondents as they could through the mail. Although it is feasible to send 250 survey forms through the mail, it may be far too time consuming to attempt to call 250 respondents on the phone. Of course, if a researcher can assemble a team of interviewers, a phone survey may be the way to go, especially if all of the interviewers can be trained to conduct the survey in a similar fashion. Training all interviewers to ask the same questions and to record responses in the same way adds to the reliability and validity of a survey. But whether a researcher chooses to do all interviews personally or assign them to a team of callers, the guidelines in Writing Strategies 7–2 will be useful.

In the City Hospital case, it is far more feasible to send a printed questionnaire through the hospital's internal mail system. A telephone survey is unnecessary.

WRITING STRATEGIES 7–2   Tips for Telephone Interviews

Interview Tip 1: Make the calls at the most opportune time for respondents, not for yourself.

Explanation: If you are calling respondents at home, you don't want to catch them when they are busy. For example, if you are expecting that your respondents are at work during the day, then call in the evening—but you do not want to interrupt their evening meals nor do you want to disturb respondents with calls after 9:00 P.M. Or, if you are calling respondents at work, try to call early in the day, when they are not in the middle of lengthy tasks or rushing to lunch or home.

Interview Tip 2: Identify yourself and your purpose immediately.

Explanation: Many potential respondents will be initially suspicious of your call. Many will wonder if you are selling something and be tempted to hang up quickly. However, given the opportunity, many people are quite willing to express their opinions publicly. Therefore, you need to have a crisp, brief opening that makes it clear that you are not selling, but doing research on a particular subject. If you are calling respondents at work, be sure to give them an accurate estimate of how long the survey will take and give them the opportunity to reschedule speaking to you at a time when it's more convenient for them.

(continued)

WRITING STRATEGIES 7–2 (Continued)

Interview Tip 3: Use a polite, businesslike tone.

Explanation: A professional, conversational tone will be the least threatening. When introducing yourself, your purpose, and the survey, do not read from a prewritten text. If your pitch sounds canned, it comes off as less personal, and some people will be less likely to respond. At the same time, you want to be sure to say the same thing to every respondent; therefore, you should have a prearranged introduction that you know well enough not to read. When you get to the questions, however, you will most likely need to read them. This is expected and acceptable to most people.

Interview Tip 4: Make the questions as brief as possible and answers easy to choose.

Explanation: Because the phone survey respondent cannot reread a question and the range of possible answers, you need to write questions and answers a listener can easily grasp. For example, because it may be too cumbersome to expect respondents to choose from five possible answers, such as "strongly agree, agree, no opinion, disagree, strongly disagree," you may want to limit the possible answers to "agree, disagree, no opinion."

Interview Tip 5: Decide how you are going to handle alternative answers.

Explanation: Many respondents may not want to be forced into any limited answer scheme. They may want to explain their opinions in some complexity. You must decide beforehand how you plan to handle that type of response. Do you throw out these types of responses, not counting them as valid? Do you listen for something that fits your answer scheme and then say, for example, "So you are saying that you disagree?"—hoping that the respondent will verify your interpretation? You can decide how you will handle this issue, but the key is for you to be consistent across every interview.

Pilot Testing/User Testing

pilot surveys
preliminary versions of a survey sent to a small, representative sample (about 3 percent) to test a survey's usefulness and validity

To increase the chances of getting useful and valid results from their survey, researchers often run a **pilot survey** to test their questionnaire's efficacy. Trying a survey out on a small, but representative, sample (perhaps about 3 percent of the respondents to be surveyed) will help to test its performance before a researcher goes to the expense of reproducing and administering the survey on a large scale. A pilot study can also help researchers refine survey questions, whether they are in written or interview form.

Determining the Results

Once a researcher has piloted a survey, refined it, and administered it to the full sample, he or she will need to organize the resulting data for a final report. With most closed-question surveys, the primary job of organizing involves tabulating the quantifiable results. Simply put, the answers are counted up: the number of respondents, for example, who answered "strongly agree" for a specific question, the number of respondents who answered "agree," and so forth. Once a researcher has counted all of the answers, he or she can tabulate the **raw scores** of the survey.

Although the raw scores may tell you something useful or insightful, usually your goal is to develop some sort of **statistical analysis** of those raw scores. Such analyses can provide even more insight into the issues you are investigating. Two types of statistical analyses can be undertaken in survey research: *descriptive* and *inferential statistics.*

Descriptive statistics provide manageable ways of presenting quantitative descriptions. There are a variety of ways to gather descriptive statistics depending on whether you want to describe single variables or the ways variables are connected. One of the most familiar kinds of descriptive statistics involves **data reduction.** If you sent out a survey with 100 questions to 200 people, you would have 20,000 answers with which to deal. Data reduction allows you to reduce mountains of data to manageable summaries.

Inferential statistics allow you to handle data in a different way. Often the research you do will involve a sample drawn from a larger population. You aren't particularly interested in the sample, but you do want to know what the sample tells you about that larger population. A number of statistical techniques allow you to make such inferences. These techniques allow you to determine the **statistical significance** of your findings. Any type of statistical analysis can become very complicated, and inferential statistics are particularly complex. For this reason, we refer readers who want to explore statistical significance further and in depth to the handbooks of research methods that we list at the end of this chapter. In the paragraphs that follow, we discuss only the simplest descriptive statistic: **frequency.**

Frequency is merely the percentage of responses to a given question. If, for example, 20 out of 50 respondents answer "strongly agree" to a particular statement and another 20 answer "agree," that means that 80 percent of the respondents agree or strongly agree with the statement.

See Example 7–14 for some of the raw scores from the survey at City Hospital. To practice tabulating other descriptive statistics, try Exercise 1 at the end of this chapter.

INTERVIEWS

Interviews are the most commonly used research method in the workplace. From informal conversations around the coffee maker to formal depositions in the presence of lawyers, more people gain information through interviews than through any other research activity. Oftentimes in the workplace,

raw scores
final count of all answers gathered from a survey

statistical analysis
a mathematical interpretation of the raw scores of a survey

descriptive statistics
a type of statistical analysis used to describe single variables or connections between variables

data reduction
process that allows a researcher to reduce raw scores into manageable summaries

inferential statistics
a type of statistical analysis used to determine what a sample indicates about a larger population

statistical significance
result of inferential or descriptive statistical analysis

frequency
percentage of responses to a given question; the simplest descriptive statistic

interview
a primary research tool, commonly used in the workplace

EXAMPLE 7–14   Raw Scores: Selected Survey Results of Study
of Med-Ease Software Use at City Hospital

Total number of respondents: 102 (of 150 surveyed)
Please circle the correct answer.

 1.   You are:

Male = 44   Female = 58
18–25 years old = 25
26–45 years old = 30
46 or older = 47

 2.   Your highest level of education?
High school graduate = 66
Associate's degree in college = 15
Bachelor's degree = 10
Master's degree = 10
Doctorate = 1

Ordinal questions on attitudes
Please circle the response that most closely reflects your opinion about Med-Ease.

 3.   The menu system enables me to move to the functions that I need to do
my job.

Strongly agree = 27
Agree = 15
No opinion = 10
Disagree = 40
Strongly disagree = 10

 4.   The help screens can be found quickly.

Strongly agree = 7
Agree = 13
No opinion = 10
Disagree = 60
Strongly disagree = 12

 5.   The help screens do not provide enough information to solve my problems.

Strongly agree = 7
Agree = 15
No opinion = 8
Disagree = 35
Strongly disagree = 37

researchers can gather information from other people quickly and informally, through a brief phone call or over e-mail. However, in this section of the chapter, we introduce readers to the most challenging situation: formal interview strategies. Some of these strategies will prove useful in more informal situations as well. Interviews differ from surveys in that the primary research instrument is not a questionnaire but the interviewer. Therefore, interviews are primarily a function of interpersonal communication.

Although both interviews and surveys are based upon question and answer strategies, there are two key differences between the methods. First, surveys are designed to be delivered to a large number of potential respondents, whereas the time demands associated with oral interviews usually allow for far fewer respondents. Second, most surveys employ closed questions, thus limiting the respondents' possible answers. On the other hand, oral interviews are ideal for probing, follow-up questions, allowing for lengthy, possibly unanticipated answers. Interviewees, in other words, are often allowed far more freedom in constructing their answers and therefore may provide unexpected insights. And because interviews enable researchers to begin investigations before knowing all the parameters of an issue, they are often employed to gather information before a survey or questionnaire is used.

This is not to say that all interviews are completely unstructured; in fact, few are. Generally, there are two types of interviews: *open* and *directed*. **Open interviews,** in theory, are much less structured, meaning that the interviewer imposes few limits on an interview. In reality, open interviews are often based upon one or more starter questions that give the interviewer a general focus. Then the interviewer allows the interview to flow more freely according to the responses of the interviewee.

Directed interviews are the counterpart to self-administered survey questionnaires. The interviewer brings a set of specific questions and attempts to stick to those questions regardless of the responses of the interviewees.

More often than not the best form of interview *combines* features of the open and directed interviews. These interviews impose a structure on the interviewee but allow for the interview to move into unanticipated but relevant areas. This process attempts to use time efficiently while allowing for unexpected discoveries.

In the City Hospital investigation, for example, you might decide that you need to supplement your survey with several in-depth, directed interviews with users in each department. These interviews would focus on the users' attitudes about the new system but still allow them to raise issues that the survey would only allow in a limited way through the one open-ended question at the end of the survey. Because an interview allows for follow-up questions that can probe a particular issue, you may be able to investigate key points in depth. When you combine this depth with the breadth of issues you cover in the survey, you can gain an increasingly complex view of the attitudes you are investigating. Example 7–15 provides an overview of possible interview goals at City Hospital.

open interview
unstructured interview in which the interviewer imposes no limits on the questions asked or on the responses of the person being interviewed

directed interview
structured interview in which the interviewer asks a set of specific questions; counterpart of the self-administered survey questionnaire

---

EXAMPLE 7–15   Interview Research for Biomed Corporation
Med-Ease Software Use at City Hospital

Research Questions

How easy is Med-Ease to use?
How well do users like it?

Research Goal

To explore in-depth the attitudes of two users in each of the five departments

Research Method

Interviews

Planning for the Interview

Before actually sitting down with interviewees, a careful researcher prepares by scheduling the interview, determining the site, choosing a recording method, and writing a list of thoughtful, precise questions. Additionally, before beginning any interview research don't forget to obtain the approval of your human subjects review board.

Scheduling the Interview

If researchers are conducting interviews in a workplace setting, usually the interview will interfere with the normal work patterns of the interviewees. Generally, a researcher should make it clear to the interviewee that he or she would like to schedule the interview at the most convenient possible time for the respondent. Because this request may be the first contact with a potential interviewee, smart researchers set a tone that encourages cooperation.

At the same time, researchers should think about scheduling a time when the interviewee is reasonably rested and alert. Interviews right before the workday ends may not yield the best responses from the interviewee who may be watching the clock and reticent to provide in-depth information because of the time constraint; typically, the earlier the time in a person's workday the better.

Researchers should be sure to inform the interviewee of the expected length of the interview. Interviews that last between 15 and 45 minutes are typical. Any interview over an hour is pushing the endurance of both researcher and interviewee, both of whom need to remain alert and focused throughout the interview.

Determining the Site

The ideal site for an interview is both convenient for the interviewee and reasonably free from distractions or interruptions. If possible, researchers should do some legwork beforehand to inspect the best interview site. The worst thing that could happen is to agree to a site—for example, the interviewee's workplace—where the interviewee is continually interrupted with telephone calls and visits by co-workers. Consider conducting the interview in a conference room or even in a quiet nearby coffee shop.

Choosing a Recording Method

A careful, ethical, and considerate researcher always asks the interviewee beforehand whether it is permissible to use an electronic recording device. Typically, interviews are tape recorded. Occasionally, they may be videotaped, even though it is the rare interview that demands a visual record. Furthermore, video cameras can be intimidating to an interviewee, so only use this technology when there is a compelling reason.

If tape recording is allowed, be sure that to test *all* aspects of the equipment. Have a reliable power source, mirophones, extension cords, if necessary, and more than enough tapes. Do a voice check to ascertain the right sequence of buttons and the appropriate volume. Arrange the microphone and tape recorder so that it clearly records the necessary voices, but do not place it so prominently that it intimidates the interviewee. Few people like to have a microphone stuck in their faces. It's always a good idea to take notes although the interview is recorded; even the best recorders malfunction at one time or another. In addition, a researcher may not have the time and resources to transcribe his or her tapes completely, so notes may be a useful adjunct to a tape recording.

If an interviewee does not want to be tape recorded, then rely on handwritten notes. In such a case, knowing shorthand is helpful. It is also a good idea to have questions written on a separate sheet or notepad. The less flipping of pages that must be done the better. If you must rely on notes, then transcribe and expand them immediately after the interview, if possible, while the interviewee's comments are still fresh in your mind and you are best able to recall what your written notes (which may be filled with notations and abbreviations) mean.

Writing Questions

Most of the questions that a good researcher asks in an interview are designed to allow interviewees to speak at length—they are not *closed-ended* questions that require only a simple limited answer. If only one word is needed—if yes/no type answers will do for the entire interview—then a questionnaire may be a preferable method. Given this characteristic of interviews, most interview questions will be open-ended to some extent. For example, if you wanted to ask employee Jane Smith at City Hospital what her general attitude was about the new Med-Ease software, you might want to avoid a closed-ended question like, "Do you like the new software?" This question merely asks someone to answer "yes," "no," or "maybe." You open up a potentially more fruitful channel by asking "What do you like about the new software? What do you dislike about it?" The interviewee's response may give you several issues to probe in more depth.

Often, researchers will go into a more open, unstructured interview not with specifically written out questions, but with a list of topics that they want to explore. The goal is to ask questions about those topics in such a way that encourages the interviewee to speak in some detail. Then a skilled researcher can move the direction of the interview by asking specific follow-up questions.

For example, at City Hospital you might decide to segment your interview into sections that correspond to the major components of the Med-Ease software system. For each of those segments, you might begin with general open-ended questions such as, "What do you like about the insurance claims manager system? What do you dislike about it?" Then follow up with more specific open-ended questions based on the interviewee's initial response (see Example 7–16).

Finally, researchers typically preplan a sequence of questions or topics so that the most difficult or controversial come either during the middle of the interview or near the end. In this way, the researcher can build up to the difficult part of an interview. Once an interviewee is in a reflective pattern of thinking, the researcher may be able to probe the tough issues better than if you hit the interviewees with the tough questions right at the start.

In the City Hospital case, for instance, you might know that one component of the software system has been plagued by bugs. If you suspect that has frustrated many of the workers, you might also decide to ask about that section last. In taking this approach, you might be able to explore the less controversial components of the system that are working

---

EXAMPLE 7–16   Plans for Interviews Regarding Med-Ease Software
Use at City Hospital

Schedule

Most interviews will be held from 10:00 to 10:45 A.M.

Site

The interviews will be held in private conference rooms within each department's offices.

Recording Methods

All interviews will be tape recorded.

Questions

The five basic sets of questions will asks users to discuss their likes and dislikes about each of the five different parts of the Med-Ease software system.

without errors, perhaps enabling the interviewee to express opinions about the other four components within a less emotional context. Of course, such a plan could backfire. It may be that you will find so much hostility about the system in general that the interviewee will want to begin complaining right from the start. There is little you can do in such a situation but try to direct the interviewee's anger toward answers that are as informative and constructive as possible for the research project.

Conducting the Interview

Regardless of how open or directed interviews are, they should have a recognizable structure: a beginning, middle, and end. As the interview begins, a considerate researcher will want to provide a brief overview of the project's purpose and the interview plan so that the interviewee knows as much as possible about what to expect. The middle of the interview comprises the sequence of questions and topics that a researcher wishes to cover. At the end of the interview, careful researchers typically ask the interviewee if there is anything he or she wishes to add before finishing. Researchers may also end the session by restating the interview and project rationale, indicating what will be done with the information gathered in the interview. Finally, they may wish to set up a follow-up interview, if necessary.

Throughout the interview, researchers need to recognize that they are engaged in a **dyadic**—a two-person—exchange that involves power and status to some extent. Interviewers have one form of power: They ask questions. If a researcher also has a hierarchical status—that is, if the researcher outranks the interviewee in the organizational hierarchy—then the researcher possesses even greater power in the dyad. Sometimes, of course, researchers will find themselves in the exact opposite situation. Either way, a dyad that is radically unbalanced in terms of power and status can negatively influence the outcome of the interview. The key is the degree of respect that the researcher shows the subject. If researchers act in a professional and respectful manner—never trying to intimidate, always displaying professional respect for the subject—then they increase their chances of acquiring valid information during the interview.

> **dyadic**
> exchange of information between two people

Clearly, one of the ways researchers give and gain respect is through their verbal communication—the way investigators describe their purpose, present their questions, and delve into issues using follow-up questions. But another important consideration is **nonverbal communication.** Careful researchers, for example, will want to dress appropriately for the interview. They will want to present an alert but relaxed posture (not so alert as to make the subject nervous but not so relaxed as to make him or her wonder whether the interviewer is paying attention). The successful researcher shows genuine interest in the words of the interviewees by looking at the individual who is speaking, by listening carefully, by providing appropriate nonverbal gestures (e.g., nodding, smiling). Never forget that in interviews the researcher embodies the research instrument. The means by which a

> **nonverbal communication**
> communication signals and cues conveyed by means other than words; includes dress, posture, facial expressions, gestures, and tone of voice

researcher gathers information comes not simply through his or her questions; it comes from the entire performance of the interviewer.

During an interview, one of the key elements will be the degree of control the researcher exerts. At one end of the control spectrum is the unstructured, open, informal interview in which a researcher exerts almost no control. At the opposite end is the self-administered questionnaire containing nothing but closed questions. Usually, interviews fall somewhere in the middle of this spectrum. Skillful researchers need to cover certain topics, but they also want to explore aspects of those topics that they cannot predict. The key here is the balance between keeping the interviewee on track but still allowing for the unexpected.

One way that researchers can maintain this balance is to allow interviewees to wander a bit, before pulling them back toward the topic at hand. Although some of those wanderings may prove fruitful, others may not. Therefore, skillful researchers always need to be ready to refocus the interviewee on the topic at hand when they sense the interview is meandering into irrelevancy. Of course, this type of problem occurs most frequently when a talkative person is being interviewed. Sometimes researchers will encounter people who are reticent, people who, for whatever reason, do not say much of anything. This type of person usually poses the toughest situation for interviewers. How does an interviewer prompt reticent subjects to speak more without determining what they will say—that is, without putting words in their mouths?

Two simple techniques may help out in this situation. First, don't be afraid to use silence. If a person provides too brief an answer, don't quickly jump in with another question. Wait a moment. Keep looking at the person as if you are still expecting him or her to say more. Silence in a dyad usually sends the message that the silent one expects the other to speak. However, don't let silence lapse into a standoff; a skillful researcher may need to go on to the next question and then circle back to the question that provoked the silence.

A second technique involves restating a couple of key words that the interviewee uses in such a way that indicates the expectations of further elaboration; usually this approach simply involves restating a phrase in the form of a question. For example, read the interview excerpt presented in Example 7-17. Notice that the interviewer has not pushed the user to say anything even though the first response indicated no problems and the last response began to pinpoint some problems. This technique can become tedious to interviewees when overused, but good researchers do restate key words when they know, with some encouragement, that an interviewee has some additional valuable information to offer.

Ultimately, researchers strive for balance in interviews: prompting people to tell more about what they think without determining what they think. Exercise 2 at the end of this chapter will allow you to test your ability to respond to a variety of interviewees, from the reticent to the hostile.

EXAMPLE 7–17   *Prompting Responses in an Interview*

Excerpt: Interview at City Hospital

Interviewer:	Can you tell what it is that you like about the spreadsheet component of Med-Ease?
User:	Hmm. Well . . . Not sure, really. Guess I like it all right. (Silence) I can use it all right once I get into it.
Interviewer:	Get into it?
User:	Yeah, you know, once I figure out which commands get me into what I want to do.
Interviewer:	Once you figure out what . . . (voice trails off)
User:	Yeah, figure out. Actually, I always forget which is the first key to hit to get the menu to appear at the top of the screen. Then I can never quite remember if I want to first go into the Workbook or into File or one of the other sections. But once I figure that out, I'm OK.

OBSERVATIONS

When a researcher interviews people or asks them to fill out a question-naire, he or she is asking respondents to step out of their natural environ-ments for a few moments and reflect on a given topic. Although these meth-ods are often fruitful, there are times when researchers may want a different kind of information. Instead of asking people to tell about their thoughts and actions, sometimes researchers want to watch subjects perform during the normal course of their duties—and listen to them talk about their actions. Researchers can perform **observation** in one of two ways: observing **natural work** or making use of **simulated work** environments.

Moreover, at times researchers need to observe things—bridges, roads, buildings, natural events, equipment, lab experiments, and so on. Often these things are found at a site or location that must be visited. Of course, people often construct, operate, or control these things. Most of the strate-gies we discuss next can be used to observe people and things.

Natural Work

One way to observe is to actually enter a group of people and try to watch what happens naturally in that group, or to visit a site and observe that en-vironment and facilities. For example, we can learn about a foreign society by asking its members to tell us about it; however, we can also learn about that society by visiting it, living in it, and recording what we see and hear. Anthropologists often do the latter. They study different cultures by ob-serving those cultures first hand. Margaret Mead lived in Samoa to learn about Samoan culture. Other anthropologists have entered and observed a variety of cultures from remote aboriginal societies to suburban and inner-city communities.

observation
a primary research tool that gathers information about people and their activities as they go about their jobs or work with products and devices or in specific environments

natural work
what happens naturally in a group, at a work site, or in the office

simulated work
work that is re-created within a laboratory setting

Just as an anthropologist studies a community, so too may a careful researcher need to study the actions of others in a working situation to understand better how they communicate, why they communicate, what the barriers to communication in that community are, and what communication strategies succeed with community members. Observational researchers may want to observe experts performing a task so that they can learn the task and, in turn, communicate it to others who wish to learn the task. Observational researchers may want to observe how people use something that has been designed to determine the worth of that design. Or, observational researchers may travel to a site to observe, measure, and visit the facilities in which products are made.

In your research at City Hospital, for example, you might want to understand better how the first users of the new software system feel about it. So far, you have surveyed and interviewed users of Med-Ease software, but you haven't actually watched users in action. If you add that kind of information to the results of your surveys and interviews, you may develop a more robust perception of the attitudes and abilities of the users and the strengths and weaknesses of your company's product.

Of course, such observational research has its disadvantages. Researchers, for example, must be careful not to assume that the few people, the small amount of work, or the specific site under observation represents all of the people, all of the work, or all the sites. Further, because researchers are entering someone else's workplace and watching him or her work, they cannot set the agenda. For example, if you are able to observe several City Hospital employees using the software as they do their jobs, you may see only a small facet of the kind of work that is done every day. Or, if you are observing the use of Med-Ease within specific facilities or locations, you may not be able to generalize what you have observed to all such facilities or locations.

As outsiders, researchers should recognize that their presence may disrupt the normal activities they want to observe. Without meaning to, they may stifle or distort the words and actions of those under observation. A careful researcher always tries to put subjects at ease so that they can go about their work in as natural a way as possible.

Researchers may also misinterpret what they see. For example, a lab experiment might seem to go quite well during one specific observation, but may deteriorate during subsequent repetitions.

Simulated Work

What can researchers do when they know that they need to observe people at work but it is impossible to gain entrance into a workplace? What can researchers do when they need to predict the production rate of a future product or the durability of a certain material or device? In such cases, researchers may be able to simulate the work that people do, the steps in production, or the material or device in use. For example, in the early 1980s a group of researchers from IBM Corporation headed by John Carroll wanted to study how people in offices learned how to use word processing programs. At the time they could not actually go and visit real offices where employees were learning how to use computers, so they constructed a simulated office, complete with all of the noises and interruptions typical in most offices. In this situation, Carroll and his team were able to observe and study people working.

Some readers of this textbook may have already studied or even created computer simulations to test the feasibility of creating and using potential products. Manufacturers, consumer interest groups, and even magazines, such as *Consumer Reports,* test the durability and safety of everything from cars to electric can openers, often through simulation.

The disadvantage of the simulation approach is that the work the participants are doing isn't their real work; instead, it is work that the researchers *envision* real people doing. Simulating the manufacture of a new product may provide researchers with an idea about the materials needed and the timing of each stage, but the simulation may not fully account for the time needed to train employees. Or there may be significant differences between the way people work in a simulation and the way they work in their natural environments. Furthermore, setting up realistic simulations can be expensive and time consuming. Therefore, in Writing Strategies 7–3 we suggest four steps that will help you conduct effective observations as a means of developing information.

In most situations researchers are allowed, at least, to write observations, known as **field notes.** One commonly used style of field notes stipulates three different types:

- **Observational notes** are hastily scribbled descriptions of what you see going on as you watch.
- **Theoretical notes** are statements that express your opinion about or interpretation of what you see.
- **Methodological notes** are reminders to yourself to do some further research work, to ask a certain question tomorrow, or to remember to talk again to a certain person (Shatzman and Strauss).

Researchers need to distinguish among these three types of notes because they frequently need to separate a description of what they see (an observational note) from an analysis of what they think it means (a theoretical note) from a statement about their future actions in the setting (a methodological note).

In the City Hospital case, for example, you might be granted entry to selected offices for a limited time on several different days. You might

field notes
written records of observations

observational notes
hastily written descriptions of what you see going on as you watch

theoretical notes
statements that express your opinion about or interpretation of what you see

methodological notes
written records to remind the observer about additional research needed

WRITING STRATEGIES 7–3   Steps for Observations

Step 1: Decide specifically what activities or facilities you want to observe.

Rationale: At City Hospital you have determined that you need to watch several individuals in each department using the Med-Ease system. You would like to observe some or all of the participants using every component of the system but you cannot stipulate that beforehand. What if no one ever uses, say, the spreadsheet component? To force someone to use it would distort the data that you seek to collect. After all, you want to study actual use and actual attitudes.

Step 2: Determine if those activities or facilities are observable in a natural setting.

Rationale: Sometimes you know what you'd like to observe, but, for whatever reasons, it is not actually feasible to do the observations. At City Hospital, you realize that all of the work you want to observe is conducted regularly, and, therefore, it is quite feasible to observe work as it is being done.

Step 3: Request entry into the site where you want to observe.

Rationale: Never assume that others will automatically understand your need to enter their worlds and observe them or their employees doing their jobs. On the contrary, most people see observers as intruders and disrupters, or worse, spies! When you request entry, you want to be prepared to explain in detail your purpose and your methods. Make it clear that you will try to be as unobtrusive as possible. On the other hand, don't overpromise. The participants should realize that you will be somewhat intrusive by your very presence. Also, remember that most people will associate observation with evaluation. If you are not there to evaluate someone's performance, then by all means make that clear to those you observe. Your presence in that case may seem less intimidating and less of an imposition.

Step 4: Choose one or more data recording methods.

Rationale: When you negotiate your entry, you should discuss how you will record what you see and hear. In some cases, you will not be allowed to bring any external recording device to the site. In such cases, be prepared to record—either in writing or on tape—what you saw and heard immediately after leaving the site.

arrange to be allowed to write field notes and tape record any discussions that go on while you are observing. In this situation, you would obtain a battery-powered, voice-activated, minitape recorder so that you could easily tape record whatever is said in your presence, with the permission of the City Hospitals and its employees, of course (see Writing Strategies 7–3 and Example 7-18).

EXAMPLE 7–18 Observation Plan: Observing Med-Ease Software
Use at City Hospital

Activity to Be Observed
Use of the Med-Ease system

Participants
Two employees from each of the five departments

Length of Observations
One hour in the morning and one in the afternoon for each employee

Recording Methods
Field notes and tape recordings

The Act of Observing

When you are conducting observational research in the field or on site, consider the suggestions in Writing Strategies 7–4.

Sometimes, just watching does not enable researchers to learn what they need to learn. In some situations, researchers may need more information; often, for example, a researcher may need participants to explain what they are doing and thinking but, in the course of their normal work, they do not articulate what they are thinking. Simply watching people work will not reveal their thoughts and explanations.

How can researchers break into the normally silent thoughts of those whom they observe? In part, this can be accomplished through questionnaires or formal interviews. But researchers may also need to hear what people are thinking as they are acting or working in their natural settings. In such cases, **think-aloud protocols**—which involve subjects in verbalizing their thought processes as they are performing some activity—can enable researchers to get at some of the thinking that goes on inside subjects' heads. This method, however, is also quite dramatically intrusive in connection with the normal activities participants undertake. Researchers also should know that participants are never able to put all of their thoughts into words and that some of their thoughts are not even available to their conscious mind. Thus, think-aloud protocols enable researchers to get glimpses of individuals' thought processes as they work, but protocols, too, are limited.

To conduct a think-aloud protocol, researchers can ask the person to state continually what is on his or her mind while reading and/or performing a task. But because speaking aloud while working is quite difficult, a focused task should be chosen, something that a person could complete within a half hour. The subject's running commentary should be recorded.

think-aloud protocol
an observation tool in which a person is asked to state his or her thoughts continually while performing a task

---

WRITING STRATEGIES 7–4   Strategies for Observational Research

Strategy 1. Make your agenda known.

Explanation: Researchers can alleviate tension and suspicion about their presence by making their purpose clear. Generally, researchers observe work because they don't have a clear understanding of how it is done. When researchers tell participants that they are there to learn from them, subjects often will cooperate freely. If researchers appear suddenly, their work may go unannounced and unexplained, arousing anxiety and hostility.

Strategy 2. Choose effective vantage points.

Explanation: Researchers should try to position themselves in places where they can observe freely without getting in the way of normal activities. This is often easy to do in offices, but in an industrial site, where researchers might be observing a new process, they must be careful not to obstruct activity.

Strategy 3. Capture as much as you can.

Explanation: First-time observers often worry that they cannot capture everything that is happening in a setting. Don't worry. No research method will ever capture the *full* complexity of any activity—not even videotaping. All researchers can do is try to record as much as possible, assuming that if they miss something important, it will happen again in some form. If it is important, careful researchers will usually create more than one opportunity to record it.

When a participant falls silent, researchers will want to prompt—questions such as, "What are you thinking now?" often get the subject talking again. Researchers should be careful that they don't interfere when subjects are really concentrating—for example, to avoid a dangerous situation, such as electrocution.

In the City Hospital case, for example, you may want to have an employee think-aloud while sitting at a computer terminal using Med-Ease and completing several specific, common tasks. Your goal will be to gain some insight into how the user perceives the task and the usefulness of the Med-Ease software.

An Important Word about Human Subjects

Researchers planning to conduct interviews, surveys, or observations as part of a primary research project should be sure to check with their company or organization about **human subjects guidelines** that need to be followed. Typically, these guidelines are written to protect the safety and rights of the human beings involved in primary research efforts. Some organizations require researchers to submit proposals for primary research involving human subjects to an oversight committee for review. Other organizations require researchers to obtain written permission from their subjects, to explain to subjects the research processes they are going through, to clarify how the data obtained through this process will be used, and to identify how subjects' confidentiality will be treated.

At the end of this chapter, we have included several Web sites that provide detailed human subjects guidelines for researchers. Because a great deal of human subjects research takes place within academic settings, most of these are university Web sites, although they span a range of professions.

human subjects guidelines
written to protect the safety and rights of the human beings involved in primary research efforts

---

■ Many technical professionals are likely to conduct research in order to develop the information needed to carry out their jobs. As Chapter 6 indicated, this research often starts with or is supplemented by secondary research, which uncovers and interprets the research of others. In this chapter you learned that primary research helps individuals discover new knowledge and includes several techniques: surveys, interviews, and observations.

■ You learned that surveys are questionnaires designed to ask the same questions of a large number of potential respondents. Longitudinal surveys allow researchers to ask questions of the same or similar groups over a period of time, while cross-sectional surveys provide a current picture of the phenomena under study. If researchers cannot question everyone involved, they may want to choose a representative sample. Often a balance of open and closed questions will provide excellent information; questions must be valid and reliable. Surveys can be presented to the respondents through self-administered forms or over the phone.

■ More people gain information through face-to-face question and answer sessions, or interviews, than through any other research activity. Interviews differ from surveys in that the primary research instrument is not a questionnaire but the interviewer. Therefore, interviews are directly shaped interpersonal communication. Interviews can be open, directed, or a combination of both. Planning a successful interview demands careful scheduling, determining the site, choosing a recording method, writing questions, and, finally, conducting the interview with careful attention to verbal and nonverbal communication and respect for the interviewee.

■ Instead of surveying or interviewing, researchers must sometimes observe people engaged in a particular task or facility used for particular processes. Such observations can be natural or simulated. Field notes in observational research should consist of observational notes, theoretical notes, and methodological notes.

SUMMARY

■ Again, whenever you conduct any type of research, you are communicating from a particular cultural, political, organizational, or gendered point of view. Likewise, the responses that you get from others are also imbued with particular points of view. Your goal in doing primary research is not only to gather information, but also, as we will discuss in more detail in the next chapter, to try to recognize your inherent biases and uncover the biases of others with whom you develop information.

ACTIVITIES AND EXERCISES

1. Consult a statistical textbook or one of the books on survey design listed at the end of this chapter and calculate the following types of descriptive statistics: mean, median, and mode. Using the raw data in Example 7-14, compute the frequency, mean, median, and mode of each item. In a memo to your classmates and instructor, discuss when it is most appropriate to use each.

2. Write follow-up questions to provoke more information or to help the interviewee focus for each of the following interview situations. (Note the interviewer is *Q,* the interviewee is *A.*) Your instructor may ask you to role-play this activity.

Situation 1

Q: Have you ever used the spreadsheet function of the software?

A: You know when I use the word processor, I never know how to print when I'm done. I mean, I have looked it up in the manual but I always have trouble finding it. That's another thing. Those manuals! Who wrote them? I can never find anything in them and then when I do I can't understand it! Anyway, I don't print out documents all that often so when I do I have forgotten how to do it and even if I remember the first print command, then I get completely confused if it shows me the screen with all of the printing choices. You know, which printer I am using, and all that.

Situation 2

Q: Do you find anything difficult about the word processing program?

A: Well, no, not really.

Q: No?

A: No.

Situation 3

Q: Would you say that you are satisfied or dissatisfied with the spreadsheet program?

A: Oh, well, I'd say that I am very satisfied with it. I have used it extensively and I would say that there is very little bad about it, other than the fact that I have trouble finding the correct embedded commands in the menus. Several times I have erased data that I had entered because I thought that I was moving data from one column to the next when actually I must have been deleting data. Stuff like that has happened to me several times. Lost a bunch of work, too. But, no . . . You know, everyone makes mistakes in this business. So, I'm pretty satisfied with the spreadsheet.

Situation 4

Q: I want to thank you for taking this time to meet with me. I don't expect this interview to take long, so if we could just . . .

A: You know, I am sick and tired of management sending their lackeys down here to snoop on us and then giving us grief about not getting our work done. You want to come in here in the middle of the day to ask me questions and then just walk out of here while I get stuck working over my lunch break because my boss wants to know why I haven't gotten my morning's work done. And because your boss told my boss that I had to talk with you. Let me tell you something with all due respect, I don't feel like giving up my lunch break for you!

3. Read the documents associated with the Year 2000 problem located at the end of this book (Appendix A, Case Documents 4). Form a team of three or four students from your class and prepare to investigate how this problem was handled on your campus. Choose one primary research method discussed in this chapter that will best help your team understand the nature of the problem. Design your research method, using the advice in this chapter, and describe in a brief memo how you would administer, conduct, or distribute that method or tool. For example, if you decide to conduct an interview, prepare not only your questions but also describe the place, the recording method, and the people whom you would interview. If you design a questionnaire, don't simply write your questions, but also determine your sample. If you wish to observe, decide not only what you are looking for, but also whether that situation should be natural or simulated and what types of notes you will need to take.

WORKS CITED

Carroll, J. M., P. L. Smith-Kerker, J. R. Ford, and S. A. Mazur-Rimetz. "The Minimal Manual." *Effective Documentation: What We Have Learned from Research*. S. Doheny-Farina, ed. Cambridge: MIT Press, 1988, pp. 73–102.

Carroll, J. M., R. L. Mack, C. H. Lewis, N. L. Grischkowsky, and S. R. Robertson. "Exploring a Wordprocessor." *Effective Documentation: What We Have Learned from Research*. S. Doheny-Farina, ed. Cambridge: MIT Press, 1988, pp. 103–126.

Debs, Mary Beth. "Reflexive and Reflective Tensions: Considering Research Methods from Writing-Related Fields." *Writing in the Workplace: New Research Perspectives*. Rachel Spilka, ed. Carbondale: Southern Illinois University Press, 1993, pp. 238–52.

MacNealy, Mary Sue. *Strategies for Empirical Research in Writing*. Needham Heights, MA: Allyn and Bacon, 1999.

Shatzman, L., and Anselm Strauss. *Field Research*. Englewood Cliffs, NJ: Prentice Hall, 1973.

HUMAN SUBJECTS GUIDELINES ONLINE RESOURCES

Panel on Human Subjects in Non-Medical Research, Stanford University, California
http://portfolio.stanford.edu/105955/text/plain

Ethical Principles of Human Subject Research, University of Michigan, Ann Arbor
http://www.med.umich.edu/irbmed/ethics/ethics.html

Ethics Guidelines—Research with Human Subjects, University of Windsor, Ontario, Canada
http://www.uwindsor.ca/research/ors/refs.html

Human Subjects Program, University of California, San Diego
http://medicine.ucsd.edu/humsub/

Research Ethics, National Academy of Sciences, Washington, DC
http://ecco.bsee.swin.edu.au/studes/ethics/

Editing and Style

Introduction ❚ Determining the Purpose and Results of Editing ❚ Doing Editorial Reviews ❚ Considering Style ❚ Making Your Sentences Work ❚ Proofreading

I am in love with language . . . It has been a curious discovery for me that working . . . with the many ways of writing English, almost all prose, has given me a strong sense of the unity behind the infinite variety of English. It is a unity you can only find through a conscious exploration of the language as it serves many different purposes. If precision and exactness are central to the writing of a poem, so are they to the framing of a business memo, the working out of an application for a job, or . . . the writing of a speech.

> Lauris Edmond, "Imagining Ourselves," in Robert Neale, ed., *Writers on Writing* (New York: Oxford University Press, 1992), p. 227.

Technical editing is an art and a skill with which a document is improved so that it better serves the needs of its defined users . . . [T]echnical editing is practiced not only in subject areas such as computer science, medicine, agriculture, insurance, environmental law, finance, and polymer chemistry, but also on occasion in such areas as painting or music . . . In all subject areas and with all types of documents, the hallmark of technical editing is a focus on the user's needs and a stress on accuracy, precision, and usability.

> Elizabeth Turpin and Judith Gunn Bronson, "Technical Editing," in Katherine Staples and Cezar Ornatowski, eds., *Foundations for Teaching Technical Communication: Theory, Practice and Program Design* (Greenwich, CT: Ablex, 1997), pp. 221–22.

Introduction

As you have learned, writing is a complex process, requiring research, analysis, and attention to many details of arrangement, sentence structure, and visual representation (illustrations, page design). It is easy to get distracted by any one of these tasks and even to lose sight of the main purpose. Thus, good writers build into the document development process opportunities for systematic review and evaluation of the work accomplished. This review and the adjustments that result constitute *editing*.

editing

the process of making a document more accurate, effective, and readable by ensuring that the grammar, usage, style, and punctuation, as well as the organization of a document, are accurate and appropriate

Editing provides an opportunity to focus on the effectiveness of writing and presentation—to stand back from the research and content development to consider whether readers will understand, whether they can find what they need, and whether they will be convinced by your data and arguments. It is a kind of quality control to confirm that the document meets the plans you have made for it and that the audience may have established for it. As you edit, you compare what you have written with what the work or academic task requires; with document specifications, such as necessary parts in a proposal; and with various standards of written communication, such as standards of grammar and spelling, organization, visual design, and style. If there are gaps between the document and the goals and standards, you need to make adjustments.

Editing may occur several times throughout the document development process but certainly will occur at the end, before you turn in your project. A well-edited document must be accurate in content, ethical in its use of information, correct and effective in verbal and visual expression, and easy to understand and use. All of these characteristics help it solve the problem or provide the information that defines its purpose. Moreover, an unedited document may be hard to use because it lacks sufficient information about the location of information. Worse, inaccuracies or incomplete information may cause readers to draw the wrong conclusions and to make the wrong decisions, some of which could affect safety or the future success of a project.

This chapter will help you understand what to do in editing and how to do it. The chapter illustrates the nature and results of editing and then provides guidelines for six types of editorial review:

1. Suitability for audience and purpose.
2. Completeness and accuracy of information; ethics.
3. Organization and consistency.
4. Visual design and usability.
5. Style.
6. Correctness of grammar, punctuation, and spelling; consistency.

The six editorial reviews ask you to consider the full range of possible opportunities for improving your document, from content and appropriateness for the audience to details of spelling. At each review, the editor is asking a different type of quality control question as shown in Example 8–1.

Editors in both the workplace and the classroom ask these questions, and a company may schedule a sequence of reviews throughout the document development process, not just at the end. In this chapter, when we use the

---

EXAMPLE 8–1   Six Editorial Reviews

Editorial Review	Editor's Questions
1. Suitability for audience and purpose	Does this document respond adequately to the task assigned?
2. Completeness and accuracy of information; ethics	Are all the required parts included? Is the content accurate and adequately developed for the intended audience? Do the content and discussion of it respect ethics?
3. Organization and consistency	Does the sequence of information reveal the relationship of ideas? Are related parts treated in consistent ways?
4. Visual design and usability	Will page or screen layout, illustrations, and searching devices (table of contents, headings, menu) help readers find what they need? Does the visual design enhance interpretation of content?
5. Style	Do sentence structures and word choices clarify meaning? Is the tone right for the audience?
6. Correctness of grammar, punctuation, and spelling; consistency	Do grammar, punctuation, and spelling conform to conventions? Are capitalization, use of numbers, dates, and abbreviations treated consistently?

---

phrase "assigned task" or a variation of this phrase, we are thinking not just of the classroom but also of the workplace. The reviews usually begin with a comparison of the document with the explicit or implied specifications for the document.

Following an example that illustrates the effect of editing, each review will be defined in more detail in this chapter. A separate section on style provides more details about sentence structure and word choice. For the other reviews you may refer to other chapters in the textbook for more information.

Some people equate editing with proofreading, but although the corrections that result from proofreading are an important part of editing, we define the purpose of editing more broadly: to make a document right for its audience and purpose and to make it complete, accurate, usable, and effective. A complete editorial review will consider content (completeness and accuracy of information), features to make the document understandable and usable, consistency, and correctness of spelling, grammar, and punctuation. Example 8-2 suggests the scope and effect of editing.

Determining the Purpose and Results of Editing

EXAMPLE 8–2   Comprehensive Editing: Methods and Results

The two versions of the same manuscript below show the scope and effect of editing. The unedited manuscript contains spelling, grammar, and punctuation errors, but it also presents confusing information for its audience of medical care providers. The edited manuscript is easier to read and understand.

Unedited Manuscript

[1] The Minnesota Obstetric Managment Inititive is a systems analyses, organized by time, consisting of a questionaire. [2] It is a research tool; not a quality assurance tool yet. [3] This study wanted to know if the cases were managed appropriately or not. [4] MOMI is the study of hospital births and medical record analysis. [5] The study is a sampling of the 37,921 births in Minnesota between 1988 and May 1989. [6] It is a random sampling of 5000 births which hopefully serves as a real reflection of this community. [7] The statistics presented here are not final or publishible, but you can see trends in terms of the study and eventually the data will be published in JAMA. [114 words]

Comment

Some typos (Managment, Inititive, analyses, questionaire, publishible) make the paragraph seem amateurish or carelessly prepared. But correcting these errors alone will not solve the more serious problems of organization and missing information.

For example, readers may wonder what system is being analyzed (sentence 1), but sentence 2 does not answer the question. Reorganizing to place the information in sentences 4 and 5 right after sentence 1 will help readers understand what the topic is. Then the comment on the information in sentence 2 will make more sense. There is some ambiguity in "reflection of this community" (sentence 6)—how does the information reflect the community?

Edited Manuscript

[1] The Minnesota Obstetric Management Initiative (MOMI) is a systems analysis, organized by time. [2] MOMI examines hospital births and medical records, using a 32-page questionnaire to gather data. [3] Designed to determine whether pregnancy cases were managed appropriately or not, the study covers a random sample of 5,000 births selected from the 37,921 births in Minnesota between May 1988 and May 1989. [4] As such, it provides an accurate reflection of pregnancy in the state. [5] MOMI is a research tool, not a quality assurance tool. [6] The statistics presented here are not final but identify trends and eventually will be published in the *Journal of the American Medical Association* (JAMA). [108 words]

Comment

The second sentence answers some predictable reader questions (What is it? How does it work?). It also shows the relationship of information—the questionnaire and the data that it produces.

The third sentence continues to clarify the connection between the instrument (MOMI) and the purpose and the relationship of the sample to all births.

(continued)

EXAMPLE 8–2 (Continued)

Sentence 4 clarifies "reflection of the community" by substituting the more specific "pregnancy in the state."

Typos have been corrected, and abbreviations have been identified.

In spite of all the clarifications of content, the edited version is shorter than the original. The writer has used language efficiently.

Because the edited document is easier to understand and more accurate than the original, it will better achieve its purpose of explaining the system analysis to medical care providers. We have numbered each sentence in brackets to make it easier for you to follow the editing process.

Revising the paragraphs requires not just a review for typos but also the ability to consider what readers need to know and in what order the information should come. The revisions reflect reviews for content and organization.

Doing Editorial Reviews

Each important document that you write requires **multiple editorial reviews.** Multiple reviews will direct your attention to specific features, and you should plan to go through your document several times, with a different focus each time.

The next few sections suggest a sequence of reviews, with the more global reviews preceding the reviews for details.

multiple editorial reviews
the various stages of the editing process necessary to produce a complete, accurate, usable, and effective document

SUITABILITY FOR AUDIENCE AND PURPOSE

Although you determined the audience, purpose, and use of your document before you or your group began writing it, writers can get diverted. You need to review what you were asked to do and consider whether the document delivers what your instructor or work supervisor requested.

Usually workplace writing begins with a problem to solve, which could be a need for new information, a process, instructions, or materials. Thus, your first editorial question will be whether you have provided the requested solution or information. You will consider the document as a whole, not its details, asking overall whether you have stayed on task and fulfilled the assignment.

A perfectly researched, thoughtfully organized, and attractive document can fail if it does not accomplish the goal that the audience has for it. This point may seem obvious, but it is easier than you might imagine to let

---

WRITING STRATEGIES 8–1   Audience, Purpose, Use

- What is the purpose of this document? Why does the audience care about it? What problem are you helping your audience solve?
- Have you put the interests and values of your audience in the forefront (as opposed to your own interests)?
- Who will use this document? Does the document allow for multiple audiences and multiple needs?
- Have you stayed on task—have you responded to the goal that was set for the document?

---

your own interests dominate the interests of the audience. For example, a proposal writer may be so committed to her organization's mission and need for money that she overlooks what the granting agency wants to accomplish with its money and does not adequately show the match between her interests and theirs. A software engineer may think of the program more than of the user's reasons for using the program and nudge the technical writer to document the program features rather than to anticipate the user's tasks.

Writing Strategies 8–1 reminds you about what you want your document to accomplish and who will be reading it. In addition to asking these questions, it's a good idea to review the assignment itself to verify that your sense of the purpose for the document matches the task you have been given.

This review is the most global of the six reviews. When you are confident that you are meeting the needs of the audience, you consider some of the specific details of the document itself, beginning with the completeness and accuracy of information.

COMPLETENESS AND ACCURACY OF INFORMATION; ETHICS

The review for accuracy and completeness of information is closely related to the audience and purpose question because "information" depends on who needs it and what they already know.

You can begin by comparing the document to any assignment specifications to make sure all the parts are there. The class assignment, a workplace specifications sheet, a request for proposals, or related documents in your organization may provide this information. If you are missing an expected part, you need to provide it before submitting the document. Also be sure that the parts of the document are arranged in the same order as the assignment specifies.

The harder question is whether each part contains the information that enables the audience to understand the document's use. Often there are

EXAMPLE 8–3   Checking for Completeness and Accuracy

Content Check	What to Consider
Background information	Even if you are sure your main reader knows the situation that initiated the document, your document should include a statement of the problem and its background. Readers are probably working on multiple projects and need the orientation of background information. Future readers may need this orientation even more. This information can go into an appendix if it would be intrusive in the body of the paper.
Terms and concepts	Unless you know for certain that the audience knows the terms and concepts of the document, you need to explain them. If explanations would be intrusive for the main audience, include them in a glossary or appendix.
Evidence	Evidence is the data you have gathered to include in the document, including facts and arguments. Gaps in content occur when a topic is incompletely researched or not all the relevant issues have been researched. Try to anticipate readers' questions, and look for explicit answers in the document.
Interpretive statements	As the writer, you probably know more about the subject than the reader even if the reader ranks higher in the organizational structure. The reader will depend on you to explain the significance of the evidence. Most facts do not speak for themselves if only because most issues are complex and require a human mind to reconcile contradictions or to show significance in a particular context.
Steps in a procedure	If a document provides instructions, the check for completeness should include a test of the instructions with a user who is typical of the prospective reader. If the user cannot complete the task following the instructions, you will need to provide more information.
Safety information	Safety information needs to be included in documents that treat processes, materials, or even concepts with safety implications.

gaps in information because a writer assumes knowledge that a reader does not have. These gaps may be in background information, definitions of terms, evidence, interpretive statements, steps in a procedure, or safety information. There is no template to prove that your information is complete (or excessive), but you can ask these questions systematically to prompt yourself to think of questions the audience may have that the document does not answer, as in Example 8-3.

WRITING STRATEGIES 8–2   Completeness and Accuracy
of Information; Ethics

■   Are all parts called for in the assignment included and in the correct order?
■   Have you anticipated and answered audience questions?
■   Have you provided background information, definitions of terms and con-
cepts, sufficient and complete evidence, interpretive statements that explain
significance, all steps in a procedure, and safety information?
■   Is the information accurately represented?
■   Have you shown the ethical consequences of the information?

The question of accuracy has to do with whether you have provided
information that can be trusted. Incomplete information is likely to be
misleading. Be careful with numbers; a misplaced decimal or reversed digits can change a meaning substantially and care is especially crucial with
financial information.

Ethical use of information is closely tied to completeness and accuracy.
Intentional gaps may invite the audience to overlook conflicts or contradic-
tions. For example, if you omit information on hazards or competing view-
points, you may invite a flawed decision or action. You need to reflect on
consequences of the information or any recommendations you make to consider whether they could result in personal or professional harm. The review
for ethics, as presented in Writing Strategies 8–2, should also consider issues
of exclusion on the basis of gender, race, class, or ethnic identity.

ORGANIZATION AND CONSISTENCY

organization
reveals the relationship
between ideas and
influences comprehension

Organization refers to the sequence of information. Organization power-
fully affects how well readers understand because it suggests the relationship of ideas. A well-organized document is also easy to use because readers
can find what they are looking for in expected places without reading the
entire document. The editorial review should consider the overall document
organization in addition to the organization of sections and paragraphs.

Although you can organize material many ways, the conventions of
the genre in which you are writing may help you make your choices. You
can also apply general principles of organization that increase reader comprehension and use of the document. Effective use of transitions creates
coherence.

Following the Conventions of the Genre

Some types of documents, such as grant proposals, memoranda, or feasibility reports, are common in the workplace. These types of documents,
or genres, have predictable components and a recognized order. These

established conventions meet readers' expectations. You can use them to understand what your readers will expect. Genre conventions, including expectations for content, organization, page design, and length, may be written down in a document assignment. If you have such a guide, you should follow it carefully. You can also obtain samples of other people's work to see what is effective in a specific situation.

Genre conventions are good guides, but they are not templates. Writers adjust these conventions all the time according to a situation. A good example of a proposal in one situation may not be a good example in another. Nevertheless, you should be familiar with the conventions for genres and use them to your advantage. If you modify the conventions, do so deliberately—with a purpose, not accidentally. The chapters in this textbook on genres will give you specific information.

Using Recognized Patterns for Organizing Materials

Research by cognitive psychologists has shown that readers use organizing principles to help understand information. For example, a chronological order prompts them to perceive that pieces of information are related by time. A general-to-specific pattern helps them to understand concepts before they learn details. By matching these organizing principles to the type of information you are presenting, you increase the opportunity that the audience will read your document accurately and easily. The structure of the document must support its meaning and your argument.

A variety of organizing principles apply to documents: comparison/contrast, spatial patterns, chronological/step-by-step, and combinations of these patterns. When framing an argument, as we discussed in Chapter 4, you might organize according to your logos or the logic of the argument: defining, describing consequence, determining value, comparing/contrasting, citing authority, and countering objections.

When you edit for organization, you consider whether the information is organized in an overall pattern that suits the meaning. You may mix patterns, but you probably have a major overall framework that ties the parts together. You also review related parts, such as sections in the body, for consistency of structure. For example, each part in the body of a feasibility study will probably follow a pattern of introduction, evidence, and conclusion related to that topic of investigation. Steps in a complex procedure may all begin with an overview that explains some concepts and goals. Once you establish a pattern, readers come to expect it in other related sections.

Providing Transitions between Thoughts

The sequence of ideas will help readers perceive the patterns of your thinking, but you also need to guide the reader through your document with **transitions,** or words that link thoughts. When a paragraph does not seem

transitions words used to link thoughts; common transitions include words that show relationships in time, place, cause and effect, contrast, and similarity

to "flow," it's likely to be because the transitions are inadequate or the ideas don't explicitly link to the surrounding sentences.

Transition words indicate relationships. These may be relationships in time (*after, before, until, while, during, since*), place (*above, below, outside, inside, to the right*), cause and effect (*because, as a result, due to*), contrast (*however, on the other hand, instead, although*), and similarity (*likewise, as*). Because these words help readers see the relationship of ideas, they are important to meaning.

Repeated words or synonyms help a reader focus on the topic. Usually a key word of one sentence echoes a word in the preceding sentence. These verbal connections create paragraph coherence. If a sentence seems out of place, check to see whether any of its words reflect words in the previous sentence.

The passage in Example 8-4 illustrates how transitional words and repetitions help a reader see the movement from sentence to sentence to form a coherent paragraph. Transition words are italicized; topic words are in bold. The transitions support an overall structure of chronology (the changeover process) and consequence. Each sentence has at least one word that repeats or substitutes for a term in the previous sentence. The subject of every sentence, except for the first, has already been named in a previous sentence. Using a familiar word in the subject position creates coherence and flow. Key terms relate to computers, training, and time. Now see Writing Strategies 8-3, which will help you check for organizational problems.

EXAMPLE 8–4   Using Transitional and Topic Words

The **new operating system** replaced **System 7.0** only after all virus protection had been updated. *When* the **system** was in place, personnel had to be **trained** on the **new codes.** These **codes** challenged even the experienced users, a challenge that management failed to understand in planning the amount of **time** needed for the complete changeover. *As a result,* the **training period** was extended for three **months.**

WRITING STRATEGIES 8–3   Organization

■ Does this document follow genre conventions for organization?
■ Does the sequence of information reveal the relationship of ideas?
■ Are related parts treated in consistent ways?
■ Do transitions within sentences make the paragraph "flow"?

VISUAL DESIGN AND USABILITY

The **visual design** of a document is its appearance as created by choices of type, spacing and margins, icons, use of headings, and illustrations. Design elements also draw a reader to particular parts of a document. **Design** also refers to navigational devices that let readers find information such as page numbers and running heads, title page and table of contents in a print document, menu of choices, topic list, and hyperlinks in a non-linear, digital document. **Usability** refers to how easily and accurately a document can be used.

Visual design makes the organization of content visible. The visible representation of content, in turn, aids comprehension and navigation. In addition to these functions, design makes a document attractive and appealing to read. Editing for visual design and usability means checking for use of established specifications, for the document divisions and their markers (especially headings), and for identifying information.

Using Established Specifications

A class assignment, conventions of a company, or a request for proposals may provide directions for page design. These directions may specify margins, number of pages, integration of visuals, double or single spacing, use of color, and placement of the page number. You probably consulted these specifications when you began writing, but editing provides an opportunity to compare what you have developed with what was assigned.

Some genres, especially the business letter, have visual design conventions that are accepted widely and do not represent just one company's preferences. Check your textbook to review those conventions, such as placement of the return address and signature. These conventions may vary across countries, and you should be aware of the practices of the country where the correspondence will be received.

Making Content Visible

When you analyze a document before editing its visual design, you are looking for ways to make it consistent, readable, and navigable. Markers, such as headings and page numbers, show the structure of the document and provide useful guides to readers on the location of information. Consistent typography and spacing for various levels of headings reveals the hierarchy of information. (See Chapter 9 on document design for more information on how to use headings.)

The design of a successful document doesn't call attention to itself. Rather, it unobtrusively guides the reader through the document. For example, the typeface of a document should not call attention to itself at the cost of readability. Warnings, cautions, and statements of danger may well be put in bold or even require the use of color in a document, but color, bold or italic letters, or unusual typefaces should never be used just for

visual design
the appearance of a document as created by choices of type, spacing and margins, use of headings, and illustrations

design
the navigational devices that let readers find information: page numbers and running headers, title page and table of contents in a print document, menu of choices, topic list, and hyperlinks in a hypermedia document

usability
describes how easily and accurately a document can be used

themselves. Too many headings look jumbled and suggest chaos rather than a thoughtful structuring of information.

In editing, make sure parallel sections (such as major steps in a process) have similar headings and that these headings use the same features of type (font, style), capitalization, and space. The verbal style should also be consistent in related headings, whether each is a noun phrase or a verb phrase or something else. Note that in this section on visual design, for example, the subheads all begin with an -*ing* word (*making, including, checking*).

Including Identifying Information and Checking Usability

Professional writing makes good use of identifying information, including a title page with a date, running header, and page number. As you think about what to include in these sections, remember that many documents have an existence over time, and group knowledge of what the document is and what issues it addresses will diminish. The record on the page has long-term as well as short-term value.

A document is usable when people can find what they need in predictable places using markers and identifying information. Of course, the necessary information has to be available to begin with, but that question relates to content and the anticipation of what readers will need to know and do. The issue of design relates to whether the content is visible so that readers can find the place they want.

Checking Hypermedia Documents

hypermedia
nonlinear digital documents that can be accessed online or over the Internet and consist of combinations of voice, data, text, still and moving images, and audio

The need to follow specifications, make content visible, and include identifying information applies to **hypermedia** or nonlinear, digital documents as well as to print. When hypermedia documents are long and complex (with many sections), people have a harder time retaining a sense of their location than they do with print, so identifying information is especially important. Menus, navigation tables, and links must be complete and accurate. Contrast between type and background should allow for easy reading. Color and graphics should be used meaningfully and consistently.

Writing Strategies 8–4 lists visual design elements to review.

STYLE

style
choices made by an author which include such elements as tone, point of view, level of formality, use of figures of speech, voice, and emphasis

Style refers to sentence structures and word choices. These choices create a sense of the writer's voice and document tone, and they also determine whether sentences seem clear or confusing. Just as you may be attracted to or put off by an individual on the basis of his or her personal style (considerate, comfortable; overbearing, whiny), so may a reader respond to the style projected by the document in a positive or negative way.

In writing, long sentences and complex words create a formal style that may be just right for an academic journal but seem pretentious in a company report. Humor may be relaxing in a user manual but seem trivializing

WRITING STRATEGIES 8–4   Visual Design and Usability

Check your document for effective, appropriate, and consistent use of:

- Genre conventions for page design.
- Style manual guidelines.
- Typeface and type size.
- Margin widths.
- Bold, italic.
- Headers and footers.
- Headings and subheadings.
- Numbers and bullets.
- Labeling of figures and boxes.
- Right, left, and center justified text.
- Measurements and mathematical formulas.

Check your hypermedia document in addition for:

- Consistent use of color and graphics.
- Readability of type (type size and style, contrast with background).
- Working hyperlinks.
- Accurate and complete navigation table.

WRITING STRATEGIES 8–5   Style

- Do the subject and verb of each sentence contain the most important information of the sentence?
- Do most sentences use verbs in the active voice and descriptive, specific verbs?
- Are nouns specific and concrete?
- Does the tone seem respectful and professional?

in the problem statement for a proposal. Writers can seem condescending or disrespectful if their choices of words seem to exclude some people, for example, if they seem unaware of an international audience. On the other hand, sentences with many qualifiers ("maybe"), weak verbs ("seems"), and conditions ("if") may make a decision maker seem uncertain. Style has to be right for the audience and purpose.

Sentence structures also affect the ease with which readers understand the meaning. A style that is "clear" uses just the right words and also places the words in the appropriate position in a sentence—subjects near the beginning, action in the verb.

The section on style and grammar in this chapter will give you some specific guidelines for managing your style so that you can achieve the effect you want. Writing Strategies 8-5 previews some principles that will be discussed there and also identifies some features of sentences to check as you edit.

CORRECTNESS OF GRAMMAR, PUNCTUATION, AND SPELLING; CONSISTENCY

correctness
the level of correct grammar, punctuation, and spelling in a document

consistency
the quality of a document that enables a user to relate items and distinguish between them; includes consistency in capitalization, spelling, abbreviations, and use of numbers

Correctness of grammar, punctuation, and spelling increases the credibility of the document as well as the ease with which it can be understood. **Consistency** in capitalization, spelling, abbreviations, and use of numbers suggests care in preparation. Using a single expression to mean one thing will minimize confusion. For example, should a computer user press, hit, or touch a key? Multiple terms suggest distinctions in meaning.

Correctness

When you examine a document for correctness, you verify the spelling, grammar, and punctuation of each word and sentence. You also make certain that all quotations are accurate, that bibliographic references are exact, and that all quantitative data (measures, weights, quantities, and duration of time) are correct.

The computer will help you check spelling and may point to grammar problems, but you can't trust it entirely. Take advantage of its help and then review the document yourself.

completeness
the internal integrity of the document; all parts the assignment called for are present

You checked for **completeness** in Writing Strategies 8–2 by considering what document parts the assignment called for, but in this review you look internally at the document details. Are all the illustrations that you referred to in the text included and labeled? If your introduction forecasts four sections, are there four sections? Does the reference list include all the items cited?

Consistency

Writers have some choices about how to use numbers, how to cite sources, and even how to spell and capitalize certain terms. Your choices should be consistent throughout the document and consistent as well with any guidelines that your organization uses. These guidelines are often contained in an organization's style guide, which records the ways that organization chooses to use numbers and so forth. Or you may use a more comprehensive style guide, such as the *Publication Manual of the American Psychological Association* or the *Chicago Manual of Style*. These references will answer many questions about how to use terms and numbers, how to label illustrations, and much more. The information that follows is very selective.

Numbers

Rules vary for when to write out a number and when to display it as a numeral. In scientific and technical documents, numbers that are 10 or above appear as numerals. Numbers preceding units of measurement, such as 2 inches, 10 centimeters, 120 miles, 5 percent, and 20 Hz, usually appear as numerals regardless of their quantity.

If you do not have written guidelines, try to achieve consistency within sentences by using numerals for all numbers in a sentence if one of those numbers is 10 or above, as in the following example:

Our personnel office houses 3 managers and 17 support staff.

When numbers and words are joined to modify a noun, called a unit modifier, hyphens help the reader understand your sentence, as in the following examples:

Seven 5-inch screws were missing from the assembly kit.
Two-car garages have become popular in the suburbs.

"Two" in the last sentence is written out—not only because it begins a sentence but also because "car" is not a unit of measurement. The hyphen helps readers understand on first reading that there are two cars, not two garages.

Hyphenation, Dates, Capitalization

Some terms are correct in both hyphenated and "solid" forms, although the trend is away from using the hyphen. In your document, is *online* one word or hyphenated, *on-line*? Consult a dictionary or your company style manual to determine best practice and then follow one option consistently.

Will you use the North American or European date style: January 1, 2000 or 1 January 2000? Or will you use figures (1/1/00)? A style guide or your company's practice will help with this decision.

Many writers use capitalization of initial letters excessively to indicate words that are important, not just for proper nouns and other established uses of capital letters. Have you capitalized unnecessarily? Check a dictionary for capitalization of specific terms. You can do a global check on the computer to replace words that do not need capitalization.

Citations, Cross-References

The list of references should follow the form specified by the style guide that your instructor or work supervisor recommends. Check for order of information, capitalization, and abbreviation style. Within the text, labels, titles, and references to figures should follow the same form (for example, in parentheses; capitalize "figure" or not; e.g. or written out "for example"). Writing Strategies 8–6 will help you check for correctness and consistency.

Putting Principles into Practice

Example 8–5 shows how the six reviews can direct a revision that makes a document more likely to work for its intended purpose. In 1981 a well-known liability case was decided by the United States Court of Appeals, Ninth Circuit. This case, *Aetna Casualty and Surety Co. v. Jeppesen and Co.,* involved technical communicators who created a set of instrument approach charts for

WRITING STRATEGIES 8–6   Correctness and Consistency

- Are sentences grammatically correct? (Subjects agree with verbs and pronouns with antecedents, etc.)
- Does punctuation reflect the structure of the sentence?
- Are all the illustrations, citations, and cross-references included?
- Are numbers, dates, terms, capitalization, hyphenation, and abbreviations used consistently?
- Do citations and cross-references follow a style guide consistently?

EXAMPLE 8–5   First Draft of Memo

DATE: 30 May 1981
TO: Communications Department Staff
FROM: Jane Lilly, Manager
SUBJECT: Lawsuit Involving Communications Department

By now you know that Jeppesen lost a major lawsuit because of the charts you produced of the airport instrument approach to Las Vegas. On November 15, 1964, a Bonanza Airlines plane crashed in its approach to Las Vegas, and all the people died in the crash caused by your faulty charts. Aetna Casualty and Surety Co. was required to pay a large wrongful death settlement. Aetna sued our company to recover the damages. The court found that the crew members were not negligent in relying on the defective chart because this chart was so much unlike other charts we've created.

You know that our products involve the setting forth of FAA specifications in tabular form accompanied by graphic approach charts. Usually, we have two views of the proper approach. One is a plan, depicted as if one were looking down on the approach segment of the flight from directly above. The second chart depicts a profile presenting a side view of the approach with a descending line depicting the minimum allowable altitudes as the approach progresses. Well, you got all the data right, but you really messed up on the charts. The plan chart showed minimum altitude at the distance of 15 miles from the Las Vegas Airport as 6,000 feet. The profile chart does not extend beyond 3 miles and shows a minimum altitude as 3,100 feet. You created two views that were the same size but the scale of the plan is 5 times that of the profile. This isn't like other charts you have created for other airports in which the profile and plan charts had the same scale.

The courts found that the information you conveyed was completely correct, but that your Las Vegas charts "radically departed" from the usual presentation of graphics on our other charts and as a consequence the graphics rendered the chart unreasonably dangerous and a defective product.

From now on we can't have any more mistakes like this. I want you all to double-check all charts making sure that not only are the data correct but that the same scale applies to all charts. I then want you to go through all existing charts and develop a revision schedule to make sure that any problems we might have with existing products are cleared up. I want these changes to be effective immediately.

Let me know if there's a problem.

(continued)

EXAMPLE 8–5 (Continued)

Editorial Review	Comment
Suitability for audience and purpose	The audience consists of experts in document design. The news is bad: their work has resulted in both a fatal accident and a lawsuit. The goal in writing the memorandum is not to have them all quit but to tell them the outcome of the lawsuit and to ask them to institute a new editorial process to make sure oversights such as the one that caused the accident don't happen again.
Completeness and accuracy of information; ethics	The review of data is appropriate because time has passed since the charts were created. These details will also set some standards for a review of chart design and revision of the chart development process. The writer is correct to take the problem seriously (though she seems more concerned about money than death), but she accuses the communications department without taking responsibility as a manager for processes and corporate policies that may have contributed to the design change in the charts. While she recommends actions, she is not specific about who and when.
Organization and consistency	Organization is conventional: the first paragraph states the problem, the second gives history and details, and the third continues with these details. The memo concludes with solutions, but the closing anticipates problems, not solutions.
Visual design and usability	Paragraph divisions alone convey the structure of the document.
Style	Style is often accusing, with many "you" statements attached to the bad news. The accusations may be factually correct, but in making the readers defensive, they do not encourage the productive response that the situation requires.
Correctness, of grammar, punctuation, and spelling; consistency	OK

pilots to use in approaching Las Vegas, Nevada. Pilots using these charts were involved in a plane crash that killed all on board. The court found that the publisher of the charts was 80 percent at fault for creating charts that, although they contained accurate data , were found to be misleading.

Example 8-5 shows a possible memorandum to the communications department informing them of the outcome of the lawsuit against Jeppesen and Co. and asking them to make changes in their editing procedures to prevent future litigation. Imagine that this memo will be read by people who work in communications, and that it's important the writer retain her credibility as their manager. Also, because the audience will be defensive, the style must be formal but respectful enough to convey support for the staff along with a clear indication that the staff will have to make the recommended changes.

The revision (Example 8-6) changes the style and tone by substituting "we" for "you." It uses headings to make the structure visible. It establishes a more explicit procedure for solving the problem. It ends by anticipating working together rather than problems.

EXAMPLE 8–6   Draft of Memo after First Editing

DATE: 30 May 1981
TO: Communications Department Staff, Jeppesen and Co.
FROM: Jane Lilly, Manager
SUBJECT: Resolution of the Lawsuit and Development of New Editing Procedures

Resolution of the Lawsuit

As many of you know Jeppesen recently lost a major law suit stemming from the charts we produced of the airport instrument approach to Las Vegas. On November 15, 1964, a Bonanza Airlines plane crashed in its approach to Las Vegas, with 100% fatalities. Aetna Casualty and Surety Co. was required to pay a large wrongful death settlement. Aetna sued us to recover the damages. The court decided that the crew members were not negligent as they had relied on a chart much unlike other charts we've created. That means that we have to bear the ultimate legal responsibility for what happened, despite the fact that our error was unintentional.

Specific Nature of the Defect in the Charts

The materials we create set forth FAA specifications in tabular form accompanied by graphic approach charts. Usually, we have two views of the proper approach. One is a plan that depicts the approach segment of the flight as if one were looking down from directly above. The second chart shows a side view of the approach with a descending line depicting the minimum allowable altitudes as the approach progresses. In this case, the data in the tabular form were accurate, but there was a problem of scale between the two approach charts. The plan chart showed minimum altitude at the distance of 15 miles from the Las Vegas Airport as 6,000 feet. The profile chart did not extend beyond 3 miles and showed a minimum altitude as 3,100 feet. The two views were the same size but the scale of the plan was 5 times that of the profile. The pilots couldn't anticipate this difference because our charts created for other airports have profile and plan charts at the same scale.

This was the basis of the lawsuit against us. The courts found that the information conveyed in the charts was completely correct, but that the Las Vegas charts "radically departed" from the usual presentation of graphics on our other charts. As a consequence, the graphics rendered the chart unreasonably dangerous and a defective product.

New Procedures

I know that the defects in our charts were accidental and that we all wish we had discovered the problem under other circumstances. However, in order to minimize the chance of this kind of problem ever happening again, we're going to need to develop some new editing procedures. I have appointed an editing committee whose responsibility will be to look at our work both in terms of accuracy and consistency. I've asked them to begin by double checking all charts to make sure that the data are correct and that the same scale is used on all charts. I would like all of you to schedule a time when you can meet with this new group and help it develop a series of procedures to guard against any other serious defects that we might inadvertently introduce into our work.

Let me know if you can think of any other procedures and safeguards we might want to introduce at this same time.

Evaluating how successfully your own document addresses its audience, how well it serves its purpose, and how well it is designed for the uses to which it will be put is difficult to do. We are all attached to our own words and don't like to change them. In collaborative writing environments, one of the people in the group who is not tied to the form and content of the current draft might do the first editing tasks. If you will edit your own work, try to let it sit a day or two before beginning your evaluation so that you can gain some perspective on what you've done.

Considering Style

Within the boundaries of correctness, writers have many choices about how to structure their sentences and what words to choose. The effect of these choices is a style. Like styles in other circumstances—dress, cars, music—styles are not so much right or wrong as appropriate or not for a specific situation. Being able to make choices to create the desired style means that writers and editors need to understand sentence structure and the effect of words. You need to have some control over language in order to create a style deliberately.

The first step to creating a successful document is knowing your audience. You use the information you gained from your audience analysis to make the best style decisions for your document. When writing for highly trained and/or highly educated audiences, you use specialized vocabulary and more complex sentences than you would select in writing for less well trained or less well educated audiences. You would not write in a condescending manner to less educated audiences or stuffy, boring prose to the more educated groups. Instead, your choices about vocabulary and complexity of sentences must fit the nature of the situation.

CHOOSING THE BEST STYLE

Style may be described on a continuum from formal to informal as shown in Example 8–7. Informal style, the kind you would use in a letter to a friend, is marked by a relaxed tone. You would probably use contractions and slang.

Professional communication is usually formal. Vocabulary in **formal style** is specialized and has precise meanings. Writers and speakers attempt

---

EXAMPLE 8–7   *Style Continuum*

Formal Styles $\longleftarrow$ $\longrightarrow$ **Informal Styles**

Formal Styles		Informal Styles
■ Grants and proposals	■ Newspaper articles	■ Letters to friends
■ Feasibility studies	■ Articles in popular	■ Personal journal
■ Job applications	magazines	■ Conversations with
■ Scholarly essays	■ Letters to the editor	people we know well
■ Reports	■ Popular books	
■ Speeches	■ Briefings	

EXAMPLE 8–8   Variations in Vocabulary and Sentence Structure

The following examples are taken from the Case Documents in Appendix A and a legal liability case involving technical communication. Each uses a formal style, but not all are alike because the documents in which they appear have different audiences and purposes.

After you look at each example carefully, jot down what assumptions you believe the authors had about their readers and how those assumptions are manifested in examples of vocabulary and sentence complexity. Determine which example uses the most specialized vocabulary and explain why that is appropriate. Which example assumes the most highly trained or educated audience? Which has the longest, most complex sentences? Which seems to be written for lay audiences and which for content specialists? Make a list of other elements that you believe characterize these as examples of formal style.

Description

AIDS stands for acquired immunodeficiency syndrome, a viral infection which causes a breakdown of the body's immune system. A person infected with human immunodeficiency virus (HIV) is vulnerable to various diseases and infections. Some of these diseases can be treated, but there is no cure for HIV infection.

Informational Release

The purpose of our bird control program on the campus plots is to protect the experimental plantings that have been initiated for development of new crop varieties. The seed produced on these plots is extremely valuable because it represents the product of genetic crosses that have been carried out over a period of years. With each plant selection there may be only a few seeds produced so it is essential that they be protected.

Legal Judgment
(Ziglar v. E. I. Du Pont de Nemours and Company)

Retail seller of insecticide/nematicide could not be held liable for failing to give warning to buyer about dangers inherent in using poison which appeared like water and was packaged in a translucent one-gallon container similar to plastic milk jug, in view of plaintiff's failure to show that retailer had reason to know or duty to discover by exercise of reasonable care product-connected danger complained of or that retailer should have known that buyer would not appreciate possible harm involved in using toxic pesticide which was packaged in clear, plastic container and looked like water.

---

to create an objective or impersonal tone, in direct contrast to the personal and subjective tone of informal styles. With formal styles, you don't use contractions or slang, and the sentences are usually longer and more complex. Still, the formal style doesn't have to be stuffy or lifeless. Indeed, documents in formal style are often interesting to read because their creators put a special premium on clarity, sentence variety, and active voice.

Example 8-8 contains passages that use a formal style but differ in vocabulary and sentence structure because of the audiences they address. The legal judgment consists of one long sentence. Formal legal writing would not be the most effective choice of style for even a very educated audience unless they were attorneys or law students. This style is marked by many prepositional phrases (*in view; in using; in clear, to plastic . . .*) and expressions familiar to those who work with the law (*could not be held liable; failing to give warning; failure to show*). A formal legal style like this one is also marked by terseness. There is no embellishment with extraneous adjectives. Examine the other passages in Example 8-8 to see how vocabulary and sentence structure vary.

Often you need to "move between" formal styles; that is, you may use a formal style to address both a well-informed and a novice audience on the same subject. Again, your vocabulary (as well as your sentence structure) will vary, as demonstrated in Example 8-9. This example illustrates that one measure of good style is the audience for whom it is written. Most people will prefer to read the revision, but lawyers will be more comfortable with the original version.

The next section will help you not only to recognize different styles but to understand the effects of sentence structure well enough that you will be able to create different styles.

EXAMPLE 8–9   Moving between Formal Styles

The legal judgment in the original passage is written for a specialized audience. It needs to be rewritten to appeal to an educated lay audience.

Original Passage

Retail seller of insecticide/nematicide could not be held liable for failing to give warning to buyer about dangers inherent in using poison which appeared like water and was packaged in a translucent one-gallon container similar to plastic milk jug, in view of plaintiff's failure to show that retailer had reason to know or duty to discover by exercise of reasonable care product-connected danger complained of or that retailer should have known that buyer would not appreciate possible harm involved in using toxic pesticide which was packaged in clear, plastic container and looked like water.

Possible Revision

The hardware store that sold a farmer insecticide, which a worker on his farm inadvertently consumed, cannot be held liable for failing to warn the buyer. There was no reason for the seller to think that the buyer would fail to understand the danger implied by using a toxic pesticide even though it looked like water and was packaged in a clear, plastic container that resembled a milk carton.

**Making Your
Sentences Work**

Style depends in part on effective use of sentence structure. Thus, this section focuses on the placement of key words in sentences.

EMPHASIZING THE MAIN POINT

Sentences have structures that help communicate the meaning of the words. One key to effective style is using sentence structure to reinforce meaning. The subject and verb are the strongest parts—they say what the sentence is about. Readers will look for the key content words in the subject and verb. The beginnings and ends of sentences are stronger than the middle. Getting your sentences off to a fast start will make the style seem crisp and clear. Subordinate clauses and modifiers are less important in the sentence than the main clause (the main subject and verb) and so should contain information that is less important.

PLACING CORE IDEAS

The main point you hope to make should be in the subject and predicate of the independent clause making up the core of each sentence. Less important parts of the sentence—subordinate or dependent clauses and phrases—should be used for additional and less central information.

In the following sentence, the subject and predicate, which represent the core of the sentence, are underlined:

A <u>program has been initiated</u> to encourage the nesting of sparrow hawks in the plot areas as a further means of biological control of birds.

If the paper is about the program, the sentence is OK, but chances are the paper is more about the nesting of sparrow hawks as a means of biological control of birds. The following sentences show two ways the sentence might be rewritten, placing the most significant elements in the central core (again, subject and predicate have been underlined):

The <u>nesting of sparrow hawks</u> in the plot areas <u>has been encouraged</u> as a further biological control of birds.

As a further means of biologically controlling birds, <u>the nesting of sparrow hawks</u> in the plot areas <u>has been encouraged</u>.

If the writer wants to put even more emphasis on biological control, the sentence could be rewritten as follows:

<u>Birds can be biologically controlled</u> if we encourage sparrow hawks to nest in plot areas.

Restructuring sentences to locate the important idea in the sentence core helps readers interpret meaning easily and accurately.

COORDINATING AND SUBORDINATING

If you use **coordination** to convey equal ideas and **subordination** to convey to the reader that one idea is less important than another, your sentence structure will emphasize your meaning. Placing the main concept in a subordinate clause will confuse readers. What should be subordinated and what should be central will depend on what you are trying to convey. Consider the following sentence:

> Al Abramson, who is a former employee of the company, owned the patent for the first model of this machine.

coordination
the linking of two equal ideas, usually with a coordinating conjunction

subordination
the linking of two unequal ideas so that the relationship between them is clear

This sentence is structured to emphasize the fact that Al Abramson owned the patent (his name is the subject of the sentence). The fact that he is a former employee is subordinated to that main idea because it is in a subordinate clause.

The sentence could be written differently, however:

> A former employee of the company, Al Abramson, owned the patent for the first model of this machine.

In this sentence, the reader is directed to a different main idea: that a former employee of the company owned the patent. That it was Al Abramson is less important and subordinated.

Sometimes your sentence includes a number of points all of equal significance. In this case your task is not to subordinate one idea to another but to coordinate them properly. The sentences below need to be revised so that the ideas contained in them can be coordinated. As they are, the sentences are repetitious, choppy, and ineffective:

> Do not install the appliance near water. Do not place it near a washbowl, bathtub, kitchen sink or a laundry tub. Do not consider putting it in a wet basement or near a swimming pool.

Below the sentences have been combined and their ideas coordinated. The sentence, with its central idea about where to avoid placing the appliance, is more effective:

> Do not install the appliance near water such as a washbowl, bathtub, kitchen sink, or laundry tub, and do not place it in a wet basement or near a swimming pool.

Coordinating the ideas of several short, related sentences can often reduce the wordiness of your document. It can also create more effective and more convincing sentences.

AVOIDING VAGUE PRONOUNS AND SLOW STARTS

Writers often give away the power in their sentences by placing a **vague pronoun** in the place of the main subject or getting their sentences off to a

vague pronouns
"it," "there," and other pronouns used to begin a sentence in place of the main idea

slow start. Sentences that begin with "It" or "There" may often be edited to begin more directly and quickly. Consider the sentences below:

It is important to wear a respirator when cleaning with solvents.

It is clear that tuberculosis has been brought under control in the United States but has not been eliminated.

There is an important reason for getting a tetanus shot: incurable lockjaw.

These sentences squander the power of the core sentence by using vague pronouns as the main subject or delay the reader getting to the important part of the sentence.

Rewriting these sentences to eliminate the vague pronouns and move the content into the sentence core makes them more effective:

Wear a respirator when cleaning with solvents.

or

You should wear a respirator when cleaning with solvents.

Tuberculosis has been brought under control in the United States but has not been eliminated.

or

Although under control, tuberculosis has not been eliminated in the United States.

Incurable lockjaw is an important reason for getting a tetanus shot.

or

Getting a tetanus shot prevents incurable lockjaw.

You do not have to remove every "it is" or "there are" in your writing. On some occasions, you would change the meaning of a sentence if you eliminated the pronoun as subject, as in the case below:

It's an imperfect system, but it's better than none.

An imperfect system is better than none.

In other cases removing the vague pronoun reference doesn't do much to make the sentence more effective, as in the following examples:

There are several reasons why you should set the timer on your VCR.

Several reasons exist why you should set the timer on your VCR.

However, your document is generally stronger when you put the main idea of a sentence in its core. One way to do that is to eliminate as many vague pronoun references and slow starts as possible. If you are working with a computer, you can use some of its features to help you identify these problems in your document, and you can edit those that are ineffective. Use the *find* or *replace* command in your word processing software to locate each "it is" and "there is" in your text and examine each one as it is identified.

PLACING MODIFIERS CORRECTLY

Modifiers are words, phrases, and clauses that describe other elements in your sentences. Modifiers can be confusing if they are misplaced or if they unintentionally modify the wrong thing in the sentence. In the sentences below, notice how the meaning changes when the modifier is moved:

modifiers
words, phrases, and clauses that describe other elements in your sentences

> <u>Even</u> the adapter may become warm when it is being used.

> The adapter <u>even</u> may become warm when it is being used.

> The adapter may become warm <u>even</u> when it is being used.

Sometimes modifiers are used in sentences but are not clearly linked to a referent. When that occurs, they are called **dangling modifiers,** and they don't say what you think they do. Dangling modifiers often result in unintentional humor. The first sentence in each pair below has a dangling modifier underlined; in the second sentence, the error is corrected:

dangling modifiers
modifiers not clearly linked to a referent

> <u>Running down the street</u>, the trash can tripped the police officer

> Running down the street, the police officer tripped over the trash can.

> <u>Looking northward</u>, the Huron Islands can be seen.

> Looking northward, you can see the Huron Islands.

The implied subject of the modifier (*running, looking*) is the subject of the sentence. But the trash can doesn't run, and the islands don't look. One way to correct a dangling modifier is to change the subject of the sentence so that it can perform the action that the modifier implies.

> <u>Although ten years old at the time</u>, my father decided that I was still too young to accompany him on the Canadian fishing trip.

> Although I was ten years old at the time, my father decided that I was still too young to accompany him on the Canadian fishing trip.

In this case the modifier also attaches to the subject of the sentence. To eliminate the dangling modifier, the modifying phrase is changed into a clause so that it is no longer a modifier.

LOOKING FOR THE BEST WORDS

Placing words in sentences according to their importance and emphasis helps to produce sentences that make sense, but you also have to choose the right words. We'll begin this section by discussing verbs. Then we will discuss nouns. In both cases, we'll recommend that you choose specific words.

Using Action Verbs

Because the verb conveys the action in the sentence and comments on the subject, a lot of the energy and precision of the sentence depend on a good choice of verbs. If you have time to edit style only minimally, focus on the verbs. They are that important to an effective style.

The following sentences show how imprecise verbs can obscure the action of your sentence. In the rewritten forms, the less precise verbs are replaced by strong, precise **action verbs:**

action verbs
strong verbs that convey precise, rather than general, meaning

An electrical surge caused the destruction of the hard drive.

An electrical surge destroyed the hard drive.

Gears can injure fingers.

Gears can crush fingers.

The *to be* verbs (*is, am, be, was, were, being, been*) as substitutes for more precise action verbs can create flabby sentences, as seen in the following sentences and their revisions:

Failure to respond to the complaint *would be in violation* of University policy.

Failure to respond to the complaint *violates* University policy.

The facts are in support of the judge's recommendation.

The facts support the judge's recommendation.

Replacing weak verbs with strong ones will make your writing both more precise and more interesting.

Using Concrete Nouns

Nouns as well as verbs affect whether the style seems clear and effective or dull and ambiguous. **Nominalizations,** for example, convert action words to "thing" words and often take some of the life out of a style. Nominalizations are verbs that have been made into nouns and can often be identified by their endings: *-tion, -ment, -ance, -ing,* and *-ence.* Not all words that end with these suffixes are nominalizations—for example, the word *nation* ends in *-tion* but isn't a nominalization. Example 8–10 gives you some examples of how nominalizations are formed.

nominalizations
verbs made into nouns; often identified by endings: *-tion, -ment, -ance, -ing,* and *-ence.*

When you find nominalizations in your sentences, try to turn the noun back into a verb and move the main idea of the sentence into the

EXAMPLE 8–10   Common Nominalizations

Verb	Nominalization
condemn	condemnation
abandon	abandonment
house	housing
perform	performance
emerge	emergence

subject position. In the pairs of sentences below, the nominalizations have been underlined.

Incorporation of the latest electronics technology makes this keyboard versatile.

This versatile keyboard incorporates the latest electronics technology.

Performance of the tape player is better with fresh batteries.

The tape player performs better with fresh batteries.

Confinement of criminals seldom results in rehabilitation.

Criminals who are confined are seldom rehabilitated.

In the second sentence of each pair, the nominalization has been replaced as the subject and has been recast as a verb. Word processing programs can help with revision on the word level by identifying nominalizations for you. When you revise have the computer search for the word endings that mark nominalizations.

You can also make your words concrete by lowering the level of abstraction. For example, the following two lists begin with the most abstract term and end with the most concrete:

Abstract to Concrete: List 1	*Abstract to Concrete: List 2*
computer program	destructive birds
word processing programs	sparrows
Microsoft Word program	2,500 sparrows per year
Microsoft Word 7.0 for Windows	2,500 sparrows between March 1993 and March 1994

Ask yourself precisely who, what, where, when, how much, what amount, what degree, what size, what percent, and what type, and reflect your answers in the most concrete way.

Using Parallelism

parallelism
similar items that are
expressed in a similar
manner

Parallelism means that similar items are expressed in a similar manner. Parallelism helps the reader interpret the information correctly by showing how ideas relate. Consider the two sentences below; the first lacks parallelism in its structure, but the revision expresses both actions—retains/relies—with verbs in the present tense:

> The memory backup power means the electronic memory retains its contents and is relying on either the AC adapter or the batteries.

> The memory backup power means the electronic memory retains its contents and relies on either the AC adapter or the batteries.

Using Positive Phrasing

Readers make mistakes interpreting and acting upon negatively worded messages. The first choice in the following list emphasizes the negative (what something is not) and the second choice the positive (what something is):

Negative Choice	*Positive Choice*
do not accept	reject
not many	few
not reliable	unreliable
not efficient	inefficient

This guideline does not mean that you should cover up negative meaning by converting it to a positive one—only that choosing words that describe what something is, rather than what it is not, will aid comprehension. You can see the effect of phrasing according to what something is in the following sentences and their revisions:

> The project will not be completed on time, and so we will not be able to stay within the budget.

> The project has been delayed, and the budget must be extended.

> Because safety tests are not complete, we cannot recommend adopting the new system.

> Until safety tests are completed, we must postpone adopting the system.

Of course, when warning your readers about potential hazards, the negative carries the strongest possible message:

> Do not operate this equipment without a surge protector.

As you will see in later discussions about warnings, other design features such as colors, boldface, and layout can help readers comprehend the messages.

USING INCLUSIVE LANGUAGE: ATTITUDES TOWARD READERS

Your style will reveal your attitude toward readers and in turn the response of those readers. As you will remember from Chapter 4, you create a persona, an image of yourself as author, to whom the reader relates in your documents. Ideally, you want to create the impression that you are competent, trustworthy, accurate, and that you have the reader's needs in mind. You do this through effective sentence structure and word choice but also by choosing language that shows respect for readers.

Language is a powerful tool that can make your audience feel included or excluded. You do not want to alienate any members of your audience inadvertently because of language that seems to exclude or discriminate against them. Such language is not only unacceptable; it is also insensitive and can constitute a form of harassment or discrimination.

In this section of the chapter, we alert you to some ways to recognize exclusionary and discriminatory language while developing a clear idea of how to use inclusive language. **Inclusive language** includes everyone in your audience and conveys an attitude of respect for them.

> inclusive language does not exclude members of the audience

Avoiding Sexism

Sexist language excludes women or treats them unequally on the basis of their gender. Sexist language will alienate some of your readers and undercut your goals in your document. We offer here some ways to make your writing more inclusive.

The "generic" pronoun *he* was once accepted as meaning all people, male and female. However, research now shows that audiences have difficulty envisioning female subjects when writers and speakers use *he*.

To avoid using *he,* you can recast sentences in the plural:

Original: An engineer must understand not only theory but also application when he works with his clients.

Revision: Engineers must understand not only theory but also application when they work with their clients.

You might also avoid the pronoun altogether:

Revision: Engineers must understand not only theory but also application when working with clients.

You could also shift the person of the pronoun to the first person (*I* or *we*) or the second person (*you*) or substitute *one*:

Revision: As an engineer you must understand not only theory but also application when you work with your clients.

Revision: As an engineer one must understand not only theory but also application when one works with clients.

You can use *he or she,* or *she or he,* but repeating these phrases too often is cumbersome:

> Revision: An engineer must understand not only theory but also application when he or she works with clients.

Terms as well as pronouns can include or exclude readers. Gender-marked terms, such as *manmade* and *chairman,* should be changed to terms that include both genders (*artificial* or *made by people* and *chair* or *chairperson* are acceptable substitutes in our examples). Also, you need to edit out suffixes that mark a noun as exclusively female or male. Feminine suffixes often imply lesser, special, or diminutive. For example, we now use *flight attendant* rather than *stewardess* and *waiter* or *server* for both genders rather than *waitress* for females. Suffixes such as "ess" and "ette" should alert you that the word might be marked as feminine and should be changed.

You should also avoid identifying jobs or activities as belonging to one gender or the other. If one gender has traditionally filled a role, avoid noting the other gender's entry into that role as an exception. Also, do not distinguish people by their gender (no more than you would do so by their race or religion). To do either makes it difficult for your audience to imagine the job or activity as being open to all. The underlined words in the following sentences and revisions show you what to edit out, and the problem presented by the word in the original sentence:

> Original: The <u>lady</u> manager introduces us to <u>her</u> staff.

> Problem: Implies that most managers are men, and calls attention to her gender as her distinguishing characteristic.

> Original: The <u>doctor</u> must <u>use</u> his talents to diagnose while the <u>nurse</u> must use hers to care for and heal.

> Problem: Implies a clear division of job and talent by gender. Only men can be doctors, and only women nurses.

> Original: I would like to introduce you to our two new employees. Dr. Collins comes to us from the <u>physics graduate program</u> at MIT, a <u>man</u> of much distinction already, and <u>Nancy</u> is <u>mother</u> to three and obviously a very busy <u>gal</u>.

> Problem: Identifies the male by his formal or professional title and the female by her first name, showing less respect for her. Also notes professional accomplishments for the male but places the female in her traditional private role. Finally, considers the male an adult and the female a youth.

In Example 8–11, we list some words and phrases that are considered exclusionary on the basis of gender and their acceptable alternatives. Although our list is not comprehensive, it should give you a good idea of what to look for as you do your final editing of your document.

EXAMPLE 8–11   Exclusionary Terms and Alternatives

Exclusionary Terms	Alternatives
fellow man	other people, humans
forefathers	ancestors
Founding Fathers	Founders
kinsman	relative, kin
layman	nonspecialist, layperson
man	human, person, individual
(to) man (verb)	to work, to staff, to operate
manmade	artificial, made by humans
manpower	personnel
spokesman	speaker, spokesperson, representative
workmanship	skilled work
workman	worker
statesman	political leader
chairman	chairperson, chair, head, moderator
clergyman	member of the clergy, cleric
congressman	member of Congress, Congressional representative
fireman	firefighter
foreman	factory supervisor
mailman	letter carrier, postal worker
salesman	salesperson, salesclerk
coed	student
poetess	poet
actress	actor, performer

Avoiding Other Exclusive or Discriminatory Language

Language can discriminate against groups other than women—perhaps on the basis of race, religion, age, special appearance, or nationality. Groups often prefer certain terms, and you show your respect by choosing these terms. For example, the terms *Latino/Latina, African American,* and *Asian American* are now the terms acceptable to those whose ancestry is so noted. People who are living with AIDS prefer to be called just that—or *people with AIDS*—rather than *AIDS' victims* or *people with the AIDS disease.* Older people prefer to be called *senior citizens* rather than the *elderly.*

People with disabilities prefer not to be called the *handicapped* ("The handicapped need ramps") but rather *people with handicaps or disabilities* ("People with disabilities often need ramps"). Those with poor or no eyesight, those who have lost their hearing, or those who get around in wheelchairs generally prefer to be called *physically challenged, hearing impaired,* or *visually impaired* rather than *handicapped, deaf,* or *blind.* If you have a question about what to call someone, it is polite to ask what they prefer and smarter than alienating readers by using discriminatory language.

Using inclusive language invites your audience to see themselves within your words.

Avoiding Jargon

jargon
the specialized vocabulary of a field or discipline

Jargon refers to the specialized vocabulary of a field or discipline. Because it is specialized, it may have particular meanings that substitute words do not have, and it may be appropriate when you know that all your audience understands these terms. But its specialization also means that jargon may have no meaning at all to many readers or listeners and may exclude them. People wanting to learn about the Internet and create their own Web pages, for example, are often alienated and frustrated by the specialized terms that programmers and designers use in manuals—HTML, cascading style sheets, URLs, tags, and other such terms which don't communicate well if the intended users don't know their meaning. If these users don't know the words, they can't read the manuals or follow the online help and do their tasks, and the document has failed—not because readers are stupid but because the writers have not adequately assessed how best to communicate with these readers. The use of jargon, like every other choice you make in writing and speaking, needs to be governed by who will read or hear what you write or say. For specialists, jargon often communicates more precisely than more familiar words, but for general audiences, it's best to avoid jargon and to define unfamiliar terms when you can't be confident that readers know their meaning.

Choosing Verbs in the Active Voice

voice
the nature of a persona in a document. Voice can be stern, friendly, condescending, formal, or gracious. Voice also refers to the relationship between the subject of the sentence and the action expressed by the verb

active voice
a grammatical construction that clearly identifies the agent of the action, e.g. the police arrested the burglar

passive voice
a grammatical construction that omits the agent of the action

Many style guides encourage writers to choose verbs in the active voice rather than the passive voice. **Voice** refers to the relationship between the subject of the sentence and the action expressed by the verb. In **active voice,** the subject performs the action expressed by the verb. In **passive voice,** the subject is acted upon or receives the action. In the first sentence of Example 8–12, the woman (subject of the sentence) does not do the attacking but receives the attack. In the active voice companion sentence, the assailant is the subject of the sentence and performs the action of attacking.

Active voice tends to make your style much more direct and explicit because it's natural to expect that the subject will perform the action. Active voice sentences generally have strong verbs and indicate who or what is the agent of action in the sentence. Passive voice sentences are usually longer and have some form of the weak verb *to be* in them. If you don't know the agent of action or if you want the reader's primary focus to be on the action itself or the object of the action, passive voice is appropriate. However, using active voice when possible will increase the clarity and energy of your writing.

Because active voice clearly indicates who is responsible for the action occurring in a sentence, its choice has ethical and practical implications. The clear assignment of responsibility in technical documents can often be the difference between safety and an accident that can lead to injury or costly litigation. In the first sentence below, the identity of the

EXAMPLE 8–12   Active and Passive Voice

The simple subject and simple predicate are underlined.

Active Voice	**Passive Voice**
An assailant attacked the woman.	The woman was attacked by an assailant.
The committee reached a decision.	A decision was reached by the committee.
Support staff must check surgical equipment at the beginning of their shifts.	Surgical equipment must be checked by the support staff at the beginning of their shifts.

persons responsible for maintaining a safe working environment is obscured by the passive construction. In the example below it, responsibility is clearly identified by the active voice:

> The steam pressure control must be turned off by the boiler room engineer before any maintenance of the turbine can be undertaken.

> The boiler room engineer must turn off the steam pressure control before any maintenance of the turbine can be undertaken.

Proofreading

proofreading
final check of a document for spelling, grammar and surface mistakes

Proofreading is often the final check of your document and occurs before you print your last draft. Your goal is to produce a clean copy ready for printing. Proofreading is your last chance to make sure that the document will look and read as you hope it will. Writing Strategies 8-7 gives you an idea of what to look for as you proofread.

WRITING STRATEGIES 8–7   Proofreading Checklist

In addition to any errors you can find in spelling and punctuation, proofread for the following:

- Hyphenation, especially at the end of lines.
- Extra spaces between words, between lines.
- Run together words.
- Failure to italicize or boldface where called for.
- Transpositions.
- Broken fonts.
- Inappropriate page breaks, especially headings that appear as the last line on a page and the last word of a paragraph on a separate page.
- Unaligned type.
- Uneven margins.
- Inconsistent indentations.

WRITING STRATEGIES 8–8   Summary Table: Reviews, Questions, and Checks

Editorial Review	Editor's Questions	Specific Checks
1. Suitability for audience and purpose	Does this document respond adequately to the task assigned?	■ What is the purpose of this document? What problem are you helping your audience solve? Have you placed your audience's interests and values into the foreground (as opposed to your own interests)? ■ Who will use this document? Does the document allow for multiple audiences and multiple needs? ■ Have you stayed on task—that is, have you responded to the goal that was set for the document?
2. Completeness and accuracy of information; ethics	Are all the required parts included? Is the content accurate and adequately developed for the intended audience? Do the content and discussion of it respect ethics?	■ Are all the parts that the assignment calls for included? ■ Have you anticipated and answered audience questions? ■ Have you provided background information, definition of terms and concepts, sufficient and complete evidence, interpretive statements to explain significance, all steps in a procedure, and safety information? ■ Is the information accurately represented? ■ Have you shown the ethical consequences of the information?
3. Organization and consistency	Are related parts treated in consistent ways? Does the sequence of information reveal the relationship of ideas?	■ Does this document follow genre conventions for the organization? ■ Does the sequence of information reveal the relationship of ideas? ■ Are related parts treated in consistent ways? ■ Do transitions between sentences make the paragraph flow?
4. Visual design and usability	Will page or screen layout, illustrations, and navigation devices (table of contents, headings, menu) help readers find what they need? Does the visual design enhance interpretation of content?	■ Does your document use the following design elements effectively, appropriately, and consistently? —Genre conventions for page design —Style manual guidelines —Typeface and type size

(continued)

WRITING STRATEGIES 8–8 (Continued)

     —Margin widths
     —Bold, italic
     —Headers and footers
     —Headings and subheadings
     —Numbers and bullets
     —Labeling of figures and boxes
     —Right, left, and center justified text
     —Measurements and mathematical formulas

- Check a hypermedia document in addition for:
  —Consistent use of color and graphics
  —Readability of type (type size and style, contrast with background)
  —Working hyperlinks
  —Accurate and complete navigation table

5. Style

   Do sentence structures and word choices clarify meaning? Is the tone right for the audience?

   - Do subject and verb of each sentence contain the most important information of the sentence?
   - Do most sentences use verbs in the active voice and descriptive, specific verbs?
   - Are nouns specific and concrete?
   - Does the tone seem respectful and professional?

6. Correctness of grammar, punctuation, and spelling; consistency

   Do grammar, punctuation, and spelling conform to conventions?

   - Are sentences grammatically correct? (Subjects agree with verbs, and pronouns with antecedents, etc.)
   - Does punctuation reflect the structure of the sentence?
   - Are all the illustrations, citations, and cross-references included?
   - Are numbers, dates, terms, capitalization, hyphenation, and abbreviations used consistently?
   - Do citations and cross-references follow a style guide consistently?

SUMMARY   Now that you have completed your document, review all the steps in the editing process to see if you are ready to present it to your audience. Writing Strategies 8-8 on pages 292-93 will help you do this.

- In this chapter you have learned that editing greatly increases the effectiveness and usability of your documents, and that editing includes a careful look at your document as a whole—its form, content, organization, and style—to make sure that it meets the needs of your audience.
- To edit your document effectively, you need to revisit your analysis of audience, purpose, and use. Editing requires a review of content, organization, and visual design in light of the audience's needs and assignment specifications.
- Editing also means checking to see that your sentences work, that they emphasize the main point and include proper coordination and subordination, that vague pronouns and slow starts have been eliminated, and that modifiers are used accurately. Editing for action verbs, concrete nouns, parallelism, and positive phrasing strengthens your document. A focused look at grammar, punctuation, spelling, and illustrations will help to ensure that your document is correct, consistent, and complete. Proofreading is the final check of your document.
- Editing and revising style are not simple tasks, but you cannot afford to ignore or slight them as you allocate time to the tasks involved in creating a polished document. Too many things—your reputation, the safety of the document's user, and your company's legal liability—depend on your careful attention. Like other aspects of the writing process, editing has ethical, political, legal, and moral considerations involved with it. Good editing also can mean the difference between whether a text will be suitable for a global audience or be narrowly restricted in its use.

ACTIVITIES AND EXERCISES

1. *Performance Appraisal.* Performance appraisals are common in business, government, and academic settings. They become part of an individual's personnel file and serve as a basis for decisions about promotion and salary increases. These documents also serve as a road map for a person's future work and are used to guide the career of the person evaluated. Therefore, performance appraisals must be clearly written. The performance appraisal below is too long, repetitious, and inadequately organized. Work with the material here, reducing it significantly and organizing it logically. Be careful not to alter the content.

> Professor Alexandra is an extremely impressive and outstanding young scholar whose progress toward tenure has been and continues to be consistently excellent. She is good natured, energetic, enthusiastic.

The people around her all respond well to her positive attitude. Professor Alexandra's abilities and skills as an administrator and leader in the department are excellent. Her desire to identify and seek out an increase in her administrative responsibilities as well as her other qualities are demonstrated in her outstanding performance as chair of the graduate committee for the past twelve months. During this time, she served as next-in-charge for the department head for two weeks as well, performing her work excellently.

Professor Alexandra continues to sharpen her scholarly and research skills by attending conferences and scholarly meetings on a regular basis several times a year. She has attended 3 conferences and 2 workshops in this evaluation period and has given a total of 5 scholarly presentations. She is extremely well read and is up to date in her field as well as up to date in a number of collateral areas. She also knows many of the more established scholars in her field. She does thorough research and is well prepared when it comes to advising students which makes her a much sought after advisor among graduate and undergraduate students. She has a keen and perceptive intellect and is especially quick at identifying and solving both research and administrative problems. She is well prepared to handle the complex tasks of administration which include making personnel decisions easily and without letting them confuse or upset her.

Professor Alexandra is very articulate in both her written and oral presentations, and she expresses herself clearly and forcefully. She is well liked and has good rapport with her colleagues, students, and with the civil service staff in the department, and she is well respected for her good judgment, easy-going nature, and fairness. Professor Alexandra continues to make a significant contribution to new knowledge in her field, publishing more articles in the last year than her colleagues in similar stages in their careers and being nominated for several scholarly awards. She was also nominated for the distinguished teaching award in both the department and in the college on the basis of her quality advising and her teaching evaluations. Professor Alexandra is a young professor of the quality we rarely find. Her scholarship, her teaching, her visibility in her field of research as well as her administrative and leadership abilities suggest that she will be ready for early promotion before others who came into the department the same year she did. She is extremely well qualified for administrative duties, and I will recommend her with great enthusiasm for selection as one of the college's nominees as outstanding young professor. Her scholarly and teaching career as well as her administrative future is extremely promising. She has the potential for a highly successful and positively regarded career in the academic world.

2. *Style Guides.* Choose two journals in your field. In the front or the back of each journal should be listed information about manuscript submissions. From this information, create a short style guide or checklist that should be used to help a person preparing a manuscript for submission to these journals. For example, the *Technical Communication Quarterly* requires manuscripts to have

an informative abstract of 50–75 words, to have titles centered on top of the first page of text, and to indicate first-level headings by boldface or all capital letters.

3. *Exclusive Language.* Identify the sexist, discriminatory, or exclusive language in the following sentences and edit to make them inclusive:

The girls in the main office always provide support for the salesmen in the field.

The groups represented in the survey were policemen, firemen, housewives, and waitresses.

Each man has to weigh the consequences of his own actions.

Engineering is a challenging occupation for any boy, or for any girl, for that matter. The math involved demands a youthful mind, the spirit of the hunter or sportsman.

We must take special care to help our handicapped clients realize the alternatives they might miss in our building designs. Those confined to wheelchairs, those who cannot see, the mentally retarded, those suffering from any abnormal condition need design features pointed out to them.

For a woman, she possesses a competitive spirit and exceptional endurance.

Every user needs to learn our graphics and word processing software through examples and exercises. Whether it's the housewife recording her favorite recipes, the young boy designing computer games, the young girl describing her dreams of romance in a new "electronic" diary, the older person writing a letter to his grandson, or the woman listing emergency babysitters to step in when she's at work, each person needs to learn the value of the software in his particular way.

WORKS CITED   Edmond, Lauris. "Imagining Ourselves." An address given at Palmerston North, Massey University. In *Writers on Writing.* Robert Neale, ed. New York: Oxford, University Press, 1992.

Turpin, Elizabeth, and Judith Gunn Bronson. "Technical Editing." In *Foundations for Teaching Technical Communication: Theory, Practice and Program Design.* Katherine Staples and Cezar Ornatowski, eds. Greenwich, CT: Ablex, 1997, pp. 221–22.

Document Design and Packaging

By revising its forms, Citibank reduced the time spent training staff by 50% and improved the accuracy of the information that staff gave to customers.

> "Plain English Pays," *Simply Stated* (Washington, DC: American Institutes for Research, February 1986).

Since the British Government began its review of forms in 1982, it has scrapped 27,000 forms, redesigned 41,000 forms, and saved over $28,000,000.

> Robert D. Eagleson, *Writing in Plain English* (Canberra: Australia: Australian Government Publishing Service, 1990).

Southern California Gas Company simplified its billing statement and is saving an estimated $252,000 a year from reduced customer inquiries.

> "Gas Utilities Switch to Plain Language," *Simply Stated* (Washington, DC: American Institutes for Research, March 1986).

Introduction

In this chapter we will teach you how to make wise choices regarding a document's design and final packaging. Design features include everything from white space—that empty space in the margins and around headings and figures—to the use of color and the choice of binding. In flipping through this textbook, you might notice that we've used examples throughout and that there is room in the margins for our definitions and your own notes. Other than that, you might not have noticed much about how the book was put together. That's because design features are supposed to be "invisible" or at least unobtrusive. Their purpose is to make reading easier, not to call attention to themselves. For that reason, you may not know some of the most common aspects of design at this point and probably little of the terminology; this chapter begins to remedy that by making you more aware of design features in general, especially those you can use to enhance the readability and effectiveness of your technical documents. This chapter focuses primarily on print design; however, you will also find some advice on designing for online documents and Web sites later in the chapter.

clarity
cluster of elements that includes freedom from ambiguity, logical development of ideas, unity, coherence, appropriate style and usage, clear transitions, and accurate word choice

Probably two of the most important reasons to choose design features carefully are clarity and consistency. **Clarity** helps your readers "see" the content of your document and to determine the meaning and importance of each piece of information. For example, the headings within this textbook are designed to help you see the hierarchy or order of the information as well as the major subjects contained in the book. Design features that support clarity help the reader distinguish between one item and the next and help the reader better understand the meaning of those items. For example, when you warn your readers of possible harm if they do not carry out your instructions to the letter, you would set aside that warning in a box, with the letters in white and the background in red. Or, to clarify a step within those instructions, not only would you *write* "Remove the four screws at the top of the printer," but also you would show an illustration of the printer top with the screws highlighted with arrows and add a caption to that illustration.

Consistency helps your readers link or relate items and to distinguish between them. For example, each time the readers of your document encounter a heading all in capital letters indented five spaces from the margin, they will know that the information that follows each of those headings will be of similar importance. Consistency in your design techniques helps you visually display the arrangement that you have decided best organizes your information. For example, each time your readers encounter an arabic number in your instructions, they realize that they must complete a new step in the process and that this step is probably just as essential as the one that precedes it and the one that follows it. Finally, clarity and consistency are related concepts. Often once you decide how to clarify an item or a piece of information, you need to be consistent in using the same design technique each time you discuss a similar piece of information. These examples again pertain to printed documents. With online documents, you would still use organizational signals, but instead of conventional headings, for example, you might achieve consistency by using similar navigational buttons.

Deciding on how to achieve consistency and clarity in your documents is the first step in helping your readers appreciate and absorb your information. You can also use document design techniques to attract and keep their attention, to help them read selectively or skim, to help them locate possible answers to their questions, and to help them remember information and to execute plans. We try to address all these challenges with this chapter but begin with the most important—clarity and consistency. You will notice that the first part of this chapter focuses on print document design. If you are involved in designing online documents and messages, you should turn to the later part of this chapter first.

Design Techniques to Achieve Clarity and Consistency

In this section we introduce you to some common document design techniques to help you clarify your information and achieve consistency throughout your document. When readers encounter a piece of information, they rely on your document design choices to help them answer three questions: What is this item of information? How is it similar to others? How is it different? Therefore, your clarity and consistency decisions help readers identify items, group items, and distinguish between them. In this section we discuss how to achieve clarity and consistency through headings and subheadings, white space, margins, visual elements for highlighting, and typography.

HEADINGS AND SUBHEADINGS

Headings and **subheadings** distinguish between various sections of your document. They also help the reader identify the topic of each section and understand which sections are similar to other sections. Efficient readers skim headings and subheadings to locate and use information. To signal relationships between major sections and subsections, use two or three levels of headings. For example, look at the three levels of headings used below:

headings and subheadings navigational devices that divide text into sections and subsections and help the reader's comprehension and selective reading

This is a Level One Heading

This is a Level Two Heading

This is a level three heading

As you can see, the level one heading is boldface and starts at the left margin. The level two heading also starts at the left margin but is not boldfaced. The level three heading is indented and set in *italics.* Your heading choices might already be determined by your company or publication style.

However, it is always important to remember that these headings signal to the reader which sections of your document are similar and which are different from other sections. For example, every time your readers encounter a level three heading, they might anticipate that you have provided them with another recommendation in your report, another step in your instructions, or another example in your memo.

WHITE SPACE

white space
empty or blank areas on a page; guides readers through structure, prevents fatigue, and creates aesthetic appeal

White space is the empty or blank areas on your page. White space guides readers through structure, prevents fatigue, and creates aesthetic appeal. You should devote at least 50 percent of a page to white space, never less. Also, as a general rule you should have more white space on the edges or margins of your text than in between items in your text. If you have too much white space between items in text, these items will seem to blow apart, and your reader will lose any sense of consistency or similarity among items. Generous white space in the margins, on the other hand, will appear to press in on the text and hold it together.

White space also distinguishes between items of information, as it sets one apart from others. White space also links items of information, as similar items might be indented the same number of spaces, for example. Look again at the headings displayed earlier—your reader will not only distinguish between level two and level three headings because one is set in italics but also will notice that level three headings are indented.

You can also add white space to your text by creating additional space between paragraphs and between lines. For documents such as letters, you can center the text on the page; that is, don't let the letter fill only the top half of the page. Adjust and add white space to enhance the letter's organization and make the content easier to read. Example 9–1 shows a letter lacking white space and one with ample white space. Each letter contains the same amount of text. The letter on the left-hand side appears more difficult to read, almost cluttered, while the one on the right looks easier to read and therefore more inviting. Also, the paragraphs in the letter on the right are separated by white space, which helps the reader distinguish between where one paragraph begins and the other ends.

margins
white space at the top, bottom, and both sides of a page

left-justified margin
all lines of text begin at the left margin; readers are used to reading left-justified text

right-justified margin
all lines of text line up along the right margin

MARGINS

Ample **margins** also help create needed white space on a page and add consistency to your document. A **left-justified margin** means that all lines begin in a straight line at the left margin. Readers are used to reading left-justified text. A **right-justified margin** means that all lines also line up along the right margin. Right-justified margins help readers identify discrete chunks of text or distinguish between them. A **ragged right margin**

EXAMPLE 9–1 Two Letters Positioned on a Page

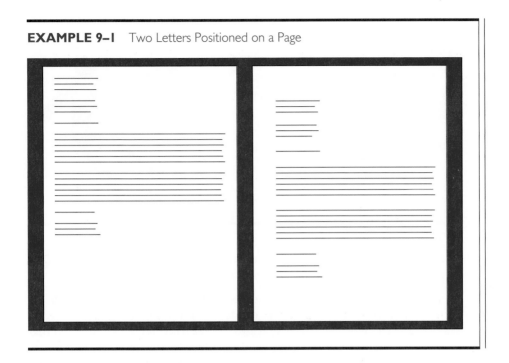

(that is, it is not justified) guides readers to continue reading and avoids unnatural spacing between words in a line of text that often appears with right-justified margins. What type of justification is used in Example 9-1? Why should this help readers?

VISUAL ELEMENTS FOR HIGHLIGHTING

Lines, borders, circles, and other shapes emphasize, isolate, and separate information to help readers locate and understand text. You can emphasize crucial information with a line or **rule** to attract attention. Use boxes to isolate and separate information. Borders and circles also help to isolate text for emphasis.

Information such as warnings, caution statements, and legal contracts should be highlighted so that readers can find it easily and consistently. Shading this information with an appropriate color (as determined by ANSI standards for danger, warning, and caution), placing a box around the information, and/or identifying it with an appropriate icon will help readers to locate important information.

Product liability guidelines require that the manufacturer must warn of a danger or hazard if the product is dangerous, if this danger is known by the manufacturer or product seller, if the danger is not obvious to the user, or if the danger is not one that arises only because the product is used in an

ragged right margin
a margin that is not right-justified; guides readers to continue reading and avoids unnatural spacing between words in a line of text that often appears with right-justified margins

rules
lines used in the text to emphasize information or to draw the reader's attention

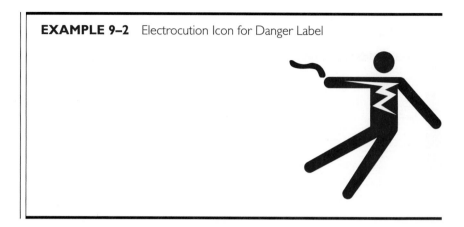

EXAMPLE 9–2   Electrocution Icon for Danger Label

unexpected way. Warnings must identify the severity and nature of the danger and clearly tell the user how to avoid this danger.

1. The word *danger* indicates an imminently hazardous situation that, if not avoided, will result in death or serious injury.
2. The word *warning* indicates a potentially hazardous situation that could result in death or serious injury.
3. The word *caution* indicates a potentially hazardous situation that could result in minor or moderate injury.

Such warning labels are usually accompanied by a picture or symbol that depicts the hazard or shows the reader how to avoid the hazard. Example 9-2 shows an ANSI standard icon symbolizing the potential hazard of electrocution; this picture would accompany a danger warning.

TYPOGRAPHY

To take the burden out of reading, you need to make conscious decisions regarding the typography of your document. **Typography** refers to the look and legibility of letters; this includes their case, style, size, spacing, and length on the page. Your choices in typography help your reader distinguish between letters and words, a point of clarity, and to group similar items of information, a matter of consistency. Your choices include how to display individual letters as well as blocks of letters or words.

Case

Perhaps you're most familiar with **case** (lowercase and uppercase or all capital letters). Lowercase letters which have any part of the letter that extends beyond an "x," for example, or below the baseline, have **ascenders** (e.g., *h* and *b*) and **descenders** (e.g., *y* and *p*) that create a distinctive shape for a word. Draw a close line around any word in this paragraph, and you'll see that your

typography
refers to the look and legibility of letters; includes their case, style, size, spacing, and length on the page

case
refers to lowercase and uppercase letters

ascenders
parts of the letter that rise above the line (e.g., *h* and *b*)

descenders
parts of the letter that drop below the line (e.g., *y* and *p*)

line goes up and down around these ascenders and descenders. These irregular outlines help readers recognize words and read with ease—they clarify.

Thus, contrary to what you might think, ALL UPPERCASE WORDS ARE NOT EASY TO READ; THEY PROVIDE NO IRREGULAR OUTLINE TO SPEED UP READING AND TO DISTINGUISH BETWEEN LETTERS. Moreover, users of online documents, those documents that appear on the computer screen, report that reading documents with all uppercase words makes them feel as if they are being shouted at. Therefore, use all uppercase letters for WARNINGS or other information that must be intimidating. Otherwise, use uppercase letters only as capital letters to begin sentences or for titles and headings. For example, each time your readers see a phrase set in all caps, they know that they will encounter similar information to the last phrase set in all caps, a matter of consistency.

Typeface

Typeface or **font** refers to a collection of type that has a unique name and set of distinctive characteristics. A typeface or font comes in two styles, **serif** and **sans serif** (French for "without" serifs). Serifs are the little "feet" and "hats" attached to the ends and tops of letters. Serifs on the ends of the letters extend or pull a reader's eye across the line of text. A typeface such as Palatino is a serif typeface; that is, it has these little extensions:

> This is Palatino.

A typeface such as Helvetica is a sans serif typeface; sans serif typefaces do not have extensions:

> This is Helvetica.

You might base your decision on what is familiar to your readers. For example, people in the United States most often use serif typefaces, although most people in Europe use sans serif typefaces (Wright). For large blocks of type, however, serif typefaces are often best because the serifs pull the reader from one letter to the next. As you will see later in this chapter, this advice is not true for online documents where sans serif typefaces are preferred.

Size or Point

You'll hear people refer to **type size, point,** and **font size** somewhat interchangeably. For your purposes, they all refer to the size of the print. Technically, a point is the unit for measuring font size: one point is equivalent to .014 inch, and 72 points equal one inch. For titles and headings, use a larger point size such as 12, 14, or 18 point. For text, standard point size ranges from 9 to 12 points and, for online design, 11 or 12 point type works best. If you're using columns, the narrower the column, the smaller point size you should use, so that your reader encounters a sufficient number of words before jumping to the next line of text. Point size will help your

typeface, font
a collection of type that has a unique name and set of distinctive characteristics

serif, sans serif
Two styles of type; serifs are the little "feet" attached to the ends of letters; *sans serif* means "without" serifs

type size, point, font size
terms often used interchangeably to refer to the size of the print in a text

readers distinguish between items of information; for example, your level one headings might not only be set in boldface but also be a point or two larger than your level two or three headings.

Spacing

spacing
the distance between
letters and between lines of
type on a page

Spacing is the distance between letters and between lines of type on a page. People use two terms to describe spacing: leading and kerning. **Leading** refers to the space between lines of text; generally, shorter lines need less space between them. Type fonts with short ascenders and descenders need more space between lines so the text doesn't look too heavy and dense. For example, in a line set in Palatino, the letters themselves might be 11 points high and the space between the lines 2 points. In Helvetica, a sans serif type, the letters themselves might be 11 points high and the space between the lines 1 point. We designate the first as Palatino 11/13 (letters set 11 points with 2 points between lines for a total height of 13 points) and the second as Helvetica 11/12. **Kerning** refers to the space between letters. Some typefaces create extra space between letters and may be harder to read. Your decisions about leading and kerning help your readers distinguish between letters, lines, and blocks of type.

leading
the space between lines of
text

kerning
the space between letters
in a word of type

Length

line length
the width of a column of
text

Line length is the width of a column of text. There are at least two ways to judge proper line length. You can count off and set about 50 characters per line. Or, you can set an alphabet and a half of your chosen type style, that is, the length of the letters *a* through *z* if set in a row. Longer lines make it harder for readers to retain information and return to the beginning of the next line of text. Online documents require much shorter lines.

Highlighting

highlighting
a way to emphasize text by
using **boldface,** *italics,* or
underlining

Just as you can **highlight** items in your document by lines, borders, circles, and other shapes, you can emphasize or highlight text by using **boldface,** *italics,* or underlining. Keep in mind that excessive use of any of these techniques wears on readers. The more you boldface words in your document, the less readers will notice and retain. Therefore, use these techniques sparingly to highlight only what really needs emphasis.

Highlighting items in your document will distinguish between them and make them stand out within your text. For example, if you are discussing a significant recommendation in your report, you might set that recommendation as follows:

> We recommend that all engineers submit **progress reports** within the first stage of an investigation. These **progress reports** must contain a complete budget, a revised schedule, a list of personnel involved, and a proposed deadline.

EXAMPLE 9–3   Instructions for a VMX (Voice Mail) System

Entering Your Mailbox

1. Call the VMX System

2. Press #

3. ■ Enter your mailbox number

4. ■ Enter your security code

❏ Single Message Queue

"You have X messages. Ready . . ."

   5 Listen to first or next message

  1 5 Listen to last or preceding message

❏ New and Saved Message Queues

"You have X new and Y saved messages. Ready . . ."

1 9 1 New Message(s)

1 9 2 Saved Message(s)

   5 Listen to first or next message

  1 5 Listen to last or preceding message

While Listening to Messages

  2 Back up several seconds

2 2 Back up to the beginning

  3 Erase the message

  4 Go forward several seconds

  5 Listen to next message

  7 Save message

  8 Time and Date message was received

❏ 1 3 Forward the message

❏ 1 4 Immediately call the sender

  1 5 Listen to a previous message

❏ 1 7 Reply with a message

  1 8 Adjust volume

  * Pause/Resume

Listening

Source: Adapted and reproduced with the permission of VMX, Inc. Copyright © 1992 VMX, Inc. All trademarks identified by the TM and (R) symbol are trademarks or registered trademarks, respectively, of VMX, Inc. All other trademarks belong to their respective owners.

The reader's eye is naturally drawn to the boldfaced words "progress reports"; this highlighting clarifies what the recommendation addresses. Also, the underlined list draws the reader's attention to what the progress report contains, and the reader notes that each item in the list is of equal importance because each receives similar highlighting.

Now that you have read how design features help you achieve consistency and clarity in your documents, look at Example 9-3 to see how consistency and clarity are achieved visually in the instructions on how to use a voice mail system. Notice that each of the primary categories—Entering Your Mailbox, Single Message Queue, New and Saved Message Queues, and While Listening to Messages—are consistently displayed. For example, each category appears in larger type size than any other information on the page. Two of the categories—Single Message Queue and New and Saved Message Queues—are further highlighted by a shaded box to the left of the heading.

The reader's eye is drawn to the three identical boxes on the right-hand side of the document that highlight common actions the reader might take to forward a message, call the sender, and reply with a message. The italicized information below the Single Message Queue and New and Saved Message Queues clarifies these categories by telling the readers what they will hear within the VMX system. Notice also that each button that the readers will elect to push—for example, 4 for "go forward several seconds" or 14 for "immediately call the sender"—are set to represent the buttons on a phone. These icons clarify to the readers that they must push these buttons and are highlighted consistently throughout the document. Finally, each line of text is set in serif typeface, and generous white space separates each item of text so that readers can easily find the steps and processes.

In the next sections of this chapter, we will discuss how your design features can help you respond to other challenges, such as engaging your readers, helping your readers locate possible answers to their questions, helping your readers read selectively and with ease, and helping readers remember what they have read.

Design Techniques to Engage Readers, Help Them Locate Answers, Read Selectively, and Remember

When all else is equal in printed materials, we know that physical attractiveness can tip the balance for readers. As technical communication scholars Martha Andrews Nord and Beth Tanner stated, "Attractive documents look as if the producers value the information enough to care about its appearance and suggest that the same care has gone into writing the text as was invested in its design" (224). Look across the texts on any bookshelf or the many handout packets that you have purchased during your college career. What packaging features motivated you to use a textbook or other document?

We also know that readers need navigational tools or guideposts to help them find the location of a potential answer. Numerous studies have found that the most effective readers are those who use guideposts such as tables of contents, indexes, and headings to aid their search for information. Thus, even in the shortest document (e.g., a one-page memo or a short form), you should use document design techniques that allow readers to locate information quickly and use that information effectively.

You can motivate your readers to open a document and locate a potential answer by focusing on the packaging as well as the larger organization of a document. Writing Strategies 9-1 presents a reader's main goals before reading as well as the document design techniques you can use to engage readers.

WRITING STRATEGIES 9–1   Reader Goals and Design Techniques I

Reader's Goals Before Reading	Writer's Document Design Techniques
To formulate a question and become motivated to open a document and locate an answer	Focus on packaging: Attractive binding High-quality paper Balanced and consistent layout Interesting graphics Legible cover and print Appropriate use of color
To find the location of a potential answer	Focus on the larger organization: Table of contents Indexes Structural headings Tab markers and dividers

---

ENGAGING READERS

You can engage readers with a document by considering your readers' needs and focusing on attractive packaging and the structural layout. These document design features create a strong first impression. Although there are numerous document design techniques for this stage of reading, here are some starting techniques and examples in each case. In this section we cover some of the external features of a document, such as binding and covers, as well as give some important advice on color.

Attractive Binding, Cover, and Paper

Generally, large documents are bound by spiral binding, by staples through the center margins, or by glue on the side or spine. If readers need to keep a large document (more than 50 pages) open to a specific page, provide a spiral binding. If possible, provide identifying information on the document's spine or binding so that readers can locate the document quickly on a bookshelf. For reports submitted to your classes in college, however, a simple plastic cover purchased in your college bookstore will dress up the document and make it look more professional. Although there are numerous materials available for creating covers for documents, choose one that showcases your document without detracting from it. Make the title on the document large, legible, and clear.

If you want your document to be noticed and appear to be produced with care, use bond or other high-quality paper. If you are printing a document on both sides of the paper, choose high quality paper so that the text doesn't bleed through to the other side.

Structural Layout and Graphics

Begin major sections of a document with a consistent layout (note the first pages of every chapter in this text). When using figures and tables, label them consistently and place them as close to their textual explanations and citations as possible. Use a common graphic to unify a group of documents such as tutorial and reference texts or field and laboratory reports.

Color

color
used in a document to link or contrast objects, focus attention, serve as a navigational aid, or express values in charts and maps

Choose a **color** scheme that works for your readers. For example, a bright color scheme for a train schedule may not work for forms or labels. You can unify or achieve consistency in a set of documents or among items of information with a similar color. However, do not overuse color or select multiple colors to be flashy; this distracts rather than engages readers. Also, use caution when you use color and avoid it if color-blind readers might be confused or endangered. Consider these basic guidelines:

- Use only four to six color codes.
- Use color conventions from the reader's work world (e.g., red for danger, blue for water), but be aware of cultural differences when you choose a color.
- Pick harmonious colors and avoid conflicting colors.
- Never use colors that compromise legibility.
- Never depend on color alone for any critical distinction (Horton, "Visual Literacy"; "Pictures Please").

Color theory can be complicated. If you are designing a long document, such as a computer software manual or a document whose success depends on attracting and keeping the reader's attention (e.g., an advertisement), here are some additional guidelines.

hue
a synonym for color (e.g., red, blue, green, purple); often used to refer to shades of color

primary colors
the dominant colors in a spectrum, such as red and blue, which provide contrast in a document

complementary colors
the opposite of each primary color on a color wheel

Colors are selected first by their **hue** (a synonym for color). White is the absence of color, and, particularly, in design and printing, black is considered a color. The **primary colors** are dominant because they appear more solid and provide great contrast in a document. For example, setting your tabs or headings in red, blue, and yellow will signal significant difference or distinction to your reader. However, there are several sets of primaries. Reflected light primaries relate to the dominant areas of the right, left, and middle of the visible spectrum and are red, blue, and green. A **complementary color** is the opposite of its primary and also provides contrast to that primary. For example, in the reflected light color scheme, magenta is opposite of green on the color wheel and consists of a blend of blue and red.

Again, complementary colors help you distinguish between items of information. On the other hand, adjacent colors, those that appear next to each other on the color wheel, such as red and magenta, are not in great contrast; instead they appear almost as shades of each other. Therefore, you might use adjacent colors to help your reader distinguish between items that do have some similarity to each other. Or, you might use one color, such as red, to make note of similar items but use a different value or tint to each item in red.

Value refers to the lightness or darkness of a hue. Different values of red, for example, can both help you to achieve consistency among items of information and to clarify or distinguish between them. All of your headings might be set in red, but those that occupy a lower place in your hierarchy or arrangement, such as level three headings, might be set in a lighter value red. Finally, the **chroma,** or the intensity or saturation, of a particular hue makes a difference in how much we notice the color. For example, red is much more intense if it does not contain much gray and can be used to highlight significant information, such as warnings in instructions.

When selecting color to attract your reader, to establish associations among items of information, and to help your reader remember, be sure to look carefully at each hue as well as the hues in context. Some colors that appear to be warm (e.g., red and yellow), are more visible than cool colors (e.g., blue and green), and appear to advance toward the reader. Therefore, if you set a red heading against a blue background, the item of information you place in red will gain more attention than the items in blue. Also, the darker the background, the lighter a color will appear against it, so much that it might wash out and lose attention. Moreover, colors that are too close in hue and not separated by white space will blend and make it difficult for your reader to see distinct items. For example, if the bars or the lines in a graph are set in the same hue but with different tints and no black lines or space between them, the reader cannot distinguish among these bars or lines.

Finally, although color is effective in helping us engage the reader and in achieving consistency and clarity in documents, it is expensive. In designing your documents, you are most likely to use **spot color** or solid or tinted colors that do not overlap with each other to create new hues. Spot color is very effective in document design and much less expensive than full or four-color processing. **Four-color processing** requires the separation of hues from the original color art and applies inks from four successive color runs on the same page to reproduce the original art. The four color runs apply yellow, magenta, cyan, and black and are screened and applied carefully to create tones and shadows. If you must rely on color photographs reproduced in a document such as an advertisement to show an item realistically and accurately, you will probably depend on four-color processing. But again, your document will most likely rely on spot color to help you clarify and achieve consistency. These suggestions vary somewhat with regard to online documents (see the last sections of this chapter).

Of course, when selecting colors to use in a document, you need to be aware of differences among cultures and settings on an international scale. In the future, designers hope to have a uniform color code that translates well in all cultures and settings. For example, they have proposed the following

value
the lightness or darkness of a hue

chroma
the intensity of a particular hue

spot color
solid or tinted colors that do not overlap with each other to create new hues

four-color processing
separation of hues and application of inks from four successive color runs to reproduce original art

scheme for road signs: red means "no"; blue means "OK"; red overrules blue; green symbolizes health services; and yellow means warning. Rather than provide here a list of what various colors represent in different international settings, we simply ask you to think about this: Suppose that a designer who was unfamiliar with color symbolization in the United States used green for stop signals and red for go signals in a major U.S. city. Imagine the chaos! Now think about the confusion you might cause in your own documents to be used in international settings if you failed to research color symbolism for those cultures.

Example 9–4 shows the cover of a document that describes how to conduct emergency medical treatment for children. This document had the following design techniques to engage readers:

- The spiral binding enables the document to stay open once the reader locates the appropriate section.
- Although you cannot "feel" this document, it is made of card stock (heavy paper) with a waxed coating. Thus, it is durable, and if it gets dirty, the reader can wipe it off easily.
- Tab markers on the right side make the structural layout of the document automatically visible from the front cover of the document.
- The visual on the front cover automatically cues readers about the purpose of the document.

HELPING READERS LOCATE POSSIBLE ANSWERS

Once readers become motivated to open and use a document, they need to locate possible answers quickly and efficiently. To aid readers, you need to focus on the document's larger organization. Essentially, you need to divide the text into manageable units and signal or clarify relationships between these units. We suggest the following techniques.

Table of Contents or a List of Headings

For larger documents, include a table of contents. Many word processing programs include a table of contents utility. Even for short documents (say two pages), providing a list of the headings at the top helps readers to determine whether they have located the appropriate section. This list also functions as a preview of the document and guides readers' under-standing of the information. When developing a table of contents or a list of headings, follow suggestions for developing lists given later in this chapter.

As an example, note the contents for the Centers for Disease Control's *Fact Book FY 1991* in Example 9–5. This example also shows this docu-ment's two-page structural organization.

EXAMPLE 9–4   Cover for an Emergency Medical Treatment Document

916363-00*$7.95*EMT, Inc.

EMERGENCY MEDICAL TREATMENT:

CHILDREN

TECHNICALLY
REVIEWED BY THE
**NATIONAL
SAFETY
COUNCIL**

A HANDBOOK OF WHAT TO DO IN AN EMERGENCY TO KEEP A CHILD ALIVE UNTIL HELP ARRIVES.

NOT BREATHING • NO PULSE • CHOKING • BURNS • BLEEDING • SEIZURES • POISONING • DROWNING • HEAD, NECK, BACK INJURY • ACCIDENT PREVENTION • BROKEN BONES

By Stephen Vogel, M.D. & David Manhoff

Source: Stephen Vogel, M.D. and David Manhoff, *Emergency Medical Treatment: Children.* Wilmette, IL: EMT, Inc. Copyright © 1984 by EMT, Inc. All rights reserved. Reprinted with permission.

EXAMPLE 9–5   Table of Contents from the Centers for Disease Control *Fact Book FY 1991*

Fact Book FY 1991

U.S. DEPARTMENT OF HEALTH & HUMAN SERVICES • Public Health Service • Centers for Disease Control

Mission

To prevent unnecessary disease, disability, and premature death and to promote healthy lifestyles.

Source: U.S. Department of Health and Human Services, Public Health Service, Centers for Disease Control.

Indexes

For larger documents, you need to include an **index** that allows readers to find specific information quickly. Readers locate information via a good index far more often than from a table of contents. Many word processing programs include an indexing utility.

Structural Headings

Structural headings identify the largest chunks of information in a document. **Running headers** and **footers** appear in small type at the top or bottom of pages and repeat information such as topic, page number, and the date. Magazines and newspapers use a **standing head**—a separate line above the text and set off by a different typeface or graphic feature—to identify regular departments such as editorials. You can also use icons as structural headings; for example, the icon below might be used to indicate a section about time limits.

You can also guide readers by using **hanging heads** or headings placed in margins so that readers can skim and skip text.

Tab Markers or Dividers

Separate sections of a larger document by using **dividers** or **tab markers.** Dividers are sheets of paper inserted between sections in your documents and may appear in color. Tab markers are made out of heavy paper or of plastic and are attached to the right edge of the opening pages of each section in a document. These function as structural headings and quickly guide readers to appropriate sections.

HELPING READERS READ SELECTIVELY AND WITH EASE

Once readers have found the location of a possible answer, you need to help them read selectively and easily in order to make a decision. We have already covered the basic design elements that help a reader read with ease in the beginning of this chapter when we discussed clarity and consistency. Writing Strategies 9-2 summarizes those elements for you again and notes how they enable the reader to read selectively and with ease when finding the information essential to make a decision.

To further help readers read selectively, you need to visualize the layout of a page of your document. For this stage, you need to focus on the overall design of your page.

---

index
navigational device at the back of documents that helps readers find information quickly

structural headings
identify the largest chunks of information in a document

running headers, footers
navigational aids in small type running at the top and bottom of a page, repeating information such as topic, title, page number, or date

standing heads
separate lines of text above the main text and set off by a different typeface or graphic feature to identify different components of a document

hanging heads
headings placed in margins to facilitate skimming through a text

dividers, tab markers
sheets of paper inserted between sections; tabs are heavy markers attached to the right edge of a page; both guide readers to sections of a document

WRITING STRATEGIES 9–2   Reader Goals and Design Techniques 2

Reader's Goals During Reading	Writer's Document Design Techniques
To read selectively	Focus on guiding readers by using: Grid layout White space Headings and subheadings Margins Color Visual elements (boxes, rules)
To read with ease and use information to make a decision	Focus on typography: Case Style (typeface) Size (point) Spacing (leading and kerning) Length Highlighting

Grid Layout or Page Design

Imagine your page as a grid, a blueprint that you can repeat on each page for consistency. Does the purpose for your document warrant a traditional one-column format, or should you think about using a two- or three-column format? Where should you place graphics or other visuals in this **grid layout?** Some companies specify a layout that they use consistently in all their internal and external publications. However, if you have the freedom to choose layouts, keep in mind the following guidelines.

A one-column format is standard for most class essays and reports you might do in school. However, with a two- or three-column format, you can reserve one or two columns for headings and for graphics and one or two columns for text, and you have more freedom to use white space to separate these elements. Look at the two different page layouts displayed in Example 9–6. Notice that with the one-column format on the left-hand side of the example, all graphics or illustrations must be placed below the text. Readers would have to rely on in-text references to "find" their way to these illustrations, and the text that describes the illustrations might be far away from them. On the other hand, the page layout on the right-hand side of the example offers greater clarity to the readers because the illustrations are placed alongside the explanatory text, text that serves almost as a caption to the illustrations. Readers also encounter the three sections—each section consisting of a block of text and an illustration. Notice also how the white space that divides these

grid layout
use of the page as a grid or blueprint upon which to make design decisions (e.g., number of columns and placement of graphics)

EXAMPLE 9–6 Two Different Page Layouts

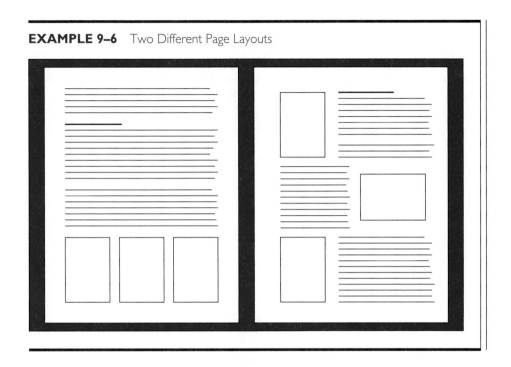

sections helps readers understand the organization of the page and choose which text block and illustration pertain to their interests and needs.

In some circumstances, you are not limited to the standard 8 1/2-by-11-inch page for your layout; you can use a two-page spread called a **folio.** Example 9-7 presents a two-page spread or a folio from the emergency treatment document. Notice all the design elements used in this page from the emergency medical treatment document, whose cover we saw in Example 9-4.

folio
the two-page spread used in page design

Notice how color is used not only to link objects but also to distinguish between them and to focus the reader's attention on important information or actions that the reader's should take. Color can then speed the reader's search through color-coding symbols.

HELPING READERS REMEMBER

After readers have read your document, they need to execute action plans or implement decisions based on what they have read. Writing Strategies 9-3 lists goals and techniques for this stage of reading. You can help readers implement their decisions by providing lists, examples, glossaries, and appendixes.

EXAMPLE 9–7   Folio from the Emergency Medical Treatment Document

Not Breathing/No Pulse

Gently tap or shake child. Ask: "Are you OK?" If child does not respond, shout for help.

1 Open airway, check breathing.
Lay child on back. If you must roll child on back, keep (support) head and neck in straight line. Tilt head back gently, lift chin slightly. If you suspect neck or back injury, pull open jaw without moving head (see inset). Look, listen, feel for breath (3-4 sec.). If not breathing, or you are in doubt, start rescue breathing (#2). If breathing, place child on side (unless head, neck or back injury).

2 Give two slow breaths.
Pinch nose. Cover child's mouth with yours. Give two slow, *gentle* breaths (1-1$\frac{1}{2}$ seconds each) into child's mouth. Allow chest to rise and fall between breaths. NOTE: *Watch chest.* If chest does not rise and fall after 2 breaths, retilt head, lift chin up and try again. If airway is blocked, go to picture #3. If chest does rise and fall, *check pulse* as in picture #5.

3 Something in windpipe.
Straddle child's thighs. Place the heel of one hand just above the navel, but well below lower tip of breastbone. Place your other hand directly on top of your first hand with fingers pointing to head. Press upward into stomach with up to 5 quick thrusts. Open mouth by grasping tongue and lower jaw between thumb and fingers, and lifting. Only if you see object, gently sweep index finger (hooking motion) deeply into mouth at base of tongue to remove from throat.

4 Repeat two slow breaths.
Tilt head back. Lift chin. Pinch nose. Cover child's mouth with yours and give two slow, *gentle* breaths. Watch chest rise and fall. Repeat #3 and #4 if necessary.

(continued)

EXAMPLE 9–7 (Continued)

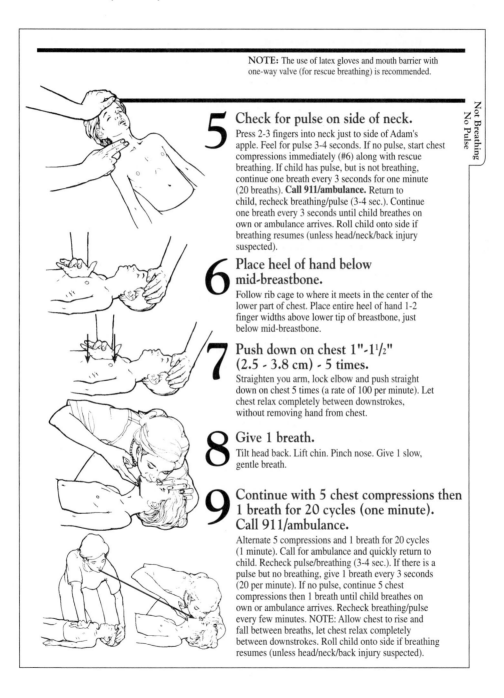

NOTE: The use of latex gloves and mouth barrier with one-way valve (for rescue breathing) is recommended.

**Not Breathing
No Pulse**

5 Check for pulse on side of neck.

Press 2-3 fingers into neck just to side of Adam's apple. Feel for pulse 3-4 seconds. If no pulse, start chest compressions immediately (#6) along with rescue breathing. If child has pulse, but is not breathing, continue one breath every 3 seconds for one minute (20 breaths). **Call 911/ambulance.** Return to child, recheck breathing/pulse (3-4 sec.). Continue one breath every 3 seconds until child breathes on own or ambulance arrives. Roll child onto side if breathing resumes (unless head/neck/back injury suspected).

6 Place heel of hand below mid-breastbone.

Follow rib cage to where it meets in the center of the lower part of chest. Place entire heel of hand 1-2 finger widths above lower tip of breastbone, just below mid-breastbone.

7 Push down on chest 1"-1½" (2.5 - 3.8 cm) - 5 times.

Straighten you arm, lock elbow and push straight down on chest 5 times (a rate of 100 per minute). Let chest relax completely between downstrokes, without removing hand from chest.

8 Give 1 breath.

Tilt head back. Lift chin. Pinch nose. Give 1 slow, gentle breath.

9 Continue with 5 chest compressions then 1 breath for 20 cycles (one minute). Call 911/ambulance.

Alternate 5 compressions and 1 breath for 20 cycles (1 minute). Call for ambulance and quickly return to child. Recheck pulse/breathing (3-4 sec.). If there is a pulse but no breathing, give 1 breath every 3 seconds (20 per minute). If no pulse, continue 5 chest compressions then 1 breath until child breathes on own or ambulance arrives. Recheck breathing/pulse every few minutes. NOTE: Allow chest to rise and fall between breaths, let chest relax completely between downstrokes. Roll child onto side if breathing resumes (unless head/neck/back injury suspected).

WRITING STRATEGIES 9–3   Reader Goals and Design Techniques 3

Reader's Goals after Reading	Writer's Document Design Techniques
To execute action plans or implement decisions	Focus on providing: Lists Examples Glossaries Appendixes
To evaluate the outcome of actions or decisions	Focus on providing: Checklists Boxes for recording information Quick reference cards

---

Lists

By their format alone, **lists** are chunks of information clearly separated from the rest of the text. Because of the additional white space and the structural order, they automatically increase comprehension. They are especially helpful for executing action plans. Document design specialists suggest that you do the following:

- Indent the list so that numbers or bullets stand out in the left margin.
- Avoid setting lists in type smaller than the body unless you want the reader to consider the information less important than the rest of the text.
- Punctuate consistently (Nord and Tanner).

You can also place cues before list items:

- ☑ A checkoff box if each item must be performed.
- A bullet if list items are equally important.
1. A number if list items should be considered in a certain order (Carliner).

Lists help you clarify relationships within your text as you identify steps in a process, recommendations to consider, forms to complete, and so on. Lists also help you to be consistent throughout your document. For example, each time readers encounter a numbered list, they realize that they must read, understand, and take action chronologically. When they encounter a bulleted list, they understand that the order might not be as important as the content of each topic.

Examples

Throughout this textbook, we have worked to provide examples of what we describe. Examples are invaluable for helping readers to understand what you mean and to execute action plans. Provide examples whenever possible, and separate these with visual elements to highlight them for readers. Your readers remember your information better if they can visualize that information within specific circumstances, times, or settings.

Glossaries

Glossaries define information presented in the main text. In the workplace, you are more likely to understand a greater number of technical details than your readers. Use glossaries to define difficult terms used in your document and to help your readers recall exactly what you wrote. Readers can refer to your glossary during or after reading.

glossaries
definitions of information presented in the main text

Appendixes

Appendixes provide supplemental information without cluttering your document. Note the appendixes used in this textbook. What chapters would this information interrupt if placed in the body of the text? You can place examples or other information in appendixes so readers can access the information to execute action plans. Often appendixes will contain more detailed information that helps readers remember the specifics of your document.

appendixes
supplemental information collected at the end of a document

HELPING READERS EVALUATE DECISIONS

The next set of design features not only helps readers to remember what you have written but also helps them to evaluate their actions or decisions based on your information. Three features are checklists, boxes, and quick reference cards.

Checklists

Checklists help readers select among listed items, especially when all items are needed to execute an action. For example, if readers need to assemble a gas grill, after reading they'll need to refer to a checklist to see if all of the parts are available or to ensure that they have taken all the steps in order.

checklists
lists that help readers evaluate the outcome of their actions or decisions (e.g., one consults a checklist of parts before assembling a device)

Boxes

Boxes separate or emphasize a portion of text. You can include key points that summarize the text and place these in boxes alongside the text. After reading, users can refer mainly to the boxes to get a gist of the text. You can also provide boxes as specific places for readers to record information after they have read the text. In this way readers have a space for evaluating the outcome of their actions or decisions.

boxes
separate material outlined by rules to emphasize a portion of text

Quick Reference Cards

You can place the most important steps, hints, or points on **quick reference cards** for readers to refer to after they have used a document. Although quick reference cards are often included with computer manuals (see Example 9-8), readers appreciate shortcuts for reading, and quick reference cards can function as a summary or reference to the most important points in a document.

quick reference cards
important steps, hints, or points for readers to consult after using a document

EXAMPLE 9–8 Microsoft Quick Reference Card

Microsoft Word Quick Reference Guide

If you want to	Do this
Add a footnote	Click insertion point where you want footnote to go. Press ⌘E. Click **OK**. Type footnote text in footnote window.
Add a header or footer	Under **Document**, choose **Open Header** … OR **Open Footer** … Type header/footer text in header/footer window. Close window.
Boldface text	Select text. Press ⌘-[Shift]-B.
Center text	Press ⌘R. Select Centered paragraph alignment icon on ruler.
Change line spacing	Press ⌘R. Select desired line spacing icon on ruler.
Change text	Press ⌘H. Type text you want to change in **Find What:** box. Type what you want to change it to in **Change To:** box. Click **Start Search**.
Change to plain text	Select text. Press ⌘-[Shift]-Z.
Check spelling	Press ⌘L. If necessary, click **Start Check**.
Close a file	Press ⌘W. If necessary, save changes to open file.
Copy text	Select text. Press ⌘C. Selected text remains in document and is copied to clipboard.
Cut text	Select text. Press ⌘X. Selected text is cut from document and is copied to clipboard.
Define styles	Press ⌘T. Make font and format selections from pull-down menus while in dialog box. Click **OK**. (Requires Full Menus.)
Find text	Press ⌘F. Type text you want to locate. Click **Start Search**.
Format a paragraph	Press ⌘M. Make desired selections in dialog box. Click **OK**. (Requires Full Menus.)
Go to page number	Press ⌘G. Enter page number in dialog box. Click **OK**.
Italicize text	Select text. Press ⌘-[Shift]-I.
Left-justify text	Press ⌘R. Select Flush Left paragraph alignment icon on ruler.
Number pages	Under **Document,** choose **Open Header** … OR **Open Footer** … Click insertion point in header/footer window where you want page number to go. Click page number icon. Close window.
Open a new file	Press ⌘N.
Open an existing file	Press ⌘O. Select file to open. Click **Open**.
Outline a document	Press ⌘U. (Requires Full Menus.)
Paste text	Click insertion point in document where you want text to go. Press ⌘V. Text from clipboard is copied to insertion point.
Print a file	Press ⌘P. Make desired selections in dialog box. Click **OK**. Follow directions on screen.
Quit Microsoft Word	Press ⌘Q. If necessary, save changes to any open files.
Save as new file	Under **File**, choose **Save As**… Enter document name in dialog box. Click **Save**.
Save an existing file	Press ⌘S.
Set margins	Press ⌘R. Move indent markers on ruler to desired settings.
Show/Hide the ruler	Press ⌘R.
Underline text	Select text. Press ⌘-[Shift]-U.
View the page layout	Press ⌘I.

Source: Ann Hill Duin and Kathleen S. Gorak, *Writing with the Macintosh ® Using Microsoft ® Word*. Cambridge, MA: Course Technology, Inc. Reprinted with permission.

Decisions about how to achieve consistency and clarity, about how to engage readers and help readers read with ease, locate possible answers, and remember what they have read must be based on your understanding of your readers' goals.

To envision your readers' goals in the technical documents you'll write and design, think of your readers as users of texts. Based on the research of Janice Redish, we know that readers use texts in order to:

- Learn something.
- Do something.
- Learn how to do something.

In this section of the chapter, we help you analyze your audience, purpose, and situation so you can select design techniques to enhance your readers' ability to read to learn, read to do, or read to learn how to do something.

Reading to learn something includes tasks where readers are using a document to extract and retain information that they will "call up" later and use again. In this type of reading, readers use study strategies such as previewing and reviewing to get the information to stay in memory. Most college reading tasks are reading-to-learn tasks (e.g., studying a textbook to prepare for a quiz). The design techniques that enhance your readers' ability to remember also support reading to learn.

Reading to do something includes tasks where readers are using documents to get information to help them complete a needed task. This is when readers look up information, use the information to complete a task, and then perhaps forget the information. Most workplace reading tasks are reading-to-do tasks because most documents in the workplace are manuals, flyers, labels, and forms (e.g., filling out a reimbursement form or ordering parts for a machine). The design techniques that help readers to locate answers and read selectively also support reading to do something.

Reading to learn how to do something includes tasks where readers are using documents to learn how to complete a task that they will likely do again or need to know how to do later. Reading to learn how to do something means that readers locate information, use it, and then record how they used it so that they will not have to spend time relearning how to do something. Many college and workplace reading tasks include reading to learn how to do something (e.g., completing laboratory procedures or assembling a spreadsheet or database). Essentially, this is the primary purpose for your reading this textbook; you read to learn how to do something. In the case of this chapter, you read to learn how to use document design techniques to help readers.

Whether your reader uses your document to learn, to do something, or to learn how to do something, you are helping your reader solve a problem—to order new parts, to get reimbursed for expenses, and so forth. In the workplace, we rarely have the pleasure of using technical documents only to learn something. Cognitive psychologists call this active use of documents **situated learning;** that is, readers participate actively in the process of understanding

Considering Document Design and Readers' Goals

situated learning
use of documents by readers who actively participate in the process of understanding a document

a document. Patricia Wright, an applied cognitive psychologist specializing in document design, has created a six-step sketch to show how users interact with technical documents before reading, during reading, and after reading:

Before Reading
- Formulate question, not necessarily a precise question.
- Find location of potential answer.

During Reading
- Comprehend text, but reading may be highly selective.
- Construct action plans or use information to make a decision.

After Reading
- Execute action plans or implement decisions.
- Evaluate outcome of actions or decisions (340).

To understand the implications of this technique, look at the document, "AIDS Public Policy in Minnesota," in Case Documents 1, Appendix A. Imagine that you are a reader asking, "Given that people living with AIDS frequently suffer from tuberculosis, what efforts is Minnesota taking to control the spread of tuberculosis in AIDS patients and in the non-HIV-infected population?"

To implement the six steps as you search for an answer, comprehend the text and execute any action plans or decisions, you probably will notice the title of the document immediately, "Tuberculosis and AIDS." The title gives you a good idea that this document will help you answer your question. As you continue to skim through the document, you probably will locate quickly the second boxed heading, "Current Status of Tuberculosis Control Efforts in Minnesota." In this section, you will focus on the table reporting the number of new tuberculosis cases, but reading below the table, you see that control efforts seem to consist primarily of seeking grants to support treatment services. You want information on more concrete efforts to control the spread of tuberculosis. Should you stop reading here and decide you need to turn to a different document? Or, should you persist and use the "Recommendations" section?

This example illustrates that a developer of texts needs to focus on document design techniques that help readers use texts before, during, and after reading. You can use the design techniques we discussed in the sections that opened this chapter to help readers access and use your information at all three stages of reading.

Developing a Style Sheet

At the beginning of this chapter, we discussed how design choices support consistency throughout your document. One of the best ways to plan for consistency in your document is to create a **style sheet** or style guide. Although a style sheet can contain the word and sentence level choices we discuss in Chapter 8, a style sheet can also be your plan for the visual look of your text. Style sheets are especially helpful whenever you contribute to a group writing project. What document design techniques will you use to help readers before reading, during reading, and after reading? What spacing, margins, justification, typeface, and point size will you use? How will you build consistency into your headings, subheadings, and lists?

Whether you're writing individually or collaboratively, if your writing task requires a lot of formatting or has several sections, it helps to design a style sheet. To begin, keep your style sheet simple. Writing in a consistent typeface with consistent margins and spacing will help.

To create a style sheet, make planning decisions about which techniques you will use and why, as shown in Writing Strategies 9–4. By creating

style sheet
device to ensure consistency in document design decisions; includes style, usage, and grammar decisions

---

WRITING STRATEGIES 9–4   Style Sheet Decisions

	Document Design Technique	Style Sheet Decisions

Focus on these techniques first	grid layout
	white space
	margins
	headings and subheadings
	color
	visual elements
	typography:
	    case
	    style (typeface)
	    size (point)
	    spacing (leading, kerning)
	    length
	    highlighting	

Focus on these techniques second or third	binding quality of paper layout graphics cover color table of contents indexes headers and footers tab markers and dividers lists examples glossaries appendixes checklists boxes for recording info. quick reference cards	

a style sheet, you will spend less time in the future making decisions about style. You can develop your style sheet on paper, but we suggest using the style feature in a word processing program to develop a style sheet. Then you can access it each time you begin a new document.

Considering Differences When Designing a Document

Keep in mind that every suggestion in this chapter may not hold true for someone from another culture. In some cases, a document design feature that one gender prefers may not be preferred by the other gender. Visual symbols or gestures that are common and well understood in one culture may miscommunicate or even offend when used for another culture.

Example 9-9 depicts three frequently used gestures and the meanings they convey in the United States. However, these gestures may be understood very differently if used in another culture:

a. Although widely used in the United States and northern Europe to signal approval or to hitchhike, this gesture has vulgar meanings in Australia, some Mediterranean countries, and much of Africa.

b. The circle gesture has a variety of meanings other than the one common in the United States. In Japan, for instance, it means

---

EXAMPLE 9–9   Three Visual Symbols

a   for   **OK, correct**

b   for   **Perfect!**

c   for   **Stop!**

Source: William Horton, "Visual Literacy: Going beyond Words in Technical Communication," *Technical Communication* 39, no. 3 (1992), pp. 447–51. Reprinted with permission from *Technical Communication,* published by the Society for Technical Communication, Arlington, VA.

money—the circular opening representing a coin. In France it means zero or worthless. In South America, it suggests that the viewer obtain self-knowledge in a carnal sense.

c. The palms-out gesture, used by traffic police in the United States and Great Britain to tell motorists to come to a halt, has a scatological meaning in Greece (Horton, "Visual Literacy," 450).

Although you may run slightly less risk of misinterpretation when making typographical decisions, assume that at some point in time—in this information age in which we live—your document will likely be read by someone from another culture. In short, don't put U.S. blinders on. Consider your larger audience.

Integrating Text and Graphics

Look back at Example 9-7 on emergency medical treatment and notice how the text and the pictures support the message conveyed. The reader first encounters the pictures so that he or she can visualize the actions that must be taken to give proper care to children not breathing and having no pulse. The text then expands upon and supports this initial visual message. Neither the text nor the pictures could readily stand alone. For example, in step one the reader sees the caregiver placing two fingers under the chin of the child, and then reads these words: "Tilt the head back gently, lift chin slightly." In designing an online message, the user's experience is based on a similar dependence of text and graphics. Look at the Web sites mentioned in Appendix A, Case Documents 4, to see this integration of text and graphics.

Although graphics can add a great deal of information to a text, they must be chosen and used carefully. For example, an extraneous picture can simply detract from your message or prevent your reader from engaging with your textual message. Or, if you depict a person of one gender—for example, a man holding a rifle that you are teaching your reader to disassemble—women users might have a hard time picturing themselves capable of or interested in this process. Finally, graphics might be powerful enough that they take over your story and convey just a part of your message.

Document designer Karen Schriver suggests five ways to integrate text and graphics successfully:

1. *Redundant*—identical content in text and graphics to repeat key ideas.
2. *Complementary*—different content in text and graphics to help readers and users understand complex information or key ideas.
3. *Supplementary*—different content in text and graphics but one dominates and provides main ideas and the other supports, reinforces, or elaborates those ideas.
4. *Juxtapositional*—different content in text and graphics as each represents a clash or tension between the ideas in each

mode; "the idea cannot be inferred without both modes being present simultaneously."

5. *Stage-setting*—different content in text and graphics because usually the graphic depiction "forecasts the content, underlying theme, or ideas" presented in the text (Schriver, 412-13).

Consider these suggestions as you select and integrate your textual and visual messages.

Creating Interactive and Online Documentation Design

Although we have thus far focused primarily on print documents, technical communicators are increasingly being asked to design online messages and Web sites for their companies and clients. Graphics, design, and format are integral parts of the user's experience with your online message. This section of the chapter focuses on the particular nature of online messages and documents.

Designer Theo Mandel offers three golden rules of interface design.

1. Place users in charge. Included in this rule are suggestions such as displaying descriptive messages and text, providing immediate and reversible actions, providing meaningful paths, and allowing users to manipulate interface objects directly.
2. Reduce users' memory load. Included in this second rule are suggestions such as relying on recognition and providing visual cues, providing real-world metaphors, and promoting visual clarity.
3. Make the interface consistent. Included in this final rule are suggestions such as maintaining consistency within and across products and providing aesthetic appeal (Mandel, 51-74).

When discussing the second golden rule, Mandel notes that users rely on visual cues to tell them where they are, what they are doing, and what they can do next in online documentation. For example, Mandel suggests when designing online documentation, you test the visual cues of your message by walking away from the computer and then coming back sometime later—can you tell where you are? And what you were doing? (66).

Although Mandel's suggestions are not remarkably different from those given throughout this chapter, the aspects of design that pertain to print documents play out somewhat differently in interactive and online documentation design. For example, because of hypertext links, Web pages, in particular, need to be designed as freestanding messages. Therefore, headings are more informative than in print documents so that the user can locate information by referring to a single Web "page" within your site. Clear but informative titles serve as text markers and are often the first thing the user sees when the page comes up. These titles also become the "bookmarks" that are added to the users' list of Uniform Resource Locators, or **URLs,** and help readers find their way back to your page. Usually pages have dates somewhere on the bottom so that the user can tell when the page was created or updated and how current the information is. These pages also indicate to the user where you are—your company and your international site. Your home URL on the majority of your Web pages also allows the user to find his or her way back to your site.

URL
an abbreviation for **Uniform Resource Locator,** an addressing system for the World Wide Web

The task of designing online documents and Web sites can be overwhelming and is often covered in semester or year-long university courses. However, we can give you some starting points.

TEXTUAL CHOICES

When selecting the font or typeface to use for your online documents, generally you should select sans serif fonts and avoid decorative typefaces for on-screen displays. If you do use serif fonts, select those with little contrast between the thick and thin parts of the letters because the thin strokes tend to disappear. Use larger font sizes such as 11 or 12 point consistently throughout your online document and avoid too much variation. Indicate changes in your message by changing the color or weight of your font or its placement on the screen. Keep your text lines much shorter than you would in print documents (perhaps one alphabet length maximum or 26 letter spaces). You don't want to force users to turn their heads to see your full line of text. Finally, avoid placing too many indicators of emphasis or highlighting in your online documents.

LAYOUT AND FORMATTING

We have discussed consistency in print documents. Online document design requires an even greater need for consistency. Each time your user encounters a procedure or step, it must look like every other procedure or step. For example, as JoAnn Hackos and Dawn Stevens mentioned in their guidelines for online design, most Web pages have a standard set of buttons on each screen, and users look in this location on each screen for these navigational aids (214). Online courses in technical communication offered at the University of Minnesota rely on a template to assure that navigational aids will appear consistently on each screen (see Example 9–10).

By exploring Web sites on the Internet (e.g., Appendix A, Case Documents 4) you can see that the placement of navigational buttons, scroll bars, and pop-up windows is all part of effective online documentation design. You must use these in a helpful and consistent way. For example, if you have a help system, or "office assistant" as in Microsoft Office for the Macintosh 98, your user needs to know how to access or hide this assistant and how to ask questions. And, your user will rely on your rulers

EXAMPLE 9–10   Navigational Aids on a Course Template Page

You can visit this site to see more navigational tools at *http://www.agricola.umn.edu*

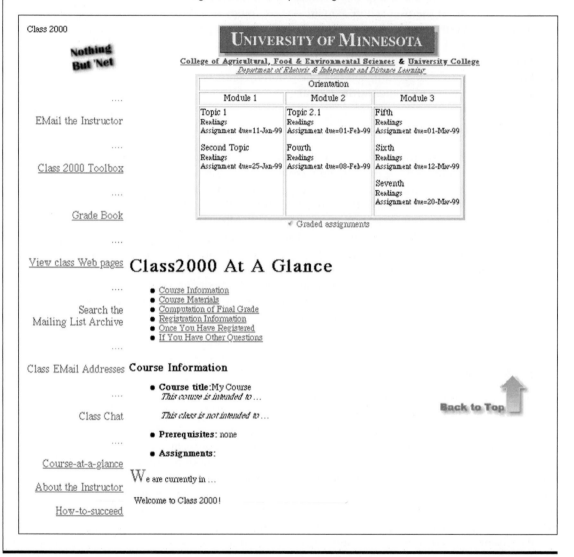

and toolbars to display all formatting choices. Finally, your Web site user needs to know how to return to your home page, go back to the previous page, select another link, or look for information under a key word.

Finally, in creating your online designs, remember that your user will generally want to move up and down your page, not horizontally. And, your window should display all information without asking your user to scroll up and down more than one or two times. All your important information should be seen when the reader opens that window, with pop-up windows seen in the center of the window and near the information they supplement (Hackos and Stevens, 218).

COLOR

Avoid selecting too many colors to aid your user in understanding your online documentation and in interpreting your Web pages.

When you design a document, you have some idea of its intended use; you have some idea about your target for readers' subjective feelings and attitudes about the document. But how do you measure reader satisfaction with your document? How do you know if your document design decisions are correct? How do you know if the techniques you've chosen have resulted in a higher-quality document?

Evaluating Document Design Decisions

The drawback of evaluating documents is that it takes time; the benefit is that the information obtained is extremely valuable for guiding revision and future document design decisions. For example:

> When a technical publications group at AT&T began evaluating documents and focused on streamlining the process of developing technical documentation, they reduced the cost of documentation by 53 percent, reduced documentation production time by 59 percent, and increased the number of projects completed by 45 percent (Edwards).
>
> In England, when the Department of Customs and Excise revised its lost-baggage forms used by airline passengers, they cut a 55 percent error rate down to a mere 3 percent ("Plain English Pays," 1, 4).

Research aimed at measuring quality in document design can be classified by two dominant approaches: direct, or "criterion-reference" measures, and indirect, or "prediction" measures (Schriver). These two approaches serve different purposes when evaluating document design.

direct measures
study how readers use, read, or rate a document and judge that activity against a criterion

DIRECT MEASURES (CRITERION MEASURES)

Direct measures are those that study how readers use, read, or rate a document and then judge that activity against a criterion. Direct measures include (1) **concurrent methods,** designed to collect data about how people think, feel, or respond as they use a document, and (2) **retrospective methods,** designed to collect data as readers reflect on their experience when reading or using a document (Schriver).

For example, if you designed instructions for assembling a gas grill and then wanted to evaluate your document, you might give your instructions

concurrent methods
designed to collect data about how people think, feel, or respond as they use a document

retrospective methods
designed to collect data as readers reflect on their experience when reading a document

to readers/users and record how they used the instructions. You would then judge that activity against a criterion such as the maximum time that it should take users to assemble the grill. Although this concurrent method would provide information about how users respond as they use your document, you might also ask users to reflect later on their experience, a retrospective method.

As another example, if you designed a formal report with the bulk of your findings in appendixes, you might give the report to readers and record how often they flipped to the appendixes to study specific results. A day or two later, you might also ask readers to record what they remembered most from your report.

INDIRECT MEASURES (PREDICTION MEASURES)

indirect measures
designed to predict a reader's ability to comprehend or use the document

Indirect measures are those designed to predict a reader's ability to comprehend or use the document (Schriver). We will cover one primary method, quality metrics, in this chapter.

Quality Metrics

quality metrics
a four-phase indirect measure of a reader's ability to use a document; ends with a regression analysis

To develop a **quality metric,** you need to carry out a four-phase plan:

1. In Phase 1 you gather evidence from existing research about how readers use a specific type of document (e.g., a tax form, a proposal, or a brochure), noting especially the document design features readers appreciate most.
2. In Phase 2 you modify a document according to the document design features that readers appreciate most. You then test these documents with readers/users to determine whether the features improve understanding, attitudes, and ease of use.
3. In Phase 3 you use the results from this testing to identify a set of features (e.g., a style sheet) that will be used for the prediction measure.
4. In Phase 4 you conduct a regression analysis to determine what weight each feature should have when predicting the comprehensibility or usability of the document (Schriver).

Although this four-phase plan may appear too complicated to use as you develop technical documents, it may be worthwhile to follow when designing documents for the corporation or other business that you may be working for.

EVALUATING A PHONE BILL

Let's apply these measures to the redesigning of a telephone bill. Technical communicator Deborah Keller-Cohen and her colleagues studied the collaborative effort as a linguist, a graphic designer, and an executive manager developed and tested new telephone bill solutions to improve customer comprehension and satisfaction with a very familiar, yet complex document. Example 9–11 shows the original telephone bill.

Based on what you have now learned about document design techniques, how might you redesign this telephone bill? This group's main concerns were to make relationships between the charges clearer, to increase legibility, and to simplify the language. The collaborative team followed this process when redesigning it:

1. They made initial changes which they felt customers desired and the company could implement. They then tested these changes through face-to-face interviews with customers.

EXAMPLE 9–11 Original Telephone Bill

```
              ◉  Midwest Bell   P.O. Box 1234   Anywhere, Midwest 01234

                                                      123  456-7890 000
                                                         DEC 10, 1984
                                                         PAGE      2

MIDWEST BELL - CURRENT CHARGES

FLAT RATE SERVICE
MONTHLY SERVICE     DEC 10-JAN 09                              19.11
DIRECTORY ADVERTISING                                          2.75
LOCAL AND ZONE USAGE-SEE DETAIL PAGE                           6.85
OTHER CHARGES AND CREDITS-SEE DETAIL PAGE                      1.92
ITEMIZED CALLS-SEE DETAIL PAGE                                 2.12
TAXES:  (FED     .87)  (STATE    1.16)  (LOCAL    1.16)        3.19


              MIDWEST BELL TOTAL CURRENT CHARGES               35.94


BILLING INQUIRIES - CALL TOLL FREE 1-123-4567
IF MOVING OR PLACING AN ORDER FOR SERVICE - CALL TOLL FREE 1-123-0000




                                                         CONTINUED

                     KEY TO CALLS ON OTHER SIDE
```

Source: Deborah Keller-Cohen, Bruce Ian Meader, and David W. Mann, "Redesigning a Telephone Bill," *Information Design Journal* 6, no. 1 (1990), pp. 45–66. Used with permission of the authors.

2. Based on the interviews in stage 1, they developed three possible designs for the bill. They used three methods with customers to test these designs against the original bill.

■ A comprehension task to assess which version was easier to understand (e.g., one multiple-choice question asked how many zone calls were made).

■ An attitude survey to see which version customers preferred (e.g., customers were asked to indicate how easy each bill was to understand).

■ **Focus groups** in which customers talked about their preferences.

3. They developed a composite telephone bill based on their findings in stage 2. They tested this bill against the original bill, again using a comprehension task, an attitude survey, and focus groups.

focus groups
a method of gathering data by bringing together groups of people around a specific topic or issue

---

EXAMPLE 9–12 Composite Telephone Bill

● Midwest Bell

P. O. Box 1234
Anywhere, Midwest 01234

**MIDWEST BELL
CURRENT CHARGES**

123 456-7890 0000
Dec 10, 1984

Page 2

		Calls	Minutes	
Monthly Charges	Local Services.. Dec 10 thru Jan 9			19.11
	Directory Advertising			2.75
Local Calls	Calls over 50 call allowance - 23 at 8.2¢ each			1.89
Zone Calls	Day Rate	9	38	3.62
	Evening Rate	3	16	1.04
	Night/Weekend Rate	2	6	.30
Itemized Calls	See page 3			2.12
Other Charges and Credits	We increased our monthly service rate by 64¢ per day beginning Dec. 7. The new rate applied to Dec 7 thru Dec 9 and was not billed at that time.			1.92
Taxes	Federal: .87 State: 1.16 Local: 1.16			3.19

MIDWEST BELL CURRENT CHARGES — **35.94**

This amount is included in the total amount due on page 1

**For questions about your bill call 555-5555
If moving or placing an order for service call 555-4444**

KEY TO CALLS ON OTHER SIDE

Source: Deborah Keller-Cohen, Bruce Ian Meader, and David W. Mann, "Redesigning a Telephone Bill," *Information Design Journal* 6, no. 1 (1990), pp. 45–66. Used with permission of the authors.

This collaborative group essentially followed the first three phases of quality metrics. Whatever method you choose—concurrent methods, retrospective methods, or quality metrics—be sure to evaluate your document's design. We cannot stress this enough. Once you do, you will be pleased with how much you learn, and you will be able to apply what you learn to your future document design efforts.

SUMMARY

- This chapter describes how you might design documents so that busy readers can quickly locate what they need, comprehend the information, and refer to it later. By envisioning readers' goals and by thinking of readers as users of texts, you can predict whether a reader wants mainly to learn something, do something, or learn to do something when using your document.
- You now have a list of document design techniques to use to motivate readers to pick up your text: packaging features such as an attractive binding, the quality of the paper, a balanced and consistent layout, interesting graphics, a legible cover and print, and the appropriate use of color. Once readers become motivated to open and use a document, they need to locate answers to their questions quickly and efficiently. You can help readers locate information by providing a table of contents and/or index and using structural headings (headers, footers) as well as tab markers or dividers.
- To help readers read selectively, you now know how to visualize the grid layout or page design of your document. Remember to create ample white space by using headings and subheadings and margins. Focus the attention of your readers through the appropriate use of color and visual elements such as lines, borders, circles, and other shapes. To help readers comprehend the information, you need to make conscious decisions regarding the typography of your document; that is, the case, style, size, and spacing of letters as well as the length of a line on the page. To highlight specific text, use boldface, underlining, and italics; however, do not overuse highlighting features. Remember that your overall goal is to achieve clarity and consistency in your document.
- After users have read your document, they need to execute action plans or implement decisions based on what they have read. You can help readers implement their decisions by providing lists, examples, glossaries, and appendixes as well as checklists, boxes for recording information, and quick reference cards.
- One of the best ways to plan for consistency throughout your document is to create a style sheet in which you make planning decisions about which document design techniques you will use in

a document and why. Focus on page design and typography first; then determine techniques for helping readers before and after reading. By creating a style sheet, you will spend less time in the future on document design decisions.

- In designing interactive and online documents, you should remember to place your users in charge, to reduce their memory load, and to make your interface consistent. Selecting sans serif fonts and clear navigational tools, avoiding too many colors, and allowing your users to move up and down your online documents all enhance your designs.

- You can use direct or indirect measures to evaluate your document design decisions. It's most important to watch real people using your document and to collect information (data) about how they use it. You then use this information to redesign your document, and in the case of developing a style sheet, to determine which document design techniques you should be using.

- Keep in mind that document design techniques appropriate for one culture may not be as appropriate for another culture. Although some techniques may work across cultures (e.g., the use of headings and lists), some cultures may prefer one technique to another.

ACTIVITIES AND EXERCISES

1. *Reviewing Design Techniques.* Take another look at the design features of this textbook or any other textbook you have available. Flip through the text and generate a list of the design features, such as the type style used, the distribution of white space, the display of lists, the nature of headings and margins, the binding method, the use of color, and the placement of figures. First, focus on the *larger message:* How is the text organized visually so that you get a clear sense of where information is located? Make a list of features that help you get a sense of the overall organization of the text. Second, focus on the *smaller message:* What smaller features, such as headings, bullets, and figures or examples, help you to find specific information quickly so you can use that information? Again, make a list of these specific features. Now, share your list with another person in your class. What did someone else notice that you didn't? What document design features do you most appreciate? What document design features are not present in this text that you wish were there to guide you?

2. *Expanding Document Design Techniques.* Go to the library and read a chapter or two from one of the following sources in which technical communicators talk about document design:

   - Barnum, Carol M., and Saul Carliner, eds. *Techniques for Technical Communicators.* New York: Macmillan, 1993.
   - Hackos, JoAnn T., and Dawn M. Stevens. *Standards for Online Documentation.* New York: John Wiley, 1997.

- Horton, William K. *Designing and Writing Online Documentation.* New York: John Wiley, 1990.
- Mandel, Theo. *The Elements of User Interface Design.* New York: John Wiley, 1997.
- Schriver, Karen A. *Dynamics in Document Design.* New York: John Wiley, 1997.
- Swann, Cal. *Language and Typography.* London: Lund Humphries, 1991.
- Any of the articles in the *Information Design Journal.*
- Any of the *Proceedings* from the yearly conference of the Society for Technical Communication, Arlington, VA.

Prepare a brief oral presentation on your findings for your class.

3. *Locating Document Design Technologies.* Explore your current word processing program to identify the document design features available to you. What document design techniques are available for you to use? Bring a list of these and compare word processing programs with other students.

Chapter 10 includes a section on computer graphics. In this section, we note that because computer tools for communicators are developed and improved so rapidly, any list of such tools is quickly outdated. At your campus or workplace, identify the computer graphics tools available for your use.

4. *Practicing on Your Own.* Add subheadings and headings to a document that you wrote for another course. Note how this process helps you to identify areas where your content and organization were weak. After you have created headings and subheadings, choose other document design techniques that would improve this document.

5. *Writing in the Workplace.* Locate a text that you or someone else has written for work. Evaluate its design using one of the evaluation methods discussed in this chapter.

6. *Document Design Dilemmas.* As you work near others, perhaps in a computer lab, ask them what they consider "visual repellents"; that is, what document design decisions annoy them as readers. Do they find certain typefaces displeasing? Do they think that college writers should avoid left- and right-justifying texts? Bring your findings to class for discussion.

7. *Working with Functional Documents.* Locate your phone bill or any other day-to-day functional document (e.g., forms, labels). Redesign and test the revised version(s) according to the steps followed by Keller-Cohen and colleagues in this chapter.

8. *Recording Your Reading Process.* Choose any of the case study documents included in this textbook and record exactly how you approach and use the document

- Before reading.
- During reading.
- After reading.

Do you follow Patricia Wright's six-step sketch as discussed in this chapter (p. 340)? Why or why not?

What document design changes would you recommend be made to the document?

9. In 1992 the University of the Midwest was informed by the Coalition of Citizens Concerned with Animal Rights (CCCAR) that the university was trapping and killing large numbers of birds in the experimental crop fields on its campus. After much discussion, the University developed the position statement titled, "Bird Control on the Experimental Plots," included in Appendix A, Case Documents 3.

   Working with a partner, redesign this entire document using Example 9-13. As you redesign the text, record your document design process. After you have redesigned the text, get opinions from outside readers regarding your changes and record these opinions.

EXAMPLE 9–13   Plan for Redesigning a Text

Document Design Decision	Specific Changes in the Document	Opinions from Outside Readers

---

10. This exercise focuses on the Exxon Company's report, "How Tiny Organisms Helped Clean Prince William Sound" in Appendix A, Case Documents 2. Make suggestions for redesigning this document for engaging readers

   ■ Before reading.
   ■ During reading.
   ■ After reading.

   Record your suggestions in the grid (Examples 9-14, 9-15, and 9-16). Bring your completed grid to the next class for discussion.

EXAMPLE 9–14   Grid for Redesigning a Report I

Reader's Goals Before Reading	Writer's Document Design Techniques

To formulate a question and become motivated to open a document and locate an answer	Focus on packaging:
	    Attractive binding
	    Quality of paper
	    Balanced and consistent layout
	    Interesting graphics
	    Legible cover and print
	    Appropriate use of color

To find the location of a potential answer	Focus on the larger organization:
	    Table of contents
	    Indexes
	    Structural headings
	    Tab markers and dividers

Changes Suggested for the Exxon Document:

EXAMPLE 9–15   Grid for Redesigning a Report 2

Reader's Goals During Reading	Writer's Document Design Techniques

To read selectively	Focus on guiding readers through:
	    Grid layout
	    White space
	    Headings and subheadings
	    Margins
	    Color
	    Visual elements (boxes, rules)

To read with ease and use information to make a decision	Focus on typography:
	    Case
	    Style (typeface)
	    Size (point)
	    Spacing (leading and kerning)
	    Length
	    Highlighting

Changes Suggested for the Exxon Document:

EXAMPLE 9–16   Grid for Redesigning a Report 3

Reader's Goals After Reading	Writer's Document Design Techniques
To execute action plans or implement decisions	Focus on providing: 　Lists 　Examples 　Glossaries 　Appendixes
To evaluate the outcome of actions or decisions	Focus on providing: 　Checklists 　Boxes for recording information 　Quick reference cards

Changes Suggested for the Exxon Document:

---

11. To test your understanding of color blindness, answer the questions in Example 9-17.

12. Evaluate the effect upon you of this textbook's binding, quality of the paper, headings, typography, color, index, table of contents, the overall layout, and so on. Does it motivate you to open the text and use it to learn to communicate effectively? To test these elements, write down the time it takes you to locate information about collaboration, ISO-9000 standards, and suggestions for designing Web pages. Then, using Example 9-18 go back and list the document design techniques that aided your search in each case. As you complete this short exercise, what document design techniques helped you the most? The least? What techniques discussed in this chapter are not used in the text, perhaps due to cost constraints or traditional textbook specifications?

13. To illustrate these three major goals for reading, complete the exercise in Example 9-19. The first column in this grid presents 10 contexts or scenarios for reading. In the second column, decide whether the reader's main goal is to learn something, to do something, or to learn how to do something. Then, working with a partner, compare your responses and list a document design technique that you have seen or would use to help a reader better understand and use a text in this scenario.

EXAMPLE 9–17   Understanding Color Blindness

Color blindness can limit people's ability to perceive information displayed in certain colors. To test your understanding of color blindness, answer the following questions:

1. What fraction of men have trouble distinguishing red or green?
   100%   80%   8%   0.8%   .04%   0%
2. What fraction of women have trouble distinguishing red or green?
   100%   80%   8%   0.8%   .04%   0%
3. As we age our ability to perceive one color diminishes. What is that color?
4. The center of our vision, where our vision is most acute and where we read text, is less sensitive to one color than others. Which color is that?

Answers are at the end of this chapter.

―――――
Source: Adapted from William Horton, "Understanding Color Blindness," *Technical Communication* 39, no. 3 (1992), pp. 447–51. Reprinted with permission from *Technical Communication,* published by the Society for Technical Communication, Arlington, VA.

EXAMPLE 9–18   Design and Location Exercise

Locate the Following:	Time to Locate the Information:	Document Design Techniques That Aided Your Search:
Information on how to initiate a collaborative writing project		
Information on writing with standards such as ISO-9000		
Information on designing Web page		

EXAMPLE 9–19   Goals for Reading

Context for Reading	Reader's Goal 1. To learn 2. To do 3. To learn to do	Document Design Technique to Help the Reader/User
Filling out a tax form		
Learning the best ways to avoid exposure to HIV		
Comprehending one's rights in a legal contract with a landlord		
Reading an advertisement about exercising equipment		
Locating an emergency phone number		
Making a decision whether to vote for or against the building of a nuclear power plant in your county		
Assessing the current status of tuberculosis control efforts in your state		
Using a manual to solve a telecommunications problem		
Following the directions to assemble a gas grill		

WORKS CITED   Carliner, Saul. "Lists: The Ultimate Tool for Engineering Writers." *Writing & Speaking in the Technology Professions: A Practical Guide.* D. Beer, ed. New York: IEEE Press, 1991.

Eagleson, Robert D. *Writing in Plain English.* Canberra: Australian Government Publishing Service, 1990.

Edwards, A. W. "A Quality System for Technical Documentation." *Proceedings of the 36th International Technical Communication Conference.* Washington, DC: Society for Technical Communication, 1989.

"Gas Utilities Switch to Plain Language." *Simply Stated.* Washington, DC: American Institutes for Research, March 1986.

Hackos, JoAnn T., and Dawn M. Stevens. *Standards for Online Documentation.* New York: John Wiley, 1997.

Horton, William. "Pictures Please—Presenting Information Visually." *Techniques for Technical Communicators*. Carol M. Barnum and Saul Carliner, eds. New York: Macmillan, 1993.

_____. "Understanding Color Blindness." *Technical Communication* 39, no. 3 (1992), pp. 447–51.

Keller-Cohen, Deborah; Bruce Ian Meader; and David W. Mann. "Redesigning a Telephone Bill." *Information Design Journal* 6, no.1 (1990), pp. 45–66.

Mandel, Theo. *The Elements of User Interface Design*. New York: John Wiley, 1997.

Nord, Martha Andrews, and Beth Tanner. "Design That Delivers—Formatting Information for Print and Online Documents." *Techniques for Technical Communicators*. Carol M. Barnum and Saul Carliner, eds. New York: Macmillan, 1993.

"Plain English Pays." *Simply Stated*. Washington, DC: American Institutes for Research, February 1986.

Redish, Janice C.; A. G. Elser; R. Thornburgh; and L. Downey. "The U-Metric Questionnaire: A Tool for Assessing the Usability of Manuals." *Proceedings of the 39th International Technical Communication Conference*. Arlington, VA: Society for Technical Communication, 1992.

Schriver, Karen A. *Dynamics in Document Design*. New York: John Wiley, 1997.

Wright, Patricia. "Writing Technical Information." *Review of Research in Education*. E. Z. Rothkopf, ed. Washington, D.C.: American Educational Research Association, 1987, pp. 327–85.

Answers to Questions About Color Blindness

1. About 8 percent of men have trouble distinguishing red and green.
2. Only 0.4 percent of women suffer from the same malady.
3. Blue, primarily because the lens of the eye yellows with age.
4. Blue. The center of the retina is covered with a yellow patch that filters out blue light.

Source: Adapted from William Horton, "Understanding Color Blindness," *Technical Communication* 39, no. 3 (1992), pp. 447–51. Reprinted with permission from *Technical Communication*, published by the Society for Technical Communication, Arlington, VA.

Visual Display
and Presentation

Graphical excellence is that which gives to the viewer the greatest number of
ideas in the shortest time with the least ink in the smallest space . . . . And graphical excellence requires telling the truth about the data.

Edward R. Tufte, *The Visual Display of Quantitative Information*
(Cheshire, CT: Graphics Press, 1983), p. 5.

Designed for pragmatic purposes, documents help people learn, use technology,
make decisions, and get their jobs done. Documents often concern topics in science and technology, education and training, government and law, economics and
finance, health and medicine, risk communication and safety, or public policy and
the environment. Since people rely on documents to make decisions that influence their safety, livelihood, health, and education, the highest ethical standards
must be brought to bear in making textual choices—in deciding what to say and
what not to say, in what to picture and what not to picture

Karen Schriver, *Dynamics of Document Design*
(New York: John Wiley, 1997), p. 11.

In Chapter 9 you learned how to design and package your documents effectively. Part of that design involves the **visual display of data,** which we look at here. When your written and oral presentations contain data, you need to think about how to interpret those data so that they are meaningful and accessible to your audience. You need to select visual displays for your Web pages and online documentation. Quite often what you are displaying visually are the "good reasons" or the evidence that supports an argument. Because visual displays of data either support or take the place of text, you should plan these visual displays as you plan your text rather than leaving them for last.

Generally, you use a visual display of data to accomplish one or two of the following tasks:

- Set off or emphasize important information or evidence.
- Make complex or detailed information accessible.
- Make the abstract more concrete.
- Make the concrete universal.
- Symbolize the structure or organization.
- Condense large amounts of data.
- Show relationships or trends among data.
- Compare and contrast the size or magnitude of data.
- Show what something looks like.
- Show what percentages or proportions are assigned to the parts of a whole.
- Demonstrate how to do something.
- Illustrate how something is organized or assembled.

To understand the **rhetorical function of a visual display,** you need to understand first the difference between a prose statement that contains data and a visual display of that data. In the following prose statement, the audience would be so busy trying to keep the numbers in mind that they would have a difficult time seeing relationships:

> The University of Minnesota Agricultural Experiment Station traps sparrows, blackbirds, starlings, and grackles to protect valuable seed in its fields. In 1985, 6,305 sparrows, 581 blackbirds, 1,086 starlings, and 163 grackles were trapped. In 1990, 2,317 sparrows, 1,468 blackbirds, 400 starlings, and 131 grackles were trapped. In 1995, 3,843 sparrows, 502 blackbirds, 1,104 starlings, and 325 grackles were trapped. Finally, in 2000 3,396 sparrows, 395 blackbirds, 2,160 starlings, and 412 grackles were trapped.

In Example 10–1 the audience can see quickly that although more sparrows are trapped than any other kind of bird, the number of sparrows trapped dropped sharply in 1990 from its high in 1985.

In the text of a document containing this information, an audience would need an explanation of that contrast and other changes, such as the increase in blackbirds trapped in the same year that fewer sparrows were trapped and the sharp increase in the number of starlings trapped

Introduction

visual display of data
placing of data in a graph, chart, or drawing to support or take the place of text

rhetorical function of a visual display
the specific purpose of a visual display, including the clarification of relationships among pieces of data and the trends they suggest

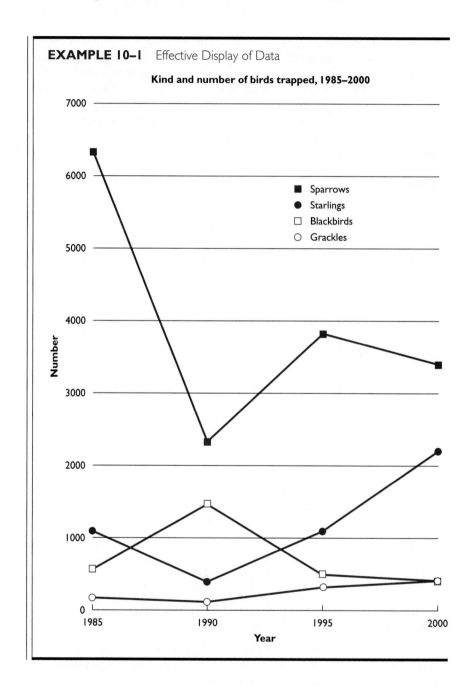

EXAMPLE 10–1   Effective Display of Data

Kind and number of birds trapped, 1985–2000

since 1990. Displaying data visually helps the audience quickly see trends the text would identify, interpret, and explain.

Although visual displays are very effective, they cannot simply be dropped into your document. More concrete visual displays of data, such as tables and photographs, require surrounding text that helps an audience decide what is essential information. More abstract visual displays of data, such as line graphs and drawings, often require less explanatory text but restrict an audience to those experienced in decoding symbols. You need to introduce all visual displays of data, relate them to your subject, and help your audience interpret them. You will need to mention the visual display by its number and, if helpful, the subject or title in the text the first time your audience will need the information referred to, and you need to place the display as close to where it is mentioned as possible.

To be effective, text and visual displays of data must be fully integrated. For example, you might refer to specific data and then "point" to the visual display of that data:

> The total number of birds trapped in university fields in 2000 was 1,591 compared to 2,034 in 1985 (see Table 4).

Or, you might call your audience's attention to the function of the display the first time they would find it useful.

> Table 4 displays the number of birds by type trapped in university fields in 1985, 1990, 1995, and 2000.

You may need to explain how or when to read your visual display. For example, you might say:

> To compare the number of sparrows, blackbirds, starlings, and grackles trapped in university fields from 1985 to 2000, see Table 4. Data are given for the first year only in each decade; see the bottom line of the table for the total number of birds trapped.

Finally, because you may be using your display of data to persuade your audience to draw some conclusions about your data, you might say:

> The data in Table 4 demonstrate that since 1985 the university has successfully decreased the number of birds trapped in its experimental fields.

Remember our advice about introducing, discussing, and interpreting your visual displays as we discuss each type of display in this chapter.

Selecting a Visual Display: Considering Audience and Purpose

Your understanding of your audience's needs will determine the type of visual display you select to present your evidence or support your argument. When considering visual displays, you can often focus on your audience's experience with your topic, their technical understanding, their language and culture, and the purpose of your document—how your audience will use your message.

EXPERIENCE

An audience that has little experience with your subject will need a simple and obvious visual display. A table with only two or three column and row headings and related data is easily understood by most audiences. A line graph with two lines plotted against familiar units of measurement such as time and dollars is a familiar display you can use to introduce an unfamiliar subject.

More experienced audiences can handle detailed or more abstract visual displays, such as exploded drawings that show how the parts of a mechanism fit together. An audience that has worked with a similar mechanism can handle the detail of a more complex or symbolic display such as a schematic.

As you think about each display discussed in this chapter, you should consider its appropriateness for your purposes in terms of how much experience your audience has had with your subject.

TECHNICAL UNDERSTANDING

technical understanding
the specialized knowledge about a product or process that an audience brings to a document

The degree of **technical understanding** your audience has also should affect your selection of a visual display. The more technical understanding audience members have, the more abstract and symbolic your display can be. General or lay audiences will recognize common or familiar symbols and units of measurement only. A lay audience would understand a pictorial graph of the number of birds trapped only if the birds were distinguished by color or tone and not by particular features such as wing span or markings.

Audiences with some degree of technical understanding, such as technicians or mechanics with a high school or two-year college education, can handle more detail and symbols or icons common to their trade or job. These audiences would also be familiar with certain types of visual displays; for example, the electrician would be educated to read schematics of circuits.

professional audiences
composed of highly trained professionals, including scientists, computer programmers, engineers, physicians, and systems analysts

Professional audiences—engineers, scientists, systems analysts, computer programmers, and others—can understand visual displays using symbols and abstractions alone. These audiences can read everything from combination graphs to flow charts easily when the information and types of display are common in their profession.

LANGUAGE AND CULTURE

When we are thinking about U.S. audiences, the audience's ability to read English and its cultural background also should influence your choice of visual displays. Generally, the less comfortable the audience is with English, the more you need to limit your words and rely on widely accepted symbols, pictures, or icons. When possible, determine your audience's ability to read English. The majority of your audience should fit into one or two of the following categories:

- Reads English well.
- Has difficulty reading English.
- Reads English as a second language.
- Must rely on a translation of English.

If your words are going to be translated, use the most common, least ambiguous words and back up these words with widely understood symbols. On the other hand, when your audience has difficulty reading or does not read English at all, you will have to rely on the symbols alone to carry the message. The audience that reads well can handle most combinations of numbers and words.

The **cultural background** of the audience also determines its ability and willingness to read your visual display. For example, as we mentioned in Chapter 9, colors in various cultures carry different meanings. In Western cultures yellow can carry a negative connotation although in some Eastern cultures it symbolizes courage. Images of parts of the body and gestures as well as symbols should be selected with your audience's culture in mind. A waving hand might represent the end or good-bye in Western culture, but in other cultures can signal the beginning or hello. Finally, a culture that reads from right to left would have difficulty reading a table designed for a Western audience that reads from left to right.

cultural background
shared understanding of various symbols, shapes, and colors within a given culture

PURPOSE

To decide which visual display to use, you also need to understand generally which visual display works best in what circumstances and what use your audience will make of your data. Use Writing Strategies 10-1 to help you select what form of visual display will best serve your audience's needs. For example, if your audience wants to know the actual appearance of an object to see how it would fit into their environment, a photograph would serve these purposes. But if they need to assemble that object, a diagram would show the parts of the object and how they attach.

Even though you can use Writing Strategies 10-1 to get started on selecting and creating a visual display of data, the purpose of your display will help you make specific decisions about the type of visual display you will need. If your audience needs to know the mechanical and chemical processes used in office copying machines, Writing Strategies 10-1 tells you that drawings "show the physical components of mechanisms, objects, or organisms." But you will still have to decide what kind of drawing you should use. Writing Strategies 10-2 gives you some additional criteria to use once you are clear

WRITING STRATEGIES 10–1   Visual Display and General Use

Tables

- Give a large amount of data or detailed information in a small space
- Show the characteristics of objects, ideas, or processes
- Aid item-by-item comparison
- Display exact numeric values
- Give exact values for comparison

Graphs

- Show the relationships, trends, and patterns in two or more sets of data
- Support forecasts and predictions
- Help the audience interpolate or extrapolate
- Present complex information symbolically
- Interest the audience in the data or add credibility to words

Charts

- Show the components, chronology, or steps of a whole organization, process, mechanism, or organism
- Display the interrelationships between the components, stages, or steps

Photographs

- Show the actual appearance of an object or organism
- Add realism of detail, tone, texture, and color to a display or discussion
- Display features of objects or organisms difficult to draw, such as emotions
- Prove something is real or exists

Drawings and Diagrams

- Show the actual appearance of an object or organism excluding unnecessary detail
- Show the physical components of mechanisms, objects, or organisms
- Show generic or nonspecific objects or organisms
- Display objects that might have existed in the past or might in the future
- Show views that would be impossible to see otherwise

Maps

- Offer geographical information on data and objects
- Show locations in relationship to each other
- Show distribution of objects over a particular location
- Prepare the audience for travel

about the specific purpose of your document. To support your office copying machine description, you might select an action drawing if you want to display the interaction of the machine parts, an exploded drawing if you want to display the parts as assembled or attached to each other, or a symbolic drawing if you want to display the chemical interactions.

WRITING STRATEGIES 10–2   Linking Purpose to Specific Visual Display

Visual Display	Purpose
Graphs	
Simple bar	Display approximate values
Multiple bar	Compare approximate values
Stacked or area	Show how values contribute to total
Pie or 100 percent bar	Represent percentages or portions of a whole unit
Deviation	Display values above and below zero (negative/positive, loss/gain, hot/cold)
Line	Demonstrate or predict trends
Combination bar/line	Relate approximate values to trends
100 percent area	Show trends in components of a whole
Pictorial	Communicate abstract or unfamiliar values through symbols
Drawing	
Line	Show the generic or selective features of an unfamiliar or complex object
Action view	Explain an action or process of an object or organism
Phantom or translucent	Offer an inside view of an object or organism
Exploded view	Show how to assemble an object
Schematic or symbolic	Symbolize a system to a specialized audience
Charts	
Block	Reveal relationships between parts or categories of a system
Organizational	Show relationships between parts of an organization
Flow	Show interrelated decisions or steps in a process

Preparing Data for Visual Presentation

Before you create your visual display, you need to make sure your data are complete and accurate, and you will want to organize them usefully and meaningfully. To do that, keep the following guidelines in mind:

- Include all data available that pertain to your topic, even the data that might contradict final recommendations.
- Collect your data from a large enough **random sample** to support your recommendations and interpretation.
- Indicate all the possible interpretations of your data.
- Make sure that the data really support what you are saying.
- Clearly indicate the **units of measurement** of your data.
- Condense the data to a manageable form (e.g., $11,000 can be expressed as 11 with a column head or axis label of "dollars in thousands").

random sample
a representative sample selected from a table of random numbers or computer program for random selection

units of measurement
universal standards by which data are measured such as inches, Hertz, meters, pints, or miles

mean
arithmetic average from
dividing a sum by the
number of items added

median
the middle number in a
series containing an odd
number of items or the
number midway between
the two middle numbers in
a series containing an even
number of items

mode
number or value appearing
most frequently in a series

■ Indicate the measures of central tendency, if needed:

Mean: The arithmetic average obtained by dividing a sum by the number of items added.

Median: The middle number in a series containing an odd number of items or the number midway between the two middle numbers in a series containing an even number of items.

Mode: The number or value that appears most frequently in a series.

Once you have compiled your data, then you can create the most effective visual display to meet the needs of your audience.

Creating Effective Tables

If your audience needs to understand large amounts of concrete information in a short period of time, a table might serve their purposes. Tables show data in a small space for comparison and interpretation; with their rows and columns of numbers or characteristics, they are particular and concrete. Audiences won't have to read symbols and deal with abstract lines and curves, but complex tables (more than two or three column and row headings) are usually reserved for audiences with experience in the topic or with a high degree of technical understanding.

As with all visual displays of data, your audience will need you to integrate your table into the text by introducing, discussing, and interpreting it. When creating a table, organize data into groups and write short descriptive labels for each group. Order the table in the most appropriate way for your audience and purpose: alphabetically, geographically, chronologically, highest to lowest, or smallest to largest. Include all your data and do not ask your audience to interpolate or extrapolate. Compare your data vertically, rather than horizontally, and place columns that must be compared next to each other. **Columns** in a table have vertical references, while **rows** or **lines** have horizontal references. **Stub heads** classify line heads and therefore have vertical reference (see Example 10–2 for the typical arrangement of a table).

Finally, if you remember the following guidelines, your audience can read your table more easily:

columns
the vertical structure of
information in a table

rows, lines
the horizontal structure of
information in a table

stub heads
the categories or
classification of line
headings in a table

■ Select standard units of measurement; use common abbreviations for them, and indicate them either in column and row or line headings or beside each number.

■ Align the numbers along the decimal point or along the units column.

■ Align prose along the left-hand margin or center within the column.

■ Round off numbers if possible.

■ Convert fractions to decimal points and limit numbers to two decimal points.

EXAMPLE 10–2   Parts and Labels for a Typical Table

Stub Head	Single Column Head[a]	Multiple Column Head[b]		Single Column Head
		Subhead	**Subhead**	
Line head	111.1	222.2	333.3	444.4
Line head	11111.1	22.2	33.3	44.4
Subhead	111.1	2.2	3333.3	44.4
Subhead	111.1	2222.2	333.3	4.4
Line head	11.1	2.2	33.3	444.4
Column average or total	111.1	222.2	33.3	44.4

Source: XXXXX

[a]Footnote

[b]Footnote

- Use parallel grammatical forms for words.
- Box the table but avoid excessive rules (seldom are vertical rules necessary). Use spaces to help cluster and separate groups of information.
- Use footnotes, indicated by lowercase letters and numbers as **superscripts,** for clarification.
- Place the source of the table below the table.
- Turn large tables 90 degrees and place the top at the binding of the document.
- Continue long tables by writing "continued" at the bottom of the first pages and repeating the full title, "continued," and full column headings on each page that follows.

superscripts
numbers or letters raised above the line used to identify footnotes

Finally, remember that displaying data is a rhetorical act. When you create a visual display, you are persuading your audience that you have gathered your data thoroughly, that you have interpreted the data correctly, and that you can logically use that data to support your argument. Examples 10-3 and 10-4 show two displays of the same data. The examples tell different stories, however, and helping an audience read them provides different challenges for the writer. The first table (Example 10-3) might persuade an audience protesting the trapping of birds that numbers have declined since the 1985, or it might help experiment station researchers analyze how effective the traps have been in capturing each kind of bird over two decades. The second table (Example 10-4), on the other hand, could be used by those protesting the trapping to emphasize the totals per year and per kind. Certainly a total of 8,135 birds trapped in one year or 15,861 sparrows trapped in four years seems more significant than 2,034 birds per kind

EXAMPLE 10–3   Kind and Number of Birds Trapped, 1985–2000: Version One

| Kind of Bird | Year | | | |
	1985	1990	1995	2000
Sparrow	6,305	2,317	3,843	3,396
Blackbird	581	1,468	502	395
Starling	1,086	400	1,104	2,160
Grackle	163	131	325	412
Average per year[a]	2,034	1,079	1,444	1,591

[a]Rounded off to the next highest number.

Source: University of the Midwest. Agricultural Experiment Station.

EXAMPLE 10–4   Kind and Number of Birds Trapped, 1985–2000: Version Two

| Year | Kind of Bird | | | | Total per Year |
	Sparrow	Blackbird	Starling	Grackle	
1985	6,305	581	1,086	163	8,135
1990	2,317	1,468	400	131	4,316
1995	3,843	502	1,104	325	5,774
2000	3,396	395	2,160	412	6,363
Total per kind	15,861	2,946	4,750	1,031	

Source: University of the Midwest. Agricultural Experiment Station.

trapped in one year. However, notice that because Example 10–3 asks the audience to read horizontally, as well as to follow the natural vertical pattern, the audience might miss the "Total per Year" column.

Creating Effective Graphs

All graphs translate numbers into pictures—or bars, lines, areas, or pictorials for your audience. The numbers are plotted as a series of points on a coordinate system. The **vertical axis,** called the **Y axis,** begins with zero and often indicates quantity; the **horizontal axis,** called the **X axis,** often indicates time, item, or category. Therefore, graphs usually illustrate changes in one amount (e.g., quantity) in relation to changes in another measure (e.g., time). In the following sections, we discuss common types of graphs you might select if your audience wants you to show trends and to compare data. Because graphs have many features in common, here are some general guidelines for making them:

- Always title the graph with the largest lettering used and center it at the top.
- Make the axis lines heavier than any grid lines but not as heavy as curves.
- Start at zero, unless a large amount of the grid is not necessary; then clearly indicate a break in the grid but retain the zero baseline.
- Avoid using two amount scales (such as temperature on a left vertical axis and humidity on a right vertical axis) but if you must, clearly label and code your lines or bars.
- Label each axis (horizontally if possible) and indicate the unit of measurement used if the unit of measurement does not follow each number.
- Place a figure every second or fourth grid ruling. Use standard intervals of 10, 20, 30, and so on.
- Use small tick marks inside the scales if grid rulings are not used.
- Distinguish between curves or lines, bars, pieces of the pie, and so forth by tone, color, or symbol listed in a key or legend.

In general, graphs, particularly line graphs, are more abstract than tables; that is, graphs are more generic, more symbolic, and more likely to show the structure or organization of the data than a table. Therefore, if your audience needs your data displayed in a graph—for example, the audience is interested in a trend or forecast—then be aware that your audience must have had some experience with your topic and with this type of visual display to understand the display and be persuaded by it. Graphs are not always interchangeable. Each type of graph lends itself to specific purposes.

CREATING BAR GRAPHS TO DEMONSTRATE MAGNITUDE AND DIFFERENCE

Bar graphs best demonstrate the magnitude or size of several items or emphasize difference in one item at equal time intervals. Each bar represents a separate quantity, and multiple bars may be grouped and displayed horizontally or vertically. Vertical bars best indicate height and depth, and horizontal bars best indicate distance, length, and time.

Although bar graphs are not the best choice when your audience needs to see a trend, they work well when an audience is interested in distinct values over a specified period of time and the values are noticeably different. Bar graphs can also show how data change from place to place, and 100 percent bar graphs can show composition as a whole. Bars can also be stacked to show totals or combined with line graphs to help an audience

vertical, or *Y* axis
the vertical axis that begins at zero and usually indicates quantity on a graph

horizontal, or *X* axis
the horizontal axis usually indicates time, item, or category on a graph

understand difference along with a trend. Be aware that the darkest bars in your display may seem larger to your audience and attract their attention; this phenomenon might help emphasize a point in your argument, but you need to play fair with your audience and make sure that your data really support your point.

The following guidelines will be helpful when you are planning any bar graph:

- Make the bars the same width and label each bar.
- Make the space between bars one-half the bar width.
- Make the vertical or *Y* scale at least 75 percent as long as the horizontal scale.
- Make the longest bar extend to nearly the end of its parallel axis.
- Consider adding exact numbers or extending the grid lines to help relate the bars to the values.
- Consider using a **legend** or **key,** and distinguish among bars by varying pattern or color in multiple bar graphs.

legend or key
information that explains devices such as color or patterns on a graph

Simple Bar Graphs

Again, although graphs are more abstract than tables, almost any audience can quickly grasp the point of a simple bar graph. The bar graph in Example 10-5 shows how many sparrows were trapped in university experimental fields over a certain period of time. An audience can quickly and easily contrast the years in which a great many sparrows were trapped (e.g., 1983 and 1985) and the years in which few were trapped (e.g., 1990 and 1998).

If your data are too close to compare by the size of the bar alone, place exact numbers over each bar (e.g., 1988 and 1995 in Example 10-5). The farther a bar is from the vertical axis, the more difficult it is for the audience to see differences accurately. However, Example 10-5 would work well if your audience needed to focus on the far greater number of sparrows trapped in 1983 than trapped in more recent years.

Multiple Bar Graphs

You can group several bars in a multiple bar graph if your audience needs to see the magnitude of different variables or items in relation to each other over distinct periods of time. However, limit the groups to three or four bars. Audiences generally need a legend that lists the variables and shows the pattern or color that distinguishes them.

In Example 10-6 the kinds of birds trapped are displayed using bars of various patterns. Because few blackbirds and grackles were trapped over this time period, the bars representing them are quite small, so to help the audience see them, they are darker. Grid lines allow the audience to gauge more easily the number of each kind of bird trapped. The great numbers of

EXAMPLE 10–5   Simple Bar Graph

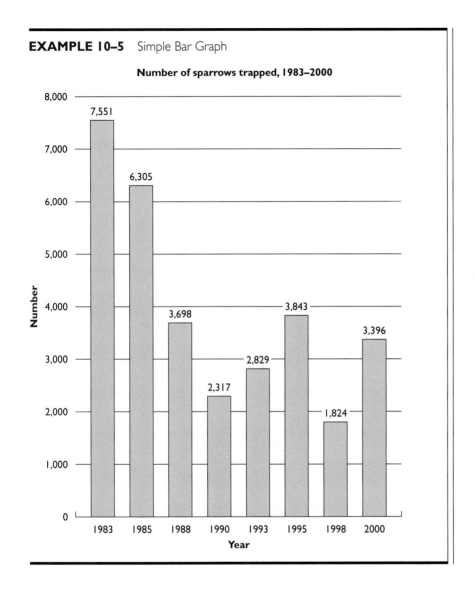

Number of sparrows trapped, 1983–2000

sparrows trapped becomes obvious immediately because of the contrast in bar height. Making a graph three dimensional helps an audience better see the smaller bars, but using this technique increases the visual size of bars and may misrepresent your data.

You make a horizontal bar graph by turning a vertical bar graph and scales on its side or 90 degrees to the right. This placement usually elimi-nates the need for a key. If possible, the longest bar should be placed at the

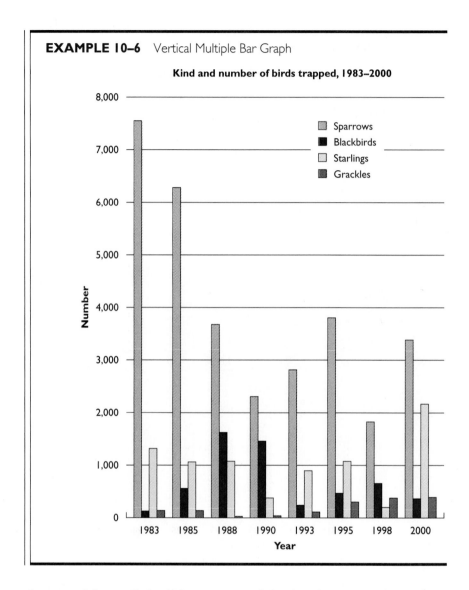

EXAMPLE 10–6   Vertical Multiple Bar Graph

Kind and number of birds trapped, 1983–2000

bottom of the graph for balance. Some of the data from Example 10-6 are displayed horizontally in Example 10-7.

In a horizontal bar graph, you have the room to label the bars themselves. However, labeling at the end of the bar lengthens the look of the bar and therefore changes the audience's impression of its magnitude. The horizontal bars in Example 10-7 could also have been labeled at the left of the *Y* axis, between the date and the axis line. A horizontal bar graph helps an audience see increasing or decreasing values more easily. However, the audience would have more difficulty seeing the differences between items—here the kinds of birds.

EXAMPLE 10–7   Horizontal Multiple Bar Graph

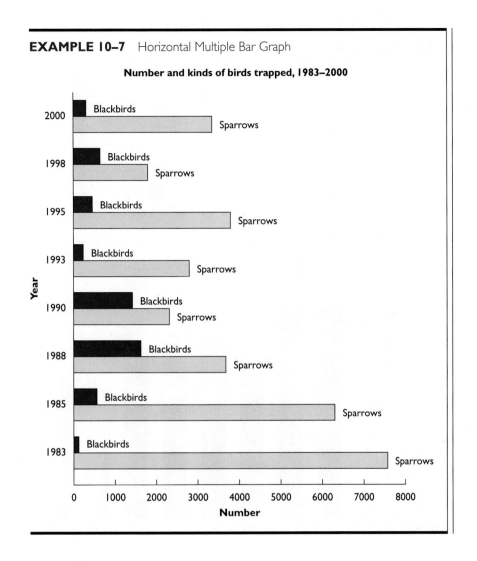

Number and kinds of birds trapped, 1983–2000

Stacked Bar Graphs

If your audience needs to see not only the size or magnitude of items but also how the items contribute to a whole quantity and how that quantity changes over time, you might stack the bars, either vertically or horizontally, in a graph. The magnitudes of the different items or components in a bar are differentiated by patterns or tones. In Example 10-8, an audience can see clearly how many grackles, starlings, blackbirds, and sparrows were trapped in a given year and also how the total number of birds trapped changed from year to year in this stacked bar graph. Because it would be difficult to read those totals, we have placed the numbers above the stacked bars. This aid is also important because in this graph the numbers

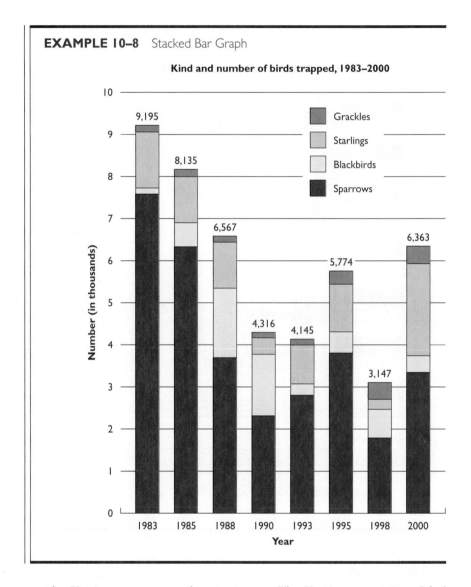

EXAMPLE 10–8   Stacked Bar Graph

Kind and number of birds trapped, 1983–2000

on the *Y* axis are compressed to save space. The *Y* axis now carries a label that reads "Number (in thousands)" so that the audience can translate 1 into 1,000, and so on.

Finally, notice in stacked bar graphs how difficult it is to compare the upper segments (the grackles, starlings, and blackbirds), as these bars have no **common baseline.** For example, it's impossible to tell how many blackbirds were trapped in 1983. Stacked bar graphs are intended to serve as a means for comparing totals rather than individual elements. Although lines could be added to link each segment, it's wise to limit the number of items to two or three.

common baseline
in vertical and horizontal bar graphs, the shared starting point of each bar; this feature is missing from stacked bar graphs

EXAMPLE 10–9   100 Percent Bar Graph

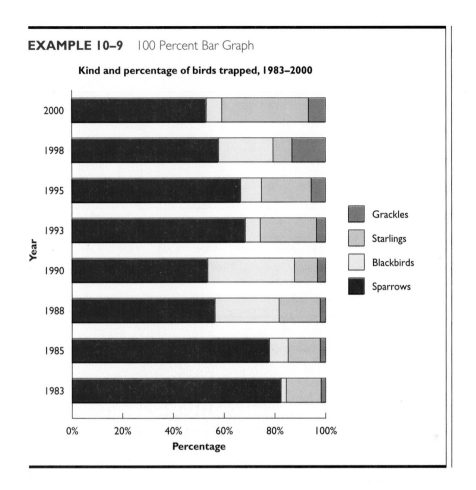

Kind and percentage of birds trapped, 1983–2000

100 Percent Bar Graphs

Components of a whole can also be displayed in 100 percent bar graphs in which each bar extends to 100 percent and the segments are measured in percentages. If your audience wants to see the relative size of each segment but not the magnitude, then display your data in this kind of graph. In Example 10-9, it is easy to figure out that the greatest portions of birds trapped were sparrows, but the numbers trapped are not displayed. This type of graph is quite similar to a pie graph, which is much easier for almost any audience to interpret. However, as with the stacked bar graph, individual segments do not line up along a common scale, so it's difficult for an audience to compare, for example, the number of blackbirds trapped for each year with any accuracy. Reading data according to their position along a common scale is often the most accurate **perceptual task** that an audience can perform (Cochran, Albrecht,

perceptual task
the intellectual effort performed by an audience to retrieve and to interpret data from a graph or other visual display

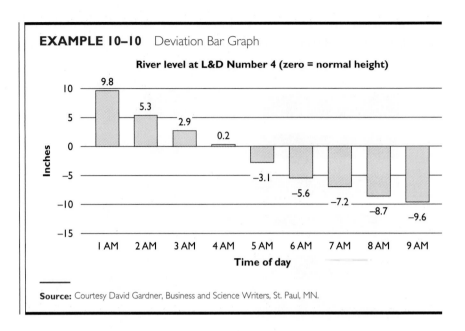

EXAMPLE 10–10   Deviation Bar Graph

River level at L&D Number 4 (zero = normal height)

Inches / *Time of day*

Values shown: 9.8, 5.3, 2.9, 0.2, −3.1, −5.6, −7.2, −8.7, −9.6

Times: 1 AM, 2 AM, 3 AM, 4 AM, 5 AM, 6 AM, 7 AM, 8 AM, 9 AM

Source: Courtesy David Gardner, Business and Science Writers, St. Paul, MN.

and Green, 27), so be aware of the difficulties your audience might have with a 100 percent bar graph.

Deviation Bar Graphs

opposite values
values such as positive/negative, gain/loss, increase/decrease that can be displayed on a deviation bar graph

A deviation bar graph indicates **opposite values**—negative/positive, loss/gain, and such—that extend on either side of a central point. For example, if your audience were interested in river levels, taken over a period of time, say from midnight to 9:00 A.M., Example 10-10 shows the river level on the hour and computes an average. Zero equals normal height, and the hourly averages are either above or below zero. Placing the exact numbers above and below the bars helps overcome the problem the audience might have with reading data not positioned along a common scale.

Combination Bar/Line Graphs

If your audience needs to see both trends and comparisons, combine a bar graph with a line graph. Notice that in Example 10-11, an audience could see from the bars the annual purchases of alternative-fuel vehicles (AFVs) by the federal government and from the line the total number of AFVs the government projects it will have purchased by 1998. The line, rather than stacked bars, gives greater emphasis to this number than a stacked bar graph would. As we will see in the following section, however, line graphs are more abstract than bar graphs; they demand that the audience interpret symbols and understand the structure and organization, rather than the particulars, of the data.

EXAMPLE 10–11   Combination Bar/Line Graph

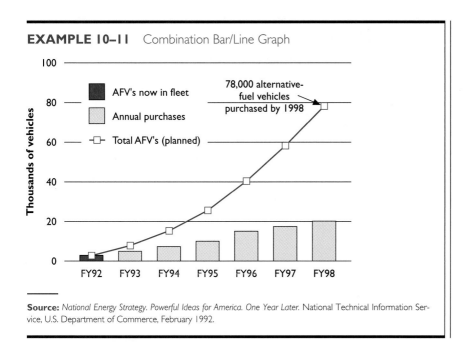

Source: *National Energy Strategy. Powerful Ideas for America. One Year Later.* National Technical Information Service, U.S. Department of Commerce, February 1992.

CREATING LINE GRAPHS TO SHOW TRENDS AND MOVEMENT

Line graphs are most useful in showing trends or movement in a long series of data or several series of data, particularly if an audience must **interpolate** (fill in) or **extrapolate** (extend) that series and it does not need to know the exact values. If your audience needs to see cause and effect—that is, what happens to one value (dependent value) as another (independent value) changes—then display your data in a line graph.

Because line graphs are more abstract than bar graphs and can accommodate and symbolize complex data, be sure that your audience has enough experience and technical knowledge to understand your display. Keep in mind the following guidelines:

interpolate and extrapolate
two means of displaying trends in data by filling in gaps (interpolating) or projecting trends (extrapolating)

- Place the dependent variable on the Y (vertical) axis and the independent variable on the X (horizontal) axis.
- Keep vertical and horizontal axes proportionate.
- Make sure the slope or steepness of the line accurately depicts the trend indicated by the data.
- Mark individual data points on each line so the audience can see how you used the data to determine the slope.
- Place no more than three or four lines on each graph.
- Use colors or symbols to distinguish your lines.
- Label each line or use a legend if necessary to avoid clutter.
- When showing positive and negative values, use the X axis as the mean.

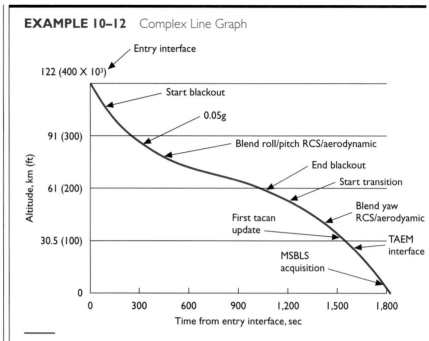

EXAMPLE 10–12   Complex Line Graph

Source: John F. Hanaway and Robert W. Moorehead, *Space Shuttle Avionics System.* National Aeronautics and Space Administration, Office of Management, Scientific and Technical Information, NASA SP-504, 1989.

Look back at Example 10-1 to see a simple line graph. Rather than comparing the numbers of birds trapped in each year, the lines in this figure show increases and decreases as well as trends and projections. A reader could be fairly certain that the number of grackles trapped would not increase in 2003, but would not be surprised if the number of starlings increased, given the trend from 1990 to 2000.

A variation on the line graph (Example 10-12) depicts events during movement or action. The points marked on the line indicate each stage of the entry phase of the Space Shuttle Avionics System measured by a decrease in altitude over a period of seconds. An audience with technical knowledge of the system would understand not only the events but also why each event appropriately follows the preceding one at that time and altitude.

CREATING AREA GRAPHS TO SHOW TRENDS AND TOTALS

If your audience needs to know both trends in data and also the sum of those data, display the data in an area graph (also called a multiple band graph). The area below each line is filled in by a pattern or color and represents a unique value. The topmost line represents the total or sum of all the data. Usually the largest area or the line with the least slope appears closest to the X axis.

EXAMPLE 10–13   An Area Graph

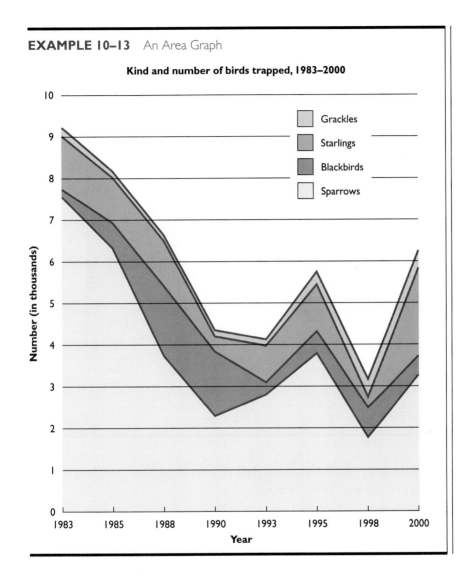

Kind and number of birds trapped, 1983–2000

Legend: Grackles, Starlings, Blackbirds, Sparrows

Y-axis: Number (in thousands) — 0 to 10
X-axis: Year — 1983, 1985, 1988, 1990, 1993, 1995, 1998, 2000

Because area graphs display two aspects of data, they are difficult to interpret, and you should use them for experienced and educated audiences. An audience seeing Example 10-13 might need to be reminded that the top line represents the total number of birds trapped in a given year, while the areas between each line emphasize the trends for that kind of bird only. In other words, the number of blackbirds trapped in 1990 is about 1,500, not more than 4,000. Add **grid lines** like those in Example 10-13 to help an audience read the data for more recent years; however, remember that area graphs are abstract and therefore are best used to demonstrate trends, not exact data.

grid lines
lines that help readers pick out data from given categories

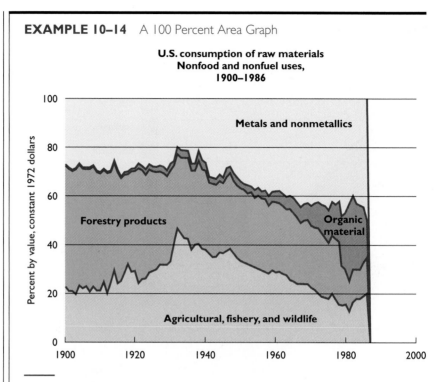

EXAMPLE 10–14   A 100 Percent Area Graph

**U.S. consumption of raw materials
Nonfood and nonfuel uses,
1900–1986**

Source: Donald G. Rogich. "The Future of Materials: Plastic Components Are Growing." *Minerals Today.* U.S. Department of the Interior, Bureau of Mines, MS2501, June 1991.

CREATING 100 PERCENT AREA GRAPHS TO SHOW TRENDS IN THE COMPONENTS OF A WHOLE

If your audience needs to see the trends in the components of a whole, you might display your data in a 100 percent area graph. An audience looking at Example 10-14 could see how the amount of organic material has increased since 1900, while the number of forestry products has gone down.

The Y axis indicates the part of 100 percent that each component makes up. Although stacked bar graphs and pie graphs show proportions of a whole unit, the 100 percent area graph is more difficult to read because it shows proportions of a whole trend.

CREATING PIE GRAPHS TO SHOW PROPORTIONS OF A WHOLE

coordinate system
describes the X and Y axes
on a graph

Pie graphs represent proportions or percentages of a whole, rather than figures plotted on a **coordinate system.** They work well with inexperienced audiences who have little technical knowledge or who read with difficulty. However, although pie graphs gain attention and convey impressions, they

EXAMPLE 10–15 Pie Graph with One Separated Wedge

Kind and number of birds trapped in 2000

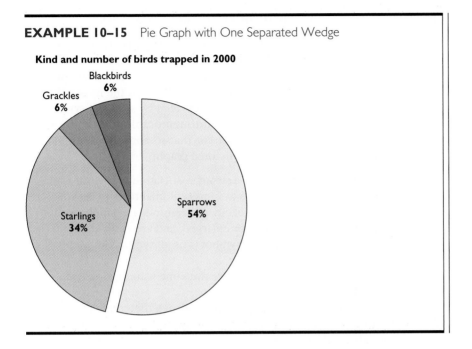

offer little help in close comparisons. Audiences have difficulty accessing the differences between wedges similar in proportion. Also, pie graphs cannot be used to compare component parts of two or more wholes. Although exact numbers can be added to the pie, experienced audiences prefer line, area, or bar graphs.

When creating pie graphs, keep in mind the following guidelines:

- Start the pie graph at the 12 o'clock position with the largest wedge.
- Make sure that wedges add up to 100 percent with each 3.6-degree segment of the pie equaling 1 percent.
- Keep wedges at least 5 percent or 18 degrees.
- Limit your pie graph to 8 wedges.
- Combine small segments under "Other."
- Separate only 1 wedge for emphasis.
- Place labels horizontally outside the pie. Avoid lines and arrows.
- Indicate exact percentages in labels.
- Avoid three-dimensional pies because the closest wedge looks larger than it actually is.

The pie graph in Example 10–15 would be best in an oral presentation to gain the attention of an audience interested in the high percentage of sparrows trapped in 2000. By separating out the sparrow wedge and adding the exact percentage, the audience can see immediately that sparrows constituted over half of the number of birds trapped.

CREATING PICTORIAL GRAPHS TO SHOW ABSTRACT CONCEPTS

icons
symbols conveying abstract
ideas or concepts

If your audience needs to understand abstract ideas quickly, use pictorial graphs (or pictographs) to communicate abstract ideas or concepts by using symbols or **icons.** These symbols can be used as numerical counting units, with each representing a specific number. Perhaps the most effective pictorial graph is a bar graph with the bars replaced by stacks of symbols. A pictorial graph is visually interesting and memorable but, if designed poorly, can distort or misrepresent data. For that reason, keep in mind the following guidelines when creating a pictorial graph:

- Make sure that symbols have a strong association with your topic, value, or idea, such as barrels to represent quantities of fuel oil.
- Make sure that symbols are readily understood, are simple and recognizable when reduced or enlarged, and are easily divisible (if necessary). If more than one symbol is used, symbols should contrast well in shape.
- Design all symbols so that they are the same value and size.
- Space all symbols equally.
- When displaying different quantities, change the number of symbols, not the size or height of one symbol.
- If possible, round off numbers to eliminate fractions (and therefore divided symbols).

In Example 10-16, the daily workload of a Data Distribution Facility (DDF) for distributing spacecraft data from various satellites to science centers is depicted in a pictorial graph. This graph was prepared to share with vendors who must demonstrate that they can provide the needed number of rewritable optical disks of the anticipated production workload. The symbols used in the graph take the place of traditional bars and represent the optical disks. The audience can grasp immediately the anticipated increase in the number of disks needed as more satellites are launched, and the audience can later study exact numbers that appear at the top of the stacked symbols.

Representing Data Accurately

distortion
the misrepresentation of
data in a visual display

Before we discuss other types of visual display, we need to remind you about accurate representation of data. Most of the common problems that might lead to the **distortion** of your data presentation occur in bar, line, and pictorial graphs. Of course, you should make sure that your sample of data is randomly collected, large enough to represent your topic, and gathered accurately. You must present all data gathered, even the unfavorable or atypical. You must be sure that your data measure what you claim and that the causal relationships you describe are really supported by your data. If your data are taken from another source, you should check the accuracy of that study carefully and credit that source for your audience.

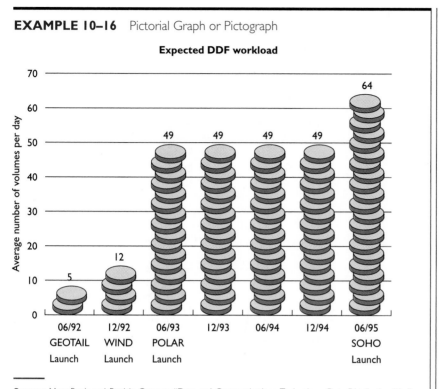

EXAMPLE 10–16   Pictorial Graph or Pictograph

Expected DDF workload

Source: Mary Reph and Patricia Carreon. "Data and Communications Technology, Data Distribution Media Study," *Research & Technology*, Goddard Space Flight Center, NASA, 1990.

Also, you need to avoid the five following common pitfalls in displaying data:

- Distortion of the *X* or *Y* axis in graphs.
- Distortion by **suppressing the zero** in graphs.
- Distortion by increasing the width or height of bars in graphs.
- Distortion in three-dimensional pie graphs.
- Distortion through improper use of icons or **isotypes** in pictorial graphs.

suppressing the zero
a distortion of data by starting graphs at other than zero

isotypes
another term for icons used as numerical counting units in visual displays

AVOIDING DISTORTION IN *X* OR *Y* AXIS

The values on the *X* and *Y* axes of graphs should be spaced equally to portray relationships and trends accurately. This rule is particularly important in line graphs. The information display in Example 10-1 is distorted in Example 10-17. In Example 10-17 the increases and decreases

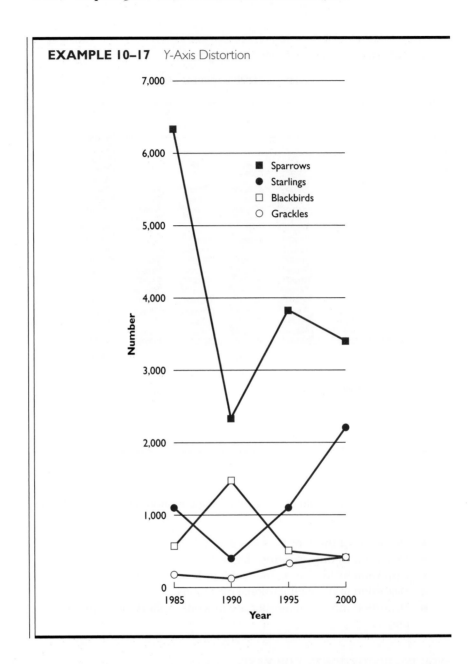

EXAMPLE 10–17   Y-Axis Distortion

in the number of birds trapped appear much greater than they actually are because the tick marks on the *Y* axis are spaced farther apart than those on the *X* axis. The same distortion would take place if the values on the *Y* axis were measured at more frequent intervals or at fewer or more intervals midway through the axis. For example, marking values

EXAMPLE 10–18   Data Distortion by Suppressed Zero

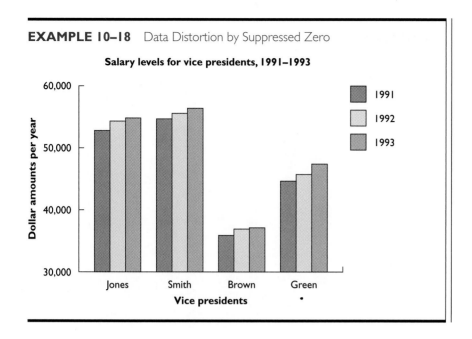

Salary levels for vice presidents, 1991–1993

after 3,000 at 3,500, 4,000, 4,500, 5,000, 5,500, 6,000, and 6,500 would cause the drop in the number of sparrows trapped to appear quite dramatic.

AVOIDING DISTORTION BY SUPPRESSED ZERO

As we discussed previously, graphs should always begin at zero. You can see the distortion in Example 10-18 because the values begin at 30,000, and therefore, the salaries of the vice presidents look rather modest.

You might be tempted to suppress the zero if none of your values come close to it. For example, every vice president gets more than $30,000 per year. However, you need to begin the graph at zero and indicate by a visual break that you have left out a portion of the bar or line (as in Example 10-19).

AVOIDING DISTORTION IN WIDTH OR HEIGHT OF BARS

If you make any changes in one dimension of a graph without making corresponding changes throughout the graph, you will introduce distortions into your visual display. Increasing the width of a bar in a graph, without making a change in the height, and vice versa, will distort the visual appearance of the data. The bars in Example 10-20 are so wide that the audience will perceive the number of birds trapped to be much greater than

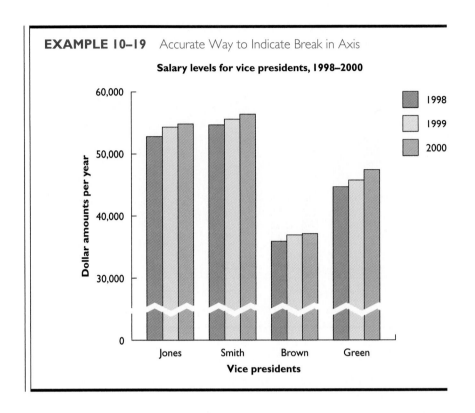

EXAMPLE 10–19   Accurate Way to Indicate Break in Axis

Salary levels for vice presidents, 1998–2000

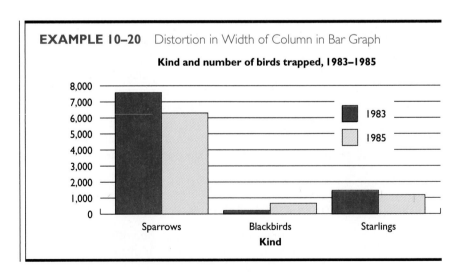

EXAMPLE 10–20   Distortion in Width of Column in Bar Graph

Kind and number of birds trapped, 1983–1985

EXAMPLE 10–21   Distortion in 3-D Pie Graph with Separated Wedges

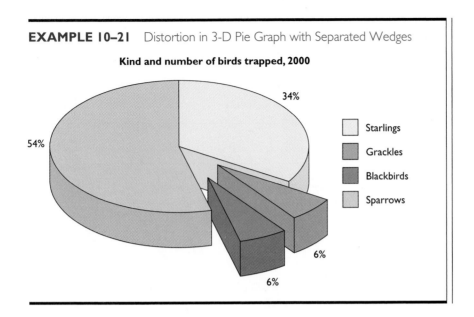

Kind and number of birds trapped, 2000

- Starlings
- Grackles
- Blackbirds
- Sparrows

34%

54%

6%

6%

---

they are. If you increase the width of a bar, you need to increase the height in proportion, and vice versa.

AVOIDING DISTORTION THROUGH 3-D

When creating a pie graph, you may be particularly tempted to make your presentation more dramatic by using a three-dimensional (3-D) approach. However, the data "nearest" the audience in the 3-D version will appear greater than it is. The data from Example 10–15 have been displayed in Example 10–21 in a way that makes the number of blackbirds and grackles appear much greater than they are (particularly because we have also separated out these two wedges from the pie).

AVOIDING DISTORTION THROUGH IMPROPER USE OF ISOTYPES OR ICONS

When you use a pictorial graph, remember that isotypes or icons are counting units. If you attempt to display your data by enlarging or shrinking one icon (as in Example 10–22), you will misrepresent your data. The audience cannot gauge how many sparrows were actually trapped because the contrast between the smallest and largest icon is so dramatic. The same data are accurately displayed in Example 10–23.

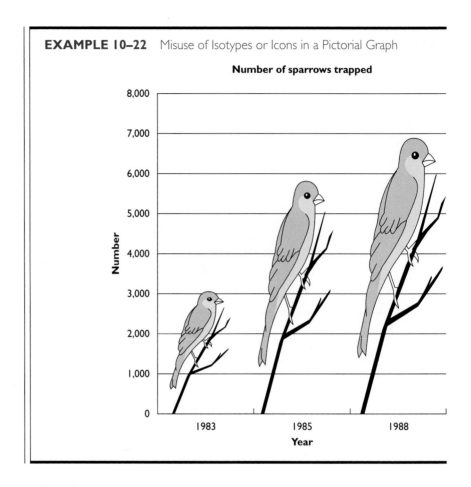

EXAMPLE 10–22   Misuse of Isotypes or Icons in a Pictorial Graph

Number of sparrows trapped

Creating Photographs

If your audience needs to see a subject in detail and in a realistic setting, a photograph, particularly one in color, will meet this need. Photographs can be reduced, cropped, and enlarged (although enlargement sometimes decreases sharpness) to enhance a subject; the angle of view can add emphasis to a particular part or approach. For audiences unfamiliar with your subject, a photograph can bring instant recognition and orientation as photographs are concrete, rather than abstract. However, so much detail may appear in a photograph that your text must tell your audience what to look for. When taking a photograph of your subject, consider the following suggestions:

- Avoid too much detail in the photograph; for example, be certain that what surrounds and provides background for your subject is necessary and not distracting.
- Consider adding callouts or small arrows to the photo to point to the main features.
- Consider adding a caption to help the audience understand what it is seeing.

EXAMPLE 10–23 Accurate Use of Isotypes or Icons in Pictorial Graph

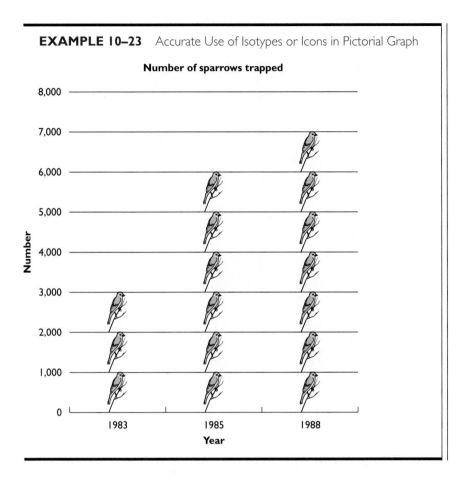

■ Consider adding a familiar object to the photograph to help the audience gauge size and dimensions.

A photograph of the "L" blast furnace complex of Bethlehem Steel Corporation comprised part of a press packet on opening day in the 1980s (see Example 10-24). Because the photo clearly shows the facility (and its dimensions in contrast to the two workers on the right), the general audience reading about the opening day of the furnace in a newspaper could visualize the process described in the caption below. However, because the photograph has so much detail, the caption must help the audience decide what to look at (e.g., the facility rather than the railroad tracks):

> Large facilities designed to clean 360,000 cubic feet per minute of blast furnace gas occupy this end of the "L" blast furnace complex at Bethlehem Steel Corporation's Sparrows Point, Maryland, plant. High-energy scrubbers, using about 10,000 gallons per minute of industrial water, help clean the gas. The water is clarified in two 90-foot-diameter thickeners and then cooled in the towers shown at left. It is then chemically treated before being recirculated. The iron-bearing slurry recovered from the

EXAMPLE 10–24   Photograph of "L" Blast Furnace

Source: Courtesty of Bethlehem Steel.

thickeners is processed and recycled through the sinter plant to the blast furnaces. Expenditures for environmental control facilities on the furnace, the largest in the Western Hemisphere, totaled $22 million.

The angle of view, a wide shot of the furnace with the camera looking slightly upward, emphasizes the size of the furnace. Again, because the photograph captures the exact image of the furnace, this visual display is most appropriate for a general audience reading a newspaper or an annual report, an audience that might never have seen such a facility before and would not be able to understand a more abstract depiction.

Creating Drawings and Diagrams

If your audience requires that you usually convey structure or organization symbolically, rather than depicting the actual appearance of a subject, you will need to create drawings or diagrams. If that's the case, make sure that your audience has the needed experience or education to decode these drawings and diagrams. Because they are more abstract than photographs, drawings and diagrams emphasize important points, focus on specific components, or convey through symbols components of an object, system, or process.

EXAMPLE 10–25 Drawing of "L" Blast Furnace

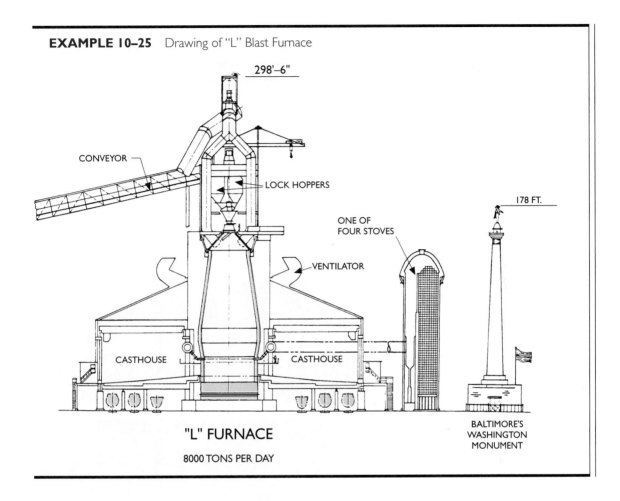

For example, the drawing of the Bethlehem Steel "L" blast furnace (Example 10–25) shows the main parts of the furnace (notice that the local audience can still appreciate the size of the furnace in contrast to the sketch of Baltimore's Washington Monument or read the exact dimensions in the drawing). However, extraneous detail has been eliminated, and the drawing comes close to that of an engineering blueprint.

The type of drawing or diagram you create depends on the education and experience of your audience and on what aspects of your subject you want to depict. In general, keep in mind the following guidelines:

- Select the amount of detail your audience will find useful and manageable.
- Label clearly all parts, steps, or stages by horizontal, parallel labels and arrows. Use a key only when absolutely necessary.
- Select symbols that are familiar to your audience.

- Give dimensions, orientation, and point of view information about your object. (For example, the object might be "seen from the top right, which is 11 feet off the ground.")
- If useful, show how parts relate or are connected or attached to the whole and to each other.

CREATING ACTION VIEW DRAWINGS TO SHOW INTERACTION OR STEPS

An action view drawing will meet the needs of audiences who must understand how components interact or how one step in the process leads to another. Example 10–26 displays the effect of a passenger thrown forward during a train crash. The display was created by a computer graphics program

EXAMPLE 10–26   Action View Drawing

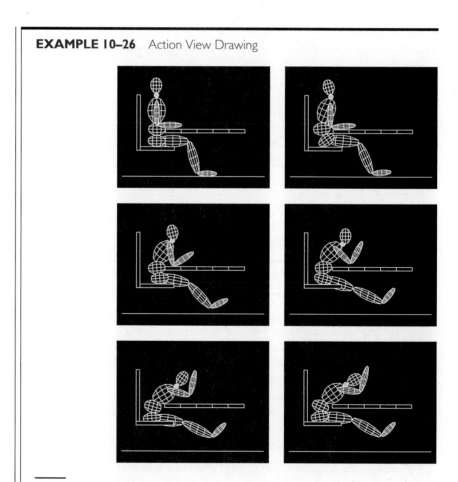

Source: *National Energy Strategy: Powerful Ideas for America, One Year Later.* National Technical Information Service, U.S. Department of Commerce, February 1992.

developed at the Department of Energy's Lawrence Livermore National Laboratory and is used to measure the effect of stress traveling through structures. Even an inexperienced audience would have no difficulty understanding the movement of the human body upon impact.

CREATING TRANSLUCENT AND PHANTOM VIEW DRAWINGS TO SHOW INSIDE STRUCTURES AND PARTS

Translucent or phantom view drawings show the inside structure and parts of objects or organisms in relationship to the outside frame or covering. A centrifugal grinder pump, a part of a sewer system, shown in Example 10–27, depicts the inside components in relation to the outside shell of the system. The audience gets an impression of the inside without being overwhelmed by details and labels.

EXAMPLE 10–27   Translucent View Drawing

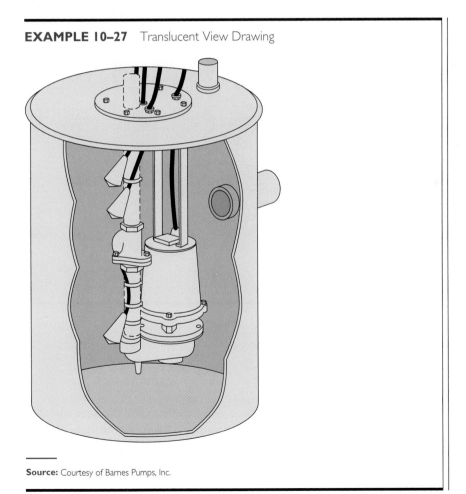

Source: Courtesy of Barnes Pumps, Inc.

EXAMPLE 10–28   Cutaway View Drawing

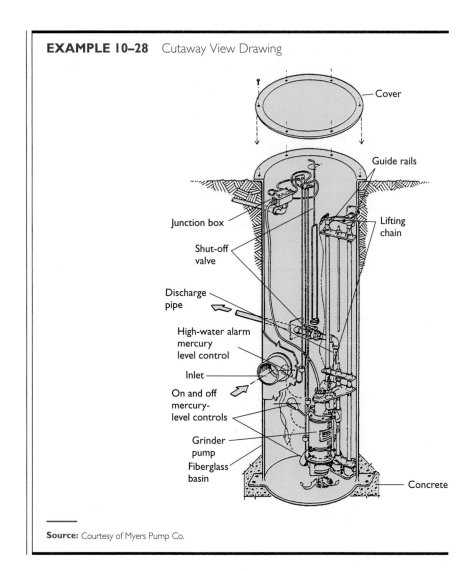

Source: Courtesy of Myers Pump Co.

CREATING CUTAWAY VIEW DRAWINGS TO SHOW
CROSS-SECTIONAL STRUCTURES

An experienced or technical audience who needs to see the structure and parts of a cross section of a device or organism can do so through cutaway view drawings. Example 10-28, a cutaway view of the grinder pump package, depicts the object as if it were sliced down the middle and opened up to show the inside.

EXAMPLE 10–29   Exploded View Drawing

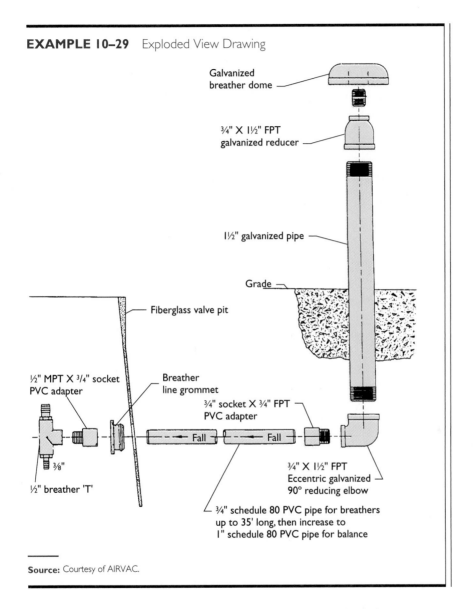

Galvanized
breather dome

¾" X 1½" FPT
galvanized reducer

1½" galvanized pipe

Grade

Fiberglass valve pit

½" MPT X ¾" socket
PVC adapter

Breather
line grommet

¾" socket X ¾" FPT
PVC adapter

Fall

Fall

⅜"

½" breather 'T'

¾" X 1½" FPT
Eccentric galvanized
90° reducing elbow

¾" schedule 80 PVC pipe for breathers
up to 35' long, then increase to
1" schedule 80 PVC pipe for balance

Source: Courtesy of AIRVAC.

CREATING EXPLODED VIEW DRAWINGS TO SHOW CONSTRUCTION OF AN OBJECT

An experienced and technical audience who must construct or assemble an object will find exploded view drawings as an even more abstract way to depict this information. In Example 10-29, the parts of the external breather dial of a wastewater treatment system are blown apart or separated so that the audience can see how the parts connect.

EXAMPLE 10–30 Schematic or Symbolic Diagram

Source: Courtesy of AIRVAC.

CREATING SCHEMATIC OR SYMBOLIC DIAGRAMS TO SHOW THE MOST ABSTRACT INFORMATION

Only technical audiences experienced in the specific subject require schematic or symbolic diagrams. These diagrams rely on specific symbols understood by those in the field and are highly abstract. Only audiences trained in reading schematics and familiar with a vacuum station could understand Example 10–30.

Creating Maps

Maps help your audience locate geographic information by showing features of a landscape, the site of an event, or data as they pertain to a location. To understand the progression of the oil spill from the *Exxon*

EXAMPLE 10–31   Map of the *Exxon Valdez* Oil Spill Location and Movement

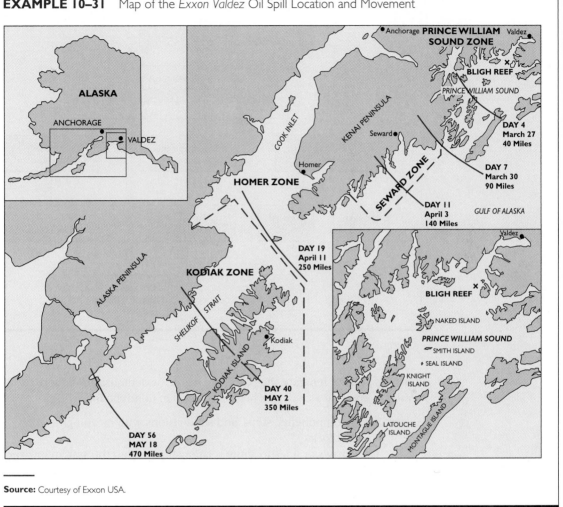

Source: Courtesy of Exxon USA.

Valdez, an audience could study the map presented in Example 10-31. The dates and mileage imposed upon the map, as well as the enlarged version of Prince William Sound and the small map of Alaska, help the audience visualize what might be an unfamiliar location and how that location was affected over time.

Creating Charts

You can create a chart for an audience that needs to understand the components of an organization or the steps in a process. Much like exploded

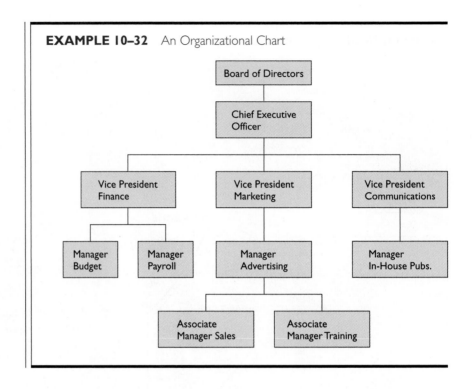

EXAMPLE 10–32   An Organizational Chart

drawings, charts separate the parts of a whole or demonstrate how parts interact. In creating a chart, it's helpful to keep in mind the following:

■ Label components, steps, and subdivisions with meaningful labels within blocks.
■ Use arrows or lines to show relationships within the system or over time.

CREATING ORGANIZATIONAL CHARTS TO DISPLAY HIERARCHIES

hierarchy
the arrangement of positions and power within an organizational structure

Organizational charts (such as Example 10-32) display the **hierarchy** (chain of command as well as formal communication links) of the positions within an organization through a series of interconnected blocks.

CREATING BLOCK COMPONENTS CHARTS TO ILLUSTRATE PARTS OR COMPONENTS

A block components chart will help your audience understand the components, parts, or subdivisions of a whole system. For example, a space shuttle avionics system, its components, and the relationship between components are shown in Example 10-33.

EXAMPLE 10–33   A Block Components Chart

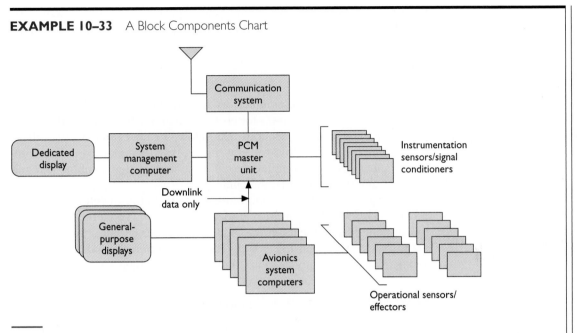

Source: John F. Hanaway and Robert W. Moorehead, *Space Shuttle Avionics System,* National Aeronautics and Space Administration, Office of Management, Scientific and Technical Information, NASA SP-504, 1989.

EXAMPLE 10–34   A Flow Chart

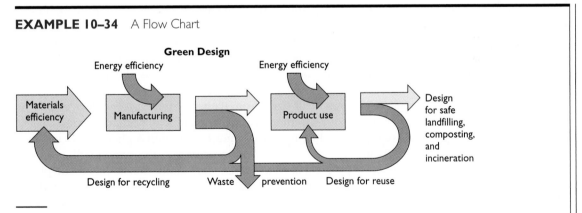

Source: *Green Products by Design: Choices for a Cleaner Environment,* US Congress, Office of Technology Assessment (Washington, DC: US Government Printing Office, 1992).

CREATING FLOW CHARTS TO DISPLAY STEPS IN A PROCESS

Flow charts show your audience the steps in a process. The time that each step takes or how one step causes another will frequently be indicated in the flow chart. Example 10-34 demonstrates how product design affects materials flow in a "Green Design." The impact on the environment is reduced through recycling, reduction of waste, and efficient use of material and energy.

Using Computer Software and Hardware to Create Effective Visual Displays

Increasingly, technical communicators are finding computers to be useful—in some cases, essential—tools for creating, manipulating, and producing visual presentations. In the sections that follow, we have listed a few of these tools. Some of them allow for graphic representations in print environments; others support computer-based presentations that combine text and graphics in multimedia environments.

computer tools
computer applications designed to create a variety of visual displays based on information stored in data files

Computer tools for communicators are being developed and improved so rapidly that our list will soon be outdated. You'll need to stay on top of new developments in the field of computer graphics by consulting graphic designers who use computers and computer-supported graphics and by reading trade publications on desktop publishing.

USING IMAGE/TEXT SCANNING PACKAGES

scanners
electronic devices that allow users to convert text and images to digital files that can be stored on the computer and manipulated with image software

Scanners are electronic devices that allow you to take a still image (e.g., a page of text, a photograph, line art, cartoons), digitize, and store it on your computer. Programs such as Adobe Photoshop allow you to take those stored images, manipulate them on the computer (e.g., changing shapes, sizes, outlines, color, and introducing visual effects), and print them—in color, gray scale, or black and white. Computer programs that support full-feature **optical character recognition (OCR)** enable you to use your scanner to import text as text rather than just as an image of text. This is a wonderful time saver because with OCR software to go with your scanner, you don't have to retype pages.

optical character recognition (OCR)
software that allows the use of a scanner to import text as text so that it can be edited

Scanners come in all sizes, from flatbed scanners (large devices that resemble a photocopy machine) to smaller handheld, portable devices. These tools are especially helpful if you want to use both a photograph and a related drawing in a document. After scanning in the photograph, you can use a software program such as Adobe Photoshop, to manipulate the photograph, eliminating extraneous detail to create a drawing. Once scanned and stored on the computer, these images can be used both in print documents and online.

Image and text scanning packages allow you to scan images electronically from books, magazines, and photographs, and to import them into a digital environment. When these images are digital—that is, converted into the electronic impulses computers use—they can be manipulated according to the rhetorical contexts for which they will be used, the audiences for whom they are designed, and the information that they are intended to convey. Make sure that the ease with which you can scan, digitize, manipulate, and use existing images doesn't lull you into violating copyright laws when you create **digital images.** You'll need permission when you use someone else's work, even if you change it electronically. Whenever possible work with material you develop yourself.

Typical image scanning products, for example, allow you to import and export images in a variety of standard industry formats, correct and balance the color or tone of images, increase or decrease brightness and contrast, capture computer screens, crop and scale images, and compress images so that they take up less memory in a computer.

digital images
an electronic form of data storage that increases the flexibility with which images can be modified to suit a variety of rhetorical contexts

USING SPREADSHEETS

Spreadsheets are computer programs that allow you to store, manipulate, and represent data in columns and rows. Most full-featured spreadsheets also provide tools for the visual representation of data in print contexts and can help you create area charts, bar charts, column charts, stacked column charts, line graphs, pie charts, scatter plots, 3-D charts, line/column charts, and volume charts. We used Microsoft Excel to create the bird trapping illustrations in this chapter.

When you create a spreadsheet in Microsoft Excel, like that shown in Example 10–35, you can rely on the software to track, calculate, and analyze your data and to create visual displays of that data. To create such a spreadsheet, you type that data into a grid of cells. You can also enter formulas that total, subtract, multiply, and divide your numbers, although no such formulas were required for the spreadsheet for our bird trapping illustrations. You can also edit the cells to change your data, select only one portion of the data to appear in a chart, and change the spreadsheet's structure by widening columns and rows, inserting or deleting rows and columns, and moving and copying data. To view your data graphically, you ask Excel to create a chart. Once Excel has created a chart, you can tailor it to meet the needs of your audience by clicking on any element and using the program's tools to edit that element. You can add or change titles, labels, legends, arrows, and

EXAMPLE 10–35   A Typical Spreadsheet

	1983	1985	1988	1990	1993	1995	1998	2000
Sparrows	7551	6305	3698	2317	2829	3843	1824	3396
Blackbirds	137	581	1630	1468	264	502	684	395
Starlings	1348	1086	1110	400	905	1104	228	2160
Grackles	159	163	129	131	147	325	411	412
Total	9195	8135	6567	4316	4145	5774	3147	6363

gridlines. You can choose and change color, patterns, and shading, and you can change the scale, labeling, and look of the axes. You must be careful not to distort your data or mislead your audience. If you want to change or edit your data, you have to do so in the spreadsheet or worksheet rather than on the actual chart or graph. If you have access to Excel, you might enter the bird trapping data shown in Example 10–35 and re-create for yourself the visual displays in this chapter.

USING MULTIMEDIA PACKAGES

Multimedia packages will help you create computer-based documents that can incorporate video, animation, voice, sound effects and music, still images, and text. The multimedia presentations that you create using these packages require computer support when they are displayed for an audience. Such packages can help you show the operation of a mechanism, the change associated with a phenomenon over time, the aging or growth of an organism, or the unfolding of a technical process or procedure, among many other temporal representations. Generally, multimedia presentations are meant to be viewed on a computer or a television screen rather than in print, but animation sequences can be depicted in print. Example 10–36 illustrates an animation sequence and "score" in a multimedia program called MacroMind Director.

EXAMPLE 10–36   Animation Sequence

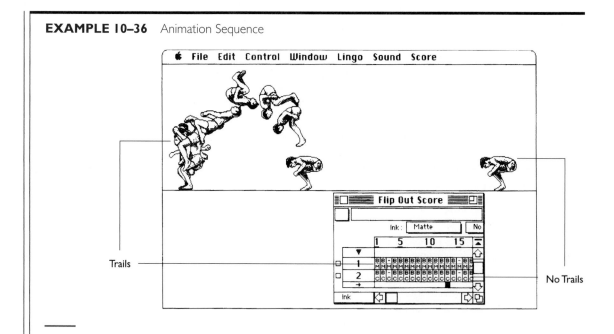

Source: MacroMind Director, Version 3.0, *Studio Manual,* 2nd ed. Reprinted with permission of Macromedia, Inc. © 1992.

USING VIDEO EDITING PACKAGES

Increasingly, technical communicators are called upon to be familiar with video used for training, sales, and public relations. Video editing packages support direct video input from cameras into computers. When video sequences are digitized, these programs allow users to select video clips, zoom in on parts of sequences; slow or speed up sequences; create various transitions including wipes, dissolves, and cuts; edit sound tracks; and key text onto video images. Example 10–37 shows how input from video editing programs—in this case QuickTime—can be imported into multimedia presentations.

USING CLIP ART AND PAINT AND DRAW PROGRAMS

If you are involved in any of your company's communication activities—flyers, newsletters, posters, multimedia presentations—you will soon become accustomed to turning to clip art packages for ready-made art work. Clip art packages contain a range of preselected images, often accessible by topic (e.g., celebrations, maps, animals, objects, business images,

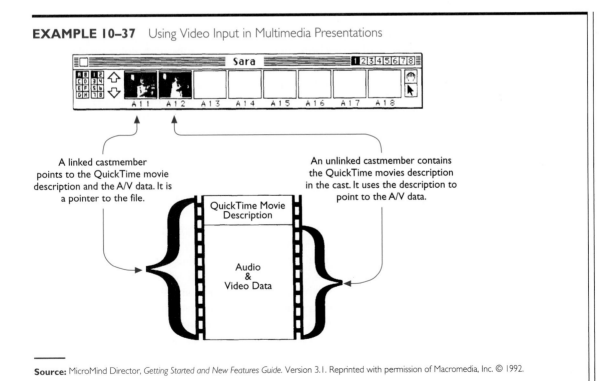

EXAMPLE 10–37 Using Video Input in Multimedia Presentations

Source: MicroMind Director, *Getting Started and New Features Guide.* Version 3.1. Reprinted with permission of Macromedia, Inc. © 1992.

sports, borders, food, or transportation) that can be selected, cut, and pasted into documents to illustrate or enhance text. If you don't have access to clip art software, you can purchase the hard copy and scan in the images.

Paint programs allow you use the computer for many of the routine tasks that previously had to be sent out to graphic artists. These packages provide you with tools similar to those used by artists: brushes, compasses, lines, erasers, spray paints, texturizers, dots, dashes, and colors, among others. With them, you can design original images or modify images imported through scanners or from the Internet.

With draw programs you can create and manipulate "freehand" art on the computer. Among the common tools available on such packages are rectangles, lines, ovals, arc, and freehand polygons. These items can be colored, filled with gray scale tones, grouped and ungrouped, shadowed, or moved. Draw programs are often used on systems that include additional support hardware: scanners, digitizing tablets and pens, and color printers.

USING PRESENTATION SOFTWARE

Sometimes you will want to prepare overhead projection sheets, color slides, or computer images for a presentation. Many companies have presentation software available to do this. Presentation software helps writers create overhead projection sheets or 35mm slides with color or black-and-white images (e.g., bar charts, graphs, still images).

Programs such as PowerPoint are actual presentation graphics programs which create charts and graphs as well as slides, handouts, and overhead transparencies. You can run the slides created by such programs on your computer screen or project them onto a screen for your audience. Such programs as PowerPoint allow you to create bulleted text slides, graphs, tables, and organizational charts and to access clip art and drawing tools. You can take screen shots of projects being designed on the computer and use them as part of your presentation. You can first develop your presentation in outline form, where you can see all of your text in one place, or you can generate slides one at a time by typing text directly into a text placeholder on the slides as you go along. You can rearrange your slides, change their overall design, and delete slides that do not pertain to a particular audience. You can create a logo and select color and type font. Example 10–38 shows what a typical PowerPoint slide looks like in print. Finally, during an oral presentation, you can display your 35mm slide and handouts as well as create an on-screen, electronic presentation with special effects such as sound and music. With such tools available, many professional audiences expect a very sophisticated and polished display of data.

EXAMPLE 10–38   PowerPoint Slide

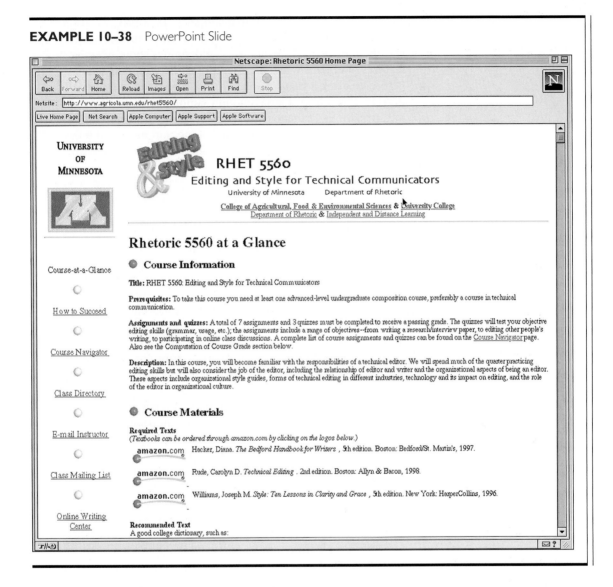

■ In this chapter you have learned that visual display of data can be used for everything from making complex information accessible to symbolizing abstract structures. However, all visual displays of data must be introduced, interpreted, and explained in your text for the data to support an argument or inform an audience. Generally you select and create a visual display of data according to your purpose and audience, particularly the experience, technical understanding,

SUMMARY

and language and culture of that audience. However, we also know that certain types of visual displays of data lend themselves to specific purposes. For example, line drawings show selective features of objects while photographs show complete detail.

■ Before creating a visual display of data, you need to gather all pertinent data from a large enough sample, indicate all possible interpretations of that data, make sure your data really support what you are saying, indicate units of measurement and measures of central tendency, and condense the data into a manageable form.

■ Selecting the proper form of graphic display is important. You might select a table if you need to present complete and specific data to an audience that understands your topic. Graphs, on the other hand, translate numbers into pictures—bars, lines, areas, or pictorials. More abstract than tables, graphs illustrate changes in one amount relative to another. Bar graphs (simple, multiple, stacked, 100 percent, or deviation) demonstrate magnitude or size of several items or emphasize the difference in one item at equal time intervals. Line graphs show trends or movement in series of data. Simple area graphs show not only trends in data but also the sum of those data, whereas 100 percent area graphs show trends in the components of a whole. Pie graphs represent proportions or percentages of a whole, and pictorial graphs use icons or symbols to represent data.

■ When creating bar, line, and pie graphs, you have to be particularly careful not to misrepresent your data by distortion of the X or Y axis, distortion by suppressing the zero, distortion by increasing the width or height of bars, distortion in 3-D, and distortion through improper use of icons.

■ Photographs depict subjects in detail and in a realistic setting. Drawings or diagrams, on the other hand, are more abstract and convey structure or organization symbolically. Action view and translucent view drawings are understood by many audiences, whereas cutaway or exploded view drawings and schematic drawings are understood only by experienced audiences in the particular field. Maps help audiences locate geographical information while charts display the components of an organization or object or the steps in a process. Both are readily understood by most audiences.

■ Finally, we covered the computer software and hardware that you might use to create your visual displays of data. These scanning programs, spreadsheets, multimedia packages, paint/draw programs, and many others are constantly being updated.

1.   *Critiquing and Revising Visual Displays.* Analyze the visual displays presented in Examples 10-39 through 10-42 and answer the following questions:

   a.   Who might be the audience for the visual display? What can you tell about their level of education, technical knowledge, ability to read English, and so on?

   b.   What did the creator of the visual display want to show? How might that person interpret the data in the display? In what type of publication might you find such a display?

   c.   What type of visual display do you see? Was the visual display successful according to the guidelines of displays in general and the type of display specifically? In what ways? Was the visual display unsuccessful? In what ways?

   d.   How would you revise the display to improve its effectiveness? Be prepared to discuss your impressions with your classmates.

ACTIVITIES AND EXERCISES

EXAMPLE 10–39   Graphics Tools from a Word Processing Program

With this tool		You can
↖	Selection tool	Select, move, and resize elements
𝒜	Text tool	Type text in a graphic
╲	Line tool	Draw lines
▢	Rectangle tool	Draw rectangles and squares
⧉	Front/back tool	Switch the order of stacked elements
◁┼▷	Flip tool	Flip selected elements horizontally or vertically
☰	Text alignment tool	Specify the alignment of text
⊢—⊣	Line width tool	Specify line width

Source: Reprinted with permission of Microsoft Corporation

EXAMPLE 10–40   Depiction of a Space Shuttle Communication System

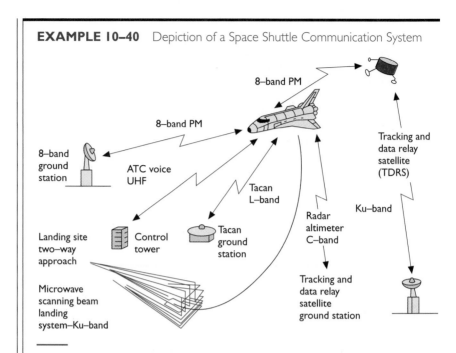

Source: John F. Hanaway and Robert W. Moorehead, *Space Shuttle Avionics System,* National Aeronautics and Space Administration, Office of Management, Scientific and Technical Information, NASA SP-504, 1989.

2. *Introducing, Interpreting, and Explaining Visual Displays.* A small diameter gravity sewer is displayed in Example 10-43 on page 395. It has the following features:

- Collector mains do not carry solids.
- Primary treatment occurs at each connection.
- Only settled wastewater is collected.
- Construction cost is low.
- Grit, grease, and similar contaminants are separated from wastewater.
- Fewer obstructions occur in collector mains.
- Interceptor tanks appear upstream of each connection.
- Grit, grease, and similar contaminants are retained in interceptor tanks.

  Write a paragraph in which you introduce the visual display, interpret it, explain the main features by referring to the display, and recommend the sewer system for consideration over conventional systems.

3. *Creating Visual Displays.* Present the following information in a visual display or displays that would be meaningful to graduating science and engineering students at your university or college.

  In 1989 the average salaries of doctoral scientists and engineers were gathered by field, sex, and years in the profession. In environmental science men earned $55,600 and women $43,600 (women's salaries were 78.4 percent of men's); in physics men earned $59,100 and women

EXAMPLE 10–41   Scenario for a Renewable Energy-Intensive Future
in the United States

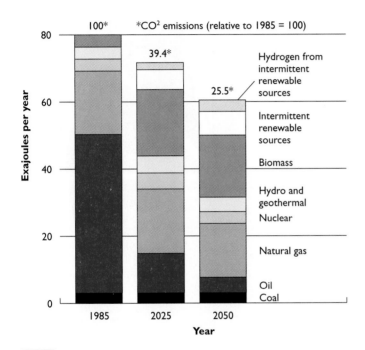

100*   *CO^2 emissions (relative to 1985 = 100)

Hydrogen from intermittent renewable sources

Intermittent renewable sources

Biomass

Hydro and geothermal

Nuclear

Natural gas

Oil

Coal

Exajoules per year

1985   2025   2050

Year

Source: T. B. Johnson, H. Kelly, A. K. N. Reddy, and R. H. Williams, eds. *Renewable Energy Sources for Fuels and Electricity* (Washington, DC, and Covelo, CA: Island Press, 1992). Granted with permission of Island Press © 1992.

$48,700 (women's salaries were 82.4 percent of men's); in computer science men earned $60,100 and women $50,000 (women's salaries were 83.2 percent of men's); in mathematics men earned $52,400 and women $43,800 (women's salaries were 83.6 percent of men's); in psychology men earned $51,300 and women $44,300 (women's salaries were 86.4 percent of men's); in chemistry men earned $55,900 and women $46,900 (women's salaries were 83.9 percent of men's); in engineering men earned $62,900 and women $53,400 (women's salaries were 84.9 percent of men's); in social science men earned $52,000 and women $44,200 (women's salaries were 85 percent of men's); in life sciences men earned $53,200 and women $43,100 (women's salaries were 81 percent of men's). In all fields men earned $56,000 and women $44,800; in life sciences men earned $53,200 and women $43,100 (women's salaries were 80 percent of men's).

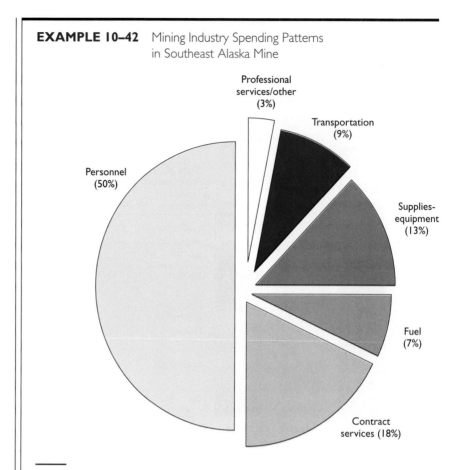

EXAMPLE 10–42   Mining Industry Spending Patterns in Southeast Alaska Mine

Professional services/other (3%)

Transportation (9%)

Personnel (50%)

Supplies-equipment (13%)

Fuel (7%)

Contract services (18%)

Source: Linda M. Daniel, "The Tongass: Getting Beneath the Surface," *Minerals Today,* US Department of the Interior, Bureau of Mines. MS2501, August 1992.

In terms of years in these professions, for less than 5 years, men earned an average of $42,500 and women $36,900 (women's salaries were 86.8 percent of men's); between 5 and 9 years men earned $48,900 and women $42,700 (women's salaries were 87.3 percent of men's); between 10 and 14 years men earned $54,800 and women $48,000 (women's salaries were 87.6 percent of men's); between 15 and 19 years men earned $60,400 and women $51,200 (women's salaries were 85.4 percent of men's); between 20 and 24 years men earned $63,700 and women $55,200 (women's salaries were 86.7 percent of men's); between 25 and 29 years men earned $67,400 and women $58,300 (women's salaries were 86.5 percent of men's); between 30 and 34 years men earned $70,100 and women $63,400 (women's salaries were 90.4 percent of men's); for

EXAMPLE 10–43   Municipal Solid Waste Management, 1960 to 2000: (Projected)

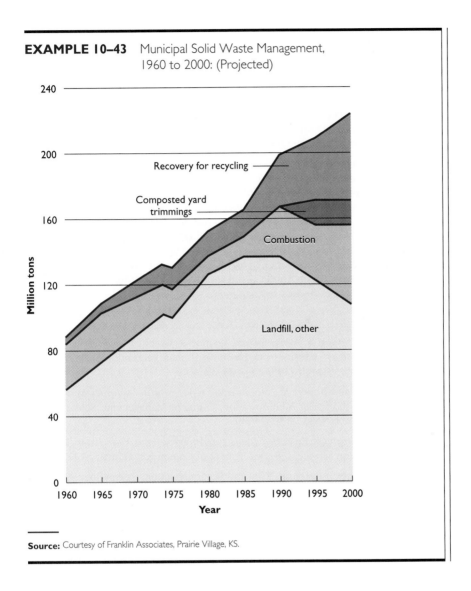

Source: Courtesy of Franklin Associates, Prairie Village, KS.

over 34 years of experience men earned $74,500 and women $62,300 (women's salaries were 83.6 percent of men's).

Source: Betty Vetter, "Ferment: Yes/Progress: Maybe/Change: Slow," *MOSAIC* 23, no. 3 (Fall 1992), p. 35.

4. Survey the visual presentation software and hardware available at your college or university. What is available in the computer labs? In the administrative offices? In the printing or communication department? Write a memo to your teacher and classmates describing available resources. Include where possible examples of visual displays produced on these systems.

5. Study the data displayed in the tables in the "Minnesota Department of Health AIDS Epidemiology Unit/ Pediatric/ Adolescent HIV/AIDS Monthly Surveillance Report, November 1, 1992" (Appendix A, Case Documents 1). Create alternative displays of these data, emphasizing aspects that you find important, to distribute to the general public.

6. In the "Fact Sheet" on AZT (Appendix A, Case Documents 1) find data within sentences and paragraphs that might be better displayed visually. Create these visual displays and explain how they would be integrated into the fact sheet.

7. The following information pertains to the bird trapping at the University of Minnesota Agricultural Experiment Station in St. Paul (Appendix A, Case Documents 3). Take a stand on one side or the other of the controversy and display the data below in a way that would support your points at the public hearing on the issue. The data represent the total number of birds (all species) trapped for the year. Notice that for some years data were not gathered. 1981—4,307; 1982—5,114; 1983—3,407; 1984—6,145; 1985—no data; 1986—4,752; 1987—9,846; 1988—no data; 1989—4,989; 1990—6,488; 1991—6,014; 1992—773; 1993—735; 1994—no data; 1995—3,648; 1996—6,322; 1997—2,205; 1998—6,187; 1999—10,380; 2000—5,951.

WORKS CITED

Cochran, Jeffrey K.; Sheri A. Albrecht; and Yvonne A. Green. "Guidelines for Evaluating Graphical Designs: A Framework Based on Human Perception Skills." *Technical Communication* 36, no. 1 (1989), pp. 25-32.

Hanaway, John F., and Robert W. Moorehead. *Space Shuttle Avionics System.* National Aeronautics and Space Administration, Office of Management, Scientific and Technical Information, NASA SP-504, 1989.

National Energy Strategy: Powerful Ideas for America. One Year Later. National Technical Information Service, U.S. Department of Commerce, February 1992.

Reph, Mary, and Patricia Carreon. "Data and Communications Technology. Data Distribution Media Study." *Research & Technology.* Goddard Space Flight Center, NASA, 1990.

Rogich, Donald G. "The Future of Materials: Plastic Components Are Growing." *Minerals Today.* U.S. Department of the Interior, Bureau of Mines. MS2501, June 1991.

Schriver, Karen. *Dynamics of Document Design.* New York: John Wiley, 1997.

Tufte, Edward R. *The Visual Display of Quantitative Information.* Cheshire, CT: Graphics Press, 1983.

Vetter, Betty. "Ferment: Yes/Progress: Maybe/Change: Slow." *MOSAIC* 23, no. 3 (Fall 1992), p. 35.

Creating Effective Documents

Creating Definitions and Descriptions

Sharing a language means sharing a conceptual universe.
    Evelyn Fox Keller, *Secrets of Life, Secrets of Death: Essays on Language, Gender and Science* (New York: Routledge, 1992), p. 27.

Since the meaning of a word helps organize the way we encounter an entire situation, misusing a word encourages false perceptions, occludes possibilities, and gives us bad habits.
    David Dobrin, *Writing and Technique* (Urbana, IL: National Council of Teachers of English 1989), p. 310.

Introduction

In Chapter 4 we stated that the most powerful kind of "good reason" to support an idea is an argument from definition. And, in Chapter 10, we discussed how visual displays, such as photographs, drawings, and diagrams, could help an audience see an object or organism that might be described in a document. In this chapter we cover in detail definitions and descriptions—two rhetorical patterns that drive arguments and convey information in a great many technical communications. Whether you're giving a progress report or deciding on the feasibility of a recommendation, writing a résumé or delivering a sales pitch, you'll need to define and describe.

In the past, technical communication—perhaps because it has been associated so closely with the concerns of scientists, engineers, and technology developers—has been portrayed as a straightforward process of representing objective reality as accurately as possible through language. Certainly, the project of "defining" words and "describing" objects or processes grows out of such assumptions.

We now understand, however, that techniques of definition and description are neither neutral nor objective, and that even those writers who use definitions with precision and care cannot always define things in a way that will avoid all confusion or eliminate the need for interpretation on the part of audiences. Some specialized fields have created lists of standard definitions for key technical terms, and different practitioners will still interpret definitions and descriptions according to the contexts of their fields, their experiences and education, their belief systems and values, the goals of the companies and organizations where they work, and their different understanding of the subject matter.

By defining and describing, you can help readers approach a topic, avoid error, reduce confusion, or reach a goal, but you can never be sure that you have conveyed an *absolutely* precise meaning or understanding. You can work to provide your audience with the clearest possible understanding of a concept, object, process, or phenomenon, however. In this chapter we provide you with some techniques that help you describe and define.

Creating Definitions

categorical proposition
a subject, verb, and predicate; also called an informational definition

In Chapter 4, you studied how to write categorical propositions. A **categorical proposition** (also called an informational definition) consists of a subject (the word to be defined), a linking verb (such as *is* or *was*), and a predicate (something to be said about the word being defined). Example 11-1 illustrates several different categorical propositions.

As practice in recognizing the elements of categorical propositions identify the subject, linking verb, and predicate in each example given in Example 11-1.

EXAMPLE 11–1   Categorical Propositions

Examples

Science and engineering curricula are usually rather demanding.
A recycling program is something that this campus badly needs.
Ms. Protevi was an ideal boss.
Nursing is a more interesting profession than most people imagine.
Certain climatic changes now in evidence are truly disturbing.

EXAMPLE 11–2   Classical Format for Definitions

Formula

name (term) + class/family/species/group + special characteristics = definition

Examples

light pen—A light-sensitive input device used to select an entry or indicate position (Szymanski et al., 738)

liquid-based solar heating system—a solar heating system in which liquid, either water or an antifreeze solution, is heated in solar collectors (Beckman et al., 172)

Although categorical propositions are useful when an audience is familiar with your terms, you may need more **formal definitions** if you are defining unfamiliar terms for an audience or if your purpose depends on the clearest possible understanding of terminology. When we mention defining a term throughout this chapter, remember that a term may convey information about a concept, process, object, or procedure.

The classical format for constructing definitions involves the formula illustrated in Example 11-2. Especially important in formal definitions of objects is the task of characterizing the **special features** that distinguish a particular item from other items in a related classification group. In defining processes, procedures, or phenomena that occur over time, you can modify this format, placing the focus on **function** or **operation,** as indicated in Example 11-3. In both patterns, you indicate through language—as accurately, precisely, and clearly as possible—what a term *is* by indicating what differentiates it from other similar terms.

This approach to definition is limited on one hand by precision and on the other by efficiency. To be absolutely precise, you would have to differentiate a term *from all other similar terms.* For example, to define

formal definitions
the term to be defined, the class/family/species to which the term belongs, and the special characteristics that distinguish the term from other terms in the same class/family/species

special features
characteristics that distinguish a particular item from other items in a related classification group

functional or operational definitions
definitions for processes, procedures, or phenomena that occur over time; constructed by using the name + the function or operation = special uses

---

EXAMPLE 11–3   Format for Functional or Operational Definitions

Formula

name (term) + *function or operation* + special uses or operations = definition

Examples

selecting—The process of retrieving only certain records in a table in a relational database management system (Szymanski et al., 745)

kerning—In text composition or typesetting, nudging letters together selectively to improve spacing gives the illusion of equal spacing between letters

---

accurately and precisely a ¾-inch, plated, hex-head lag screw, you would have to differentiate it from every other type of screw: from bolt screws; from slotted-head screws and Phillips-head screws; from decking screws and brass furniture screws; and from ½-inch hex-head lag screws and 1½-inch hex-head lag screws. Fortunately, most readers bring a great deal of contextual knowledge to situations that require definitions. A customer putting together a set of bookshelves, for example, probably knows that 6-inch, galvanized decking screws are inappropriate. She also knows it is customary for manufacturers to include the appropriate hardware in a little plastic bag that comes with shelves requiring assembly, and she knows the approximate size of the hardware from observing the size of the predrilled holes. As a result, such customers need only a picture or diagram of the appropriate screw to identify a hex-head lag screw and not a fully fleshed-out definition. Therefore, you often need to mention only the key characteristics that differentiate a term from other similar terms customers or users might encounter in the same context.

Thus, you need to analyze your audience's expertise and the context in which your definition appears. See the various definitions of the word *gas* listed in Example 11–4. Some of the definitions are more appropriate to a general audience and some to a specialized audience. Also, the particular meaning of the word *gas* depends on the context in which it is used and defined.

The level of specificity required by a definition and the nature of the definition itself depend on your analysis of the audience's knowledge, experience, and interest. Definitions can be incorporated within the text through modification at the word, phrase, clause, sentence, or paragraph level; incorporated within the text through example, metaphor, analogy, negation, or comparison-contrast; covered in informational footnotes, endnotes, or captions; illustrated in figures or tables; or included in an appendix or glossary.

EXAMPLE 11–4   More Specialized and Less Specialized Definitions

Examples

Gas—A shortened term for *gasoline*. The fuel used to power most internal combustion engines.

Gas—In the intestinal tract, oxygen, nitrogen, hydrogen, carbon dioxide, methane, and in decomposition of proteins, phydorgen sulfide, indole, skatole, ammonia, and such.

Gas—The action of treating chemically with gas. Because of its expansion capabilities, gas immediately penetrates large areas. For example, during World War I many soldiers suffered the ill effects of poisonous gas spread over the trenches and battlegrounds. Or, certain insects are best destroyed or controlled by spreading gaseous pesticides over the area in question.

Gas—The state of matter, as opposed to solid or liquid, characterized by low viscosity and density. Gases respond with relatively great expansion and contraction to changes in pressure and temperature. Gases diffuse easily and distribute uniformly in a container.

Gas—The control device for an engine powered by gas. For example, to move the car forward one must step on the gas.

Gas—The occult principle thought to be present in all bodies. From the Flemish. Named by chemist J. B. van Helmont (1577–1644) and taken from the Greek word *khaos,* meaning chaos or chasm. The gases within dead bodies cause swelling and decomposition.

DEFINITION BY MODIFICATION

Depending on the amount and the extent of definition that you think your audience may need, you can define your terms through **modification,** using four different kinds of strategies as indicated in Writing Strategies 11-1. Using these strategies, choose a piece of equipment from your favorite sport or hobby and write a series of definitions for that equipment that you define through modification.

    Extended definitions, as indicated in Example 11-5, may involve a paragraph, several paragraphs, or even an entire document.

modification
process of adding words, phrases, clauses, or sentences to clarify or hone a definition

DEFINITION BY SYNONYMS AND ANTONYMS

Definitions often include or use **synonyms** (words that mean the same thing) for key terms or **antonyms** (words that mean the opposite). Both synonyms and antonyms help a reader understand a definition by providing terms that may be familiar to them or that expand their frame of reference for a concept, as indicated in Example 11-6.

synonyms
words that mean the same thing

antonyms
words that mean the opposite thing

WRITING STRATEGIES 11–1   Definition Through Modification

Strategy 1: Modifying individual words
Example

The *spring-operated* closing mechanism is located on the inner surface of the oven door.

Strategy 2: Modifying phrases
Example

Double poling—*used primarily by skiers in racing and skating situations*—is characterized by identical poling motions on both sides of the body.

Strategy 3: Modifying clauses
Example

The antenna, *which is the device receiving the signal,* should be repositioned to avoid interference.

Strategy 4: Modifying sentences
Example

Cumulonimbus clouds produce what people call "thunderstorms." *To be termed a "thunderstorm," a storm must include visible lightning, audible thunder, and precipitation of some kind.*

DEFINITION BY EXAMPLE AND ENUMERATION

examples or enumerating items
ways of extending definitions by including items that fall within the scope of the word being defined

You can frequently extend definitions by including **examples** (often indicated by "e.g.," which stands for the Latin words *exempli gratia*) or **enumerating** items that fall within the scope of the word being defined. Example 11-7 provides several definitions using example and enumeration.

DEFINITION BY METAPHOR AND ANALOGY

metaphor
a comparison without the use of the words *like* or *as;* e.g., "The wind was a bulldozer in the forest."

analogies
comparisons of two items to illustrate elements of similarity; one of the items must be familiar to the audience

Writers often use **metaphor** in definitions to explain unfamiliar items by implying comparison to familiar items (recall our discussion of metaphor and analogy as comparisons in Chapter 4). The effective use of metaphor depends on cultural knowledge shared by you and your audience.

In a similar fashion, writers may use **analogies** to compare two things and illustrate elements of similarity. As you can imagine, the effective use of analogy also depends on your having a sense of the wide variety of experience audiences bring to reading tasks. In Example 11–8, for instance, the authors of a bicycle maintenance and repair manual count on their readers to have a broad knowledge of other sports activities and the equipment that goes along with them. Example 11–8 shows definition by metaphor and analogy.

EXAMPLE 11–5   Extended Definition Through Modification

1.   Nature of radiant heat transfer. In the preceding sections of this chapter we have studied conduction and convection heat transfer. In conduction heat is transferred from one part of a body to another, and the intervening material is heated. In convection the heat is transferred by the actual mixing of materials and by conduction. In radiant heat transfer the medium through which the heat is transferred usually is not heated. Radiation heat transfer is the transfer of heat by electromagnetic radiation.

Thermal radiation is a form of electromagnetic radiation similar to X rays, light waves, gamma rays, and so on, differing only in wavelength. It obeys the same laws as light: travels in straight lines, can be transmitted through space and vacuum, and so on. It is an important mode of heat transfer and is especially important where large temperature differences occur, as, for example, in a furnace with boiler tubes, in radiant dryers, and in an oven baking food. Radiation often occurs in combination with conduction and convection . . .

In an elementary sense the mechanism of radiant heat transfer is composed of three distinct steps or phases:

1.   The thermal energy of a hot source, such as the wall of a furnace at T_1, is converted into the energy of electromagnetic radiation waves.
2.   These waves travel through the intervening space in straight lines and strike a cold object at T_2 such as a furnace tube containing water to be heated.
3.   The electromagnetic waves that strike the body are absorbed by the body and converted back to thermal energy or heat.

Source: Christie J. Geankopolis, *Transport Processes and Unit Operations,* 2nd ed. © 1983, p. 258. Reprinted by permission of Prentice Hall, Inc., Upper Saddle River, NJ.

EXAMPLE 11–6   Definition Using Synonyms or Antonyms

Examples

Assembler—A computer program that converts (i.e., translates) programmer-written symbolic instructions, usually in mnemonic form, into machine-executable (computer or binary-coded) instructions. This conversion is typically one-to-one (one symbolic instruction converts to one machine-executable instruction) (Giesecke, 635).

[I]n this molecular orbital, the likelihood that an electron will be found in the region between the two nuclei is high. Electrons in such a molecular orbital tend to be in the region where they can hold the nuclei together. This molecular orbital is called a *bonding orbital* . . . Another molecular orbital is obtained by subtracting the 1s orbital on one atom from the 1s orbital on the other . . . the resulting values in the region of the overlap are close to zero. This means that in this molecular orbital, the electrons spend little time between the nuclei. We call it an *antibonding orbital* . . . (Ebbing and Wrighton, 258–59).

EXAMPLE 11–7   Definition by Example and Enumeration

Examples

Noise has been defined as unproductive sound. Sometimes the noise level may be so high that permanent aural damage may result—e.g., from continued use of a chain saw or considerable exposure to a musical group with amplifiers (Hook, 57).

cathode ray tube (CRT)—The principal component in a CAD display device. A CRT displays graphic representations of geometric entities and designs and can be of various types: storage tube, raster scan, or refresh. These tubes create images by means of a controllable beam of electrons striking a screen (Giesecke, 635).

dedicated—Designed or intended for a single function or use. For example, a dedicated work station might be used exclusively for engineering calculations or plotting (Giesecke, 637).

A *computist* is a mathematician who specializes in computing such things as time of high and low tides, eclipses, dates, business accounts, and the like in which specific answers useful in practical applications are the goal (Hook, 243).

EXAMPLE 11–8   Definition by Metaphor and Analogy

Example of Definition by Metaphor

Inside the CRT is a "gun" that "shoots" electrons out to a screen. The electrons hit a phosphorescent coating on the inside of the picture tube. When the phosphorous on the screen gets "hit," it glows (Brecher, 27).

Example of Definition by Analogy

clipless pedals—"road bike pedals that use a releasable mechanism like that of a ski binding to lock onto cleated shoes . . ." (*Bicycling Magazine's Complete Guide*, 298).

DEFINITION BY NEGATION

negation
describes an object, concept, or term by what it is not or identifies characteristics it does not have

Defining by **negation** involves describing an object, concept, or term by what it is *not* or by identifying characteristics it does not have. The danger of negative definition is that there are so many *nots* to choose from. Poorly constructed negative definitions, therefore, can combine long strings of negatives without really helping readers come to a clear definition of the term. In composing negative definitions, writers should be careful to choose only the most salient or telling negative characteristics. Example 11-9 provides examples of both poor and helpful definitions by negation.

DEFINITION BY COMPARISON-CONTRAST

Definition by comparison can extend the range of your audience's understanding by connecting several objects or concepts that are related in some

EXAMPLE 11–9   Definition by Negation

Example of Poor Definition by Negation

Sprong, the new caffeinated soft drink, is neither a cola nor a citrus-based product.

Example of Helpful Definition by Negation

Collateral publications—Ad agency term for printed pieces, such as brochures and annual reports, that are not directly involved in advertising (Beach et al., 197).

EXAMPLE 11–10   Definition by Comparison or Contrast

Examples

Stride rate—The number of times the basic movement or cycle is completed in a given time period. Average race pace is 2.2 strides per second for hill running and 1.6 strides per second for diagonal (Hall, 234).

direct-view storage tube (DVST)—One of the most widely used graphics display devices, DVST generates a long-lasting, flicker-free image with high resolution and no refreshing. It handles almost an unlimited amount of data. However, display dynamics are limited since DVSTs do not permit selective erase. The image is not as bright as with *refresh* or *raster* (Giesecke, 637).

digital computer—A high-speed programmable electronic device that stores processes and retrieves data by *counting* discrete signals, as opposed to measuring a continuous signal, as is done in an analog computer (Szymanski et al., 732).

What is damp haze?—Damp haze lies somewhere in the scale between dry haze and light fog. It consists of microscopically small water droplets or particles which attract water (hydroscope nuclei) suspended in the air. It differs from dry haze in that it has a grayish appearance and occurs with high relative humidities. It differs from light fog since its droplets are most widely dispersed and smaller in size (Forrester, 61).

---

way, either by similarity or difference. Example 11–10 provides several examples of definition by **comparison** or **contrast.**

comparison or contrast
connecting several objects or concepts by similarity (comparison) or by difference (contrast)

DEFINITION BY VISUAL DISPLAY AND PRESENTATION

There are times when a visual presentation, such as a photograph or a drawing, can do a much better job of defining an object or concept than can words, given the nature of language and the economy of visual images. In Example 11–11, various kinds of bookbindings are defined through illustrations. Although the writer could have tried to convey the

EXAMPLE 11–11   Definition Through Visual Presentation

These eight bindings are commonly available at printing companies and trade binderies. Each has advantages and disadvantages, depending on page count, kind of paper, and how the product will be shipped and used. To assure correct imposition, the binding method must be known before stripping and platemaking.

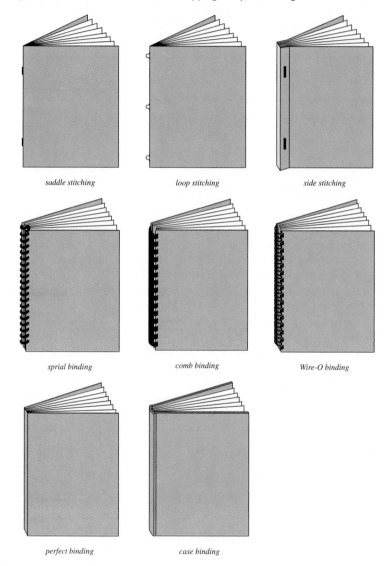

saddle stitching   *loop stitching*   *side stitching*

sprial binding   *comb binding*   *Wire-O binding*

perfect binding   *case binding*

Source: Mark Beach, Steve Shepro, and Ken Russon. *Getting It Printed: How to Work with Printers and Graphics Arts Services to Assure Quality, Stay on Schedule, and Control Costs* (Portland, OR: Coast to Coast Books, 1986), p. 120.

differences between saddle stitching and loop stitching through words, the illustrations were much more economical and informative.

As we discussed in Chapter 10, you need to understand your audience's experience, technical expertise, language, and culture as well as your purpose when creating a visual presentation. In Example 11–12, for instance, the writers are addressing an audience of engineering students. The writers are defining a device somewhat unfamiliar to the students, but they can rely on the students' experience reading technical diagrams. Given this

EXAMPLE 11–12   Definition Through Visual Presentation and Text

THE CHEMICAL ROCKET ENGINE

The advent of missiles and satellites has brought to prominence the use of the rocket engine as a propulsion power plant. Chemical rocket engines may be classified as either liquid propellant or solid propellant, according to the fuel used.

The figure shows a simplified schematic diagram of a liquid propellant rocket. The oxidizer and fuel are pumped through the injector plate into the combustion chamber where combustion takes place at high pressure. The high-pressure, high-temperature products of combustion expand as they flow through the nozzle, and as a result they leave the nozzle with a high velocity. The momentum change associated with this increase in velocity gives rise to the forward thrust on the vehicle.

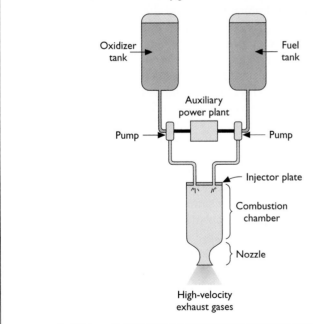

Source: Gordon J. Van Wylen and Richard E. Sonntag. *Fundamentals of Classical Thermodynamics,* 3rd. ed. SI Version (New York: John Wiley and Sons, Inc. 1990), p. 14. Reprinted by permission of John Wiley and Sons, Inc.

context, they choose to define the liquid propellant rocket with a line drawing. The drawing is abstract because only important details that distinguish this rocket from others (e.g., the solid propellant rocket) are included. Also, the writers know that their audience will be familiar with the technical terms that help define the rocket, such as *oxidizer* and *combustion chamber*. Finally, because the drawing defines the liquid propellant rocket according to its parts, the writers can reserve their text for a description of the rocket's functioning.

Including Contexts in Definitions

To create useful and clear definitions you need to consider context. Adding contextual information—information about the social, cultural, environmental, systemic, ideological, or historical contexts within which terms are used or understood and within which technology is set—will enrich your definitions.

SOCIAL AND CULTURAL CONTEXTS

Considerations of the social and cultural settings within which technological terms exist can provide your audience with a way of understanding the terms in connection with human beings and the institutions and organizations human beings create. These **social and cultural contexts** are particularly important when the purpose of the document is to solve problems for people. Definitions that include some reference to social contexts often focus on human agency, social institutions, or cultural practices.

The definitions of *data entry* and *pathnames* in Example 11-13, for instance, illustrate definitions placed in social and cultural contexts. The definitions convey information about the culture that makes and uses technology and the relationships that exist between humans and the technology.

Defining terms in their social and cultural contexts differs in a fundamental way from defining technical terms as existing apart from the social norms that shape them—as consisting of assemblages of parts without references to the social purpose that inspired their use or as mechanisms that run without references to the humans that operate them. Defining objects and concepts in terms of how human beings make, design, and use them can help your readers envision the ways that they and other people relate to technology. This practice can help both readers and authors focus on the human uses of technology rather than simply on the technology itself.

ENVIRONMENTAL OR SYSTEMIC CONTEXTS

Consideration of the **environmental or systemic contexts** within which technological terms are used can expand your audience's understanding. Such definitions assist people in troubleshooting and problem solving when

social and cultural contexts
human settings in which technological terms exist

environmental or systemic contexts
the environments or systems in which technical terms can be placed

EXAMPLE 11–13   Definition Through Social and Cultural Context

Examples

Data entry—Entering data into a computer in a timely manner, at a reasonable cost, and with minimal error (Szymanski, 731).

Stack, document, and education pathnames—"In a Macintosh desktop world of folders and icons, the three pathname cards in the HomeStack are as mysterious as an MS-DOS C> command prompt. If you never let the user get within sight of these cards while setting up your stack, all the better" (Goodman, 28).

[A] *stack developer* is anyone who designs a HyperCard stack that one or more people will be using. That includes corporate stacks developed for in-house use, perhaps as training vehicles or as the basis for departmental information systems. . . A stack developer is a computer consultant whose charter is to create information tools for clients, whether the tools are for time and money management or a freestanding kiosk of trade show exhibitors and products (Goodman, 3).

EXAMPLE 11–14   Definition Using Environmental or Systemic Contexts

Examples

Spokes are regularly damaged in two ways. Something gets caught in them or the chain overshifts low gear and lands on them. Common sense can minimize the chances of either problem occurring. . .

The rule with *low-gear shifting* is conservatism. Always shift gently into low gear when riding after a wheel change or when riding a bike whose derailleurs may be improperly adjusted. Test the gear before you reach a steep section of the road where your shift will be rapid and pedaling force extreme. And when adjusting the derailleur, do not allow it to move closer to the spokes than necessary. . . Once scratched, driveside spokes are much more likely to break when they are subjected to a lot of pressure (*Bicycling Magazine's Complete Guide*, 36).

mechanisms fail to work as they should. In a manual of bicycle maintenance, for example, the definitions of *spoke damage* and *low-gear shifting* include the information contained in Example 11-14.

ECONOMIC CONTEXTS

In definitions, considerations of the ways that terms are connected to cost, supply, and demand, or systems of exchange can expand an audience's understanding. The examples of *VHS* and *compatibility* in Example 11-15, illustrate information about **economic contexts** that may be useful to audiences in specific technical environments.

economic contexts
the environments of cost, supply and demand, and systems of exchange in which terms can be placed

---

EXAMPLE 11–15   Definition Using Economic Context

Examples

VHS (S-VHS)—"A trademarked name for the first popular videotape cassette recording, storage, and playback format whose hallmark is consumer acceptance and the resulting economies of scale" (*MacroMind Director,* 373).

compatibility—The ability of a particular hardware module or software program, code, or language to be used in a CAD/CAM system without prior modification or special interfaces. *Upward compatible* denotes the ability of a system to interface with new hardware or software modules or enhancements (i.e., the system vendor provides with each new module a reasonable means of transferring data, programs, and operator skills from the user's present system to the new enhancements) (Giesecke, 636).

EXAMPLE 11–16   Definition Using Ideological Context

Example

[Computer] scientists seem to work under two assumptions that lead to a fallacy par excellence: that *human information processing* is optimal, and that programs fare better by imitating and emulating human intelligence. It is important to note that this contention leads to a fallacy of "faulty hypostatization." The use of human behavior as a criterion for success or failure of emulation procedures should always be regarded in relation to the qualitative differences between man and machine (Obermeier, 158).

IDEOLOGICAL CONTEXTS

ideological elements focus on the relationships between objects, phenomena, and concepts and the human beliefs that shape them

Sometimes, definitions can be enhanced, clarified, or focused by making implicit **ideological elements** explicit. Such definitions may focus on the relationships between objects, phenomena, and concepts and the human beliefs that shape or influence them. The passage in Example 11–16 has been taken from a text that defines "human information processing" in relation to computer-based information processing, with a specific reference to the ideological belief system of computer scientists.

HISTORICAL CONTEXTS

historical information elements that allow events to be placed in a temporal environment

Some definitions include **historical information** that expands an audience's understanding of the temporal context within which technology is necessarily designed, used, and modified. Historical information helps in-

---

EXAMPLE 11–17 Definition Using Historical Context

Examples

database query language—A fourth-generation language used in conjunction with a relational database; acts as an interface between a user and a relational database management system to facilitate easy access without use of complex programming code (Syzmanski et al., 731).

gears—toothed members transmitting rotary motion from one shaft to another, are among the oldest devices and inventions of man. In about 2600 B.C. the Chinese are known to have used a chariot incorporating a complex series of gears. . . Aristotle, in the fourth century B.C., wrote of gears as though they were commonplace. In the 15th century A.D., Leonardo da Vinci designed a multitude of devices incorporating many kinds of gears (Juvinall and Kurt, 550).

---

struct or train an audience, shows change or timing, or illustrates a sequence. You might use historical information to provide the **etymology,** or word history, of a term. The definitions in Example 11–17 contain historical information.

etymology
history of a word and its formation

Deciding When and Where to Include Definitions

The placement of definitions differs according to the purpose of the document, its layout and design specifications, the ways that the audience will use the text, and the conventions of a given discipline.

IN-TEXT DEFINITIONS

Some texts identify key words with **highlighting** (bold or italics) and provide a definition immediately after the term, generally in an **appositive, noun phrase,** or **noun complement.** Some writers might prefer, especially if a definition is rather long or complicated, to devote an entire sentence or more to the definition, generally, following the key term. Examples of both key words and in-text definitions are illustrated in Example 11–18.

FOOTNOTE/ENDNOTE DEFINITIONS

In some texts, and for some disciplines, accepted placement of definitions involves **footnotes** (located at the bottom of the page) or **endnotes** (located at the end of a chapter or section). The definition of the computer program ELIZA, as shown in Example 11–19, is presented in an endnote form.

highlighting
identification of key words by the use of bold or italic typefaces

appositive, noun phrase, or noun complement
grammatical forms that usually follow the statement of the key word in a definition

footnotes
definitions, comments, or references placed at the bottom of a page

endnotes
definitions or references placed at the end of a chapter, section of text, or book

---

EXAMPLE 11–18   Definitions in Texts

Examples of short in-text definitions

Gravure, a process of photomechanical printing [appositive definition], used in newspapers to provide visual references for readers.

An *acid* is any substance other than a salt that increases the concentration of hydrogen ions in aqueous solution [noun complement] (Ebbing and Wrighton, 283).

Example of longer in-text definitions

Warning!
Kickback may occur when the nose or tip of the guide bar touches an object, or when the wood closes in and pinches the saw chain in the cut. This contact may abruptly top the chain saw and in some cases may cause a lightning fast reverse reaction, kicking the guidebar up and back toward the user . . . Kickback may cause you to lose control of the saw (Inside cover, *Instruction Manual/Owner's Manual,* Stihl, 28).

EXAMPLE 11–19   Definition in Footnotes or Endnotes

[61]ELIZA mimics a Rogerian psychotherapist, whose technique consists largely of echoing utterances of the patient; it therefore uses very little memory, and arrives at its "answers" by combining transformations of the "input" sentences with phrases stored under key words (Kurzweil, 491).

VISUAL DISPLAY/SIDEBAR DEFINITIONS

In some documents, especially those that support immediate reference to technical materials or processes, you might define key terms or words in an accompanying visual display or a **sidebar** of text. Example 11-20 shows a portion of a page from a chemistry text. On this page the alloy bronze is cued by a text flag (a small square) and defined in a nearby sidebar marked with the same flag.

GLOSSARY DEFINITIONS

Glossaries are collections of terms located either at the beginning of a text (if readers need to know the information contained in the descriptions before

sidebars
margins of a text where definitions can be placed

glossaries
collections of terms located at the beginning or at the end of a document, depending on the needs of the readers

EXAMPLE 11–20   Definition Within a Side Bar

Group IVA Elements: Valence-Shell Configuration ns² np²

As mentioned earlier, Group IVA elements show a distinct trend from nonmetal (carbon C) to metalloid (silicon, Si, and germanium, Ge) to metal (tin, Sn, and lead, Pb). Carbon exists in two well-known allotropic forms: graphite, a soft, black substance used in pencil leads, and diamond, a very hard, clear, crystalline substance. Both tin and lead are metals that were known to the ancients. Tin is mixed, or alloyed, with copper to make bronze. ■

These elements form oxides of general formula RO_2. Examples are CO_2, a gas used to make Dry Ice (solid CO_2) and carbonated beverages; SiO_2, a solid that exists as quartz and white sand (particles of quartz); and SnO_2, the mineral cassiterite, principal ore of tin. Lead forms the dioxide, PbO_2, but the monoxide, PbO, is more stable. Carbon also forms a stable monoxide, CO.

■ Bronze, one of the first alloys (metallic mixtures) used in history, contains about 90% copper and 10% tin. It melts more easily than copper, but is much harder.

Source: Darrell D. Ebbing and Mark S. Wrighton, *General Chemistry,* 1st ed. Copyright © 1984 by Houghton Mifflin Company, p. 189. Used with permission.

EXAMPLE 11–21   Definition Within a Glossary

abstracts   (1) A standard part of a scientific report and usually precede the introduction, (2) reference texts that include listings of citations and brief summaries, and (3) a condensed version of a longer document that summarizes main points.

bit   The smallest piece of information that can be stored and processed digitally. A bit may have only one of two values: 0 or 1 (i.e., on/off or yes/no).

HTML   HyperText Markup Language, a programming language used to create Web pages.

hue   A synonym for color (e.g., red, blue, green, purple) that is often used to refer to shades of a color.

ISO   Acronym for the International Standards Organization, which makes quality control of supplies and manufacturing processes possible. Its Web site is <http://www.iso.ch/>.

---

proceeding through the text) or at the end of the document (if readers can refer to definitions as they work through a text). Example 11-21 shows the glossary from a book on computer science.

Writing Strategies 11-2 contains questions you can ask in preparing an effective definition.

WRITING STRATEGIES 11–2   Creating Effective Definitions

Ask and answer the following questions:

1. What is the purpose of this definition, or set of definitions? Are there several purposes? If so, how are they related?
2. For what audience is this definition intended? What is the audience's expertise in this area? Experience? What do they need to know or do?
3. What is the term being defined? To what class/group/family does it belong? What special characteristics differentiate it from other things in this group?
4. What conventional definition strategies best lend themselves to this task? Among those frequently used are the following:

   - Modification
   - Synonyms and antonyms
   - Example and enumeration
   - Metaphor and analogy
   - Negation
   - Comparison-contrast
   - Visual display and presentation

5. What special contextual information will help meet the audience's needs? Information about:

   - Social or cultural contexts?
   - Environmental or systemic contexts?
   - Economic contexts?
   - Historical contexts?

6. Where should the definition be placed?
   - In text?
   - In a footnote or endnote?
   - In a visual display or a sidebar?
   - In a glossary?

Creating Descriptions

Technical descriptions can be found in all sorts of communication settings. Frequently included in reports and proposals, descriptions help audiences make decisions about designs, manufacturing, or purchasing. Descriptions of processes, such as chemical processes, organizational procedures, or safety measures, are found in manuals that accompany a specific product or direct activities. Literature written to sell or explain products includes technical descriptions of products or processes, as do documents written for technical training.

Like definitions, descriptions written for differing purposes and audiences differ in length, content, detail, tone, vocabulary, and format. Some technical descriptions of simple mechanisms, for example, are limited to a single visual display with accompanying text labels. Other technical descriptions may continue for pages. Some technical fields assemble long lists of standard definitions for commonly used terms. Depending on the purpose of a document, a technical description may include information about social and cultural contexts, environmental or systemic contexts, historical background and economic factors, and ideological assumptions. Generally, technical descriptions fall into two main categories: descriptions of objects/mechanisms/phenomena and descriptions of processes/procedures/activities.

The exact nature of technical descriptions, as well as the scope and depth of these answers, depends on your audience—their education, experience, attitude, responsibilities, among other factors—and the purpose for which the description is written, distributed, and read. Your description also depends on the nature of the information involved in the description itself and the contexts that surround your creation of the description—the time allocated for writing, the corporate contexts and purposes for the writing, the historical factors influencing the descriptive act, and such.

Example 11-22 shows two different technical descriptions of the same process, boiling. One comes from a chemistry textbook and the other from a cookbook; one addresses specialists—chemical engineers with an interest in the processes of steady state heat transfer, evaporation, and distillation—and the other is written primarily for nonspecialists with a general interest in those chemical processes associated with cooking. Notice that the description for specialists is supported by a visual display.

Look at Writing Strategies 11-2. What other characteristics does the description in Example 11-22 contain? Using Writing Strategies 11-2 as a guide, create a pair of definitions like those found in Example 11-22. Choose a term or a process, like boiling, that can be defined for both specialists and nonspecialists.

In the rest of this chapter, we offer suggestions on how to create the two main types of technical descriptions (descriptions of objects/mechanisms/phenomena and descriptions of processes/ procedures/activities).

DESCRIPTIONS OF OBJECTS, MECHANISMS, OR PHENOMENA: PARTS, WHOLES, AND SYSTEMS

Writers can describe an object, mechanism, or phenomenon by focusing on the parts or elements that make up the whole—sometimes an entire system. The technical descriptions in Examples 11-23 and 11-24 illustrate two ways of representing parts, wholes, and systems. Example 11-23 emphasizes the structural parts of a whole, and Example 11-24 emphasizes the function of parts within a system.

EXAMPLE 11–22   Technical Description for Specialists and Non-Specialists

Example of Description for NonSpecialized Audience

Boil—To cook in liquid at boiling temperature (212° at sea level) where bubbles rise to the surface and break. For a full rolling boil, bubbles form rapidly throughout the mixture.

———

Source: *Better Homes and Gardens, New Cook Book* (Des Moines, IA: Meredith Corporation. 1976), p. 408.

Example of Description for Specialized Audience

4.8A Boiling

I. Mechanisms of boiling. Heat transfer to a boiling liquid is very important in evaporation and distillation and also in other kinds of chemical and biological processing, such as petroleum processing, control of the temperature of chemical reactions, evaporation of liquid foods, and so on. . . .

Boiling is a complex phenomenon. Suppose we consider a small heated horizontal tube or wire immersed in a vessel containing water boiling at 373.2 K (100°C). The heat flux is q/A W/m^2, $\Delta T = T_w - 373.2$ K, where T_w is the tube or wire wall temperature and h is the heat-transfer coefficient in W/m$^2 \cdot$ K. Starting with a low ΔT, the q/A and h values are measured. This is repeated at higher values of ΔT and the data obtained are shown in Figure 4.8–1 plotted as q/A versus ΔT.

In the first region A of the plot in Fig. 4.8–1, at low temperature drops, the mechanism of boiling is essentially that of heat transfer to a liquid in natural convection. The variation of h with $\Delta T^{0.25}$ is approximately the same as that for natural convection of horizontal plates or cylinders. the very few bubbles formed are released from the surface of the metal and rise and do not disturb appreciably the normal natural convection.

In the region B of nucleate boiling for a ΔT of about $5 - 25$ K ($9 - 45°$F), the rate of bubble production increases so that the velocity of circulation of the liquid increases. The heat-transfer coefficient h increased rapidly and is proportional to ΔT^2 to ΔT^3 in this region.

In the region C of transition boiling, many bubbles are formed so quickly that they tend to coalesce and form a layer of insulating vapor. Increasing the ΔT increase the thickness of this layer and the heat flux and h drop as ΔT is increased. In region D or film boiling, bubbles detach themselves regularly and rise upward. At higher ΔT values radiation through the vapor layer next to the surface helps increase the q/A and h.

The curve of h versus ΔT has approximately the same shape as Fig. 4.8–1. The values of h are quite large. At the beginning of region B in Fig. 4.8–1 for nucleate boiling, h has a value of about 5700–11400 W/m$^2 \cdot$ K, or 1000–2000 btu/h $\cdot$ ft$^2 \cdot$ °F, and at the end of this region h has a peak value of almost 57000 W/m$^2 \cdot$ K, or 10000 btu/hr $\cdot$ ft$^2 \cdot$°F. These values are quite high, and in most cases the percent resistance of the boiling film is only a few percent of the overall resistance to heat transfer.

———

Source: Christie J. Geankopolis, *Transport Processes and Unit Operations,* 2nd ed. © 1983, p. 258. Reprinted by permission of Prentice-Hall, Inc.

EXAMPLE 11-23   Technical Description Representing Parts as Structural Elements

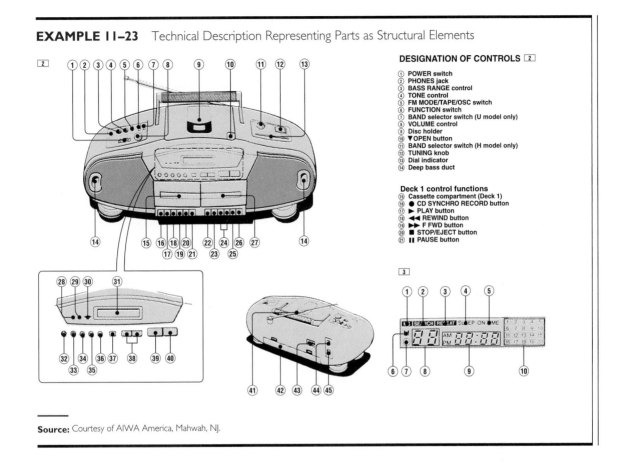

DESIGNATION OF CONTROLS ⎡2⎤

① POWER switch
② PHONES jack
③ BASS RANGE control
④ TONE control
⑤ FM MODE/TAPE/OSC switch
⑥ FUNCTION switch
⑦ BAND selector switch (U model only)
⑧ VOLUME control
⑨ Disc holder
⑩ ▼OPEN button
⑪ BAND selector switch (H model only)
⑫ TUNING knob
⑬ Dial indicator
⑭ Deep bass duct

Deck 1 control functions
⑮ Cassette compartment (Deck 1)
⑯ ● CD SYNCHRO RECORD button
⑰ ► PLAY button
⑱ ◄◄ REWIND button
⑲ ►► F FWD button
⑳ ■ STOP/EJECT button
㉑ II PAUSE button

Source: Courtesy of AIWA America, Mahwah, NJ.

In the technical description of the AIWA compact stereo in Example 11-23, the graphics and text identify various parts (e.g., switches, jacks, controls, knobs, buttons) of the stereo that users should know how to locate. Except for hints provided by the names of the individual elements, however, these labels describe the components as *structural* elements rather than detailing their *function*. The description of the fractionating column contained in Example 11-24, on the other hand, gives information not only about the various structural parts of the apparatus, but also about the function or purpose of these component parts within the system of the entire fractionation column. Thus, we learn that the glass beads in this system "provide a surface on which the less volatile part of the vapor condenses." Often technical descriptions of objects, mechanisms, or phenomena include both structural and functional information to provide readers with as much data as possible.

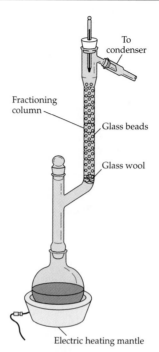

To condenser

Fractioning column

Glass beads

Glass wool

Electric heating mantle

Figure 2.11
An apparatus for fractional distillation. The temperature varies up the length of the fractioning column, being hotter at the lower end and cooler at the top. The most volatile component passes over into the condenser, whereas less volatile components condense on the beads in the column or remain in the distillation flask. As the most volatile component distills over, the temperature of the boiling liquid increases until the next most volatile component distills over.

Fractional distillation is useful when the solution consists of two or more volatile components. The temperature at which a liquid boils is known as its *boiling point*. When a solution containing liquid substances of different boiling points is distilled, the substance with the lowest boiling point normally distills over first. ■An apparatus for fractional distillation is shown in Figure 2.11. A *fractioning column*, containing glass beads, is placed at the top of the distillation flask. The glass beads provide a surface on which the less volatile part of the vapor condenses. The temperature varies along the length of the column, being hotter at the bottom and cooler at the top. The substance with the lowest boiling point passes over into the condenser from the top of the column, whereas substances with higher boiling points condense on the beads in the column or remain in the distillation flask. As the substance of lowest boiling point is removed from the flask, the temperature of the boiling liquid increases and the next substance, now the one with the lowest boiling point, begins to distill.

Fractional distillation is employed commercially to separate crude oil or petroleum into useful products. For this, the distillation is operated continuously, various fractions being taken off at different heights up the fractionating column. Thus, the lowest boiling fraction (20°–60°C), called petroleum ether, is obtained near the top of the column. Light naphtha or ligroin is a fraction with a slightly higher boiling point (60°–100°C), and it is taken off the column just below the petroleum ether. Gasoline boils between 50°C and 200°C, and kerosene between 175°C and 275°C; these come off even lower on the column. Furnace and diesel fuels boil at still higher temperatures.

■ Although solutions of volatile liquids usually increase in boiling point as the more volatile components distill off, some solutions exhibit a constant boiling point. For example, a solution of 95.6% ethanol (grain alcohol) and 4.4% water by mass boils at the constant temperature of 78.15°C; pure ethanol boils at 78.5°C. Such a constant-boiling mixture is called an *azeotrope*. It is impossible to separate ethanol and water completely by fractional distillation of simple ethanol-water mixtures.

WRITING STRATEGIES 11–3   Steps for Creating Effective
Technical Descriptions

Step 1.   Identify the rhetorical context of the description.
Step 2.   Decide on a useful scheme for identifying important parts of the whole.
Step 3.   Present the parts of the description in a way that is most congruent with
information about audience, purpose, and context and the "whole" you are
presenting.

EXAMPLE 11–25   Description Topics Listed According to Purpose

Operating	Troubleshooting	Repairing
Light soil cycle	Spots and filming	Motor
Normal wash cycle	Cloudiness	Pump
Pot-scrubber cycle	Chipping of china	Hub connection
Delay start cycle	Suds in the tub	Soap cup
Energy saver cycle	Water won't pump	Washing tower
Rinse and hold cycle	Leaking	Spray arm

To create useful descriptions of objects, mechanisms, or phenomena,
writers begin with the three steps identified in Writing Strategies 11-3.
Each of these steps is explained more fully in the short descriptions that
follow.

1. **Identify the rhetorical context of the description.** Begin by
thinking about the *purpose* of the description and the needs of the
audience who will read it. What do these people need to know
about the complex whole you are describing or the parts that comprise this whole? Why do they need to know these things? This
rhetorical information will help you make subsequent decisions
about how best to represent your topic. Example 11-25 demonstrates three ways of describing a dishwasher based on different
purposes—operating, troubleshooting, and repairing. Imagine that
these lists represent topics to be covered in the print documentation that accompanies the dishwasher when it is sold to a customer.
The first list, detailing the various cycles the washer offers, would
help individuals interested in understanding the machine's features
or how the machine functions. The second list, detailing the problems that might occur in connection with the machine's use, would

help people trying to solve a problem with the machine's performance. Such people may be everyday users of the machine trying to
cope with a minor problem or repair people called in to cope with
a more serious malfunction. The third list, detailing the various
parts of the dishwasher, would help a repair person or an owner
who needed to understand the structural components of the larger
mechanism.

2. **Decide on a useful scheme for identifying important parts of
the whole.** Writers' decisions about what the "important" parts are
will depend on their knowledge of the audience and their needs. Note
in Example 11-25, that the first list is an identification of dishwasher
cycles by function. These cycles are important "parts" of the dishwasher for customers who will have to know how to use it properly.
Similarly, consider the list of potential problems—these items represent symptoms that are important "parts" of a whole troubleshooting
diagnosis.

3. **Present the parts of the description in a way that is most congruent with information about audience, purpose, and context
and the "whole" you are presenting.** Once these parts have been
identified, the ways that writers present parts of a whole will differ according to their understanding of the audience and their needs. The
order of parts, the grouping of like items, the presentation of parts,
and the use of visual display, for example, will change according to
rhetorical context.

For instance, the repairing list in Example 11-25 could be composed
in two different ways depending on whether a writer perceives his or her
audience mainly as experienced mechanics or weekend troubleshooters.
For experienced mechanics, the various structural components of the dishwasher could be presented in an order determined by their listing in the
parts manual. (Because motor parts and subassemblies within the motor
are listed first in the technical repair and parts manual, they could be listed
first in the owner documentation, along with references to parts numbers
in the technical repair and parts manual). These parts could also be represented by a technical diagram to indicate their structural relationship to
one another.

For less experienced problem solvers, the list of components might be
presented in a reverse order—perhaps by the component's frequency of repair record—in an effort to facilitate locating the most likely problem areas
(motors require repair less frequently than do spray arms). A visual display
might help here as well, but parts could be identified with more common—
and less technical—names, perhaps linked to troubleshooting directions.

You can use the questions in Writing Strategies 11-4 to develop descriptions of objects, mechanisms, or phenomena. The answers to these
questions would appear in your description.

WRITING STRATEGIES 11–4   Questions Answered by Descriptions of Objects, Mechanisms, and Phenomena

What is it?

- What family or group of things does it belong to?
- What is it called?
- What is its model or part number?
- Does the object/mechanism/phenomenon have a historical background that helps people understand it? A cultural, social, or economic context that would help people understand it?

Where is it located?

- What are its surroundings, setting, environment? Why are these important to an understanding of it?
- How does one get to it within that location?

What are its sensory characteristics?

- What does it look like? What is its size? Shape? Color?
- What does it smell like?
- What does it feel like? Weight? Texture? Density?
- What does it taste like?
- What does it sound like?

How is it put together or constituted?

- What material is it made of?
- What are its characteristics? State (gas, liquid, solid)? Stage (immature, mature, elderly)?
- What parts or elements is it composed of? What are the distinctive characteristics of each part? How are these related?
- How does one put it together or take it apart? Does one?
- Is this object/mechanism, phenomenon part of a larger system? An environment?

How does it change/act/function over time?

- How long does it last or live?
- What special characteristics (dangers, precautions, benefits) are associated with its use, care, or maintenance?

How do people relate to it?

- Who uses/observes/attends to this object, phenomenon, mechanism?
- Who does not use/observe/attend to it? Why?
- What is its function? Within what settings/environments?
- What assumptions do people have about this object/mechanism/phenomenon that might help people understand it?
- How can people use this product/follow this mechanism safely?

EXAMPLE 11–26   Description of Food Digestion

Digestion

A biochemical process that all people experience is the digestion of foods. Certainly, a failure in digestion is a memorable experience for everyone. Digestion breaks down the complex molecules in foods to the simple molecules that can be used in metabolism for the generation of energy and the biosynthesis of cellular components. Digestion of food is moderately well understood in biochemical terms for human beings and animals. Digestion in biochemistry extends to the breakdown *within cells* of complex compounds to simple molecules that can be either excreted or used in cellular metabolism. Both of these digestive processes are essential to the well-being of complex animals.

Digestion of Food

The major nutrients in food that require digestion are proteins, starches, and fats, all of which are large molecules that are digested by hydrolysis into smaller molecules. Proteins are long-chain polymers of x-amino acids, and starches are polymers of sugars, whereas fats are *triglycerides* or *phospholipids,* which are glycerol esters of fatty acids. (The chemical structures of triglycerides and phospholipids are given in Chapter 11 and those of sugars and starches are in Chapter 15.) The digestive reactions are catalyzed by enzymes in the digestive tract. Enzymes are themselves proteins and are often defined as biological catalysts.

The first enzyme encountered by food during ingestion into the mouth is *salivary amylase.* This enzyme initiates digestion by catalyzing the partial hydrolysis of starches into shorter oligomers. Its main importance may be to protect teeth from decay by digesting adhering food particles into soluble molecules that flow easily into the stomach, thereby removing potential nutrients for bacteria.

Source: Robert H. Abeles, Perry A. Frey, and William P. Jencks. *Biochemistry* (Boston: Jones and Bartlett, 1992), p. 16. Reprinted with permission.

DESCRIPTION OF PROCESSES, PROCEDURES, OR ACTIVITIES: STEPS AND OVERVIEWS

Technical descriptions of processes or procedures also generally break down complex wholes (in this case, processes that happen over some period of time) into parts (designated as events, steps, activities, sequences). Examples 11–26 through 11–28 provide three different types of technical descriptions that focus on processes, procedures, or activities.

The technical description of food digestion in Example 11–26, begins with an *overview of the process* ("Digestion breaks down the complex molecules in foods to the simple molecules that can be used in metabolism for

---

EXAMPLE 11–27   Description of Norton Utilities "ASK Command"

Description

The Norton ASK command provides an easy way to create batch files that execute different commands, depending on user input at the time the batch file is run.

When the ASK command is invoked, it displays the prompt text, then awaits a response. The prompt will generally list the keys the user may type in response to the ASK command (any of the keys in the key-list) and explain the results of choosing each.

The user responds to ASK by typing any one of the keys in the key-list. The key-list itself is not displayed on screen; only the prompt is displayed.

The key-list is optional. When no key-list is given, any key the user presses is accepted.

After the user chooses one of the keys in the key-list (or any key, if no key-list is specified), ASK returns control to the batch file. ASK passes the key that was chosen by the user as an ERRORLEVEL code. The first choice (key) in the key-list corresponds to ERRORLEVEL 1, the second to ERRORLEVEL 2, the third to 3 and so on (if no key-list is supplied, ASK returns ERRORLEVEL 0 in response to any key being pressed). The batch program can then branch to different labels in the batch file, in accordance with the ERRORLEVEL.

Source: *The Norton Utilities Reference Manual.* Advanced Edition, Version 4.5 (Santa Monica, CA: Peter Norton Computing, 1988), p. 41.

---

the generation of energy and the biosynthesis of cellular components"). This overview is followed by a narrative of sorts—a record of digestive events that we recognize as smaller successive *events within the larger process,* as these events occur in chronological order.

The description of the Norton Utilities "ASK command" (see Example 11-27) starts with an *overview of the procedure* invoked by the command and the function of this particular procedure ("The Norton ASK Command provides an easy way to create batch files that execute different commands. . ."). The description then proceeds with the various *subprocesses within this procedure* (e.g., user invokes the command, command invokes the prompt list, user responds to list, ASK returns control to the batch file, batch program branches to "different labels"). For ease of reading and comprehension, the description is marked with key words that signal the temporal order in which the various steps happen (e.g., "when," "then," and "after").

The technical description of mouse cleaning from the *Macintosh Reference* manual in Example 11-28 begins with an exceptionally brief *overview of the procedure*—"To clean the mouse"—and then breaks the

---

EXAMPLE 11–28   Description of Mouse Cleaning

Power switch

1 **To clean the mouse, first turn off the Macintosh.**

You should always turn off electrical equipment before disassembling any part.

2 **Turn the mouse over and open the plastic ring on the bottom that holds the ball.**

Macintosh models use two different types of rings on the mouse; one is opened by turning the ring counter-clockwise, the other is opened by pulling the ring straight down until it snaps.

3 **Remove the ring and the ball beneath it.**

You may be able to lift off the ring while holding the mouse upside down; if not, turn the mouse over while holding one hand underneath it to catch the ring and ball.

4 **Clean the three small rollers inside the mouse with a cotton swab moistened with alcohol.**

Rotate the rollers to clean all their surfaces.

5 **Wipe the mouse ball with a clean, soft, dry cloth.**

Don't use any liquid to clean the ball.

6 **Gently blow into the mouse case to remove any dust that has collected there, then replace the ball and ring.**

Source: *Macintosh Reference* (Cupertino, CA: Apple Computer), pp. 295–96. Used with permission.

---

process down into discrete *steps within the procedure* that are labeled so that the audience can follow them easily. Visual displays provide additional information.

Let's examine more closely the stylistic techniques that the authors of these passages have employed to describe processes. These techniques can best be explained in relation to the concept of **performance.**

Look at the description of digestion in Example 11–26. This process is obviously not meant to be performed (at least in a conscious way) by a user. Note that the sentences focus on the process of digestion, and that no human agent, or actor, is represented as *doing* the process. As a result, the purpose of this passage is not to explain how to accomplish some task, but what happens and when. The words the writer chose supports this purpose. In the first paragraph there are six sentences; the writer has made sure that the first **noun cluster** in each of these six sentences ("A

performance
the use of the product or process being defined or described; often includes the tasks a user must perform

noun cluster
one or more nouns used at the beginning of a definition

biochemical process," "a failure in digestion," "Digestion," "Digestions," "Digestion," "Both of these digestive processes") focuses readers' attention on the process of digestion.

In contrast, look at the stylistic features of the mouse cleaning passage in Example 11-28. The purpose there is to describe a procedure that manufacturers of the product expect customers to perform: cleaning a mouse. As a result, the text and graphics focus on the users' actions—the text then constitutes directions for performing a task as well as a description of a procedure. The **imperative verbs** in many sentences ("turn off the Macintosh," "turn the mouse over and open the plastic ring," "remove the ring") support this focus. The subject of these phrases—understood, by readers of English, to be "you"—is clearly intended to perform the actions described by the text and graphics.

> **imperative verbs**
> the form of the verb used when giving commands; for example, "Unplug the machine before servicing"

To create effective technical descriptions of processes, procedures, and actions, you will want to follow the steps identified in the sections below.

1. **Identify the rhetorical context of the process or procedural description.** Begin by thinking about the *purpose* of the description and the needs of the *audience* who will read it. What do these people need to know about the process or procedure you are describing or the steps and activities that comprise this process? Why do they need to know these things? What parts of the process are they familiar or unfamiliar with? This rhetorical information will help you decide how best to represent the steps and subprocesses you identify.

2. **For a chronological framework, identify the steps/ activities/subprocesses that make up the entire process and the appropriate order of these subparts.** One good way to accomplish this task is to observe a situation that closely parallels the context in which your audience will be operating. If you are describing a complex procedure that must be performed safely and precisely, for example, observe an expert performing it and take careful notes or videotape the process. If your intended audience is a novice in a field with which you are not familiar, you may simulate the conditions by performing the task yourself. Given your knowledge of the task and the audience's needs, you should also determine the level of specificity and completeness that your audience must achieve. Novices, for example, may require a description that identifies even the smallest event or step. An audience with experience may be able to complete subprocesses without detailed explanations and may need only safety precautions.

3. **Group the steps/activities/subprocesses in useful chunks so the audience can use them efficiently. Relate these chunks one**

EXAMPLE 11–29 Description of Paper Making

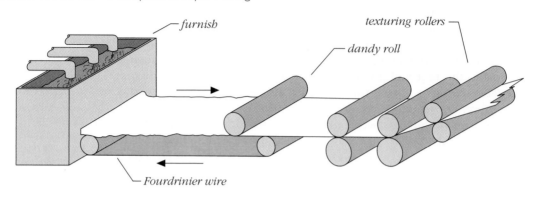

Source: Mark Beach, Steve Shepro, and Ken Russon. *Getting It Printed: How to Work with Printers and Graphics Arts Services to Assure Quality, Stay on Schedule, and Control Costs* (Portland, OR: Coast to Coast Books, 1986), pp. 95–96.

EXAMPLE 11–30 Safety Features in Descriptions

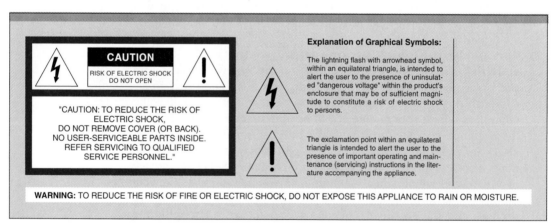

Source: AIWA Compact Stereo Radio Cassette Recorder. CSD-XW505U (2 Band) and CSD-XW505H (4 Band). Used with permission.

to another for the same purpose. Simply listing the steps of a procedure or process may not help an audience conceptualize the entire process or its major stages. Often audiences need to see steps chunked into meaningful groupings to understand the "steps" as a coherent "whole." Example 11-29 describes a coherent procedure for making paper by chunking steps of the process into logical paragraphs. Each paragraph represents a related group of subprocesses (e.g., "cleaning," "beating" and "sizing" the pulp, mixing the "furnish" and draining the furnish, applying a "finish," trimming, drying, pressing). Note also that the writer demonstrates how the various steps in this process are related to one another with phrases such as "After the pulp is beaten, cleaned, and sized, it is mixed at the rate of one pound of pulp to 100 pounds of water" and "When the furnish reaches the end of the wire."

4. **Describe safety factors.** Increasingly, companies and manufacturers are being held legally responsible for safe use of their products (Helyar). As a result, warnings and risk descriptions such as those represented in Example 11-30 are becoming common. In Chapter 9 we looked at warnings and labels from the perspective of document design. Here, we consider them in terms of creating adequate risk descriptions.

    Be sure to make health and safety warnings prominent in technical descriptions: Use the typefaces and font sizes we described in Chapter 9, include familiar warning icons (e.g., a lightning bolt for electrocution warnings, a skull and crossbones for poison), and follow standards for color and position (usually at the very beginning of a section that is subject to a warning). Treat these warnings carefully and respectfully; make sure they are accurate and checked by experts: Others depend on you for their safety and health. Also think about identifying recovery strategies for procedures that might go awry. Such strategies might come in very useful to a reader of your prose.

You can use the questions within Writing Strategies 11-5 to develop descriptions of processes, procedures, or activities. Choose a process about which you are personally familiar (e.g., waxing your skis, canning tomatoes, clipping the dog's toenails, changing the oil in your car, bathing an infant, brewing beer) and write a description of that process following the suggestions in Writing Strategies 11-5. The answers to these questions would appear in your description.

WRITING STRATEGIES 11–5   Questions Answered by Descriptions of Processes, Procedures, or Activities

What is it?

- What kind of action or process is involved?
- What is the name of the process, action, procedure?
- What is the focus? People? Mechanical process? Biological process? Natural phenomenon? Chemical process?
- Does the process/procedure/activity have a historical background that helps people understand it? A cultural, social, or economic context that would help people understand it?

Where and how does it happen?

- What are the surroundings, settings, or environments for the action?
- How does one observe/take part in/do these actions? Why?
- Who does not observe/take part in/do these actions? Why?

What are the sensory characteristics of the action, procedure, process?

- What does it look like when it happens? Before? After?
- What does it smell like when it happens? Before? After?
- What does it feel like as it is happening? Before? After?
- What does it taste like as it is happening? Before? After?
- What does it sound like as it is happening? Before? After?

How is the process, procedure, activity put together?

- What steps or stages can be identified? Would different people identify different steps or stages? If so, why?
- What are the distinctive characteristics of each? How are these related?
- Can steps be taken out or added? If so, by whom and why?
- Who determines the steps of this process/procedure/activity? Why and how? Who does not? Why not?
- Is this process, procedure, or activity part of a larger system? An environment? A context?

How does the process, procedure, activity change over time?

- What is the function?
- What characterizes the beginning state?
- What characterizes the ending state?
- What is the input? What is the output?
- What special characteristics (dangers, cautions, benefits) are associated with the action, procedure, or process?
- How do people relate to it?
- Who does/observes/attends to this process/procedure/activity? Why? Who does not do/observe/attend to it? Why not?
- What assumptions do people have about this process/procedure/activity that might help them understand it?
- How can people use this process/follow this procedure safely?

- This chapter began with a discussion of formal definitions. A formal definition usually contains the term to be defined; the class, family, species, or group in which the term belongs; and the special characteristics that distinguish the term from other terms in the same class, family, species, or group. Functional or operational definitions stress the *use* of the term as the special feature. In differentiating the term you are defining from the others in the same group, you need to consider the precision and efficiency of your definition, both of which depend on your rhetorical situation.

- When creating a definition, you can define your term by modification, by synonyms and antonyms, by example and enumeration, by metaphor and analogy, by negation, by comparison or contrast, and by visual display and presentation. You also need to place your definition within an appropriate context for your audience. You might explain the social and cultural, the environmental or systemic, the economic, the ideological, and the historical contexts of the term you are defining. Finally, you must decide where to place your definition in your document. You might put it in your text immediately after you mention your term and signal the definition by a highlighting cue such as boldface or italics. You can place the definition in a footnote or endnote, a glossary, a visual display, or a sidebar.

- The second part of this chapter focused on descriptions, which are often developed using techniques similar to those used in definitions. Descriptions, too, may be placed in social, cultural, environmental or systemic, historical and economic, or ideological context.

- Generally, technical descriptions fall into two main categories: descriptions of objects, mechanisms, or phenomena; and descriptions of processes, procedures, or activities. When creating descriptions of objects, mechanisms, or phenomena, you should consider the parts and the whole of your subject; that "whole" may be an entire system. Often, the parts of this type of description are depicted by their structure and their function. To create an effective description of an object, mechanism, or phenomena, identify the rhetorical context, decide on a useful scheme for identifying important parts of the whole, and present the parts of the description in a way that is most congruent with available information about audience, purpose, and context and the whole you are presenting.

- To create an effective description of a process, procedure, or activity, you should consider the steps and overviews required in your description. Many of the techniques used in this kind of description pertain to performance—actually doing the process or what happens at what time during the procedure or activity. When describing a process, procedure, or activity, identify the rhetorical context and work with a chronological framework to identify the steps, activities, or subprocesses that make up the entire process and the appropriate order of these subparts. Then, group the steps, activities, and subprocesses in useful chunks so the audience can use them effectively and relate these chunks one to another for the same purpose. Finally, describe any safety factors required for your audience, purpose, and context.

ACTIVITIES AND EXERCISES

1. Consult the AIDS case documents (Appendix A, Case Documents 1). Using this material and additional information you can get from the library or from a search on the Web, write the copy for a brochure on AIDS that is appropriate for eighth-grade students enrolled in a sex education class. Make sure the brochure includes a definition of AIDS, a description of how AIDS is acquired, and instructions about how to avoid HIV infection. Make sure that the material you write is appropriate for a middle school audience. If possible, get permission to test the brochure copy with several individuals of the appropriate age group. *Do not test this document* unless you have obtained the appropriate permissions from parents, educators, and students.

   Write the copy for another brochure about AIDS for the parents of the students described above. This brochure will be distributed during a parent-child AIDS information night. Make sure the brochure includes a definition of AIDS that is appropriate for parents of children who might soon be sexually active, a description of how AIDS is acquired within this population, and instructions about how to help children avoid HIV infection. Make sure that the material you write is appropriate for parents. If possible, get permission to test the brochure copy with several individuals of the appropriate age group. *Do not test this document* unless you have obtained the appropriate permissions from parents and educators.

2. Working collaboratively in teams of four, write a description of the registration process on your campus that can be sent to enrolling first-year students so that they get an accurate picture of the complex procedures involved and the role they must play in these procedures.

   In this description, make sure to identify all of the major stages within the registration process, and the subprocesses within these (e.g., planning a schedule, meeting with your advisor, choosing classes, dealing with time conflicts, paying the bill). Also make sure to define the key terms associated with the registration process (those specialized terms that students learn only after they have a semester or two of experience—registrar, bursar, recitation or discussion versus lecture section, teaching assistant, advisor). You may also want to include warnings of registration dangers, tips for alert students, shortcuts for the adventurous, and ideas for preparation. Be sure to check with the registrar's office about the accuracy of your document, but send only one member of your group to the registrar so you don't overwhelm the staff.

3. Read the Year 2000 case at the end of this book and review the Web sites associated with this case (Appendix A, Case Documents 4). Using the strategies we identified in this chapter, write a description of the Year 2000 problem for the following audiences:

   - Sixth graders who are in a school computer club.
   - A group of senior citizens who are well read in many areas, meet regularly to discuss current events, but who have very little technical knowledge.
   - Readers of the local Rotary or Kiwanis Club newsletter.

   Submit a copy to your teacher for review.

4. Find an example of a poorly written definition. Rewrite and revise this definition using the strategies we have identified in this chapter. You might want to try rewriting the definition for a particular audience and situation. Submit a copy of both the original and a copy of the revised definition to your teacher for review.

5. Choose a small device or machine with which you are familiar. It is best to choose something simple (e.g., a manual pencil sharpener, an egg beater, a manual can opener, or manual grass clippers), something with moving parts, and something that allows you to see all the parts of the device. Study the device by taking measurements, learning what materials make up the parts, and discovering how the parts are attached to each other. Observe how the parts move and how that movement might cause other parts to move in turn.

   Once you have completed your observations, imagine that you are writing about the device for an audience who needs to assemble the device. First, write a definition of the device. Next, describe the device and how the parts relate. Include the measurements and materials of the parts and their method and location of attachment. Use Writing Strategies 11–3 to help you write the most effective description.

6. Find a term that has taken on a new meaning for a particular audience or is used to describe a new technology within your own major field of study. For example, in the computer field, words such as *disk, network,* and *mouse* have come to mean different things than they did originally. Using Writing Strategies 11–2 for inspiration, create a definition of the term that fully explains and illustrates it. Imagine that your definition will be part of a manual that accompanies a device or a textbook for high school students. You may include a visual illustration if you think it appropriate.

7. Assume that you and other members of a collaborative team must present a brief *oral* presentation on the bioremediation process used to clean up Prince William Sound in Alaska. Read over the case documents in Appendix A that pertain to this issue and create a definition and description of bioremediation. You may use visual displays if you like. Choose one of the following audiences:

   - The Environmental Protection Agency.
   - The citizens of Prince William Sound.

- The stockholders of the Exxon Corporation.
- The viewers of a major network's evening news.

8. Gather warning, caution, and danger labels from appliances and products you have at home. Try to determine from them how risks are defined. Write a memo to your instructor discussing what you've learned about defining levels of risk.

WORKS CITED

Abeles, Robert H.; Perry A. Frey; and William P. Jencks. *Biochemistry.* Boston: Jones and Bartlett, 1992.

Beach, Mark; Steve Shepro; and Ken Russo. *Getting It Printed: How to Work with Printers and Graphic Arts Services to Assure Quality, Stay on Schedule, and Control Costs.* Portland, OR: Coast to Coast Books, 1986.

Beckman, William A.; Sanford A. Klein; and John A. Duffie. *Solar Heating Design: By the f-Chart Method.* New York: John Wiley, 1977.

Bicycling Magazine's Complete Guide to Bicycle Maintenance and Repair. Emmaus, PA: Rodale Press, 1990.

Brecher, Debra L. *The Women's Computer Literacy Handbook.* New York: New American Library, 1985.

Dobrin, David N. *Writing and Technique.* Urbana, IL: National Council of Teachers of English, 1989.

Ebbing, Darrell D., and Mark S. Wrighton. *General Chemistry.* Boston: Houghton Mifflin, 1984.

Forrester, Frank H. *1001 Questions Answered about the Weather.* New York: Dover Publications, 1981.

Geankopolis, Christie J. *Transport Processes and Unit Operations.* 2nd ed. Boston: Allyn & Bacon, 1983.

Giesecke, Frederick E.; Alva Mitchell; Henry C. Spooner; Ivan L. Hill; Robert O. Loving; and John T. Dygdon. *Principles of Engineering Graphics.* New York: Macmillian, 1990.

Goodman, Daniel. *HyperCard Developer's Guide.* New York: Bantam Books, 1990.

Hall, Marty. *One Stride Ahead: An Expert's Guide to Cross-Country Skiing.* Tulsa, OK: Winchester Press, 1981.

Helyar, Pamela S. "Products Liability: Meeting Legal Standards for Adequate Instructions." *Journal of Technical Writing and Communication* 22, no. 2 (1992), pp. 125–47.

Hook, Julius Nicholas. *The Grand Panjandrum: And 1,999 Other Rare, Useful, and Delightful Words and Expressions.* New York: Macmillan, 1980.

Instruction Manual/Owner's Manual, Stihl 028. Waiblingen, Germany: Andreas Stihl, n.d.

Juvinall, Robert C., and Kurt M. Marshek. *Fundamentals of Machine Component Design.* 2nd ed. New York: John Wiley, 1991.

Keller, Evelyn Fox. *Secrets of Life/Secrets of Death: Essays on Language, Gender and Science.* New York: Routledge, 1992.

Kurzweil, Ray. *The Age of Intelligent Machines.* Cambridge: MIT Press, 1990.

MacroMind Director, Version 3.0: Interactivity Manual. San Francisco: Macro-Mind, 1991.

Obermeier, Klaus K. "Computers and Their Frame of Mind—The Linguistic Component." *Humans and Machines.* Proceedings at the 4th Delaware Symposium on Language Studies, October 1992, University of Delaware. Stephanie Williams, ed. Norwood, NJ: Ablex Publishing, 1985, pp. 154-65.

Szymanski, Robert A.; D. P. Szymanski; N. A. Morris; and D. M. Pulschen. *Introduction to Computers and Information Systems.* New York: Macmillan, 1991.

Van Wylen, Gordon J., and Richard E. Sonntag. *Fundamentals of Classical Thermodynamics.* 3rd. ed. SI Version. New York: John Wiley, 1990.

Instructions, Specifications, and Procedures

Introduction ▌ Planning Procedures, Instructions, and Specifications ▌ Gathering Information about the Steps and Substeps, Activities, and Subprocesses Involved ▌ Creating Effective Instructions and Procedures ▌ Writing Specifications ▌ Writing Special Types of Instructions ▌ Offering Instructions Online ▌ Testing Instructions, Procedures, and Specifications

Technology presumes there's just one right way to do things and there never is. And when you presume there's just one right way to do things, *of course* the instructions begin and end exclusively with the machine. But if you have to choose among an infinite number of ways to put it together, then the relation of the machine to you, and the relation of the machine and you to the rest of the world, have to be considered, because the selection from among many choices, the *art* of the work, is just as dependent upon your own mind and spirit as it is upon the material of the machine.
Robert Pirsig, *Zen and the Art of Motorcycle Maintenance*
(New York: Morrow, 1974), p. 166.

Instead of running usability tests at the end of your programming cycle, think of optimizing your applications from the get-go. Call it usability design. . . Tweaking with a bad design isn't going to help any.
Rachel Parker, cited in Theo Mandel, *The Elements of User Interface Design*
(New York: John Wiley, 1997), p. 101.

In Chapter 11, you learned how to describe objects, mechanisms, and phenomena as well as processes, procedures, and activities. This chapter builds on what you learned there. Here we focus on the **performance** aspects of processes, procedures, and activities, in particular users' actions. You will learn how to use descriptions in the course of teaching people how to do various operations. You will learn how to write instructions, specifications, and procedures, all of which focus on performance and frequently *support* technology. Indeed, they might be considered parts of technologies themselves.

Instructions, specifications, and procedures serve as important interfaces between people and technology. These documents provide the information necessary for people to use products, complete procedures, and follow professional codes and standards. **Procedures** describe how to coordinate an activity in which many people may be involved. For example, procedures might be used to describe how operators of a printing press would coordinate their activities while the press is running; some people would apply the appropriate ink, others feed the blank paper into the press, and still others monitor the final folding and cutting of the printed material.

Instructions (a kind of document that is closely related to procedures) detail the specific steps that an individual would follow to complete a task. For example, the printing press operator responsible for folding and cutting would adjust certain mechanical parts of the press for each job. **Specifications** state what standards must be followed for a task to be completed safely and according to professional codes. For example, the press operators along with the customer would specify the paper weight and quality of ink for each printing press run, and unions would specify safety measures that operators must follow to avoid injury.

When you write procedures, specifications, or instructions, think of yourself as a **user's advocate**—someone who is on the user's side, helping the person use technology effectively, efficiently, and safely. Take it as a challenge to help users to feel confident, safe, and comfortable as they use the technologies and processes that you are describing. Ideally, you can begin serving as a user's advocate even while the product or service is being developed. For example, the people assigned to write instructions for how to use a computer graphics package might serve on the team that creates the computer code for the package. That way, the communicators can ensure that the codes are designed for the convenience of users as well as manufacturers—by suggesting, for example, that a 3-D option be added or the spreadsheet simplified. Even if you are not part of a design team, however, you should think of yourself as *working for the users* when you write instructions, specifications, and procedures because you have both a legal and an ethical obligation to provide users with accurate and complete information that will allow them to work with technology efficiently and safely. And as we mentioned in Chapter 3, being an advocate is also a matter of *ethos*—of creating a reliable narrator for your documents that your audience can trust.

Introduction

performance
the focus in instructions, specifications, and procedures on the user performing a task

procedures
a description of how to coordinate an activity in which many people are involved

instructions
details of the specific steps an individual must take to complete a task

specifications
standards that must be followed for a task to be completed safely and according to applicable codes

user's advocate
the obligation a technical communicator should take in creating instructions, procedures, and specifications: to represent the user's needs and interests

In this chapter, we're going to lead you through the planning process you can use to create effective instructions, procedures, and specifications. By now, some of this information should seem familiar. We'll take you through your analysis of the rhetorical context, through the identification of the process and product information you will need to create your documents, and through the steps needed for effective organization of your documents. In addition, we will discuss some of the textual and design conventions that have come to be associated with these documents. And we will suggest a way of testing and revising documents during the composing process so that you can be sure that the finished products will work the way you want them to.

Planning Procedures, Instructions, and Specifications

As you know, you should start all writing processes by planning. You prepare documents most effectively if you begin by thinking about the rhetorical context of your work. In this sense, creating instructions, specifications, and procedures is not much different than other kinds of writing. You need to begin by thinking about audience and purpose.

IDENTIFYING YOUR AUDIENCES AND PURPOSES

Identifying the audience and purpose of instructions, specifications, and procedures is absolutely essential. If you provide instructions that users of a product cannot understand, or if the instructions are not suited to your audience's particular needs, the users may damage the equipment, become frustrated, or even endanger their lives. If you leave out important details on how to maintain a product, parts might need to be replaced too frequently. If you fail to envision everyone who needs to follow a procedure, you might address only one group of readers, leaving others out or confused. If you fail to anticipate trouble, you may leave users frustrated and powerless. As you know from Chapters 2 and 3, your documents must always proceed from a clear sense of purpose that emerges from your assessment of your audience's needs, attitudes, and technical knowledge and expertise. You must also consider your readers' situation and environment (see Example 12-1).

Spend enough time determining everything you can about your audience—about what they will bring to their task and about what you will need to provide for them. Be attentive to the various audiences who might encounter your document. Some may be novices, and others may have an insider's knowledge of the process you are describing. Given today's global workplace, not all the users of your document will have English as a first language, so you will need to think about your audience in broad terms. The techniques you learned in Chapter 3 should equip you to determine whom your document is designed specifically to serve.

After you determine who your audience will be, you will need to answer a number of questions about their needs and backgrounds before you

---

EXAMPLE 12–1   Planning Procedures, Instructions, and Specifications

1. Determine your audience.
2. Determine your audience's needs, attitudes, and prior knowledge.
3. Consider your audience's reading environment and how your document will be used.

WRITING STRATEGIES 12–1   Assessing the Needs and Background of an Audience

What are my audience's needs?

- What does my audience need to know how to do? (Remember, there may be many things.)
- What tools and materials will my audience need to follow the instructions or processes or to meet the specifications?
- What exactly is involved in the process or activity my audience wants to learn about?
- How can my audience get out of trouble if something goes wrong?

What is my audience's background knowledge?

- How much prior experience does my audience have with the process and the technology, or related technologies and processes?
- How much detail does my audience require?
- What level of education and technical expertise does my audience have?
- What technical terminology can my audience handle?

---

can begin writing. Writing Strategies 12-1 provides you with a set of questions you can use to get started.

To assess your audience's attitudes, think for a time about the feelings your audience will bring to your document. Are your readers apprehensive or are they confident (or even overconfident, so that they might be tempted to read superficially)? Do they look forward to performing the tasks you are describing, or are they reluctant users? Do they have prior experiences that build confidence, or are they nervous newcomers?

CONSIDERING THE PURPOSES OF PROCEDURES, INSTRUCTIONS, AND SPECIFICATIONS

Once you have answers to the more general questions about who your audience is, what their needs are, and what background they have, you can focus on the particular ways you will have to meet your audience's needs as you design procedures, instructions, and specifications.

Procedures frequently describe how a group of people are to cooperate in various situations or operations, so your purposes in writing procedures will be related to helping a group to work together to accomplish a common goal. If you think about how engineers cooperate to design and build highways, how companies deal with the removal of asbestos from buildings, or how kidney transplants are coordinated in a hospital, you can imagine how complicated it can be to achieve a good set of procedures. Your audience often consists of individuals who have been trained to do specific jobs, all of which are an important part of the overall task. The procedures you create tell people how to accomplish the overall goal of the activity. Your document may be used to clarify, standardize, or modify existing procedures, or it may be used to train new employees who must see how their work affects others.

Often a procedure outlines who is responsible for what so that a larger goal can be reached. For instance, Example 12–2 describes how a plan administrator, an insured person, an attending physician, and a specialist all participate in a common goal, making sure that the cost of a second opinion is justified and therefore covered by an insurance company. To create a procedure that all readers could understand, the writer of the example had to describe the chronological interaction between each person involved.

Unlike the procedures audience who needs to know its special role in a larger operation, the instructions audience may be a single person who will complete an entire task on his or her own. Aside from this difference, your aim in writing instructions will be the same as it was for procedures: to teach someone how to accomplish some technical task (or tasks) in an efficient and safe manner. Your audience for instructions may be the general consumer or the expert who wants to use your document to install, assemble, operate, maintain, or repair devices—or to do all of those things. Your instructions must not only guide your reader through the activity but also help the reader when problems develop. Example 12–3 offers an excerpt from a set of instructions on how to use a cellular phone. Note that the writer assumed that the audience had little prior knowledge of the device but that the user would understand from previous sections of the document certain technical terms, such as *memory system* and *storage methods.*

Specifications are documents that describe the required methods and materials for a projected work, such as plans for creating a landfill or constructing a synagogue. Specifications are often used to provide information to independent firms when bids are being solicited for construction projects or for purchases from vendors. Audiences for specifications are almost always experts. When you create specifications, you must follow standard specification formats because specifications are legally binding.

EXAMPLE 12–2   Second Opinion Procedure

When a program is established, it is not necessary to publish a list of the procedures on which a second opinion will be required. Instructions to the insured simply state that when he or she is told by the attending physician that surgery of a nonemergency nature is required, the insured should promptly contact the plan administrator in person or by phone. The insured will be asked to furnish the administrator with the name of the attending physician, his or her own version of the medical problem involved, the date surgery is scheduled, the insured's address, phone number, and a choice of two preferred appointment dates and most convenient times.

On the basis of this information, the plan administrator decides whether a second opinion is necessary. If the administrator feels that it is, he or she consults the directory of specialists, picks the appropriate specialists, and makes an appointment for the examination. The administrator then notifies the claimant of the appointment time and sends him or her a presurgical screening form. The presurgical screening form must be completed promptly by the attending physician and the insured. The insured is required to authorize release of necessary information to the specialist. The attending physician is asked to indicate the surgical procedure, diagnosis, pertinent history, X rays, and laboratory findings, where the procedure will be performed, expected date of surgery, date of hospital admission, expected length of hospital stay, and total surgical fee, including aftercare.

This form is submitted to the specialist who will make the second examination. He or she will consider not only the need for the surgery but also the qualifications of the specialist who plans to do it. If hospitalization is planned, the specialist will also determine whether such hospitalization and expected duration are necessary. The specialist must have access to the X rays and lab work done by the attending physician. Prior approval of the plan administrator is needed before additional tests can be made.

Source: Excerpts adapted from Mary M. Lay, *Strategies for Technical Writing: A Rhetoric with Readings,* copyright © 1982 Holt, Rinehart and Winston, Inc., reprinted by permission of the publisher.

EXAMPLE 12–3   Excerpt from Instructions for a Cellular Telephone

Memory Dialing

Your telephone provides an alphabetic and numeric memory system capable of storing up to 99 frequently called names and phone numbers. Before setting up your memory, we suggest that you read the rest of this section to become familiar with the ways in which you will access and use it.

Enter Memory (Alpha Entry Mode)	Telephone numbers can be entered into the memory three different ways through the Alpha Entry Mode. Choose one of the following storage methods and use the directions given to enter the information you wish to store.

Source: Motorola Inc., Pan American Cellular Subscription Group.

EXAMPLE 12–4   The Audience of Specifications

The concrete in areas to be refinished shall be removed to a depth of approximately ¾ inch below the adjoining surface. A cut ¾ inch deep shall be made with an abrasive saw along the perimeter areas prior to removing concrete.

Concrete removal shall be done by abrasive blast cutting, abrasive sawing, impact tool cutting, machine rotary abrading, or other methods. The method used to remove concrete shall be approved by the engineer. Cut areas shall be cleaned free of dust and other loose and deleterious materials by brooming and high pressure air jets.

Epoxy adhesive conforming to the adhesive required for bonding dowels shall be applied to the surfaces to be refinished before placing Portland cement concrete filling.

Source: State Highway Construction, Department of Transportation, State of California.

Example 12–4 shows specifications used for state highway construction. You can see that the audience must be familiar with the construction materials and methods.

IDENTIFYING OTHER ITEMS RELATED TO RHETORICAL CONTEXT

Before you finish planning your processes, specifications, or instructions, think for a moment about the conditions under which your documents will be read. Seldom are such documents read all at once from cover to cover. More often they are read in pieces. Audiences take up one or another part of the procedure, read the document related to that part, and then set the document aside until that part is completed; they then return to the document when they are ready to do the next step in the task. Often procedures or instructions describe activities that take some time to accomplish; one day or several days can separate one part of the task from another. A set of instructions for using a computer, for instance, will help a reader to install and begin to use the machine; at a later time the user may return to the instructions for troubleshooting advice, advanced applications, or add-ons; still later, the user may consult the instructions for disassembly.

The point is that your sense of purpose must accommodate not only a range of uses but a range of uses over time. Thus, you may well have to think about how to design your documents to help readers manage them in pieces. The guidelines you learned in Chapter 9 about layout and page design, and about helping readers to manage a document with the help of tables of contents, indexes, headings, and other such navigational aids, will serve you well when you write procedures and instructions.

After you have a sense of the audience for your document—your readers' needs, background knowledge, attitudes, and any special conditions under which the documents might be used—you will need to gain a lot of additional knowledge about the product or processes involved. This knowledge will prepare you to write about the activity with attention to detail and comprehensiveness. Often you will have considerable expertise about the activity that you are writing about; indeed, your expertise might be the reason that you were chosen to write the procedures or instructions. If you are not especially expert, however, that can be to your advantage because you will more easily identify with your audience's lack of expertise. Whether you are an expert or not, remember that your job is to serve as an advocate for the user, not just the manufacturer.

How can you learn in detail about the procedures you must describe? Sometimes a product's **functional specifications** will provide you with essential information. Functional specifications describe in detail exactly what the product is designed to do and how it will perform. In some cases you might be asked to write these specifications, but in other cases, you'll be able to find those specifications, developed by designers, before writing instructions or procedures for users.

Even if no functional specifications are available, you can find out what the product can do and what readers hope to learn from you. Proposals, diagrams, schematics, and marketing information commonly offer information about what the technology is designed to accomplish and what it cannot. In other words, study what has been written already about the product or process before you begin your own writing. In many cases, you will have to find the written information to study. This is part of the secondary research you learned about in Chapter 6. Use your skills for creating effective searches to gather what already exists about the technology or process with which your documents will deal.

In addition, you may well want to interview potential users to determine their hopes and expectations. You can learn from them not only what they expect from the technology involved but also what they expect from you through your instructions. Interviews and surveys can offer important information as well about your readers' level of background knowledge and their attitudes about the technology. Put into action some of the techniques for doing primary research you covered in Chapter 7 to gather this first-hand information.

Completing a **task analysis** of the product or process is a good way to make sure that your instructions, specifications, or procedures will ultimately be complete. A task analysis systematically analyzes product performance. It identifies in detail what the product will do, though not necessarily what tasks a user might be doing when using the product. For instance, Example 12-5 lists some questions that one writer answered in the course of making a task analysis of a computer graphics program. You

Gathering Information About the Steps and Substeps, Activities, and Subprocesses Involved

functional specifications
detailed descriptions of what a product will do and how it will perform

task analysis
systematic analysis of product performance

EXAMPLE 12–5   Task Analysis Questions for a Specific Graphics Program

1. What materials or information is required before the software can be effective for the average users (e.g., should information on the kind of numerical data needed for a spreadsheet be indicated in the documentation)?

2. What tasks must be completed before this program will become operational, (e.g., boot up computer? open systems software? open hard drive menu? Will our documentation contain these instructions)?

3. In what order must these initial tasks be performed? Will our documentation contain these instructions?

4. In what situation will the user be depending on the program (e.g., will the user be maintaining complex spreadsheets? Importing tables and graphs into text? Creating visual displays for oral presentations)?

5. What specific skills will the user need to complete the task? Will other aspects of our software have taught the users these skills (e.g., mouse operation? Menu use? Will the user need to learn how to format a spreadsheet within our documentation or is that knowledge assumed)?

6. What knowledge of visual display will be needed (e.g., how to choose between tables, charts, and graphs? Will our documentation contain these principles)?

7. What steps are need in the creation of a spreadsheet? In the creation of the visual display? Are these steps separate or do they overlap (e.g., will the design of a specific spreadsheet determine which visual display is best for the data)?

8. What will the final product look like (e.g., will the program include color options? What textual elements can be added to the visual display—titles, axis labels, arrow labels)?

9. What warnings and cautions should be added (e.g., will the user's personal computer need a certain amount of memory to run the program? Will the user be able to run the graphics program and a word processing program at the same time? How often should the user save the spreadsheet while choosing between visual display options)?

10. How much time will be required to complete both the spreadsheet and the visual display (e.g., should the documentation contain information on how much time will be needed to create the spreadsheet so the user need not stop in the middle of the task)?

11. With the suggested options for the program, how much cost will be involved in developing the documentation? How much will the final product cost the user?

should typically develop a set of similar questions to determine what a product will do when you write instructions for it. Notice that the questions have to do with matters like these:

■ What tools, materials, and background does the user need to accomplish the task?

- What must a user be able to do in order to accomplish the instructions or processes? What special skills are necessary? What preparatory steps are required?
- What steps and substeps are needed to accomplish the activity?
- In what order will the tasks be performed?
- In what circumstances will the user be working?
- Are there alternative ways that a user might proceed?
- What warnings and cautions are necessary?
- What can go wrong—and how can users get out of trouble if something does go wrong?
- How much time will typically be needed to accomplish each task and subtask?

To get a full picture of the product or process you are documenting, you need to ask questions not only about the specific tasks involved but also about the environment in which the tasks will be performed, the equipment or materials needed to carry out the task, the skills or knowledge needed to complete the task, and the time and cost constraints on the tasks.

If you develop your information as the product or process itself is being developed, you will be constantly updating your information as the product or process changes. If you become involved after design is well along or after the product is completed, you will have to work harder to get information because many of the experts involved in development will be unavailable to you. In that case, you'll need to schedule extra time for hunting down information.

PLANNING THE DESIGN OF YOUR DOCUMENTS

As you develop information for your instructions, specifications, or procedures, you will also have to think about the physical layout that will best accomplish your goals. What will your document look like? What will its dimensions be (think about where users will store and use the documents)? How many visuals will be necessary and who will prepare them?

Although you know a great deal about design considerations like these already from previous chapters, sometimes you will find that decisions about design, organization, and structure will be made for you. A number of companies and funding agencies (such as the U.S. government) have style manuals or specification guidelines that dictate what your instructions, specifications, and procedures must look like. Example 12-6 shows military specifications for the general style and format of technical manuals. These specifications dictate everything from page sizes to acceptable symbols.

Many companies have style manuals that serve as a form of quality control and that ensure a consistent "look" to company materials. All documents written for the company must conform to those style guides. For

---

EXAMPLE 12–6   Military Specifications for Design

3.1.4 Manual outline. When specified in the contract or order, a manual outline shall be provided and shall contain the following:

    a. A text plan that shall be in accordance with the requirements of the technical content specification, showing paragraph titles to indicate the intended coverage of the various aspects of the equipment or system. Each paragraph title or notation shall be followed by a brief statement outlining the information to be presented.

    b. An illustration plan and a table plan that shall be keyed to the text plan. Each illustration and table listed in the plans shall be described. The illustration plan shall contain figure numbers, title, information, intent, approximate size and nature of illustration (halftone, schematic, line drawing). The table plan shall describe the tables by table number and information content. . . .

Source: Military Specifications M-M-38784A Technical Manuals.

---

instance, many organizations observe the ISO 9000 Series of International Standards to maintain quality and consistency in the processes used to develop products and services and also in their supporting documents.

Creating Effective Instructions and Procedures

user friendly
concept of creating instruction manuals (and other documents) oriented toward the user's needs

task-oriented instructions
instructions designed to help users perform the tasks they wish with a product

It is difficult to offer general advice about instructions and procedures because they are written for so many different situations. In the past 20 years or so, the concept of **user friendly** manuals has demanded that instruction manuals be designed according to what a user does with the product; the instructions are **task-oriented.** Consequently most instructions and procedures are written simply, are well designed visually, and contain the following conventional features:

- Introduction.
- A description of the technology or process involved, including an account of the theory of operation that animates it.
- A list of materials and equipment required.
- Directions, arranged chronologically.
- Troubleshooting guide.

Let's consider each of these conventional features in turn.

INTRODUCTION TO PROCEDURES AND INSTRUCTIONS

Introductions to instructions and procedures do the same work that introductions do in other documents: They name the subject, purpose, and sometimes the audience of the document, offer necessary background information, and forecast the contents of the document.

Often the first sentence or paragraph of a set of instructions or procedures will explain its subject and purpose:

> This manual explains how to tune your new Karhu cross-country skis.

> This manual explains how to manage the deer population at Valley Forge National Park.

> This guide provides explanations and step-by-step instructions for installing and using your 56Kbps modem.

You may wish to identify the specific product with such information as the product name, model number, and trademark. Often the purpose of your document will be simple or apparent, but sometimes your introduction will have to anticipate readers' questions about the aim of the instructions or procedures, or specify exactly what the instructions will permit a user to do.

Often the introduction lets the audience know what they will need to have on hand to carry out the instructions. In the case of instructions for assembling a piece of furniture, the introduction will list tools needed to complete the work. For installing a 56Kbps modem, for example, the introduction might specify minimum system requirements:

- Macintosh II computer (68020 processor) or any later Macintosh model.
- System 7.1 or later.
- 4 MB of RAM for System 7.1.
- 8 MB of available disk space.
- 640 × 480 pixel monitor or larger; color preferred.

This section of your instructions is really a set of specifications, so you are creating a document within a document. For the product to work correctly, in this case a 56K modem for the Macintosh, your specifications must be accurate and unambiguous.

Oftentimes the audience of a document will also be obvious and general—a manual for tuning skis will address anyone who wants to perform the tune-up—but sometimes you will want to let readers know whether or not a set of instructions or procedures is meant for them. A sentence like "You don't need any special knowledge of skis to perform this tune-up; any beginner can do it" can build a reader's confidence and make a reader feel specifically included in the implied audience. On the other hand, you may on occasion want to slow down overly motivated readers through your introductory comments:

> You will want to begin using your computer as soon as possible, but please take the time to read this manual first. It will save you time and possibly some frustration over the long run, and ensure that you make the best use of your computer.

Just as you identify the specific product with which you are dealing, your introduction should define the assumed audience of the instructions, the background knowledge the user is expected to have, and any requirements the user needs to have (such as licenses, certifications, and so on).

Describing the organization of your document in the introduction will reassure your readers that everything they need is included in your document and offer them a road map for managing the task:

> This guide will first give you an overview of the tune-up process and make you familiar with the concepts involved. It lets you know what you'll need in the way of supplies in order to get started. The next section will take you through the steps of the tune-up, one by one. Finally, you will find a troubleshooting guide that will help you out in case something goes wrong.

Most readers appreciate forecasting statements because they create expectations about the document that it will systematically and reassuringly satisfy. You might also take the time to explain the conventions you are following and how any special information such as warnings and cautions will be displayed in the manual itself.

An introduction to a set of instructions or procedures is not always explicitly named "Introduction." It may go under the heading of "General Information," "About This Manual," "Welcome to Your New Machine," or the introduction may have no heading at all. Do not worry about that or spend too much time thinking about what to call your introduction. Focus on what your introduction must do for your reader. Example 12–7 shows the introduction to a set of instructions, one that contains subject, purpose, and forecasting. (Another set of instructions that illustrates the principles described here and elsewhere in this chapter is included in Appendix A; see "Taking Care," the manual on caring for AIDS patients.)

DESCRIPTION OF THE TECHNOLOGY OR PROCESS

We mentioned that introductions conventionally contain background information. In the case of instructions and procedures, the background information is typically an account of the theory of operation that guides a process or technology and possibly a description of the mechanism. To be able to operate or repair a piece of equipment, readers need to know the names, function, and location of parts; therefore, many sets of instructions or procedures contain descriptions of mechanisms that are like the ones you learned to write in the previous chapter. Example 12–8 on page 450 illustrates a typical description of the parts listed in the instructions for unpacking a vacuum cleaner.

Similarly, many readers find that knowing the principles of operation behind a machine or process will help them to perform the process or operate the machine. In that case, provide such a description in your instructions or procedures. Most readers find that knowing *why* something is done helps them to know *how* to do it. Example 12–9 on page 450 contains a typical the-

---

EXAMPLE 12–7 An Introduction to Instructions

Chapter 1: Installing the TelePort Modem and Software

This guide provides explanations and step-by-step instructions for installing and using your TelePort 56 modem. Once you have installed the TelePort modem and its software, you can use the modem for all your communication needs, including faxing and connecting to online services and the Internet.

This chapter provides installations instructions and discusses the following topics:

- Overview of the TelePort 56 modems.
- Installation requirements.
- Using the online documentation.
- Following basic safety precautions.
- Installing the TelePort modem.
- Installing software for the TelePort modem.
- Registering your purchase.
- Using the Setup Helper.
- Installing Adobe Acrobat Reader.
- Internet access.
- Finding more information.

Source: TelePort™ 56, *User's Guide.* Global Village TelePort for Macintosh, p. 1.

---

ory of machine operation—in this case a diamond-bladed rocksaw—written for a nonspecialist user. Although the paragraph uses technical information, a novice user would be able to get a general idea of what to expect. More technical audiences, of course, would be impatient with such information and prefer to hear more about the theoretical principles of the product's internal operations. For electronic equipment, for example, expert audiences would prefer to see schematics of circuitry and formulas or algorithms upon which its operation is based.

LIST OF MATERIALS

Usually instructions and procedures properly include an account of what readers will need in the way of time, materials, and other resources to carry out the process described. You probably know from your own experience that you can help readers greatly by anticipating their material needs *before* they begin their work. Early in your document, insert a list of what people will need, and you will avoid the kinds of frustrations associated with having improper tools or inadequate help. In the case of instructions, tell readers how much time they should leave for the job in order to do it properly and completely.

EXAMPLE 12–8   Illustration of Parts, with Callouts

Unpacking

Unpack cleaner from carton and identify the parts shown. Remove and dispose of cardboard packing.
A. Canister Cleaner
B. Power nozzle
C. Paper bag (installed in cleaner)
D. Extension wands, hose and cleaning tools

The cleaner assembled will look like the drawing below.

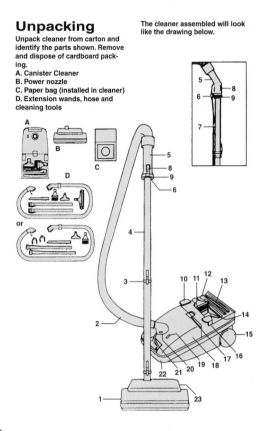

1. **Power nozzle**
2. **Lightweight hose**
3. **Spring latch:** locks extension wands together or locks wand to nozzle or tool.
   Plastic latch ring on plastic wands: see item #7
4. **Extension wands:** used with rug and floor nozzle. One or both wands may be used with cleaning tools. (Some models have **plastic wands;** some models have **metal wands.**)
5. **Hand grip**
6. **Receptacle:** connects power nozzle cord underneath hand grip.
7. **Cord clip:** fastens nozzle cord to extension wands. On some models a separate cord clips snaps onto wands.
8. **Suction regulator:** close for maximum suction; open if power nozzle is difficult to push on some carpets or if lightweight rugs or draperies are being cleaned.
9. **Plastic latch ring:** located on hand grip and plastic wands; rotate to lock wands and cleaning tools to hose.
10. **Cord hook:** rotate hook inward for quick cord release.
    On some models:
    Cord rewind pedal: step on pedal to rewind cord for storage. **Hold plug until cord is rewound.**
11. **Dusting brush**
12. **Crevice tool**
13. **Wall/floor brush**
14. **Furniture nozzle**
15. **Large wheels:** provide easy towing when cleaner is pulled over carpet, area rugs and door sills.
16. **On-Off pedal:** conveniently located on top of cleaner.
17. **Bag lid latch:** lift to open bag lid.
18. **Bag lid**
19. **Check bag indicator*:** shows red when paper bag should be checked for fullness.
20. **Hose connector**
21. **Receptacle:** connects power nozzle cord to canister cleaner.
22. **Carrying handle**
23. **Headlight***
*Available on some models

This cleaner is intended for household use.

Operate cleaner only at voltage specified on data plate on bottom of cleaner.

Source: Hoover Futura Cleaner, *Owner's Manual.* Used with permission of the Hoover Company.

EXAMPLE 12–9   Theory of Operation for a Rocksaw

Diamonds are the hardest substance on earth and can cut anything. Each rocksaw blade has many tiny, industrial-grade diamonds embedded in its edge. A sliding carriage, mounted beside the sawblade, holds the rock securely. The carriage is part of a passive counterbalance system. A 7-lb. counterweight pulls the carriage and rock past the blade. The friction of the blade against the rock produces tremendous heat. Therefore, a nonvolatile cutting oil bathes the blade to remove the heat. The oil also carries the debris away from the cut.

DIRECTIONS

Detailed directions are obviously the central part of any set of instructions or procedures and they will take up most of the space. As you already know, the conventional way to write instructions and procedures is to guide readers chronologically through the process with the help of named (and often numbered) steps and proper illustrations. Break the process into its component parts, and discuss each part of the process fully and in patient detail.

Sometimes the directions for performing a task amount to a sort of recipe: First you do this, then you do that, and finally do the last thing. If there is an established recipe for the activity that you are describing, by all means go ahead and offer it—one, two, three. But in many cases there may be several ways of performing the action, several ways of getting the job done. If there are alternatives and options, be sure to cover them when you write your instructions or procedures. You not only will be producing a more flexible and resourceful document, but also will be showing your readers that you respect them and appreciate their abilities. That bit of ethos will make your documents all the more effective.

Here are some further details of the conventions that typically guide writers of effective manuals:

Indicate the Overall Structure of the Task with Headings and Subheadings

Most instructions and procedures are long and complicated enough that readers appreciate being guided through them carefully. Those guides can come in the form of explicit headings (indicating the major steps in the process) and subheadings (indicating the substeps required to complete each major step). A simple set of instructions that uses headings and subheadings appears in Example 12-10.

EXAMPLE 12-10   Instructions with Headings and Subheadings

Modem Problems

The TelePort modem disconnects when you try to dial a telephone number.
Verify that your telephone cables are securely connected.

When the TelePort modem tries to connect with another modem, you receive a "No dial tone" message.
Make sure that you have set up the modem according to the installation instructions in Chapter 1, "Installing the TelePort Modem and Software." Make sure that the modem is not connected to a digital telephone line. Also make sure that the line from the telephone jack is connected to the modem's line connector.

Source: TelePort™ 56, *User's Guide.* Global Village TelePort for Macintosh, p. 35.

Present Steps in List Format

As you know from your own experience, most people use instructions in pieces. They read about one step in the process and then perform it. Then they read the second step and do that. You can accommodate this reading habit by helping readers find their place each time they turn back to the instructions if you offer information in lists. You can help your reader even more if you highlight the items in your list with *bullets.*

Be Complete and Detailed

The amount of detail in your instructions depends on your audience's experience with the process involved. In general, however, you should attend to details and be patient with readers. Start your instructions with the earliest possible action and include every action, no matter how obvious it might seem to you. You will want to avoid creating a patronizing tone with too much inappropriate detail, but you will want to guide the reader patiently and completely through the process. Remember this: it is far more common for readers to think, as they read, "Slow down! I'm getting lost!" than it is for them to think, "Hey, quit giving all this detailed instruction! I can get along without it!" When in doubt, include more information, not less.

Use the Imperative Mood

When you are telling your reader what to do, offer your comments to readers directly in the form of *commands;* tell them just what to do by using *imperatives.* Examples: "Fill the tray with half a quart of fluid." (Not "The tray should be filled with half a quart of fluid.") "Clean each part carefully before you use it." (Not "Each part should be cleaned before it is used.") "Insert the CD-ROM into your CD-ROM drive." (Not "The CD-ROM should be used next.") Imperatives are simple and easy to read, and they involve the reader directly in the activity that is being described. Notice how readers are addressed by means of imperatives in Examples 12–10 and 12–11.

Use Illustrations to Clarify Things

As you can also see from Example 12–11, visual displays in instructions help readers orient themselves to the product before them. Plan the visuals for your instructions as you plan your content. Depending on your audience, you will be using visuals to augment your writing at ratios as high as 60 percent prose to 40 percent visuals. Some instructions, especially those aimed at international audiences, are 100 percent visuals. (Most airlines, for example, present their safety and evacuation procedures visually, making them understandable to a nonreading audience.) Creating illustrations is

EXAMPLE 12–11   Simple, Detailed Instructions for the General Consumer

Setting up the computer

Step 1: Plug in the power adapter

Plugging in the power adapter recharges the computer's battery while you work. You should plug it in now in case the battery has drained during shipping and storage.

▲ **Warning:** Use only the power adapter that came with your PowerBook computer. Adapters for other electronic devices (including other portable computers) may look similar, but they may damage your computer. ▲

■ **Plug the power adapter into a standard electrical outlet or power strip. Then plug the power adapter cable into the power adapter port (⎓) on the back panel of the computer.**

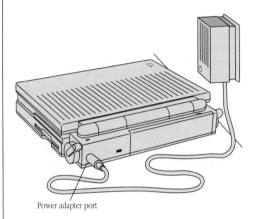

Power adapter port

Step 2: Open the display

■ **Slide the latch to the right and lift up the display.**

Position the display at a comfortable viewing angle. You can adjust the angle of the display at any time.

Source: *Macintosh User's Guide for the Macintosh PowerBook 160 and 180 Computers,* pp. 2–30. © Apple Computer, Inc. Used with permission.

covered in-depth in Chapter 10, but we would add here that you should think of illustrating not only the device but the user's interaction with it in your instructions. Get some people or parts of people (their hands, eyes, and so forth) into your illustrations whenever possible. When planning pictures of people, remember to be inclusive, representing both genders as well as ethnic diversity.

callouts
illustrations and definitions pulled out from the main text and magnified to let the user see some aspect of process or product in detail

If many parts are included in an illustration, you can use **callouts** to label each part of a product, as shown in Example 12–8. If you use this technique, call the reader's attention to it by referring to the numbering system in your prose. You would, for example, tell your reader when describing the plastic latch ring, "The plastic latch ring (labeled 9)" in Example 12–8 or, if you have already told the reader that you have labeled parts by numbers, you can then just refer to the number itself: "The plastic latch ring (9). . . "

Provide Warnings at the Right Time

Your readers will be depending on you to warn them about things that might endanger them, damage the equipment, or ruin the results. It is of course important to offer such warnings at the right time—*before* trouble might arise. You also may wish to highlight your warnings visually to give them special emphasis. Note how the warning is placed and symbolized by the triangle in Example 12–11.

Providing warnings is called for on grounds of efficiency, convenience, and safety. It is also a matter of law. Legally, companies are responsible for the safety of users of their products. For that reason, many companies require a legal review of their manuals to assure the safety of product users and to minimize their legal liability.

For example, Westinghouse designed a product safety label program. In their design, color (red for danger, orange for warning, yellow for caution, and blue for notes) helps classify the seriousness of the hazard. These colors are compatible with American National Standards Institute **(ANSI)** standards for warnings and cautions. Icons (symbols and pictographs) are used in Westinghouse documents to convey warning information to nonreaders.

ANSI
acronym for American National Standards Institute

In Chapter 9 we considered warning labels from the point of view of design issues. Here we would like to remind you about what must be included in warnings, which are an essential part of instructions and procedures. When you provide warnings, you must indicate the seriousness of the hazard, explain the consequences of the hazard, and provide information about how to avoid the hazard. The terms *danger, warning,* and *caution* are not interchangeable and must be used accurately in your work. ANSI has developed standards for developing effective hazard alert messages, which you should follow (see Example 12–12).

EXAMPLE 12–12   ANSI Standards for Product Signs and Labels

DANGER indicates an imminently hazardous situation which, if not avoided, will result in death or serious injury.

WARNING indicates a potentially hazardous situation which, if not avoided, could result in death or serious injury.

CAUTION indicates a hazardous situation which, if not avoided, may result in minor or moderate injury.

Source: "American National Standard for Product Safety Signs and Labels," Z535.4. American National Standards Institute, 1989.

ANSI specifies that all alert messages must have seven basic elements in them to be effective:

1. A hazard alert symbol.
2. A signal word.
3. The appropriate color: red, orange, yellow.
4. The pictograph.
5. The hazard identification.
6. The consequences of ignoring the hazard.
7. The description of how to avoid the hazard (Kemnitz, 71).

Many manuals include sections that deal exclusively with safety. Usually these are set off graphically to call attention to them—surrounded by black or colored borders, for example. Warnings within the text are usually set off from surrounding material inside or outside the margins of the text. Placing the warnings in borders alerts readers to their presence. Be sure to remind your audience to heed these warnings by putting a cautionary statement to that effect on the cover of your instructions. Example 12-13 shows a warning and a caution developed to meet ANSI Z535 standards.

Tell Readers How to Get Back on Track if Something Goes Wrong

Readers not only appreciate knowing how to avoid trouble. They also appreciate it if the instructions recognize that not everything may go smoothly. Try to anticipate where things might go wrong. Your interviews with potential users, or the tests that we will tell you about in a moment, will give you examples to work with and offer guidance about what to do in such a case. If it isn't obvious to users how to get out of trouble, tell them how to do so. Sometimes you should put some advice into the text itself or you can use a troubleshooting guide for this purpose.

EXAMPLE 12–13   Sample Warning and Caution Meeting ANSI Z535 Standards

⚠ WARNING: MOVING PARTS can cut off hand or fingers. DO NOT TOUCH.

⚠ CAUTION: HOT PARTS can cause burns. DO NOT TOUCH UNTIL COOL.

Source: Charles Kemnitz, "How to Write Effective Hazard Alert Messages," *Technical Communication* 38, no. 1 (1991), pp. 68–74. Reprinted with permission from *Technical Communication,* published by the Society for Technical Communication.

TROUBLESHOOTING GUIDE

In a troubleshooting guide, users get a separate section that contains summary information on a variety of things that might go wrong with a product or process. The section also offers advice for rectifying the problem or getting out of trouble. For the convenience of readers, who often need ready reference, troubleshooting guides are usually included in a separate section located in the back of the instructions.

Often you can provide troubleshooting information more conveniently if you use a table format. Example 12-14 illustrates a portion of a troubleshooting guide for a vacuum cleaner. A variety of formats are possible;

EXAMPLE 12–14 Troubleshooting Instructions for a Vacuum Cleaner

Problem	Possible Cause	Possible Solution
Cleaner won't run	1. Not firmly plugged in 2. No voltage in wall plug 3. Blown fuse/tripped breaker	1. Plug unit in firmly 2. Check fuse or breaker 3. Replace fuse/reset breaker
Cleaner suction low	1. Bag full 2. Obstruction in nozzle connector, hose, or wand 3. Secondary filter dirty 4. Suction regulator improperly set 5. Hose not properly connected to cleaner	1. Replace paper bag 2. Remove obstruction 3. Clean filter 4. Reset suction regulator 5. Ensure hose is properly connected

Source: *Hoover Futura Cleaner Owner's Manual.* Used with permission of the Hoover Company.

---

many feature the problems that might arise in the left-hand column, and the solution to the problem in the right-hand column. A middle column often gives the cause of the problem. Others list problems by subheadings and place answers below the subheads (see Example 12–10).

Your instruction manual may also include things that are similar to troubleshooting guides—suggestions for maintenance and care, for example. Navigational aids, such as tables of contents and indexes, help users find what they are looking for. If your instructions contain terms that may be unfamiliar, your manual should have a glossary.

Writing Specifications

Most of the information you learned about writing good procedures and instructions can be applied to writing specifications. Your biggest concern in writing specifications should be that your document is complete and accurate. Specifications lay out the requirements for materials and processes that are critical to the success of a particular endeavor. Whether you are writing specifications for the strength and durability of street paving materials, for the flexibility and R-factor of the insulation in walk-in restaurant refrigerators, or the type and strength of drugs that will be used during anesthesia, it's essential that you have accurate information, that your specifications are expressed unambiguously, and that your work is double-checked by experts. Once you have followed those guidelines, you can feel as if you have met the standards for safety and clarity that specifications require.

Writing Special Types of Instructions

As we indicated earlier, various circumstances call for a great many kinds of instructions. The advice we provided you already should serve you well as you contemplate those situations and respond accordingly. In addition, you might be called upon to employ special instructional supports of one kind or another.

For example, instructions for many products include **templates** that fit over keyboards or on equipment control panels to remind users about the functions of a device. Most computer users are familiar with the templates that fit over the computer keyboard for some word processing programs. Example 12–15 illustrates a template explaining the use of voice mail that fits right over the face of a touch-tone telephone.

Quick reference cards provide enough information to get users started and to keep users performing basic tasks. **Tutorials,** on the other hand, are extended instructions that help beginning users understand a product's functions. When online, tutorials are sometimes called **computer-based training (CBT).** The tutorial shown in Example 12–16 takes users step-by-step through

templates
overlays that fit on keyboards or control panels to remind users about the functions of a device

quick reference cards
single-page reference documents to get users started and to provide basic information about a product

tutorials
extended instructions to help beginning users understand a product's functions

computer-based training (CBT)
step-by-step training exercises completed at the computer

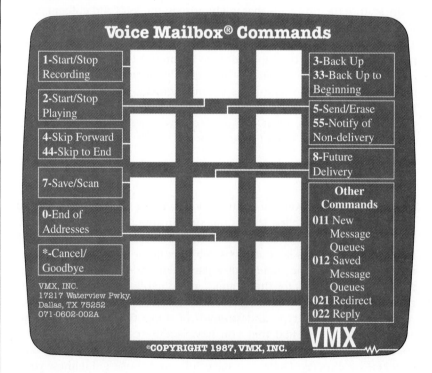

EXAMPLE 12–15   Template for Voice Mail

Voice Mailbox® Commands

1-Start/Stop Recording

2-Start/Stop Playing

4-Skip Forward
44-Skip to End

7-Save/Scan

0-End of Addresses

*-Cancel/ Goodbye

VMX, INC.
17217 Waterview Pkwy.
Dallas, TX 75252
071-0602-002A

3-Back Up
33-Back Up to Beginning

5-Send/Erase
55-Notify of Non-delivery

8-Future Delivery

Other Commands
011 New Message Queues
012 Saved Message Queues
021 Redirect
022 Reply

©COPYRIGHT 1987, VMX, INC.   VMX

Source: Adapted and reproduced with the permission of VMX, Inc. © Copyright 1992, VMX, Inc.

EXAMPLE 12–16 Tutorial

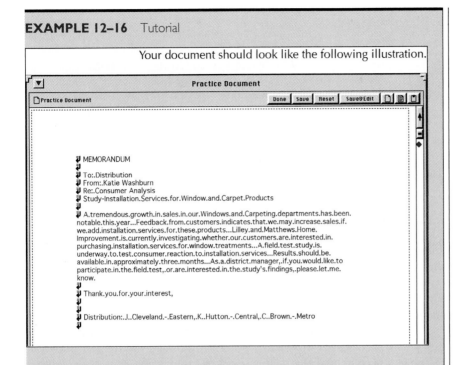

Your document should look like the following illustration.

Practice Document

Practice Document   Done | Save | Reset | Save&Edit

¶ MEMORANDUM
¶
¶ To:.Distribution
¶ From:.Katie Washburn
¶ Re:.Consumer Analysis
¶ Study-Installation.Services.for.Window.and.Carpet.Products
¶
¶ A.tremendous.growth.in.sales.in.our.Windows.and.Carpeting.departments.has.been.
notable.this.year...Feedback.from.customers.indicates.that.we.may.increase.sales.if.
we.add.installation.services.for.these.products...Lilley.and.Matthews.Home.
Improvement.is.currently.investigating.whether.our.customers.are.interested.in.
purchasing.installation.services.for.window.treatments...A.field.test.study.is.
underway.to.test.consumer.reaction.to.installation.services...Results.should.be.
available.in.approximately.three.months...As.a.district.manager,.if.you.would.like.to.
participate.in.the.field.test,.or.are.interested.in.the.study's.findings,.please.let.me.
know.
¶
¶ Thank.you.for.your.interest,
¶
¶
¶ Distribution:.J..Cleveland.-.Eastern,.K..Hutton.-.Central,.C..Brown.-.Metro
¶

Inserting Text

You can insert additional text at any time.

1. **Position** the pointer on the O in the word MEMO and **click** the Select mouse button.

   The caret appears to the right of the point where you clicked the mouse button.

2. **Type:** *RANDUM* and **press** RETURN.

   The word "MEMORANDUM" is now completely spelled out. Characters are always added, and the caret moves down to appear after the paragraph character.

3. **Position** the pointer on the Paragraph character after the word "Distribution" and **click** the Select mouse button.

4. **Add** the following text to complete the memo. Press RETURN only where indicated.

   Type: *From: (Your name)* and **press** RETURN.

   Type: *Re: Consumer Analysis Study—Installation Services for Window and Carpet Products* and **press** RETURN twice.

   Type (without pressing RETURN): *A tremendous growth in sales in our Windows and Carpeting departments has been notable this year. Feedback from customers indicates that we may increase sales if we add installation services for these products. Lilley and Matthews Home Improvement is currently investigating whether our customers are interested in purchasing installation services for window treatments. A field test study is under way to test consumer reaction to installation*

(continued)

EXAMPLE 12–16 (Continued)

services. Results should be available in approximately three months. As a district manager, if you would like to participate in the field test, or are interested in the study's findings, please let me know.
Press RETURN *twice.*
Type: *Thank you for your interest, and* **press** RETURN *three times.*
Type: *Distribution: J. Cleveland—Eastern, K. Hutton—Central, C. Brown—Metro and* **press** RETURN.

Source: Examples of instructions from XSoft software manuals are copyright of Xerox Corporation and are reprinted here with permission.

an exercise that teaches them how to insert text while they are editing documents. Notice that the tutorial not only directs users' actions—"position," "type," and "add"—but also creates sample documents for users to practice on. Finally, notice too that the user is shown exactly how the document should appear if the users have followed the tutorial correctly.

user guides
step-by-step instructions for completing all tasks associated with a product or process

User guides provide step-by-step instructions for completing all the tasks associated with a product or process. Although the tutorial shown in Example 12–16 helped the *beginning* user insert some *sample* text into a document, the passage from the user guide shown in Example 12–17 tells *any* user how to add *any* text to any document.

reference guides
a kind of glossary containing information arranged alphabetically that users will need to know to continue using a product

Reference guides are a kind of glossary. They list in alphabetical order the information users will need to know to continue using the product. Usually a reference guide describes functions of the product rather than tasks users will perform; that is, a reference guide is reference-oriented rather than task-oriented, and although most reference guides include some step-by-step instructions, these instructions are usually listed by product function rather than user task. Example 12–18 shows a page from a reference guide that explains the autohyphenation checker in a word processing program. Reference guides generally contain much more technical information than user guides. For example, they may describe the internal operation of a product and the technical information needed for its maintenance and repair.

online help
information available on the computer; it can be built into a computer application, available from a Web site or stored on a CD-ROM

Offering Instructions Online

Online help, a particular kind of instruction that is becoming more commonplace and important each day, provides on the computer screen the information found in other types of instructions. Some online help systems are simply an electronic version of what you would find in a paper manual. Their chief advantage is that they can be updated frequently to reflect changes in procedures or improvements of various kinds; and they can accommodate small differences in similar products that would otherwise

EXAMPLE 12–17   User Guide

To add text to existing text:

1. Move the mouse pointer to the position in the text where you want to enter new text.
2. Click the Select mouse button to place the caret.
3. Type the new text.

You can use the AGAIN key to repeat text entry operations, including inserting paragraph, new line, and tab characters, and backspacing over characters. The system "remembers" up to 100 keystrokes entered since the last time you made a selection. This feature is useful for repetitive text entry in tables or forms. It works within a document and between documents.

For example, if you want to enter a name and address in several different places on a form, or on different forms, you can type the name and address once and use the AGAIN key to enter it in the other places.

To repeat a text entry operation:

1. Type the text or characters.
2. Move the pointer to a new location for the text or characters and click the Select mouse button.
3. Press AGAIN.

EXAMPLE 12–18   Reference Guide

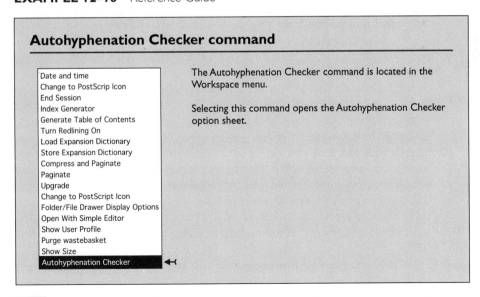

Autohyphenation Checker command

Date and time
Change to PostScrip Icon
End Session
Index Generator
Generate Table of Contents
Turn Redlining On
Load Expansion Dictionary
Store Expansion Dictionary
Compress and Paginate
Paginate
Upgrade
Change to PostScript Icon
Folder/File Drawer Display Options
Open With Simple Editor
Show User Profile
Purge wastebasket
Show Size
Autohyphenation Checker

The Autohyphenation Checker command is located in the Workspace menu.

Selecting this command opens the Autohyphenation Checker option sheet.

---

require expensive separate instructions. Although there will be print versions of such procedures and instructions in circulation, organizations can keep customers and employees up to date without having to print out and resend new instructions each time a minor change is called for. Users also appreciate online documents because they can be easily located and require no storage space.

Such online instructions should observe the conventions of any other set of instructions or procedures, but you should remember that putting print instructions directly online will not result in a satisfactory set of online instructions. There are physical constraints associated with reading from a computer screen. Users easily lose their place in the online document in the absence of the visual cues (e.g., tables of contents to turn back to, headers and footers) that are often found in a paper manual, and users cannot take computer screens with them when they perform tasks.

Therefore, if you are putting a conventional set of instructions or procedures onto a Web site, you can take one of two approaches. If users need to take the instructions with them to perform some job, you may want to design the instructions to be printed as well as to be used online. For example, you can create **PDF files** that retain all of their formatting when printed. If the material can be read on screen, then you will have to design for the screen. Look back at the design advice given in Chapter 9 to learn whether you have actually created a Web document and not simply tried to put paper under glass. Include consistent and adequate navigational aids so readers can find what they need easily and return to it as needed.

PDF files
Portable Document Format files that can be printed from a Web site and preserve the look and all formatting of the original Web page

context sensitive
online help systems that present on screen information related to the procedure being performed at the time the help is called up

Sometimes online help systems are **context sensitive;** that is, information presented on the computer screen will be related to the exact computer operation that the reader is performing at the time of calling up the help screens. For example, a person confused by a particular icon on a computer screen can ask for information online about that function. The instructions are designed to be used on the computer screen, not printed separately. In this case you should still prepare your instructions by being attentive to the matters we have just covered: conscientious planning that is both audience and task oriented; writing that is both patient and complete; document design that anticipates problems and offers warnings and troubleshooting tips; and testing that guides revision. Online help offers this key advantage: A user can ask for help at particular times that are especially relevant to specific needs; at other times, the instructions are hidden. See Example 12-19 for an online help screen for students taking an online class.

When you prepare instructions online, be aware of the special needs and expectations of users. Such users often come to online instructions with specific questions in mind: What can I do with this program, and what can't I do (goal questions)? What does each feature of the software do, and how do I make it do it (procedure questions)? What do various features

EXAMPLE 12–19   How to Send E-mail from an Online Class

 UNIVERSITY OF MINNESOTA

College of Agricultural, Food & Environmental Sciences & University College
Department of Rhetoric & Independent and Distance Learning

Rhet 8210: How to Succeed
Step 2

E-mail the
Instructor

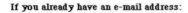

Step 2: Send me an e-mail message to get started

Your next step in this course is to send me an e-mail message.

If you already have an e-mail address:

Please send me an e-mail message
(Rhet8210.Instructor@mail.agricola.umn.edu) that tells me a little bit about yourself and why you're taking the course. I will reply to your e-mail and send you the login and password that you'll need in order to access the modules for this course on the World Wide Web.

The Course Map

If you do NOT have an e-mail address:

Check the pages in the pocket of this study guide for the information you need to activate your University of Minnesota e-mail account, use the account, and keep your account open. When your account is active, send me an email message (Rhet8210.Instructor@mail.agricola.umn.edu) that tells me a little bit about yourself and why you're taking the course.

> **NOTE:** All new University of Minnesota students experience a delay of 7 to 14 days after initial registration before they can activate their University e-mail accounts. If you already have an e-mail account, you may use that address for this course. You are not required to use a University of Minnesota e-mail account.

Class E-mail Addresses

> **NOTE:** If you have any problems as you proceed, feel free to send Paul Brady (pbrady@mail.fsci.umn.edu), our college's instructional computing coordinator, an e-mail message with your questions. Be sure to indicate that you are a student in Rhet 8210.

Electronic interaction throughout the course

I expect to hear from you at least three times a week as you communicate with me and with your classmates. In some cases you may want to use e-mail, and in other cases you can communicate via the WebChat. The thing I find most rewarding about my classes is my opportunity to interact with students, to hear your ideas and insights as well as your questions. Please share your perspectives and respond to the perspectives of your colleagues. Update us on what you do and how this information transfers to your workplace.

Assignment Submission &
GradeBook

> **NOTE:** I will make an effort to answer all of your Rhet 8210 e-mail within 24 hours except on weekends, holidays, or when I'm out of town on business.

mean (interpretive questions)? How can I get to the next part of the instructions (navigation questions)? As you design online instructions, you must be sure to answer these questions. If possible, build search capabilities into your online instructions. Then, a user can search for the answer when he or she needs it. And, don't forget to add an e-mail address where users can write to request answers to their specific questions.

Testing Instructions, Procedures, and Specifications

Before you consider your instructions, specifications, or procedures finished, you need to test them on users. The result of user testing will indicate whether your documents are complete and satisfactory or need to be revised. When you are planning your documents, leave plenty of time for testing because you really cannot be much of an advocate for users unless your documents help them accomplish whatever task they are trying to carry out. Schedule your user testing early enough in your writing process to make the results of the test helpful. Incorporate what you learn from the test into your next drafts.

Even though tests cost time and money, it's both a rhetorical and an ethical imperative to test instructions and procedures and to make sure specifications are complete, understandable, and accurate. In the long run, user testing is time-efficient and cost-efficient because it helps users operate more effectively and more safely—and it keeps complaints and claims down. Many writers of instructions overlook testing not only because it takes valuable time and money but also because the activity they are describing seems so obviously clear and simple to them. But your familiarity with the procedure that you are describing could actually be a problem for you: You know the procedure so well that you might accidentally leave out a some fundamental information absolutely critical to a novice. Because confusion about a small matter, especially early in a set of instructions, can significantly delay a user, it's essential for you to learn as much as possible about whether your draft instructions in fact work well.

Testing can take place any time in the design or development stage of a product or document. When you test early in the development stage, you can use the development team itself along with the test users to provide feedback on the draft instructions that you write even before the product is completed. If you test your materials later—after development is complete but before the product is finally released—be sure to allow for enough time to test your instructions and make revisions. Test your documents with actual users and revise them on the basis of what the user testing shows.

As you know from Chapter 9, you can test your draft document on potential users by means of direct measures (those that study how readers read, use, or rate a document), or by means of indirect measures (those that predict a reader's ability to use or comprehend a document). The following paragraphs describe some testing methods you can employ once you have produced a draft document. All of the methods depend on your employment of "test users" who are similar to the people who will actually use your finished instructions.

user edits
a kind of user testing whereby a "typical" user reads a document aloud while following the steps to see if it is possible to follow the instructions to carry out a task

USER EDITS

In a **user edit,** the simplest kind of test, a "typical" user of a document reads aloud the document while attempting to follow the steps or points. The test user does not actually try to use the instructions in this case; rather, the test user simply reads the document out loud as you listen. Meanwhile you witness this reading activity (that's why the user reads it aloud) to see if your user can manage the instructions step-by-step, point-by-point, illustration-by-illustration. In essence, this kind of test is actually a test of the readability and understandability of your draft. If the user misreads or stumbles, then you obviously need to revise that step. A user edit works well for short instructions, but it may not work so well for longer procedures and instructions, especially those that involve visual aids and complicated series of steps.

PROTOCOL-AIDED TESTS

A **protocol-aided revision,** sometimes called a "think-aloud protocol," is similar to a user edit, except that this time the test user reads the document aloud and talks to you while trying to *perform* the activity involved. As you watch the test users perform the instructions and listen to them report on how they are managing the activity, you will pick up valuable information about the adequacy of the instructions. The advantage of protocol analysis is that it gets at readers' seemingly "invisible" internal thinking processes, so listen carefully as users talk about how they are using the instructions and the product.

> **protocol-aided revision**
> type of user testing in which users perform tasks and talk aloud as they do so to reveal how they are interpreting a text

This type of testing works best when your users must find as well as follow your instructions; it reveals their problems, their misinterpretations, their apprehensions, and their questions. A disadvantage is that it requires your presence, a presence that users might find awkward and distracting. One way to overcome this awkwardness is to tape-record or videotape the process so that you can do the analysis after the test user is finished.

FIELD TRIALS

Often the most effective testing occurs when you run a **field trial** in the actual setting in which your users will need your document. Your users then report the problems they encounter, either by means of an interview or a written response to your questions. (Some companies, such as IBM, have information cards that users return with descriptions of problems, so in essence the company receives constant feedback and can revise its manuals.) Arrange to observe your test users (either in person or by means of videotape) as they try to perform the physical activity of using your draft instructions. Later you can interview test users as well.

> **field trials**
> type of user testing of a document in an actual setting to determine what problems users face

Videotaping has important advantages. Not only does it seem less obtrusive if you are not physically present during the test, but a videotape can provide a more permanent record, one that can be replayed many times, if necessary, to get at a problem. Sometimes the test user can even be questioned about the videotape to clarify where things might be going wrong in your draft.

LABORATORY TESTING

laboratory testing
effective way to test a
lengthy document in the
laboratory by video- or
audiotaping users who are
following the
documentation while using
a product

Sometimes **laboratory testing** is the most effective way to find out how well a lengthy document works. Users are video- or audiotaped while following the documentation and using the product. Data logging programs are used to record the amount of time it takes typical users to read documents and perform actions. Although the number of actual users involved in the test may be small, the value of data obtained is great, particularly if the users are given realistic problems to work on while the testing is conducted (Redish and Schell, 68–71).

SUMMARY	

- In this chapter we focused on instructions and procedures, two related kinds of documents that focus on performance and frequently support technology. We also briefly examined specifications. Procedures generally describe how to coordinate an activity in which many people are involved, and instructions detail the specific steps that an individual would follow to complete a task. Specifications provide audiences with standards necessary for completing a task safely and adequately.

- Ideally, instructions and procedures are developed at the same time as product development and design, rather than after product development. Even when that is not the case, instruction, specification, and procedure writers should function as an advocate for the user. The process of developing instructions and processes is similar to the process of writing any other user-oriented document.

- Writers should plan the document, considering the functions of the product (by means of a task analysis); the audience's needs, attitudes, and prior knowledge (by means of an audience analysis); and the physical conditions under which the document will be read. Such planning will suggest the detailed contents that you will need to include and an overall design for the document, as well as details about vocabulary and visual illustrations.

- Instructions conventionally include certain textual features. An introduction typically names the purpose and audience for the document and forecasts its contents. The introduction might also include information about materials and tools required, a description of the mechanism, and an account of its theory of operation. The body of the instructions usually describes step-by-step instructions, though you should avoid offering a simple "recipe" for users to follow strictly unless it is clearly called for. Present steps in list format, use headings and subheadings to direct the reader, be complete and detailed, employ illustrations to good advantage, and use the imperative mood and active voice. Often the style and organization of a support document will be dictated by your company, your professional code, or ISO 9000 standards. Instructions also often include warnings and troubleshooting guides to help readers avoid trouble and to tell them how to get out of trouble.

- Some special types of instructions include templates, quick reference cards, tutorials, user guides, reference guides, and online help. If you must provide online help (to accommodate revisions in procedures or to reduce distribution expenses), you should take the same general approach to writing, with two exceptions: First, online users are especially interested in getting answers to specific questions—about goals, procedures, interpretations, and navigation—so you need to anticipate those questions. Second, online users must negotiate electronic space, not printed pages, so screen design principles need to be applied. Sometimes you can design context-specific online help; such help can be hidden from sight until it is needed and tailored to meet specific needs.
- Regardless of the type of instructions you create, instructions should always be tested, either by user edits, protocol-aided tests, field trials, or laboratory testing.

ACTIVITIES AND EXERCISES

1. Locate some instructions that have accompanied a product you have recently purchased or acquired. Using the suggestions within this chapter and your own field test, critique those instructions. Describe in your critique how you tested the instructions.

2. Once you have completed your critique for question 1, *(a)* write a revised and improved set of instructions that avoids the shortcomings you spotted, or *(b)* write a set of instructions, procedures, or specifications for another product or process you know well, or *(c)* find a procedures manual that has become somewhat dated and revise it. Whatever your choice, be sure to follow all the suggestions in this chapter and be prepared to have a classmate "test" your document.

3. Arrange to interview an employee at a company in your immediate vicinity about the documentation that supports the tasks the company engages in or products the company produces. Gather information on what the documents are designed to do: how they are planned, drafted, and tested; who their users are; and any other pertinent information. Give the results of your interview, with samples if you can obtain them, in a written or oral presentation to your classmates.

4. Do some primary or secondary research on how ISO 9000 series standards are affecting a company for which you would like to work in the future. Report your findings in a memo to your instructor and classmates.

5. Read carefully all the documents that pertain to the *Exxon Valdez* oil spill in Appendix A, Case Documents 2. Assume that, as a company engineer for

Exxon, you need to write a set of instructions for the bioremediation process. You might need to do some library research to complete this assignment, but you can get off to a good start with the case documents. Plan, design, and write the instructions so that crews involved in oil spill cleanup efforts in the future can follow your step-by-step instructions.

6. Analyze the "universal precautions" for caring for a person with AIDS (Appendix A, Case Documents 1). Rewrite these instructions for a poster to be displayed in all examining rooms in your local hospital. Be aware that your users will have to follow your instructions while treating patients and therefore have only a few seconds to remind themselves about these steps. You will need to add appropriate danger, warning, or caution labels.

7. "Shareware" is basically free software available from the World Wide Web (sometimes a modest fee is requested). Most shareware is offered with little or no documentation for users, so do three things for this assignment: *(1)* Find a piece of undocumented shareware that interests you (it might be a game, an educational module, or financial product); *(2)* Write the instructions for it (you might also wish to create a Quick Reference Guide for it); *(3)* Contact the developer of the shareware to see if that developer might be interested in including your documentation with the product.

WORKS CITED   Doheny-Farina, Stephen. *Rhetoric, Innovation, Technology: Case Studies of Technical Communication in Technology Transfers.* Cambridge, MA: MIT Press, 1992.

Kemnitz, Charles. "How to Write Effective Hazard Alert Messages." *Technical Communication* 38, no. 1 (February 1991), pp. 68–73.

Mandel, Theo. *The Elements of User Interface Design.* New York: John Wiley, 1997.

Pirsig, Robert. *Zen and the Art of Motorcycle Maintenance.* New York: Morrow, 1974.

Redish, Janice C., and David C. Schell. "Writing and Testing Instructions for Usability." *Technical Writing: Theory and Practice.* Bertie E. Fearing and Keats W. Sparrow, eds. New York: Modern Language Association, 1989, pp. 63–71.

CHAPTER 13

Reports and Studies

Introduction ▮ Written Reports: Purposes and Audiences ▮ A Process for Planning and Developing Reports ▮ Types of Reports ▮ Conventions of Report Writing ▮ Parts of Reports

The reader of an industrial report does not possess the writer's knowledge and has no interest in doing so. In industry, the writer is the expert, not the reader, even if the reader is the boss. An industrial report should not focus on what the writer knows, but on what the reader needs to know, what the information means for the reader, and why it matters to the reader. The industrial writer's task is not to prove that he or she has learned something, but to produce reports that convey the significance of information to a reader in a particular context of social interactions.

Peter Hartley, "Writing for Industry: The Presentational Mode versus the Reflective Mode," *Technical Writing Teacher* 18, no. 2 (1991), pp. 162–69.

Introduction

reports
documents that contribute
to significant decision
making; along with
business letters and memos,
the most common form of
workplace writing

Reports accompany and justify or predict all significant activity in the
work world whether the work is research and development, manufacturing, government, law enforcement, medicine, or finance. Along with
business letters and memos (which often report activities or decisions), reports are the most common type of writing in the workplace. No matter
what career you plan, you will undoubtedly write and read reports.

This chapter provides an overview of reports, with emphasis on the different types of reporting situations and ways to respond to the needs these
situations present. In Chapter 14 we'll look in more detail at how reports
are used to make decisions and solve problems. Here, we offer a process
for planning and developing reports, and we summarize the contents and
purposes of various reports, including literature reviews and progress reports. We also suggest strategies for organization and style to increase report effectiveness. Finally, this chapter provides information about the formal features of reports.

Written Reports: Purposes and Audiences

Reports answer the questions "What happened?" or "What was done?" and
many times ask as well "What should be done?" Report writers, often observers or participants in what was done, report their observations to someone else who can use them. As writers, they "carry back" (re-port) information. Even though the activities reported have already occurred, the act of
reporting assumes that the information can or should be used in the present or future.

Reports may be delivered orally, but they are written when the investigation yields too much information for an audience to retain from listening,
when some written evidence is required to verify what happened, or when
the information will have value beyond the present time. Technicians write
lab reports so researchers can verify the relationship between the procedures and results. Written records provide the evidence that allows researchers to make claims about causes and effects. The secretary of an organization keeps minutes of meetings so that members will know who is to
do what and to preserve a record of decisions. Written records allow people to refer to their decisions so that their follow-up actions can be consistent. In this chapter, we focus on written reports.

Reports all have practical purposes: to preserve records and provide
information to others who can use it. Reports have a persuasive purpose
as well. Your persuasive purpose will be direct and obvious if you make
recommendations—often based on the "good reasons" offered in your report. Whether you write specifically to persuade, your reports will influence decisions, actions, and policies and, in the case of a lab report or literature review, future research. A good starting point in report writing,
therefore, is to think about what should happen as a result of your report
and how you can encourage the right things to happen.

Your reports will also influence the views other people form about your professional competence. As a report writer, you want to persuade your readers that you have performed your work competently. The information you present, your means of gathering the information, and the way in which you present it should convey your competence. You invest time and energy in researching and writing the report in part to suggest that you would like readers to take your ideas (and you) seriously.

Readers of reports are generally a select group because reports have a specific focus and purpose. Some reports may have only one reader, such as your immediate supervisor. Often reports are forwarded up the organization to the supervisor's manager or above. They may be filed and used by readers in the future who are trying to reconstruct some decisions or who are building on the information in the report to do a follow-up study. The possible uses by future readers may influence your decisions about content and presentation.

Conventions of organization and visual design follow from the varying purposes of reports. However, even within categories of reports, there are variations. Different reporting goals may require different strategies of organization and presentation. Furthermore, all organizations have their own preferences and conventions for reports. Workplace reporting situations are never exactly like textbook descriptions. Thus, in this chapter we offer guidelines for report writing based on a process of conceptualizing, researching, and managing the reporting process. We do not pretend to offer templates for reports that will work in every situation. Your task in learning to write effective reports is not so much to memorize conventions of reports as to learn a process for developing reports that can help you plan effectively for each reporting situation.

A Process for Planning and Developing Reports

The process for writing any report includes the following tasks:

1. **Conceptualization:** conceiving of the purpose in writing and the reader's need for information. Think back on this process as described in Chapter 2.

2. **Research:** researching the problem in ways that produce credible and useful information. Think back on this process as described in Chapters 6 and 7.

3. **Presentation:** choosing the best strategies of organization, document design, style, and use of visual displays. Think back on these decisions as described in Chapters 8, 9, and 10.

4. **Management:** managing the research and writing tasks to ensure a high-quality report that is finished on time.

To conceptualize the report being used and acted upon, consider the completed report in use and work backward to plan the steps that will get

conceptualization
process of conceiving of the purpose in writing and the reader's need for information

research
process of finding credible and useful information

presentation
process of choosing the best strategies of organization, format, style, and visual display

management
the research and writing tasks needed to produce a high-quality report

you to this end. You will write a better report if you can imagine the report used not only in the present but also in the future. You will write your reports more efficiently if you can imagine, from the beginning, the document itself—how it will be organized and formatted, the sections and supplements it will contain, and the visuals that will enhance the readers' understanding. Knowing something about the common types of reports and specific strategies for organizing and presenting reports will help you conceive of your writing tasks. Printed guidelines or directions from a supervisor may establish some specific expectations for content, presentation, and length. You can and should supplement what you are told by examining sample reports written within your organization.

Research often means keeping good records of activities and results during the course of a project. The evidence that forms the substance of the report comes from notes made while the work is in progress. Anticipating the report required helps you determine what kinds of records to keep and what research methods match the kind of information that the report will require. Again, the information in Chapters 6 and 7 will help you choose and carry out your research plans. Also, you can use the information in Writing Strategies 13-1 in any reporting situation to help you conceive and plan your report.

Types of Reports

Reports are called by different names, such as *recommendation report, progress report,* or *literature review.* These names reflect different purposes and expected outcomes. Various types of reports influence action in different ways. In planning, it's smart to consider the ultimate uses of the report and recognize that the report's work is not finished once the words are inscribed on the page or screen. This knowledge helps you determine content, research methods, and presentation in order to achieve the report's goal.

The first three types of reports listed in Example 13-1—literature reviews, lab notes, and scientific articles—are used primarily in research. They share the purpose of pointing to and enabling further research. The other reports are used mostly in business, to support planning and decision making. Progress reports are written in both settings. The purposes of reports in research and management settings overlap because all reports require investigation and sharing of information, and all reports invite readers to make decisions. Nevertheless, one purpose usually dominates. Some projects require a series of reports. For example, a research project may require a proposal, literature review, several progress reports, and a scientific article.

Other types of reports could be added to the list, but you can see from these examples that the classification of reports by type depends on their situation and intended outcome—what actions or decisions should result from them. Completing the report is often only the start of the action. Reports

1. *Conceptualization: Think of the report as it will be used when it is completed.*

   Purpose
   Why is a report needed? (Consider the uses that will be made of it and problems it attempts to solve.)

   Readers
   Who will read the report?
   In addition to the primary reader, will the report be forwarded to others within or outside the organization? Will the report be filed and perhaps read by readers in the future? Who will these secondary readers be? What will the readers do as a result of reading the report?

   Content
   What questions will the readers have about the report topic?
   In what order will the readers be likely to ask the questions?

2. *Research: Plan methods of gathering information.*

   What are the best ways to get answers to the questions the reader will have? (Consider interviews, surveys, site inspections, library research, financial analysis, and other suggestions in Chapters 6 and 7. Also plan for the records you will need to keep during the course of the project that is reported.)

3. *Presentation: Imagine the document itself, and choose the best strategies of organization, format, style, and visuals to make it effective.*

   How should information be arranged in the report to make the information easy to comprehend and to find? What will be the overall pattern—e.g., comparison, chronology, topics in order of importance?
   What arrangement may increase its persuasiveness? (Is there some information that should appear early or be delayed?)
   What information should be displayed in tables, graphs, or illustrations in order to increase comprehension or to persuade?

4. *Management: Identify research and report preparation tasks, and set a schedule for completing them.*

Research Tasks	*Completion Date*
a.	
b.	
c.	

Writing, Report Preparation Tasks	*Completion Date*
Writing	
Introduction	
Body	
Conclusions and recommendations	
Summary or abstract	
Revision and editing	
Preparation of visuals	
Preparation of preliminary pages	
Preparation of supplements	
Final proofreading	

EXAMPLE 13–1   Reports and Resulting Action

Type of Report	Purpose and Action That Results
Literature review	Reports current knowledge to determine best practice or to justify new research.
Lab notes	Describe procedures and results meticulously to establish a clear link between a cause and effect. The notes provide the data for the lab report or scientific article.
Scientific article	Details a research problem, methods of investigation, and results of research so that other researchers can replicate experiments and verify results or follow up with additional research. Practitioners may apply the results.
Progress report	Documents work accomplished during a specific period to justify ongoing support or recommend some change in direction.
Feasibility study	Analyzes the likely outcome of possible future actions and recommends actions.
Annual report	Summarizes accomplishments and forecasts future work in order to encourage ongoing support and, in some cases, to satisfy legal requirements.
Environmental impact statement	Predicts possible consequences to land, wildlife, and people of a proposed project and helps decision makers determine the feasibility of the project.
Personnel report	Evaluates applicants' qualifications or employees' performance in order to determine job assignments, promotions, and pay schedules.
Financial report	Evaluates the balance of assets and liabilities to determine taxes or investment options.
Trip report	Presents records of events and observations when people are away from the office to enable a change in a policy or procedure or a decision about reimbursement.
Medical report	Describes symptoms, treatment, responses, and recommended actions so that medical personnel can make sound follow-up decisions about patients.
Police report	Records who-what-when-where information from accidents or crime scenes so that police investigators can determine responsibility and degree of damage.
Minutes	Summarize discussion and decisions to provide a history and to encourage follow-up actions.

---

also share the common feature of resulting from some type of investigation, whether that investigation represents scientific research or recording the details of a car accident.[1]

---

[1] M. Jimmie Killingsworth and Michael K. Gilbertson have developed a theory of technical discourse and the action that results from it in *Signs, Genres, and Communities in Technical Communication*. The book and an article by Carolyn Miller, "Genre as Social Action," have influenced our discussion in Chapters 13, 14, and 15 of genres as distinguished by the actions that result from them.

The formal features of reports, including their appearance on the page and content divisions, vary according to the type of the report and the requirements of the organization for which the report is written. That is why we cannot provide a template for a report format (though similar reports share similar parts and structure). Rather, a good writer develops an appropriate format according to the assignment requirements and by thinking through the reasons for the report and the needs of the readers.

This section continues with guidelines for writing and thinking about common types of reports that you might write in class or on the job. Our discussions of progress reports and literature reviews provide the most detail.

PERIODIC REPORTS AND COMPLETION REPORTS

Reports are frequently scheduled to appear periodically and at the completion of a project. These reports help decision makers determine whether and how the project should continue. At the completion of a project, a report establishes what has been accomplished and whether the goal has been achieved.

Investment companies send **periodic reports** to investors monthly, quarterly, or semiannually, as well as annually, summarizing account balances and investments and explaining the performance of the investments. Investors depend on this information to make decisions about maintaining, increasing, or canceling an investment. You get one version of a periodic report in your bank statement. You use the information in the statement to guide spending in the next reporting period.

> **periodic reports**
> reports comparing the activities of the current reporting period with that of a previous period

Employees and divisions within an organization are evaluated annually, and an **annual report** indicating achievements and goals for the coming year is often the basis for the evaluation. Likewise, divisions within organizations and the organization as a whole produce annual reports summarizing achievements and goals.

> **annual reports**
> reports summarizing and evaluating annual activities for a company

All periodic reports compare the activities of the current reporting period with that of the previous period or with goals that were established at the beginning of the project or with some other benchmark of performance. Achievements are meaningful only in comparison with expectations. The argument of a periodic report—the claim the writer makes—is based on comparison. The argument may also show causes and consequences to explain the results reported.

The basis for the information you use in a report of your own achievements or progress on a project will probably be notes you take throughout the project with the addition of correspondence, e-mail, financial analysis, or other evidence of the work being done. This type of reporting requires you to keep good records of your work and to ask others who may work with you to keep records. When you write any kind of periodic report, you will be able to follow the suggestions for organizing a progress report described in the next section.

Example 13–2 illustrates part of a semiannual report to the shareholders of the Vanguard Balanced Index Fund, a mutual fund with investments in the

EXAMPLE 13–2   Periodic Report to Mutual Fund Shareholders, Front Matter

The letter format personalizes the report, a strategy that makes the shareholder feel important and connected to the people who are responsible for the investments.

The first paragraph reports good news and a "thesis statement" that shapes the reader's interpretation of the rest of the report. The basis for a positive evaluation is a comparison of the fund's performance against a benchmark—performance of comparable funds.

The table emphasizes the good news by drawing the reader's attention. Its factual information supports the claims.

Whereas the table emphasizes return on investment, this factual information explains the increase in value.

The letter continues by placing fund performance in context of a robust U.S. economy, a severe slump in the Asian markets, and good Vanguard management.

The letter concludes (these sections are omitted here) with a caution to investors to maintain a balanced portfolio and to invest for the long term because these good returns cannot continue indefinitely. The writers sign and thereby personalize the letter and commit to its claims.

CHAIRMAN'S LETTER

Fellow Shareholder:

Vanguard Balanced Index Fund earned an excellent return of +10.8% during the six months ended June 30, 1998, as the stock market continued its remarkable advance and the bond market made a solid showing.

As shown in the adjacent table, the fund's total return (capital change plus reinvested dividends) for the first half of the year exceeded that of the average balanced fund and precisely matched our Balanced Composite Index benchmark. This hypothetical portfolio is weighted 60% to the Wilshire 5000 Index, comprising essentially all actively traded U.S. stocks, and 40% to the Lehman Brothers Aggregate Bond Index, a proxy for the U.S. taxable bond market. The table also presents the six-month returns for those two "total market" indexes. Balance Index Fund's annualized yield as of June 30 was 3.23%.

	Total Returns Six Months Ended June 30, 1998
Vanguard Balanced Index Fund	+ 10.8%
Average Balanced Fund	+ 9.0%
Balanced Composite Index	+ 10.8%
Wilshire 5000 Index	+ 15.4%
Lehman Aggregate Bond Index	+ 3.9

The fund's return is based on an increase in net asset value from $16.29 per share on December 31, 1997, to $17.80 per share on June 30, 1998, with the latter figure adjusted for dividends of $0.20 per share paid from net investment income and a distribution of $0.

THE PERIOD IN REVIEW

The U.S. economy grew vigorously, inflation was subdued, and interest rates declined during the first half of 1998. Strong consumer spending, triggered by high employment and rising wages, was the economy's propellant and was more than enough to offset the negative effects of Asia's severe economic slump. . .

1

stock and bond markets. The 48-page report contains factual data, including lists of companies in which the fund invests and graphs of investment distribution. But a letter from the two senior Vanguard executives at the beginning of the report interprets the data by placing them in the context of performance by comparable funds and worldwide financial markets. This combination of factual and interpretive information creates strong support for the claim in the first sentence that the return over six months has been "excellent."

Investors (owners of fund shares) will no doubt be happy with this report and are likely to continue to hold their shares or to add to their holdings. Had the news been bad (returns lower than the benchmarks or lower than returns from other investments), the writers would have had to earn the readers' trust by pointing to such strengths as a long-term performance record and the prospect of future gains. Being truthful and straightforward about reasons for declines and articulate about plans to reverse them would provide credibility.

The periodic report is a standard feature of the financial world. Its purpose is partly information sharing, but it is also meant to be persuasive. In addition, the writer also uses the report to influence readers' actions, achieving these goals through good research, good reporting, and effective rhetoric.

PROGRESS REPORTS

Long-term research and development, manufacturing, or contracting projects (lasting six months or a year or more) require progress reports written to managers or project sponsors. Your instructor may also require a progress report on a long project. The **progress report** describes work accomplished during the reporting period and identifies the work remaining on a project. Progress reports are a type of periodic report specifically written for a project that ultimately will end (unlike the investment that is expected to continue indefinitely). The progress report compares the project as it is unfolding with the project as it was proposed. The objectives and schedule of the project assignment identify what will be reported and specify where the project should be at the time of each progress report.

progress reports
reports describing work accomplished during a reporting period and identifying work remaining

The progress report is a good management tool: It is an incentive to keep work progressing on schedule. A progress report also allows periodic assessment of project goals and methods so that they may be revised, if necessary, according to the preliminary results of the project. Typically, progress reports are due every three months or every six months on a long-term project. The terms of the project assignment should specify the frequency and content of progress reports.

Three outcomes can result from a progress report:

1. The project may continue as originally scheduled.
2. The project may be expanded or otherwise changed to accommodate new situations.
3. The project may be canceled.

Because the progress report may shape project directions, you will have a persuasive as well as an informative purpose in writing the report. Indirectly you will make the claim that you are doing competent work that will achieve the established objectives. The best way to do that is to have some progress to report. Excuses for failing to accomplish work will not impress the supervisor or project sponsor. On the other hand, being on schedule or even ahead and having good results to report will mark you as an effective worker.

Your progress report could recommend change. For example, if work is not progressing on the Year 2000 (Y2K) project according to the established schedule (see Appendix A, Case Documents 4), a manager may decide to hire additional workers. A project might be canceled because the work seems unsatisfactory, the benefits are not as great as had been predicted, or a number of external situations unrelated to the project take place, such as company reorganization.

You might occasionally find yourself in the position of convincing superiors to abandon a project. Although we focus on presenting your progress on a project in the best possible light in the rest of this chapter, you could find that a project is not feasible after all. In such a case, your reader will appreciate your honesty and wisdom because changing the plans can prevent a wasted investment.

As for any other periodic report, the information for a progress report comes from the records you keep of your work. These records may include data from experiments, regular site inspections, analyses of figures, or a log book of recorded activities. Keeping good records makes it easier to write the report.

A progress report has a three-part structure:

1. An introduction that reviews the project and its purpose.
2. A middle section that cites specific accomplishments and work remaining.
3. Conclusions that assess the progress.

We look at each part of the progress report separately.

Introduction

introductions
early sections of a report that describe the problem the project will solve

The **introduction** should describe the project and the problem the project will solve. This description may appear in the same words in each of the progress reports written on a project. Even though the reader of a progress report is supervising the project and presumably is familiar with the project description, include a project summary in each progress report to orient the supervisor, who may oversee at the same time a dozen or more such projects, with your specific project. The introduction also helps to establish your credibility as investigator because your description of the project demonstrates your comprehension of its goals.

You may also preview the rest of the report and its significance by including a statement indicating whether work is on schedule. Although the details of the accomplishments will appear in the body of the report, you might identify an accomplishment of particular significance. You should prepare readers for any surprises or changes in the work—something unexpected that might cause delays in completion or recommended changes in the project.

Body of the Progress Report

The **body of the report** presents the specific accomplishments in concrete and often numeric statements. This section of the report provides the evidence that will let you assess the overall progress in the conclusion. Generally, you should avoid evaluative statements in the body of the progress report. Instead of saying "the work is going well," you should say "we completed 14 of the 16 planned trials" or "the site has been excavated and the foundation laid." Readers look for facts more than for evaluation in this section of the report. You should, however, provide interpretive statements that show the significance of the findings, such as "the level of participation is 17 percent higher than in previous studies." These statements place the facts in a context so the reader will know how to respond to them.

body of the report
main section of a report that describes the problem the project will solve

The body of a progress report can be organized chronologically, by task or by topic. You will probably use a combination of patterns, with one dominant pattern and others embedded into the main pattern. For example, your main sections may be organized chronologically (work accomplished, work remaining), with subsections arranged by task or topic. Alternatively, your main divisions might be ordered according to task or topic, and the subdivisions ordered chronologically.

Your instructor may ask to you to write a report about your progress on a major assignment. Tasks for such a project include investigation tasks, such as a literature review, interviews, and a site inspection. They also include preparation tasks, such as writing the introduction, designing visuals, and creating the preliminary pages. Topics for such a project include the particular subjects you are investigating through a literature review, interview, or site inspection. Topics might include cost, size, legal issues, fertility rates, or other objects of your investigation.

Example 13-3 presents the first of three progress reports (here called status reports) on a senior design project by electrical engineering students. The team writer has arranged the report by task (hardware development and software development). The report is specific about accomplishments. The visuals aid understanding but are also persuasive because they suggest that the writer's own concept of the task is clear. Work remaining is specified in the conclusions of the report.

EXAMPLE 13–3   Progress Report

FROM: S. David Silk
   TO: Mark Greer
DATE: October 24, 1998
   RE: Status Report on ISU/Motorola ISDN Senior Design Project

This report summarizes all the concurrent tasks performed on the ISU/Motorola Project from September 1 through October 24. The purpose of this project is to build and implement an ISDN TA demonstrating Motorola's ISDN chip set. This status report is being sent prior to the actual proposal for the project so that we may incorporate, in the proposal, some suggestions made by Eric Crane in a memo dated September 12, 1998, that we received the week of October 22, 1998.

The project has been divided into two teams consisting of four students each. One team works on the hardware and one team works on the software.

As you recall, we have discussed hardware quite extensively and the team understands what is required of us. Therefore, this status report treats hardware briefly and software more extensively.

1. HARDWARE DEVELOPMENTS

Further work on the schematic began where the preliminary research left off (the preliminary work on the schematic only included the ISDN chip set). The schematic is currently being designed on a CAD system which when completed will allow us to go straight to fabricating a printed circuit board for the prototypes. Listed below are the areas that have been addressed within the last month pertaining to the schematic.

1. In addition to the ISDN hardware, the schematic has been refined to incorporate the following components:

   1. A 16 line by 20 character LCD
   2. A dial keypad
   3. The eight function keys
   4. A ringer

2. Since the preliminary work on the schematic, details on the memory requirements have been refined. Because this is the first attempt at our design, we will not be too concerned with managing the memory efficiently. Time permitting, we will fine tune the memory requirements. So for now, we intend to add additional memory as required while we wire wrap the design for initial testing. Listed below is each type and amount of memory that will be incorporated into the first wire wrap design.

   1. EPROM - 1M
   2. RAM - 64K
   3. Battery backed RAM - 64K
      (Note: This memory has been allocated for storage of the users' personal preferences.)

For your convenience, a printing of the schematic has been enclosed with this report.

2. SOFTWARE DEVELOPMENTS

Developments over the last month have been quite extensive in defining the software for our device. Within the last month, several issues have been resolved. The issues identified relate to how Q.931 and Q.921 will be developed, how the software architecture of our device should be structured, and how code can actually be produced for the MC68302 with the equipment available to us.

(continued)

EXAMPLE 13–3 (Continued)

2.1 IMPLEMENTING Q.931 AND Q.921 ON OUR TA

Within the last month, we have conceptualized the TA's software architecture. Because of the extent of the work involved, it has been decided that we will not write Q.931 or Q.921 code. We have decided to focus our attention solely on writing the software to control our device. To acquire the necessary code, we have solicited the assistance of a group of graduate students who are currently developing Q.931 and Q.921 code. In return for their code, we will provide them with a test bed, our device.

Essentially, the graduate students will be providing us with standard code that is portable and is not designed to operate on any particular hardware. As a result, we will be required to write an interface to their code in order to run it on our TA. To expedite this procedure, a student from our design team has been assigned to work with the graduate students from the start in order to familiarize himself with the protocol and to understand how we will integrate it into our TA.

2.2 SYSTEM SOFTWARE FOR OUR TA

The architecture for our device will be built around the EDX code that will be made available to us by Motorola in the near future. EDX's flexibility to perform as a cooperatively multitasking operating system will lend itself to our application.

The TA's architecture has been designed to be event-driven and object-oriented. The software for our TA will be broken down into self-contained modules that will consist of a main routine which does nothing but loop continuously getting messages and calling its appropriate local procedures. No module can call a procedure that exists in another module, instead, a message must be sent to that separate module. Hence, via EDX, messages will be passed around the system to the appropriate modules executing the desired tasks. In a sense, EDX will act as a type of message center responsible for the execution of all TA functions (refer to Figure 1 for a breakdown of the major system software modules).

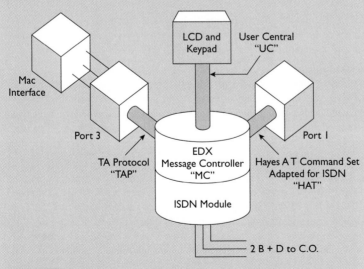

FIGURE 1

Software Architecture

(continued)

EXAMPLE 13–3 (Continued)

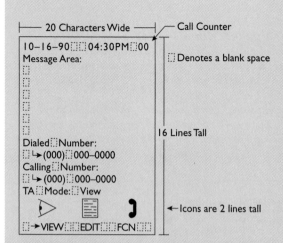

FIGURE 2

A Layout of the Internal Procedures of User
Central.
In this example, View mode has been selected.
An arrow indicates the selection.

2.2.1   System Software Modules

Each of the major modules depicted in Figure 1 has been assigned to individual members within the team. Currently, team members are preparing some preliminary documentation defining their modules' functionality and how it will interact within the entire system. Within the next week, we will begin generating code.

2.2.2   A Look at the User Interface Software

From the perspective of the user, the user interface is perhaps the most important aspect of our device. User Central is the module that is responsible for providing an efficient and user friendly means to operate the TA. The preliminary work that has been completed regarding the user interface is shown in the next two figures. Illustrated in Figure 2 is a layout of the procedures within the User Central Module. Also, illustrated in Figure 3 is the graphic layout of the LCD.

2.3   DEVELOPING CODE FOR THE MC68302

Within the next month, a new cross compiler will be installed on the University's HP System providing us with the proper environment to do our MC68302 development. The package includes a 302 Cross Compiler, Linker, Assembler, and Simulator. The majority of the software development will be done on the HP System and then downloaded to a Mac II, which in turn will be connected to the ADS Board and our prototype.

(continued)

EXAMPLE 13–3 (Continued)

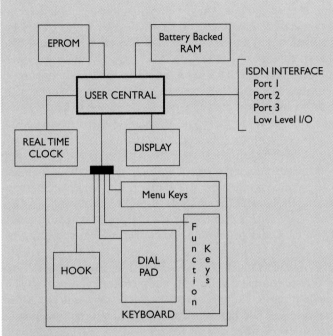

FIGURE 3
Projected Design for LCD Graphics.

3. SUMMARY

The following areas have been addressed: refinement of the schematic, acquisition of Q.931 and Q.921 code, and conceptual organization of the framework of the system architecture. Within the next few weeks we will address the following areas: complete the schematic so we can proceed with wire wrapping the design, continue working with the graduate students to evaluate what will be required to get the Q.931 and Q.921 running on our hardware, and continue fleshing out the system software for our device.

A few possible bottlenecks could occur due to delayed arrival of the following equipment: the development system for the University's HP workstations, equipment for the ADS Board, and additional ISDN components. If any one of these setbacks should occur, we will still be able to proceed with the project.

Source: Used with permission of the author, with special thanks to Patricia Goubil-Gambrell.

Conclusions

The conclusions of your progress report provide an overall assessment of the work to date. If everything is going well, you should be able to report that the work is on schedule or ahead of schedule. You will certainly indicate any areas in which you have exceeded the objectives for the project. If you have failed to achieve objectives, you will have to offer a reasonable explanation and state your plans for catching up in the next reporting period. The conclusions are also the place to propose any changes in the project that result from unexpected findings in the work so far. If you do propose changes, ask for some written confirmation that the changes are approved.

Overall, try to maintain an upbeat tone by emphasizing progress rather than problems. If there are legitimate problems, be honest about them, and point to possible solutions, but don't get trapped into complaining about how far behind you are because of all kinds of problems beyond your control. Manage your work with a mental eye on the progress report, and accomplish what you have said you would accomplish so that you can report real progress.

Progress reports usually don't have to be very long. You shouldn't have to provide all the data you have collected, but give enough detail to show that you have made progress. Report in special detail any results of particular interest or importance. Use tables and graphs for important quantitative information. As with every other type of writing you do, you can help judge what is enough information by imagining what questions your readers will ask as they read and what format (e.g, paragraph, table, or bar graph) will help them read accurately and quickly.

LITERATURE REVIEWS

literature reviews
summaries and evaluations of published material on a topic

Literature reviews report on published material on a topic. Often the quickest and most appropriate way to conduct research is to survey the literature that reports what other researchers have learned. Thus, a literature review is an essential first step in most original research. A literature review establishes what is already known about a topic. If someone has already researched your question, you won't have to conduct expensive and time-consuming original research. On the other hand, when the literature review reveals an absence of information—a gap in our current knowledge—it may show the need for new research by identifying a problem that needs attention.

One reason to review the literature is to learn best practice. For example, a technical publications group may be uncertain about whether to use frames in constructing a Web site. Through a search of the literature, they can determine what others have learned through experience or study. A written review is a report on findings and may become the basis of policies or a style guide.

A literature review may precede a specific design project. For example, an engineer trying to determine whether a particular material will work in a new application will study reports of tests on the material to determine the likelihood that the material will be suitable for the new application. The literature review eliminates the need for duplicate research and helps the engineer assess whether and how to proceed with the design, confident about whether the material will satisfy requirements.

Literature reviews are also common in academic writing. They are likely to be part of any proposal for funds to support research. The limits in existing research are an argument for the proposed research. Scientific articles typically include a literature review as part of the introduction or in a separate section immediately following the introduction. The review shows how the new research is connected to what has been done before, that is, how it extends or refutes the existing research.

Sources for a literature review may be located in the library or on the Internet. To locate them, you use the processes we described in Chapter 6. When you have written term papers, you have also used some of these searching and review processes. But the literature review differs from a term paper in some significant ways. One purpose in writing a term paper is to demonstrate to your teacher your knowledge of a subject as well as your ability to navigate in the library and cite sources. Most writers of term papers choose a "topic" and organize their papers to provide evidence for a thesis. The citations and quotations serve as evidence. The completion of the paper is probably the end of the project; the consequence of doing it well is getting a good grade and passing the class.

A literature review almost always begins with a question, and its completion points to new research or project development. The purpose for the literature review is not so much to prove your knowledge (although citing the literature does strengthen your credibility as a writer) as to determine how the existing research applies to a current or future project. Like good term paper writers, the writers of literature reviews not only summarize published articles but also evaluate and comment on the relevance of the information to the problem at hand.

A literature review is written in essay or paragraph form. In this way, it differs from an **annotated bibliography,** which is a list of sources on a topic with brief summaries or descriptions. The literature review may be organized by source or chronologically (early studies to the latest research), but when it is part of another report, it is more likely to be organized by topic. A topical pattern best shows the relationship of various sources to one another and to the research problem. Topics in an engineering materials study, for example, might include the ability of the material to withstand stress, the effects of temperature and humidity, and strength. The technical

annotated bibliography list of sources on a topic with brief summaries or descriptions

communicator putting together the review might cite several sources for each of these topics and might cite the same source in the discussion of more than one topic.

Readers of literature reviews are always asking the question "what is the relevance of this information to the current problem?" Your evaluation and interpretation answer these questions in ways that mere summaries may not, by connecting the existing research to the current problem.

Brandie McKay is an undergraduate student hoping to become a technical writer. She knows she will spend a lot of time working with a computer and wonders if there are health hazards from the radiation emitted from the monitor. Because of her question, she chooses to research electromagnetic fields for her literature review assignment. If she were already employed, she might have the same question, or her company might assign her to begin investigating the issue. Thus, the literature review she will write for her class has a workplace parallel.

What can she accomplish in her literature review?

- She can get specific information from good sources. This information is likely to be more complete and credible than hearsay.
- The research may help to define her question further. She may uncover issues that she did not know when she began investigating.
- The written review provides a starting point for new research. In the workplace it could become part of a feasibility study to determine whether the company should invest in special monitors or testing.
- If the review convinces her that she needs some protection as an employee, she has credible evidence to make her claim.

Brandie's literature review appears in Example 13–4. The search has been limited by the class assignment, but it does suggest the structure, use of sources, and interpretive statements of a more comprehensive review.

Brandie organizes her review by topic: health (including specific possible consequences) and the state of the research. The topical structure (as opposed to source-by-source structure) raises implicit reader questions (What might happen to me? How good is this research?). She refers to some of the sources in more than one section. Brandie's own voice is obvious in the review, especially in the ending statements of each section. Although she provides factual information, she interprets what those facts mean in light of her research question.

EXAMPLE 13–4   The Effect of Electromagnetic Fields on Computer Users: A Review of Research

Technical writers spend eight or more hours each day in front of a video display terminal (VDT), more commonly known as a computer monitor. Most monitors in use today are equipped with cathode ray tubes (CRT), the same picture tube found in television sets. Why is this important to a technical writer? The CRT-style VDTs emit two types of electromagnetic radiation: Extremely Low Frequency (ELF) and Very Low Frequency (VLF). Both ELF and VLF have been linked to health problems such as cancer, miscarriages, and changes in brain chemistry. In addition, the costs of research, litigation, and treatment may affect individuals and companies but they certainly affect taxpayers.

This report will provide a background on the technology of electromagnetic fields, identify the theories that propose a link between electromagnetic radiation and health problems, discuss the economic aspects of EMFs, and suggest ways in which technical writers should deal with possible health hazards in their work.

Electromagnetic Fields

Understanding the technology of electromagnetic fields requires knowledge of their origin, radiation, and measurement.

Origin

An electromagnetic field (EMF) results from an alternating (AC) electric current. This current produces two distinct fields, an AC electric field and an AC magnetic field, which together are called the electromagnetic field. AC electric fields result from the strength of the charge, while AC magnetic fields result from the direction of charge. "People can sense an electric field of more than about 20 kilovolts/meter. . . most people cannot feel the presence of AC magnetic fields. . . " (Lechter 1996).

Radiation

According to Lin (1996), electromagnetic fields produce two potentially harmful types of radiation. Technical writers are exposed to this radiation from the electromagnetic field of a computer monitor. Extremely Low Frequency (ELF) radiation comes from the vertical deflection coils wound around the CRT inside the monitor, whereas Very Low Frequency (VLF) radiation comes from the vertical coils. These coils control the electron beam that is used to light the pixels of the monitor. ELF radiation frequency ranges from 0 to 1,000Hz, while VLF can range from 1,000Hz to 500,000Hz. This frequency is important in measuring absorption by the human body. Electromagnetic energy absorption varies both with frequency and body size.

Measurement

The origin of electromagnetic fields and their ability to produce radiation are known. However, methods to accurately measure these EMFs are still being perfected. Methods for measuring the properties of magnetic fields use both

1

The introduction states the problem and it gives reasons why the audience (technical writers) might care to know about EMFs.

A forecasting statement identifies the topics of the literature review (LR) and their sequence. This forecast helps readers develop a framework for comprehending the paper and establishes its scope.

This LR is appropriately organized by topics, not by sources. The topical structure foregrounds the issues that concern the readers—definition and specific health concerns.

Explanation provides a basis for understanding the problem. Often readers retain concepts better than they retain facts. Here the main concept to remember is that a magnetic field originates in electricity and that it cannot be felt.

The writer chooses topics in part by anticipating questions of readers. Here readers might ask, "If a person cannot feel magnetic fields, how would one know they are there? Could I measure the radiation from my monitor?"

(continued)

EXAMPLE 13–4 (Continued)

Throughout the LR, the writer provides interpretive as well as factual statements, answering the reader's question, "What does this mean to me?"

A new reference to the technical writer keeps the reader's interests foregrounded in this discussion.

A forecasting statement for the section helps readers negotiate what will follow.

Each of the subsections of the report follows a pattern that the writer establishes early on: problem definition, factual information, and comment suggesting how to interpret the information.

theoretical equations and instrumental gauges. Scientists use formulas associated with Faraday's Law and Ohm's Law to calculate the magnitude of an ELF magnetic field. In addition, a Gauss meter measures the strength of AC magnetic fields. However, neither of these methods is very practical for everyday use by the technical communicator. Scientists have attempted to develop a personal measurement device, called a dosimeter, to record exposure to radiowaves (Lin 1996).

Development of a meter to detect magnetic fields is unlikely due to technical difficulties that forced the termination of the dosimeter project. However, one thing is certain: electromagnetic fields *do* exist and produce VLF and ELF radiation.

Health Risks Associated with EMFs

Why should the presence of electromagnetic fields and radiation be a concern to technical writers? Researchers have discovered a correlation between the use of video display terminals and reproduction problems and cancer in the users. Studies have also shown correlation between EMFs and other health problems such as stress, alteration of brain chemistry, and Alzheimer's disease. This section includes an overview of the research on reproduction and cancer, and it summarizes the theories that back the studies.

Reproductive Problems

Early studies on the effects of low levels of EMFs (like those found in VDTs) on chicken embryos showed a correlation between increased EMF levels and birth defects. Such defects were abnormal birth weight and slow outflow of calcium levels in the brain. In addition, a 1988 study by the Kaiser Permanente HMO found that women who used CRT-style VDTs were 73 percent more likely to miscarry. However, research on the subject of miscarriages is contradictory. A National Institute for Occupational Safety and Health study concluded that "The use of VDTs and exposure to the accompanying electromagnetic fields were not associated with an increased risk of spontaneous abortion" (Lechter 1996).

While none of these studies proves that EMFs cause miscarriages, birth defects, or other reproductive difficulties, most researchers are hesitant to rule out the possibility.

Cancer

A number of studies have assessed the risk of cancer associated with EMFs. A 1991 study by the University of Southern California found an increased rate of leukemia with children who watched black-and-white televisions. These televisions use the same type of picture tube as computer monitors. Additional studies have pointed to EMF exposure as the cause of breast cancer in men and women, brain tumors, and prostate cancer. As with the reproduction risks, cancer research cannot prove that EMFs are the cause.

2

EXAMPLE 13–4 (Continued)

Theories and Limitations of Research on Health and EMFs

Some scientists attribute the biological effects of EMFs to the resonance, or oscillation, of cells caused by EMFs. Researchers observed disruption of calcium flow through cell walls due to resonance. According to Lechter (1996), calcium helps to regulate muscle contractions, heartbeat, and development and division of cells. Still another theory suggests that altered calcium flow reduces the cell's ability to fight cancer.

Most theories represent diachronic study that examines correlations of occupations and cancer rates by analyzing databases of vital statistics. These databases are used as an inexpensive method for discovering links between EMFs and health hazards encountered on the job. However, correlation studies raise questions more than they offer proof of cause and effect. In addition, the studies often fail to account for multiple hazardous exposures in the occupational setting. In addition to magnetic fields, workers can be exposed to "metal fumes, solvents, fluxes, chlorinated biphenyls, synthetic waxes, epoxy, and chlorinated naphthalenes" (Wright 1982).

Such uncertainties in research—based more on correlations than on clear establishment of cause-effect relationships—make it difficult to determine if EMFs present reproductive and cancer hazards. However, researchers caution VDT users not to rule out any risks.

Conclusions

Research linking EMFs and health hazards is inconclusive. Nothing in the research to date provides grounds for hysteria, but the seriousness of the health hazards linked with EMFs suggest that technical writers may wisely monitor ongoing research on EMFs and health, especially looking for results based on facts, not simply on correlations. In addition, one criterion for the purchase of monitors should be the level of radiation emission. Companies or individuals may wish to investigate the effectiveness of monitor screens that reduce radiation in an effort to prevent possible problems.

Works Cited

Lechter, George. 1994. EMFs linked to health problems. http://www.milli-gauss.com/info.html (1996, Sept 25)

Lin, James, and Om Gandhi. 1996. Computational methods for predicting field intensity. In *Handbook of biological effects of electromagnetic fields,* ed. Charles Polk, 337–55. New York: CRC.

McGregor, Alan. 1996. WHO investigates electromagnetic fields. *Lancet* 7: 605.

Wright, W. E., J. M. Peters, and T. M. Mack. 1982. Leukemia in workers exposed to electrical and magnetic fields (letter), *Lancet* 7:1160.

Source: Used with permission of the author.

Conclusions answer the research question, here implicitly posed in and in the first paragraph: What effect do EMFs have on the health of a technical writer who works with a computer all day? The answer is briefly that there is no sure answer, but the writer suggests caution in behavior until researchers learn more.

The writer provides interpretive statements, but they are credible because of the preceding research.

Brandie has cited sources according to The Chicago Manual of Style author-date format.

SCIENTIFIC REPORTS AND ARTICLES[2]

**scientific reports,
articles**
the results of experiments
or other exact methods of
research

Scientific reports and **articles** report the results of experiments or other exact and systematic methods of research. Scientific articles attempt to report an accurate description of reality without the interference of bias or personal opinion. They report the test of a hypothesis and discuss whether the test results confirm or deny the hypothesis.

Scientists and engineers conduct much of their research in laboratories. The research requires tests and careful observations of the results. The researcher or a technician carefully observes and records what happens in controlled conditions. Ultimately the records provide the data for a report or article that may be published.

lab notes
notes describing methods
and results of lab
procedures in meticulous
detail

Lab notes provide the records from which the report or article can be written. The writer of lab notes must be meticulously accurate in describing procedures and results. The notes will include descriptions of procedures and materials, details about time and other conditions that may affect the experiment (e.g., temperature), and specific notes on observations. Photographs may be inserted to complement verbal or quantitative descriptions of results. A lab technician works with notebook at hand, recording results as they occur rather than trusting memory. Multiple people may record the observations, especially if the test requires checks around the clock, and they use the same notebook so that the records are contained in one place. The notebook has numbered, bound pages, to prevent any suspicion of data being inserted after the fact. These notes provide the data for the report. The notebook remains in the lab, where it is always available for reference in case the results are questioned or to verify a method or result.

If you are a student in a lab science, your lab notes are helping you learn to observe accurately. They also are the basis for a lab report that you may be asked to write. This lab report parallels the scientific article that a research scientist may write for publication.

Errors in lab notes call into question the entire experiment; moreover, they can result in the development of a dangerous product or procedure. The accurate gathering and transmitting of data is one of your legal and ethical responsibilities as a writer of lab notes and reports.

Scientists, including social scientists, may conduct their research in the field as well as in a laboratory. They, too, would keep a careful record of their observations in a notebook.

**lab reports or scientific
articles**
lab or field notes used to
make a claim extending or
challenging existing
knowledge

Lab reports or **scientific articles** use the lab or field notes to make a claim extending or challenging existing knowledge. They report in detail the experimental procedures, specific results, and discussion of the results. Their pattern of procedures, results, and comments reflects the pattern of

---

[2]We used and appreciate the insights of Dr. Marilyn A. Houck in developing this section.

scientific reports, whether they are lab reports (mostly written by students) or articles (reports for publication by researchers).

You can recognize a scientific report or article by its standard structure:

- An introduction defines the problem.
- A methods section describes in detail the methods of research.
- A results section presents the findings in factual (often quantitative) terms.
- A discussion section explains the significance of the results.

This structure is called the "**IMRaD**" structure for "*i*ntroduction, *m*ethods, *r*esults, *a*nd *d*iscussion." Often a literature review follows or is part of the introduction.

An **abstract** is a standard part of a scientific report and usually precedes the introduction. The abstract is the part that most readers are likely to study and represents the work and its significance in brief form. Because it is important and difficult to write because of length constraints, it should not be approached as an afterthought but as a major part of the article.

Much of the credibility of a scientific report or article derives from good research methodology; that is, readers believe the results of experiments when they think the experiments were conducted according to sound methods of gathering information. Thus, an important section of a scientific report is the one on methodology, which describes the methods of research in enough detail that another researcher can repeat the experiment to verify the results. It includes details about the sequence of steps, the number of trials, and the "conditions" of the trials (i.e., how the various trials may have differed from one another).

The results section includes factual and specific consequences of the procedures. Tables and graphs may be used to show measurements. The discussion section interprets the results to explain their significance in terms of the goal of the procedure and to explain any surprises.

The scientific method, and therefore the scientific report, is used in the social sciences and business as well as in science. Many academic journals consist of scientific reports.

IMRaD
organizational structure for a scientific report consisting of an **I**ntroduction, **M**ethods, **R**esults, and **D**iscussion

abstract
standard part of a scientific report that usually precedes the introduction

---

Conventions of Report Writing

Reports can vary not just in their purposes and outcomes but also in their organization, style, and visual design. Your choices about these report features can assist or hamper readers in locating and comprehending information, and they also influence a reader's response to the content. The conventions you will need to follow may be specified for you. For example, the National Information Standards Organization (**NISO**) and American National Standards Institute (**ANSI**) have developed and adopted a standard that specifies report parts, organization, format, typography, and the use of visuals in technical reports. The level of detail specified by NISO is shown by

NISO
acronym for the **N**ational **I**nformation **S**tandards **O**rganization

ANSI
acronym for American
National Standards Institute

this example: "Primary and secondary headings are aligned flush with the left column of text, and other headings are run in with indented text" (17). The Department of Defense and other government units have adopted NISO standards with the idea that uniformity in reports makes it easier to retrieve information. Organizations may develop their own style guides that specify how documents are to be prepared. The discussion about organization, style, and format that follows assumes that the writer has more options than the NISO standards allow.

organization
reveals the relationship
between ideas and
influences comprehension

Organization reveals the relationship of ideas to one another and thus influences comprehension. Some types of reports, notably the scientific article, are consistent in the use and order of parts, and substantial variation from this pattern invites readers to question the credibility of the information. Organization also influences reading patterns. Research by James Souther showed that managers do not read reports from the first word to the last. Instead they pay more attention to certain parts (most notably the executive summary) and may ignore other parts completely (the appendix). To accommodate the managerial reading style, good writers are especially careful in writing the abstract. They may also reverse the conventional beginning-middle-end structure of many written documents and place the end—conclusions and recommendations—at the beginning so that managers can find the key information easily. The discussion sections of the report then follow the conclusions, almost like appendixes. (NISO standards for technical and scientific reports would not allow this managerial organization.)

Thomas Pinelli and others, however, found that managers and scientists read NASA reports with close attention to all parts of the report. This research does not support a rearrangement of report parts for managers. When advice like this is contradictory, it is prudent to find out the preferred organization strategies where you work. It is also wise to realize that report writers do have options, and the choice of options will influence readers' responses. We will cover organization more thoroughly in Chapter 14.

Because of the variety of reports and different organizations in which they are written, there is no uniform style for all reports. Some organizations require writers to write in the third person, using the pronouns "they" and "it" rather than "I" or "we." This requirement contributes to formality in style. Some organizations encourage writers to use first person ("I"). Either style can be satisfactory, but you should check to see what the preferences in your own organization may be. Before you write your first report, you'll need to review the advice on using style effectively in Chapter 8.

format
the conventions required
for a report; may include
standards for typeface,
margins, headings, spacing,
length, and binding

Some reports are very short—handwritten memos or e-mail—although others are many volumes long. Some reports are letters; some are double-spaced in traditional typescript; others are typeset and bound to look like books. Nevertheless, within organizations and for specific types of reports, such as progress reports, there are **format** conventions. Consistency among reports means readers don't have to figure out the conventions for each new report. A sample report of the same type that has been written previously can show you the expectations about format, such as single spacing or double spacing, margins, and page numbers, as well as the expected sections and their order.

There is no need to begin each main division on a new page. Indeed, it's undesirable to do so. Blank space at the bottom of the page signals *end.* You can fill each page with text unless tables and graphs interrupt and need to be on *one* page. The exception occurs when a heading for a new division appears at the bottom of the page. Unless a heading is followed by two or more lines of text, you should force a page break to place the new section on a new page where the lines of the paragraphs can follow immediately.

According to Pinelli and his colleagues, readers like to see the illustrations and other visual presentations integrated with the text rather than placed at the end, unless there are large numbers of visuals. Readers look at the visuals as they read the text, and if they have to flip to the back of the report, they are less likely to read them. Elaborate tables, however, that are useful more for reference than for primary information should be placed in an appendix.

Headings help readers in two ways: comprehension and selective reading. They announce changes in topics and thus serve as transitions. They also reveal the structure of the information, much as an outline does, which helps readers interpret how topics relate to each other and to the whole report. Headings should at least announce the major sections, such as introduction, discussion, and conclusion. When the structure of the report is topical, headings for each topic are useful. Headings with a few words are likely to be more informative than headings with a single word (e.g., "maintenance costs" tells more than "costs").

headings
navigational tools that divide text into sections and subsections and help the reader's comprehension and selective reading

Headings provide good signals, but be careful not to use too many. If you use headings more often than every three or four paragraphs, you are probably signaling too much and your headings won't work as effective maps for the reader.

Running heads state the title of the report or a major section of the report, sometimes in a condensed version. Running heads in short reports are valuable as a source of identification if a single page is photocopied. In long reports with major divisions, a running head that names the division helps readers locate parts of the report for selective reading.

running heads
title of a report or major section of the report that appears in a condensed version at the top of the page

Page numbers are essential for cross-references and to enable selective reading from a table of contents. Preliminary pages, which we describe below, are numbered with lower case roman numerals (i, ii, iii) so that the text of the report can begin with the arabic number 1. You don't put a number on the title page and contents page, but you do count the pages when you number the others. Pages can be numbered at the top (usually on the right), at the bottom (usually in the center), or on the outside margins (left for an even-numbered page and right for an odd-numbered page) when the report is printed front and back.

Parts of Reports

Because reports are a genre widely used in the workplace, they have come to have standardized components. In addition to the body of the report, these include the following:

1. Letter of transmittal.
2. Title page.
3. Table of contents.
4. Executive summary.
5. Abstract.
6. Glossary.
7. Appendixes.
8. References or works cited.

Not all reports contain all these parts, and long, formal reports generally have more parts than short, informal reports. If the conventions of the organization where you work offer you some choices about what parts to include, choose parts that will be useful to readers and omit parts that readers will not need. In school, follow the assignment directions that your instructor provides.

Preliminary pages precede the text of the report, identifying the report and its contents and giving credit to the people who produced it.

preliminary pages
pages that precede the text of a report, identifying the contents and giving credit to the people who created it

LETTER OF TRANSMITTAL

A **letter of transmittal** or a cover letter is not a part of the report itself but often accompanies it. The primary purpose of a letter of transmittal is to announce what the attached document is and what project it relates to. You use it primarily to orient your reader to the report's purpose and topic. Letters of transmittal may accomplish other purposes as well. They can function like prefaces to books, and you can use them to acknowledge the assistance of individuals in a more personal way than through citation of sources. You can accomplish a sales purpose, too, by indicating how this report is significant. You can prepare a reader for surprising or disappointing results and start to shape the response to them. We'll cover letters in more detail in Chapter 16.

letter of transmittal
cover letter that is not part of the actual report; announces the attached document and project discussed

TITLE PAGE

The **title page** identifies the subject of the report as well as its recipient, author, and date. Its obvious purpose is to identify the document. (Example 14-5 in Chapter 14 illustrates a title page.) Other information on the title page depends on what else readers need to know. You can use it to provide other identifying information, such as a project number or grant number. If your readers need to get in touch with you and they do not have your address, the title page should include your telephone number

title page
page at the front of a text that identifies the subject of a report and its recipient, author, and date

(including FAX number) and address (including e-mail address), along with those of your coauthors. Thus, the title page may not only identify the report but make it easy for readers to communicate with the writer. Some title pages include a routing list when a single copy of the report will be distributed among several readers, and some include a descriptive abstract (defined in the next section). Often copyright material or circulation restrictions will also appear on the title page.

The arrangement of the information on the title page should correspond with expectations of your readers. If the organization does not prescribe a particular format, follow the convention of placing the title about one-third from the top of the page followed by the name of the recipient, author, and date. You can make the title prominent by using a type size bigger than the text size, perhaps 14-point type or larger.

The main purpose of the report title is to inform. A good **report title** gives two types of information:

- The topic.
- The approach to the topic.

report title
title identifying the topic and the approach taken toward it

The topic defines the subject matter, such as radiation emitted from a computer monitor. The topic alone, however, doesn't say why and how the writer has considered the topic. For example, the writer may have analyzed the feasibility of purchasing shields for monitors. Or the writer may describe the process of emission or report the results of a scientific measurement. The title for the report in Example 13–4 indicates both approach and topic:

> *The Effect of Electromagnetic Fields on Computer Users: A Review of Research*
> **topic**   **approach**

The information orients a reader to the report topic and purpose, but it also helps a reader decide whether to read the report at all. Perhaps the company files contain three reports on radiation from monitors, one analyzing the feasibility of purchasing shields, one describing the process of emission, and one analyzing data from measurements over time. The approach information lets a reader pull the right report from the file.

A report title would be a boring title for a magazine article, but it does the job of informing readers what the report is about. Reports provide information that readers need, and readers have an inherent motivation to read the reports. Cleverness may be desirable for titles of magazine articles because it attracts readers who have choices about reading, but it may trivialize a report.

TABLE OF CONTENTS

A **table of contents** (often called simply "contents") lists each main division of the report and the page on which it begins. Its primary function is

table of contents
list of the main divisions of a report and the page on which each starts; appears at the beginning of a document

to let your readers find specific information easily. Its secondary function is to give your readers an overview of the content and structure of your report. Look ahead to the contents for Example 14–5, an example we examine in-depth in the next chapter.

The main divisions of long reports may be called chapters, but the label "chapter" would be pretentious for a section of a few pages. Instead, the contents page will probably list the level-one headings (and frequently level-two headings). The page number indicates where the section begins.

The first item in the contents will probably be the abstract or executive summary. Normally, you won't list preliminary pages unless you have a preface, an unlikely part of a report unless the report is very long. However, the contents page will list the back matter, including the appendixes, references, and glossary. Don't mix a list of tables and figures with the contents page. The visuals are not sections of the report, so they are out of place in a list of report sections.

LIST OF TABLES AND FIGURES

You may list tables and figures (visual displays of data) on a page following the contents page. This list is most useful when there are many visuals and your report is long. (The NISO standards require this list when there are five or more figures and tables and make it optional for fewer.) There's less need for a tables and figures page in a 10-page report because readers can easily skim to locate the visuals. If you do include the list, format it the same way you formatted the contents page. Include the label and number (for example, Figure 1), the title of the visual, and the page on which it appears.

ABSTRACTS AND EXECUTIVE SUMMARIES

executive summaries
abstracts at the beginning of reports that give a condensed version of the entire document

A report often contains either or both a descriptive or informative abstract. An **executive summary,** a type of informative abstract, is common in managerial reports. The executive summary follows the conventions of the informative abstract in giving a condensed version of the whole report, including its conclusions and recommendations. An executive summary appears at the beginning of the report and stands alone. Readers begin the body of the report knowing the outcome, the main arguments, and the key facts. Readers can read the details in the report with an understanding of their overall significance.

TEXT OF THE REPORT

All reports have a beginning that introduces the issue, a middle that presents the results of investigation, and an ending that applies the results to the problem to show whether and how the problem is solved. The introduction describes the problem or purpose that has occasioned the report. It answers the questions of who, what, when, where, why, and how in order to establish the context for the problem and its significance. It also forecasts the rest of the report by indicating how the report will develop and what its major sections are. A literature review may be part of the introduction as a way of establishing the problem.

The ending of the report may include one or more parts: summary, conclusions, recommendations. Each of these parts has a different function and needs to be labeled accurately.

A **summary** reviews the main points of the report. The report may end with a summary if the purpose of the report has been to describe an existing situation or the current state of knowledge. The summary thus wraps up the report and reinforces the main points. The summary also can prepare the readers for your conclusions and recommendations. A review of the main points can be especially useful in a long report that asks readers to recall a lot of information.

summary
a review of the main points of a report

Conclusions answer the research question. Conclusions go a step beyond summary because answering the question requires evaluation and interpretation, including the reconciling of contradictory results. In a complex project, there may be multiple conclusions to draw and thus the section name is plural.

conclusions
report section answering the research question, assessing the overall work, and going beyond the summary in providing evaluation, interpretation, and reconciling of contradictory results

Recommendations direct action. They advise what should be done based on the conclusions. Managerial reports usually contain recommendations. For an example, look ahead at the recommendations section of the report in Example 14-5. The recommendations appear in a separate, labeled section and are listed. The recommendations do not need to provide much explanation because that has already appeared in the conclusions.

recommendations
advice designed to direct action

BACK MATTER

The **back matter** following the report's conclusions and recommendations provides additional information intended to supplement the main body of the report. Page numbering continues with arabic numerals.

back matter
additional pieces of information that are useful but not essential to a report's argument

Appendixes include material that is useful for reference but not essential to the argument of the report. A copy of a survey form, detailed tables, or the text of relevant laws would appear in an appendix so that it would not interrupt reading the report. Material that the primary reader of the report does not need but which might interest a secondary reader also could be placed in an appendix. Readers are much less likely to read material in an appendix than material in the body of the report, so do not place material there that your primary readers need. Refer to each appendix at the point in the text at which consulting the appendix would provide the reader with useful but not essential information.

appendixes
material that is useful for reference but nonessential to the argument of a report

If there is more than one appendix, each appendix should have a label and title, much like the label and title of a visual. The label identifies the appendix by a capital letter or numeral (Appendix A, Appendix B); the title identifies the content. The sequence of appendixes corresponds to the sequence in which the appendixes are mentioned in the text of the report. Each appendix should begin on a separate page and be listed on the table of contents.

references
sources used in a report and listed in the *Works Cited* section

As you learned in Chapter 6, lists of **references** or "Works Cited" include those works that have been cited in the text. These items are listed and formatted according to the documentation style that has been adopted for the report: MLA, APA, Chicago, or some other style.

glossary
a dictionary of terms that apply to the topic of a report

A **glossary** is a mini-dictionary of terms that apply to the topic of the report. Include in it only those terms that will be unfamiliar to your readers. If you believe that most of your readers will need the glossary to read the report effectively, you might move the glossary to the front of your report.

SUMMARY

- A report can be only as good as the investigation that is reported. Thus, one requirement for writing a good report is to do good research. That means being thorough in your investigation and analysis, being accurate in your observations and notes, collaborating effectively and efficiently with those whom your work touches, keeping your work on schedule, pursuing the goals that have been established for you or your project, and performing at a high level of quality, competence, and integrity.

    Good work, however, does not automatically mean that you will write a good report. A good report orients a reader to the project and anticipates and answers the reader's questions. It considers not only the primary reader but also readers to whom the report may be forwarded and future readers. The data in the report must be credible and complete, and the interpretation must explain their significance. Furthermore, a good report is well organized and formatted to help your readers find information if they read selectively and to understand the relationship of ideas or topics. A style that emphasizes accomplishment and projects confidence will enhance the report's persuasiveness and increase the chances that the report will influence decision making and actions in the way you think will be best.

- Successfully bringing together good work and writing for an effective reporting outcome requires planning at the beginning of the project and management throughout it. A lot of the work will be mental: conceptualizing the report in use and the readers' needs for information, planning research to produce that information, selecting strategies of organization and format to enhance comprehension by readers and thus the persuasiveness of the report, and scheduling tasks so that you will have a chance to complete them without cutting corners. Managing these processes involves subdividing the reporting into smaller tasks and completing the tasks in stages. This advice can work in all reporting situations.

■ In this chapter we provided you with many good work suggestions for report writing, suggestions that you'll need to remember as you study the next two chapters. Writing Strategies 13-1 will prove useful in planning your reports. In particular, we provided you with information on progress reports and literature reviews. You will need to refer back often to the conventions of report writing, as you plan the style, format, and organization of your reports. Finally, we ended this chapter with a discussion of report parts; use this section as a checklist as you prepare the reports assigned throughout this textbook.

1. Some of the Web sites in Appendix A, Case Documents 4, on the Y2K problem are reports. With a small group of classmates, choose a site. First define its purpose—that is, what problem does the report itself aim to solve? Then brainstorm the reports—both managerial reports and research reports—that might have been written in conjunction with this case. Think of who would have written a report, to whom, and for what intended action. For example, what report might follow up the report you chose? Brainstorm as many possible reports as you can that might have been written in conjunction with this case. You can use Example 13-1 to help you name reports, but it's fine to define reports according to purpose rather than by a specific name. Present your findings in a memo to your instructor or in a brief oral report to the class.

2. How does the writer of the report in Example 13-2 convey an attitude of respect for readers? How does that attitude serve the writer's persuasive purpose? How does the writer accommodate different levels of financial sophistication among readers? Be prepared to share your answers during class discussion.

3. Find a report written where you work, in the government documents section of your library, or on the Web. (Many organizations, including environmental and financial organizations, publish reports on the Web.) In a memo to your instructor, conduct the following analysis of the report you selected:

   ■ Determine its likely readers and purpose.
   ■ Analyze the writer's strategies of argument. How is the report arranged (chronologically, topically, scientifically, or by another method)? How does the writer establish credibility? What are the features of format, organization, and style? What devices, such as a descriptive abstract and title page, does the writer use to orient readers to the report topic? What devices, such as headings and running headers, does the writer use to make the report easy to use?
   ■ Finally, evaluate the report. How effective are the writer's choices? What might the writer have done to have made the report easier to understand and use?

ACTIVITIES AND EXERCISES

4. Write a progress report, addressed to your writing teacher, on your progress on a writing assignment. Report specific tasks accomplished and tasks remaining, and include a completion schedule. Consider research and report preparation tasks.

5. Anticipating an assignment of a report for decision making (Chapter 14), conduct a literature review on a topic of interest for the report that you can later incorporate into that report.

6. Assume that you have been assigned to write a literature review on one of the following topics:

    a. Possible financial consequences of the Y2K problem (see Appendix A, Case Documents 4).
    b. Genetic engineering of plants, animals, or people.
    c. Recent liability cases involving products and accompanying documentation.

    Work in the library to find at least eight sources on the topic you select (the references at the end of the appropriate documents in Appendix A may give you a start). The purpose of your report is to inform your reader on what is known and to suggest gaps in the knowledge.

WORKS CITED   Hartley, Peter. "Writing for Industry: The Presentational Mode versus the Reflective Mode." *Technical Writing Teacher* 18, no. 2 (1991), pp. 162-69.

Killingsworth, M. Jimmie, and Michael K. Gilbertson. *Signs, Genres, and Communities in Technical Communication.* Amityville, NY: Baywood, 1992.

Miller, Carolyn. "Genre as Social Action." *Quarterly Journal of Speech* 70 (1984), pp. 151-67.

National Information Standards Organization. *Scientific and Technical Reports—Elements, Organization, and Design* (ANSI/NISO Z39.18-1995). Bethesda, MD: NISO Press, 1995.

Pinelli, Thomas E. et al. "Report Format Preferences of Technical Managers and Nonmanagers." *Technical Communication* 31, no. 2 (1984), pp. 4-8.

_____. "Report-Reading Patterns of Technical Managers and Nonmanagers." *Technical Communication* 31, no. 3 (1984), pp. 20-24.

Souther, James. "What to Report." *IEEE Transactions on Professional Communication* PC 28, no. 3 (1985), pp. 5-8.

CHAPTER 14

Reports for Decision Making

In and beyond the classroom, the approach to decision-making. . . shapes the quality of decisions made. Shortsighted decisions result in part from inadequate procedures of inquiry and communication. . . . Flaws in decision-making processes and communication have been dramatically revealed by studies of the *Challenger* disaster. . . . Short-term and measurable criteria, especially cost, often influence decisions more than long-term impacts on health, safety, the environment, or on political, economic, and social systems, though these social criteria may ultimately determine the effectiveness of the decisions.

Carolyn Rude, "The Report for Decision Making: Genre and Inquiry," *Journal of Business and Technical Communication* 9, no. 2 (April 1995), pp. 170–205.

Introduction

All organizations must make decisions. These decisions result from problems, such as those we discussed in Chapter 2, or from opportunities. Sometimes the decisions are simple: An office manager discovers that paper for the laser printer is running short and decides to order a three-month supply. Sometimes the decisions are far more complex: An apartment manager who wants to resurface the tennis courts needs to consider material, cost, durability, safety, soil conditions, maintenance, and reputation of contractors.

People base some decisions on hunches or intuition, experience, preferences, and the desire to please others. For example, if your experience in a chemistry class has resulted in good grades, you are more likely to take an additional chemistry class than a student who has failed the course. These methods are legitimate in some situations, but when a decision is complex, these methods may not provide good enough reasons for choice. The decision may require information available only through research.

The person with the authority to make a particular decision may not have the time or expertise to investigate. Thus, the decision maker assigns the investigation to a staff member or a group of staff members. In a collaborative investigation, a lawyer may investigate legal aspects, an accountant may investigate financial aspects, and an industrial engineer may investigate production methods and rates. When the investigation is complete, the staff member or group presents the results of the investigation with a recommendation for action in a report to the decision maker. The researcher-writer probably reports to a superior in the organization.

Even as a new employee, you may have to write or collaborate in writing reports for decision making. Your manager will assign you to investigate and will depend on your recommendations in reaching his or her own decisions. (At this time, you might want to review the transmitter, decision maker, advisor, and implementer roles in the section "What Are Your Audience's Needs?" in Chapter 3.)

In this chapter you will learn a procedure for investigation when the expected outcome is a recommendation for action, and you will also learn more about how to arrange your findings in a report. We begin the chapter with a discussion of the kinds of problems that require decisions. We continue with methods of investigation and ways of interpreting findings, and offer suggestions for arranging the report and for handling difficult situations in decision making.

Making Decisions About Action: A Procedure

The procedure for investigating a problem and recommending an action differs somewhat from other research you may have done, including library research, lab experiments, and statistical analysis, although these methods may provide some of the necessary information for your report. The investigation procedure must match the type of problem that needs to be solved.

In Chapter 2 we recommended that you compare the *actual situation* against the *ideal situation* to find a solution (the *purpose* of your communication). We also suggested expressing your goal in a statement, adding the connecting word *but* or *however,* describing a condition that stands in the way of your goal, and then describing some action that would resolve the difficulty (Example 2–2). Finally, in Chapter 2 we contrasted procedural problems (problems of "know-how"), informational problems (problems of "know that"), and decision-making problems (problems of "what to do"). In this chapter we focus again on decision-making problems, problems that are solved by taking action. Because these decisions involve future action, there is no sure way to know that one decision is the best. But a good method of inquiry can increase the chances that you will think through the possible outcomes and make the decision based on good reasons.

The procedure for the investigation that leads to a recommendation includes these steps:

1. Defining the problem.
2. Forecasting possible solutions.
3. Identifying the research question.
4. Establishing selection criteria.
5. Establishing research methods.
6. Researching and interpreting information.

DEFINING PROBLEMS

As you learned in Chapter 2, decisions about action become necessary when there is a conflict, discrepancy, or inconsistency between an actual or existing situation and the ideal situation (often called a **goal**). A company may wish to increase sales on a product. This discrepancy between what is (current sales) and what is desired (e.g., a 10 percent increase in sales) represents a problem if the company does not know how to increase sales. Some impediment, such as a lack of marketing information, may prevent the achievement of the ideal situation. The solution to such problems is action; the possible actions include doing nothing.

goal
another name for the ideal situation

Problems don't necessarily indicate trouble: The discrepancy between the actual situation and the ideal one can represent opportunity. A company doing well in its present operations may see the possibility for expansion. The discrepancy in that situation can be seen in two ways: between the present rate of return on investments and desired rate or between the recognition of an untapped market niche and the uncertainty about the feasibility of expansion into that niche. Uncertainty about the course of action requires an inquiry.

Problems that require decisions are socially defined; that is, they do not exist in nature. Conflicts or discrepancies exist when someone identifies ways in which the present situation seems unsatisfactory. No biological or physical necessity requires an apartment to provide tennis courts. The decision results from setting priorities about how to spend available resources and how to attract renters. The laws that direct a state to meet emission standards in registered vehicles are created by a group of legislators that society elects. Because these problems are socially defined, good solutions are also socially defined. People with authority must agree that the decision is the best one of a number of possible choices.

Problems are unique in some respects to the individuals or groups that have them. An apartment manager probably cannot choose the tennis court resurfacing material based on what a friend chose in another town because of variations in soil conditions and contractors. Thus, the problems have meaning within a particular context.

Unlike mathematical problems, problems that require decisions do not have single, correct solutions that can be reached by calculation. Decision makers cannot turn to absolute rules for solving problems, nor can they be sure there is a right answer out there waiting to be discovered. The decision that is right in one context may not be right in another. Complex problems usually offer a choice in solution. (John Hayes and Michael Carter explain these types of problems in more detail.)

The social, unique, and ill-defined nature of problems that require decisions indicates the importance of defining the problems carefully and using a systematic procedure for solving them. The first step is defining the problem. Problems can be defined in an "*A* but *B*" structure with *A* representing the ideal situation or goal, *B* representing the actual situation or the impediment to achieving the goal, and "but" signaling the conflict. Sample problems are summarized in Example 14-1.

Complex decisions require a study and a recommendation report because multiple factors influence whether the decision is good, and the decision affects the company as a whole, not just one individual. Before the decision can be made, a solution must be forecast, and a study of the potential solution must be completed.

EXAMPLE 14-1   Problem Statements, Brief Form

Goal or Ideal Situation	Signal Word	Actual or Existing Situation or the Unknown
1. Laws require the vehicle fleet to achieve low emission goals.	*but*	**1.** Our gasoline engines exceed emission standards.
2. The tennis courts are deteriorating and unsafe.	*but*	**2.** We don't know whether asphalt or concrete is safer and more durable.

FORECASTING SOLUTIONS

Once a problem is defined, possible solutions must be forecast to give the investigator some direction in research. As John Hayes observes, "If people can't think up any approaches, then they can't solve the problem" (xiv). There may be more than one possible course of action in addition to the present course.

Brainstorming, as we discussed in Chapter 2, yields *possible* solutions. The apartment manager who considers both asphalt and concrete will not overlook a good option. Usually the simpler and less disruptive action is preferred over a complex action, assuming that the simpler action will achieve the desired goals.

Decisions regarding purchases often present multiple choices of brands, models, and features. In such a case, the forecasting task is to identify some possible choices. This task may require a needs assessment. A **needs assessment** is a detailed analysis of the uses for the purchase. The problem statement may identify a need for new office computers, but a needs assessment will go further to determine whether the individual will use the computer only as a stand-alone machine or also in a network. These needs will restrict the number of good choices. Thus, the decision about purchase will be better if the needs are assessed before choices are limited. We will return to the task of creating a needs assessment later in this chapter.

Forecasting also requires some selection. Some of the possible solutions that brainstorming reveals can be ruled out as impractical. The investigator chooses the option or options with the greatest promise for further study. To use a personal example, if you are investigating the problem of lack of transportation to a job, your brainstorming may yield these possible solutions: using public transportation (bus, subway), carpooling, and purchasing a car. If your job requires you to work later than public transportation is available and no one works the same hours that you work, you do not need to study the public transportation and carpooling options further. The investigation then will test whether and how the purchase of a car will solve the problem. Forecasting has been important, however, because it has let you consider less drastic options before you conclude that you will investigate only the option of purchase.

IDENTIFYING THE RESEARCH QUESTION

Forecasting identifies possible solutions, but it does not provide the information to determine which decision is best. That is the task of research. The step that follows forecasting is identifying a research question. The **research question** focuses the research by stating its objective.

Three basic types of questions focus the research for decision making:

1. *Is the project feasible?* This question points to a study of one solution that forecasting identified. The study will ask whether the individual

---

brainstorming
early stage in the writing process when one tries to create an extensive but unedited list of ideas or solutions

needs assessment
information addressing the contingencies that affect how a problem might be solved

forecasting
stage in report writing when one selects some possible solutions from a list of all possible solutions

research question
a step following forecasting in creating reports for decision making; focuses research by stating its objective

or company can and should follow a particular course of action. The *can* part of the question relates to measurable aspects of choice: money, equipment, technical know-how. The *should* part of the question relates to values.

2. *Which is better, option A or option B?* This question points to a comparison of options and is the usual question in the case of a purchase. If you know, for example, that the purchase of a car is feasible, you might compare two or more brands to determine which better meets your needs.

3. *Why does this situation occur, and what can be done about it?* This question points to an analytical study, a more detailed examination of the problem. An automobile manufacturer may ask, for example, why the labor costs are 7 percent higher at plant 1 than at plant 2 when both plants are building the same model car. When this research question focuses the study, the forecasting of the decision-making procedure will take place after the question is phrased. The researcher will brainstorm some hypothetical answers to the question (perhaps in the categories of management, training, and facilities) to forecast the possible solution.

The problem statement and forecast will indicate how the question should be phrased. The same problem might lead to different research questions and different types of study, depending on the need for information. If forecasting reveals multiple solutions with equal potential, the question will be the second one. But if one of the choices seems more attractive at the outset, a study to determine the feasibility of that choice makes sense. If the answer turns out to be no, you can explore another of the choices.

At the beginning of a study, phrase a question rather than make a statement or proposition. As an investigator you do not simply choose a solution and defend it in the report to the decision maker. Such a procedure would deny the value of information gathering and reasoning in problem solving. Instead it would base the decision on the hunch or preference and use research to justify the decision, not to explore options. If the best course of action really is unclear, your investigation should be open to various solutions and to the possibility that an unexpected or surprising solution may be the best one. The question encourages that kind of thinking.

IDENTIFYING SELECTION CRITERIA

criteria
standards for making a good decision in a particular instance; when buying a car, selection criteria might include safety, gas mileage, and dependability

Before the research begins, you should identify the **criteria** for a good decision in the particular circumstance that requires the research and report. Criteria define the information to gather in research and provide the standard for assessing the information. The apartment manager investigating tennis courts may consider cost, safety, maintenance, durability, soil and weather conditions, and perhaps marketing issues. Different people will have different criteria and may rank them differently. For example, maintenance is a different issue for companies with an on-site maintenance staff than it is for companies that ship their products to another town for service.

Because most people want a measure of certainty in decision making, they may think first of criteria that can be measured. Or they may assess only whether an action is possible. For example, if you want a new car but do not have the money for it and cannot get a loan or gift, the purchase is impossible no matter how well it would solve transportation problems. If you would like to open a plastics recycling plant but cannot hire a person with knowledge of the chemical processes used in plastics recycling, the project is not possible. If you want to design a car that uses composite material in the bumper, but the bumper built with this material will not meet government crash test specifications, the material is not a feasible choice and the action is impossible.

Even if a project is technically possible, however, it may not be a good idea if it does not meet other criteria related to project management and the values of the people who will be affected by the project. If a project would jeopardize human safety beyond reasonable limits, it is not feasible even if it is technically possible. If the city council considers bidding for a state prison in the city to bring in jobs but the citizens are adamantly opposed to a prison, the prison may not be feasible even though there is enough land and labor to build and staff the prison. **Feasibility** means that a project is both possible and desirable.

feasibility
a project that is both
possible and desirable

Example 14–2 lists possible criteria for decision making in three categories:

- ■ Technical criteria determine whether something can be done.
- ■ Managerial criteria relate to day-to-day operations.
- ■ Social criteria relate to values and the impact of a decision on people.

The categories for the criteria are not fixed. Social criteria could become technical when regulations govern them. In some situations, for example, choices regarding the environment are matters of values, but in other cases government regulations establish limits on activities that could harm the environment. Gender, race, and age could be technical issues if compliance with the law is a question. And the impact of a new project on existing ones and on the people involved is both a social and a managerial concern.

Social criteria may be hard to measure, but they affect the success of a decision. Failure to consider them invites future challenges if the action that results from the decision alienates people. These criteria are often the measure of whether a decision is ethical.

EXAMPLE 14–2 Criteria for Decision Making

Technical Criteria	Managerial Criteria	Social Criteria
Technology (availability of)	Costs: materials,	Human safety:
Physical space, structural	equipment, renovation,	manufacturers,
requirements	maintenance, financing,	operators, bystanders
Dimensions, capacity	transportation, disposal,	Gender, race, age
Materials: strength,	salaries, licenses	Quality: product, service
corrosion, weight,	Income: fees, grants,	Environmental impact: long
availability	interest	term, short term;
Parts: availability,	Market: demand, need,	manufacture, use,
accessibility	interest, competition	disposal, habitat
Material handling	Taxes	Human impact: long term,
requirements	Consistency with	short term; jobs, morale,
Compatibility with existing	organizational goals	employment benefits
systems	Organizational impact:	Access: who benefits, who
Ability to be upgraded	personnel, morale, image,	is excluded
Adaptability, flexibility	distribution of resources,	Social and cultural issues
Ease of manufacture, use	effect on other projects,	(e.g., availability of
Maintenance, service,	changes required in	schools, the arts)
technical support	existing operations	Ethical issues (e.g., conflict
(availability and quality)	Staffing: number,	of interest)
Reliability, longevity, repair	qualifications	Values: personal,
record, warranty	Training	organization or
Legal restrictions,	Policies (e.g., affirmative	community, society
standards, codes,	action)	Convenience, comfort
precedents (e.g.,	Time required: building,	Aesthetics
environment, Americans	training, performing	
with Disabilities Act)	procedures	
	Schedule	
	Quality of written	
	instructions	

Criteria for a specific decision may differ from those listed in Example 14–2. Example 14–2 will help you to anticipate the consequences of a decision in a comprehensive way and therefore to identify and select appropriate criteria for your own decision. The example also suggests that a good decision may require you to evaluate issues that go beyond the letter of the law and technical specifications.

Criteria for decision making need to be identified early in the process of investigation so that you know what to research. They may change as the research develops and you learn more about what matters in this project, but they give you a plan for starting the research. A good first step in identifying criteria is to consult with the person who assigned the research task to you and anyone else in the organization who knows the problem. You can interview these people to determine the history of the problem, such as solutions that have been tried previously, specific needs, and operating procedures that may affect the outcome of a proposed solution to the problem. This history will reveal issues that may become criteria for determining the solution. You can also use Example 14-2 to help you brainstorm relevant criteria for this project. The criteria are the measure of whether the decision is a good one.

The criteria you select reflect your ability to anticipate and assess possible future consequences of actions. A decision about a future action cannot be proven right or wrong before the action happens, but thoughtful anticipation of the future provides information that can be used in planning and in preventing mistakes.

Criteria also need to be ranked to establish which are the most important or which need to be researched first. Identifying and ranking criteria and imagining the future were important tasks, for example, when the library of a medical school outgrew its space in its location on the second floor. Some space on the fifth floor had been planned for a fitness facility but had never been developed. The library administrator wondered if the problem of limited space could be solved by moving the library from the second floor to the fifth floor.

These are some of the criteria in the library case:

- *Space*—Will the fifth floor provide sufficient room for the library, now and in the foreseeable future?
- *Access*—Will users be able to get to the fifth floor easily and conveniently? Will existing elevators and stairs accommodate the number of users who would visit the library each day? Are the elevators accessible for users in wheelchairs? Are the elevators convenient to the building entrance?
- *Support facilities*—Will available restrooms be sufficient for the anticipated number of users? Can a snack bar be installed in a convenient location?
- *Building structure*—Was the fifth floor built to withstand the weight of the books and equipment in a library?
- *Impact of displacing the fitness facility*—How important to employee health, morale, and even recruitment is the planned fitness facility? Is another location suitable for it?
- *Costs*—What are the costs of renovation and moving?

To identify criteria, an investigator must anticipate the completion of the move. The investigator visualizes the fifth-floor facility as a library to discover what information a good decision requires. Visualizing the future also helps the investigator and managers anticipate the consequences before they approve the move and start packing books.

The criteria are not equal in importance, and the order in which they emerge in brainstorming does not necessarily reflect their importance.

Ranking them determines the order of the investigation and also helps you interpret results. In the medical school library example, the two most important criteria are space and building structure. Questions of cost and access are irrelevant if the books will not fit in the space or if the floor would collapse under their weight. These criteria should be researched first.

If the space and structure are adequate, research can proceed on the other questions. If space and structure are inadequate, the research may cease. The research question has been answered: The move is not feasible. The research would proceed only if some arguments for this particular solution are so compelling that they warrant an exploration of modifications to the building to remedy the deficits. Perhaps, for example, the structure could be reinforced or the space expanded. In that case, the direction of research would change to include the new options (though the criteria for decision making would not change).

In the library case as in many decisions, cost is a criterion. It is not, however, necessarily the first criterion to research. For one thing, the cost analysis cannot be completed until structural and renovation requirements are determined. There's no point in researching costs until the primary criteria have been researched.

Sometimes studies require two or more phases. Through the establishment and ranking of criteria, you plan a preliminary study to address the most important criteria. The report recommends either to continue the study or to stop it. The division into phases may spare the time and cost of unnecessary research. Brainstorm all the possible criteria, however, even if you don't address them. Doing so will help you grasp the entire situation and to consider several criteria within the context of the whole problem.

ESTABLISHING RESEARCH METHODS

Criteria establish *what* needs to be researched; the methods indicate *how*. Many projects require several methods. Sometimes the method is a scientific test or experiment, as in a question about materials—for example, whether a storage container would corrode with particular contents. Sometimes the method is library research (as discussed in Chapter 6), although this method is likely to provide only some of the necessary information. Often the methods include surveys, interviews, and site inspections (as discussed in Chapter 7). Your task is to pull together information from various sources.

One way to determine methods is to list the criteria you come up with and then to match each with appropriate methodologies. Example 14-3 shows the planning for the medical school library example we discussed above.

Although you will probably organize your report by criteria or alternatives, you should organize the research itself by methods. The architect,

EXAMPLE 14–3   Matching Criteria and Methods

Criterion	Method(s)
Space	1. Interview the librarian to determine linear feet of stacks needed through the next 10 years and the square feet of space needed for offices, study area, and equipment. 2. Measure the fifth floor area.
Structure	1. Interview the architect who designed the building. Request an estimate of book weight and a structural analysis in light of that weight.
Access	1. Interview the librarian to determine estimated number of users. 2. In the main library, consult standards for handicap access. 3. Review the building floorplan and walk through the path a user might take from parking lot to fifth floor to judge how a user might perceive ease of access.
Displacement effects	1. Interview a personnel officer to determine the importance of the fitness facility on employee recruitment and retention. 2. Conduct a survey of employees.
Restrooms, snack bar	1. Site inspection: count the number of restrooms and stalls; measure for handicap access and grab rails; evaluate the convenience of the location with respect to the library entrance; note possible areas for expanding the restrooms or snack areas (such as adjacent storage areas that could be relocated). 2. Consult the architect or blueprints about existing plumbing and wiring and ease of expanding restrooms or creating a snack area.
Costs	1. Consult the architect for estimates on renovations. 2. Consult the building manager for estimates on the cost of moving.

for example, is a source for three of the criteria. Rather than plan three separate interviews, you can plan one interview to address three separate topics. However, if the study is a preliminary study that addresses only space and structure, you will not need to gather information on expanding the restrooms.

As you proceed with your research, your methods may change. A person you interview may suggest that you consult a source that you had not considered, or you may discover information that changes your perception of the problem and solution. If you learn from the architect, for example, that the present building structure is inadequate for the library but structural reinforcements could remedy the deficit, you begin to inquire about the nature and cost of the reinforcements. The process of investigation is flexible because it is a learning process. The planning simply gives you a place to begin.

INTERPRETING RESULTS

Decision makers look for good reasons for making a choice. Reasons, to be credible, include both facts and interpretations of the significance of those facts. Claims need the support of facts to be credible. Readers will not be convinced that the building structure will support the fifth-floor library unless the investigator cites the architect's information about the number of load-bearing walls, the weight they can support, and the weight of the books. Likewise, facts alone are often ambiguous. If the investigator cites only the figures on load-bearing walls and weights, the readers may respond by saying "so what?" The readers want a plain English response to the question that has been posed: Will the building structure support the weight of the books? A statement that answers the question interprets the facts.

Paradoxically, an abundance of facts increases the need for interpretive statements. Readers may not be able to determine the significance of all the numbers from an elaborate table, for example; the facts do not speak for themselves because there are so many. As investigator, you are closer to the facts and more expert in their meaning; consequently, you have an obligation to report to readers your judgment about the significance. Doing so makes your reasons for your conclusions explicit, and it also gives readers a basis for additional interpretations. Writing Strategies 14–1 will help you write effective reports for decision making.

WRITING STRATEGIES 14–1   Reports for Decision Making

Problem Definition
What is the problem that requires a decision about action? Express the problem with an A but B structure (Example 14–1). Include details that are available (who, what, where, when, why, how). Do not ask the research question yet.

Readers
Who has the authority to make the decision? (Perhaps there will be more than one person at different levels of authority.) What do readers already know about the problem? What are their attitudes to it? Will they be likely to resist or welcome change? What personal values may influence the decision they make? Will some future readers depend on the report to establish the history of a situation?

Reader	Knowledge	Attitudes/Values
a.		
b.		
c.		

(continued)

WRITING STRATEGIES 14–1 (Continued)

Research Question

Pose your question as a version of one of these: Is the project feasible? Which is better, option A or option B? Why does the situation occur?

Selection Criteria

What technical, managerial, and social criteria will determine that a decision is good?

Criterion	Significance
a.	
b.	
c.	

Additional Useful Information

Will readers benefit from a description of a mechanism or process? needs assessment? history and background?

Research Methods

What research methods will give you the data you need for each criterion?

Criterion	Method(s)
a.	
b.	
c.	

Resources

What people, books, or existing reports may be helpful?
What research instruments (survey forms, questionnaires) will you need to develop?

Visual Display of Data

What visual displays may help in the presentation of data?
What type of data do you have? What type of visual display do you need?

Schedule of Tasks

Task	Completion Date
research	
a.	
b.	
c.	
writing, editing	
a.	
b.	
c.	

Planning and Organizing the Report for Decision Making

The overall structure of the report makes it coherent, but the report consists of a number of small segments. The arrangement of these segments will influence your reader's response, but recognizing the segments will also help you manage your writing time. You can write these segments as you complete them, not necessarily in the order in which they appear in the report. You may, for example, write the problem statement early. You might also complete the section on one of the selection criteria before you even investigate the second.

OVERALL STRUCTURE

The report's structure of introduction, body, and conclusions emphasizes the reasoning process, from problem definition to analysis to conclusions. As we discussed in Chapter 13, reports in the workplace are often organized with the conclusions at the beginning and the details following. This order emphasizes the bottom line: the results. It also acknowledges the reading practices of managers, who are likely to read selectively.

Example 14-4 presents two report outlines, the conventional and the managerial. The items in parentheses are optional altogether or optional for that section of the report. For example, you may explain your methods of research in the introduction, or you may explain them where they apply in the body of the report. The problem statement itself may provide sufficient background information, eliminating the need for a separate background section.

Some expectations for organizing reports will probably exist at the place where you work, and you should find out what they are. If managers expect the **conventional structure,** you should provide it. If they expect the **managerial structure,** you organize accordingly. You can ask to see sample reports, or you can ask your supervisor about the preferred structure. You might also choose the managerial structure when readers are likely to agree with your recommendations. On the other hand, the conventional structure might be persuasive when your conclusions and recommendations differ from readers' expectations and preferences. That structure would emphasize the reasoning that led you to your conclusions. Refer to the sample report in Example 14-5 at the end of this chapter as you read next how decision making affects the typical structure of a report.

conventional structure standard format for organizing a report that includes an executive summary, introduction, selection criteria, conclusion, recommendations, and appendixes in that order

managerial structure a standard format for organizing a report that includes an executive summary, introduction, conclusion, recommendations, appendix, and selection criteria in that order

The Introduction

As you learned in Chapter 13, the purpose of the introduction is to orient readers to the report. They need an introduction to the topic (the problem the report addresses) and to the report itself (its purpose, scope, and order). The problem statement indicates the significance as well as the topic of investigation. The problem statement establishes the conflict that makes a decision necessary.

The sample problem statements in Example 14-1 are summaries or kernels of the problem statement. The report should provide more details

EXAMPLE 14–4   Outlines for Two Report Structures

Conventional Structure	Managerial Structure*
Executive Summary	Executive Summary
Introduction	Introduction
Problem statement	Problem statement
(Background)	(Background)
(Methods)	(Methods)
Plan of development	Plan of development
(Description of a mechanism)	Conclusions
(Needs assessment)	Recommendations
Criterion One	Appendixes:
Definition, significance	(Description)
(Methods)	(Needs assessment)
Findings	Criterion One
Interpretation of findings	Definition, significance
Criterion Two	(Methods)
Definition, significance	Findings
(Methods)	Interpretation of findings
Findings	Criterion Two
Interpretation of findings	Definition, significance
Conclusion	(Methods)
Recommendations	Findings
Appendixes	Interpretation of findings
(Glossary)	(Glossary)
(Supplementary tables)	(Supplementary tables)
(Survey instruments)	(Survey instruments)

*James W. Souther recommended this structure as a result of his research on how managers read reports; see "What to Report" and "Identifying the Informational Needs of Readers: A Management Responsibility."

about the problem. It should answer relevant questions, including who, what, where, when, why, and how.

Even if the problem statement repeats what readers already know, you should still present the details of the problem. First, the statement orients readers. Readers may have other projects and problems on their minds. The problem statement efficiently directs their attention to the specific problem that the report addresses. They can scan the problem statement if they know the problem in detail, but they cannot provide missing details if they do not know them.

Second, the problem statement convinces readers that you know the problem and its significance, and it establishes your ethos. If readers are to be convinced by your recommendations, they must first be convinced that you have been competent in researching and analyzing the problem. You make a good first impression by establishing the key details of the problem clearly and succinctly.

Third, the report will have a future. Readers in the future may consult it for a follow-up study. Those future readers will not be as well informed about the problem as present readers. For future readers, the report records the history of the problem.

Background Information

Although the problem statement will generally provide sufficient background, in some cases, additional information will help readers interpret the information in the report. Your analysis of readers will guide you in choosing whether to include **background information.** Include only what will help readers understand the information in the report and your interpretation of it. Readers will not appreciate a report that is longer than it has to be, but they will appreciate your ability to anticipate and answer their questions. Ask yourself what readers need to know. Background information can include

- History.
- Description of a process or an object that is relevant to the problem or solution.
- Definitions.
- Needs assessment.

background information
additional data to help readers interpret a report, including history, description of a mechanism or process, definitions, and needs assessments

When any of these sections becomes longer than a page or so, the topic should be a separate section (with an appropriate heading) of the body of the report rather than part of the introduction. If only the secondary readers need the information, detailed background information can be placed in an appendix.

Information on the origins of the problem, previous attempts to solve it, and the results of those attempts will place the present information in its context or **history.** If this is the second phase of a multiphase study, you should refer to the report from the first phase.

history
information on the origins of a problem, previous attempts to solve it, and the results of those attempts

When the report recommends a change in equipment, procedure, or location, you can orient readers to what is unfamiliar by a **description** (Chapter 11 will help you with this part). For example, an apartment manager may not understand how soil type affects the cracking of materials on a tennis court. A nontechnical description of the processes of soil expansion and contraction in combination with a description of properties of the materials used to build the court can help the reader understand the analysis that follows.

description
information about the principles of operation or the working of a process

If unfamiliar but essential terms will be used throughout the report, the introduction is an appropriate place to define them, although you cannot expect readers to memorize a list of unfamiliar terms. Alternatively, **definitions** may be placed in the section in which they are used (Chapter 11 will again help you with this part). The terms will have more meaning in the context in which they are used. When the primary readers are familiar with

definitions
explanations of terms unfamiliar to an audience

the terms, definitions may be placed in a glossary at the beginning or at the end of the report.

The problem statement demonstrates that something in your department or company should be changed—perpetuated, modified, or eliminated. Because the needs assessment addresses all contingencies that might affect how you solve this problem, this section sometimes becomes too long. Thus, you may want to create a separate section for it after the introduction to your report. The needs assessment addresses issues such as the following:

- We need to solve the problem by December 31, 1999.
- We need to stay within a budget of $10,000.
- We need to keep at least 80 percent of our projects on task while we solve the problem.
- We need a capacity of 18 gigabytes to store the database.

Methods

At some point in the report, you will need to tell how you got your information. Information is credible only if its source is credible. As with definitions, you may explain your methods either in the introduction or within the body of the report where they apply. You would favor explaining methods in the introduction if you used only one or two methods that applied to all the topics of investigation. You would favor explaining methods in the body of the report if the methods were unique to each topic. If methods would be repetitive and distracting to readers in the body of the report, you would place them in the introduction. If explaining methods in the introduction would require readers to evaluate methods apart from results, you would place them in the body. The criterion for choice of placement for this information is usefulness to readers.

Plan of Development

Readers appreciate an orientation to the report as well as to its subject. The forecasting statement reveals the plan of the report by naming the topics of the report and the order in which they are discussed. (Notice the plan of development in the last paragraph of Example 14-5.) If you are completing one part of a multiphase study, you should also include a scope statement—topics that this report will cover and reasons for omitting others.

The Discussion Section (Body)

The discussion section includes the results of research and interpretation of significance. The most substantial part of the body (or perhaps its entirety) will consist of a discussion of each of the criteria. If some of the options listed previously under "background" are lengthy, they will also be part of the body.

The body of a report for decision making will probably be organized according to the criteria. This structure places the reasons for the decision in the forefront. An exception might be a decision based on a scientific test. If your method is the experiment, your sections would consist of methods, results of the test, and discussion of the results.

Discussion of each criterion is like a minireport, with introduction, results, and conclusion. The discussion may include an explanation of methods for gathering the information. The introduction to the section defines the criterion and explains its significance (what the criterion means and why it is important in this context). It also includes the specific findings and interpretive statements. Interpretive statements are important because readers will remember them when they get to the conclusion of the report. Readers cannot be expected to hold numerous facts in their memories, especially if they are not sure of the meaning of the facts. In the library report, the discussion of the criterion of space might conclude with a statement such as this:

> The library requires 18,000 square feet of space. The available area on the fifth floor includes 21,050 square feet. This space would accommodate the library and allow some flexibility in arrangement of stacks, reading area, and offices. On the criterion of space, the fifth floor is a feasible location for the library.

Although the statement concludes that the space is adequate on the fifth floor, it does not claim that a move should be planned. The other criteria still need to be discussed.

You have some choice about arranging the criteria. The fundamental guideline is to use arrangement as a tool to persuade and inform readers. Usually this means placing the most important criteria first. This placement helps prioritize the issues and will shape the readers' interpretation of the information. Also, readers will lose interest if you begin with relatively unimportant topics and work up to the most important topics—the order in a suspense novel but not the proper order for a report. Because your readers are reading for information to make decisions, they will be impatient with delays.

Within the overall structure of most important to least important, try to group related criteria. Grouping aids interpretation by showing relationships. You don't want to make readers leap mentally from technical to social issues and back.

At times you may intentionally violate these guidelines because of persuasive goals or because of long-term goals. For example, one feasibility study investigated whether it was feasible to replace gasoline-powered vans with battery-powered ones to meet the standards of the Clean Air Act; the investigator found that cost precluded the change. However, on the other criteria—range, power, capacity—he learned that battery-powered vans were a good idea. Anticipating that costs would drop, he discussed the determining criterion last as a way of showing decision makers that the idea should not be ruled out for the future. If he had put cost first, they would not have read about the other features.

Conclusions and Recommendations

You answer the research question in the *conclusions.* You should be able to claim that the project is (or is not) feasible, option *A* is better than option *B,* or the situation happens because of *X,* which points to the solution of *Y.* To draw these conclusions, you interpret the results from each of the body sections in relationship to one another. The interpretations will explain why you have answered the question as you have.

The results of your analysis of each criterion may conflict with one another. If you have three criteria, two of which point to one conclusion while a third points to another, you will have to decide whether the conclusion based on the third outweighs the conclusion based on the first two. Your ranking of criteria at the outset of your investigation will help. The criteria are of different weights, so you cannot simply decide on the basis of the majority. In the library case, for example, if the space and access are adequate but the structure is not, you have to conclude with a negative answer despite the fact that the conclusion is negative on only one of the three criteria. Be sure to base your conclusions on the stated criteria and not to introduce a new criterion.

Recommendations differ from conclusions. Instead of explaining reasons for answering the research question in a certain way, recommendations direct the particular action that follows from the conclusions. A conclusion statement in the library case would be this: "Even though the fifth-floor space will accommodate the library, moving to the fifth floor is not feasible because the building was not designed to support the weight of books on the fifth floor." Recommendations might include:

recommendations section of a report that directs the particular action, unlike conclusions which answer a research question

1. Investigate the feasibility of reinforcing the building structure to support the weight, considering cost, completion time, and effect on other areas.
2. Investigate the feasibility of relocating the library to the first-floor neurology wing and moving the neurology wing to the space presently occupied by the library. Neurology occupies a space of the size required by the library, does not use all of its current space, and could be moved to a smaller location.

Note that the recommendations are phrased as directions and listed. This phrasing is not necessary, but it is appropriate because the purpose of the recommendations is to tell the decision maker what to do next. Note also that the recommendations do not include explanations. The explanations have already been covered in the conclusions.

Managing the Report Development Process

Good reports require effective investigation and writing. The third requirement for a good report is good management. Time has to be managed so that all the tasks can be completed well and on time. Management is also important because many people dread report writing. They would rather do the hands-on work and leave the paperwork to someone else. The intimidation of writing increases when the report will be long. Intimidation invites procrastination, and one cure for procrastination is management.

One way to make any project easier is to conceive of it as a series of small tasks. These tasks include research tasks, such as conducting interviews and completing lab procedures, and report developments, such as preparing visual displays, writing the executive summary, and editing. You can apply the same divisions to the actual writing. If you write small sections as you conduct your research or other project, you will soon be able to assemble a great deal of the whole from the small parts.

It may be easy to write the introduction right away because it usually consists of a problem statement that explains why investigation took place at all. In other words, the project that results in a report begins with a problem—some inefficiency or lack of information. You know the problem at the outset of your work. Writing it down early not only gives you a head start on the final report but also helps you clarify your thinking about the nature of the problem and the work it requires.

The informal notes you make as your project progresses can become the data for the final report. If the notes are good, it is easy to write the report from them. If you are conducting a lab experiment, you will certainly keep notes on your procedures and results, but even with other types of reports, you can jot down observations and details as you work. Maybe you will find it helpful to keep a notebook just for these notes.

Some sections of reports are consistent from report to report. For example, each progress report during the course of a project will include a description of the project problem and goals. When certain sections of reports become standard, they are called **boilerplate** material. You can store these sections as computer files and then paste or copy them into the new report. This process saves not only writing time but also review time. If the sections have already been approved as accurate by the organization, you don't want to risk introducing inaccuracies by rewriting unnecessarily. If someone else authored your boilerplate materials, acknowledge them according to your company policy.

boilerplate
standard sections of reports that can be stored as computer files and reused

Encountering Difficulties in Making Recommendations

So far this chapter has presented an idealized procedure for investigating possible solutions to problems and for writing them in a report. However, the workplace is never quite as tidy as textbook procedures. You will sometimes feel uncomfortable with your task and with the recommendations that you believe should be made. The following sections describe some of those situations and suggest ways to increase your ability to deal with difficult situations.

MAINTAINING AN ATTITUDE OF OPENNESS IN THE REPORT

The early part of this chapter stressed the importance of beginning the investigation with a question rather than a proposition. The reason for the question is to keep the options for choice open and not close off a good option with a premature decision or not feel compelled to defend an inappropriate option. Even though the report will end with a recommendation and is a persuasive document, its purpose is to report an investigation. Instead of establishing a thesis and defending it, the report writer asks a question and explores the answer.

At the time of writing the report, however, you will have reached a decision, and you may feel strongly about it. You have an obviously persuasive purpose as well as an informative one. And, the executive summary will reveal your conclusions and recommendations up front. The readers—the decision makers—will know the ending before they know the details and the reasoning. Yet they may benefit from approaching the issue with open minds just as you do at the beginning of an investigation. How can you accomplish your persuasive aims while still giving the readers the freedom to make up their own minds?

Sometimes the most persuasive writing is indirect. It persuades because it seems not to, and readers therefore trust it. One way to be indirectly persuasive is to delay specific conclusions until you have presented some evidence. Even though the report conclusions will be available to readers in the executive summary and, in the managerial report structure before the body of the report, you can maintain a descriptive and inquisitive attitude in the introduction. (This is another reason to write the introduction before you complete the research.) You can save your evaluative statements for the body and conclusion, letting these follow the presentation of evidence in each of these sections. It is persuasive to show readers the progress of your reasoning from questioning to evidence to conclusion.

Even though it will be your responsibility to draw conclusions, you should write with respect for the readers' right to draw their own conclusions from your evidence and reasoning.

RECOMMENDING A COURSE OF ACTION TO A SUPERIOR

When you are a new employee, you will probably feel especially uncomfortable advising a superior in the organization. Nevertheless, recommendations typically move from lower levels to upper levels in the organizational hierarchy. The subordinate officer recommends; the superior officer directs. Your manager assigns the research task and requests a recommendation, assuming that you will be conscientious in your investigation and that you have the right education and experience to interpret what you find.

When you are uncomfortable advising a superior, remember that you are advising, not directing. The decision maker has the ultimate authority to implement the recommendations and has the privilege of rejecting your advice. You are not usurping your superior's privileges, but you do have a responsibility to provide the facts and the reasons to show why you recommend one action over an alternative.

SAYING NO

It's easier to recommend yes than no. You will probably anticipate what your manager wants to hear, and you will prefer to provide it. Even posing a question of feasibility implies a motivation to pursue that course of action, and you will feel most natural in agreeing. But what if the investigation points to the opposite conclusion?

The purpose of studying the options rather than simply plunging into a course of action is to make a good decision and to avoid one that you will later regret. There's nothing wrong with saying no if the reasons for doing so are good; for example, recommending against an action with a high risk of failure will ultimately save the organization time, money, and stress. Even if the decision maker is initially disappointed, the investigation has served its purpose well: to guide the decision.

If you recommend action that conflicts with the expected recommendation, your facts and reasons have to be especially trustworthy and convincing.

DECIDING UNDER PRESSURE

Some emergency situations require quick decisions. Time limits the amount of research that can be completed. Pressures of office politics may also limit the options. You may feel that a superior has already committed to a decision and merely wants your investigation to confirm what is already decided. If your investigation points to a conflicting recommendation, you will feel pulled between the course of action that the study supports and the will of your superior.

The events preceding the explosion of the space shuttle *Challenger* in 1986 reveal the difficulties of making decisions under pressure. Both time and politics pulled against the caution of engineers who questioned whether the O-rings in the solid rocket boosters would seal at the low temperature expected during the launch. Previous experience had raised some questions about the effect of low temperatures on the sealing capacity of the O-rings. But the facts were somewhat ambiguous because no shuttle had been launched at the 36° Fahrenheit temperature that was expected for the launch day. The engineers had serious doubts, but the managers felt that proof of likely failure was insufficient. The shuttle was launched in the cold weather, the O-rings did not seal, and the explosion killed the seven people aboard.

Dorothy Winsor observes that the recommendation of the engineers in the *Challenger* disaster was insufficient to convince the managers for two reasons: the news was bad; and several different organizations were involved (the manufacturer of the boosters, Marshall Space Center, and NASA), increasing the difficulty of communication ("Communication Failures"). Bad news, multiple organizations, a shortage of time for making the decision, and criticism from the press regarding delays all interfered with reasonable decision making. In addition, the engineers were asked by the managers to prove a negative hypothesis—that the shuttle would fail. Asking the wrong question invited the wrong answer. The context in which a decision is made will influence the decision, sometimes for the worse.

To help balance the pressures of time and politics, you can rely on the ideas that social as well as technical criteria determine what makes a decision good and that judgment as well as calculation represents good thinking. The managers in the *Challenger* case wanted to depend on numbers to predict whether the O-rings would hold. But the numbers were misleading because the experience in similar weather conditions was limited and because the numbers regarding temperatures had not been sufficiently interpreted. Winsor also observes ("Construction of Knowledge") that the "facts" of the case seem much firmer to us now than they did at the time the decision was being made. The managers wanted to "know" with more certainty than was possible. Perhaps in the quest for facts, they neglected judgment. The judgment of the experts was more reliable than facts in that case.

SUMMARY

- Writing and reasoning are closely related activities. Good writing depends on good reasoning, but it can also be argued that good reasoning, at least for complex problems, requires good writing as well. The process of analyzing a problem, researching solutions, and making decisions is guided by the need to prepare the report that will present and interpret facts and make recommendations. Writing about the problem and arranging the results of your research will help clarify your thinking. Thus, writing well will make you a more valuable employee and more effective in shaping responsible decisions.

- When preparing your reports, always define your problem, recognize your readers, pose your research question, identify your selection criteria, and use research methods that match the question. Your technical, managerial, and social criteria will be ranked or weighted depending on your situation and research question. This planning will be reflected in how you organize your report. Finally, because researching and writing a report takes a great deal of time, you must manage your project closely so you can be confident that your conclusion and recommendations will effectively and safely guide the decisions and actions of your readers.

EXAMPLE 14–5   Sample Report

The title provides a lot of information—what, who, where, how.

Evaluation of Foundation Materials on Expansive Soil for English Aire Apartments' Tennis Court*

Identifying information is especially important for possible future readers. Some title pages include addresses and phone numbers to enable the reader to contact the writer.

Prepared for

> Dr. Hillary Hart
> University of Texas

By

> Russ Bynum and Eddie Eskola

March 15, 1997

The descriptive abstract expands on the title and suggests the scope of the report, but it does not provide specific findings.

This report compares concrete and asphalt as materials for reconstruction of the tennis court at English Aire apartments according to the criteria of cost, safety, durability, and maintenance.

(continued)

*This report won an Award of Merit in the 1997–98 Technical Writing Competition for College Students sponsored by the Austin Chapter of the Society for Technical Communication. It was written for a course taught by Professor Hillary Hart.. Reprinted with permission of the authors.

EXAMPLE 14–5 (Continued)

Contents

Preliminary pages are numbered with Roman numerals. The text of the report begins on page 1.

No number appears on the title page, but that page is counted in the numbering. Thus, the contents page is page ii.

The Contents page helps readers search for particular information, but it also gives a conceptual overview of the report.

The appendix is not included in this textbook because of space limitations, but the Contents page should identify materials that are appended to the main report.

(continued)

EXAMPLE 14–5 (Continued)

The executive summary provides key details from all the sections, including the problem statement and conclusions. It includes specific findings, such as cost.

The reader will know the essential findings and conclusions after reading the summary. The complete report provides the details.

Executive Summary

The tennis courts at English Aire apartments are deteriorating and need to be reconstructed. The Urban land and Ferris soil beneath the courts has a high potential for expansion and creates a heave force on the court surface.

The initial cost of asphalt is $46,300, but because of low tensile strength and low resistance to graveling, asphalt is more likely than concrete to crack and create an unsafe surface. It requires weekly maintenance to remove debris and repainting every 3 years. Its life expectancy is about 18 years; its annual cost is $2,572.

Post-tensioned concrete costs $70,500 to install, but its superior tensile strength makes it less likely to crack on swelling soil. Maintenance includes bi-weekly cleaning and repainting every 5 years. Its life expectancy is 30 years; the annual cost is $2,350.

In spite of higher initial cost, concrete provides superior safety, durability, and maintenance features and a lower annual cost.

iii

(continued)

EXAMPLE 14–5 (Continued)

Introduction and Problem Statement

The tennis court at the English Aire Apartment complex is deteriorating rapidly. The surface has many severe structural cracks, and the northern edge has foundation erosion problems (see Appendix figures A1 through A3). Ruddy's Consulting Group (RCG) has concluded that the current court problems are directly related to the properties of the soil and the soil environment.

The structural cracks indicate that major changes have taken place in the soil beneath the tennis court. A soil survey map (Figure 1) indicates that the court is built on Urban land and Ferris soil (UvE). UvE is 35% Ferris soil, a clay with a high potential for expansion. Ferris clay expands when water infiltrates the existing clay structure and causes the dry compacted clay particles to move in relation to one another. The movement of the particles to their new equilibrium positions increases the overall volume of the clay.

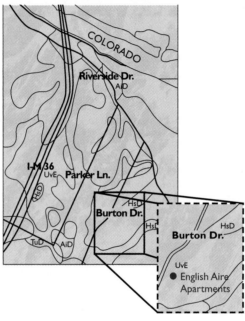

Figure 1: Soil Survey Map

When the clay below the tennis court at English Aire gets wet and expands, the volume change creates a heave force on the tennis court surface. This force exceeds the shear strength of the asphalt surface and creates structural cracks.

The environment of the soil beneath the tennis court at English Aire also contributes to the cracks and to the foundation erosion along the northen edge. The soil survey map confirmed our results from an initial site inspection concerning the slope of the tennis court. The court was constructed on an excavated portion of a hill with a slope of 10% to 15%. The north side of the tennis court has soil shifting and erosion problems beneath the court's edge due to its relative location

1

The problem statement links the effect (cracks) to a cause (soil) and thereby establishes a direction of the investigation.

Illustrations may be more common in the body of a report than in the introduction, but here the illustrations clarify and confirm the nature of the problem.

The detail in this paragraph and the previous one elaborates on the causes of the problem. The explanation also establishes the competence of the writers.

(continued)

EXAMPLE 14–5 (Continued)

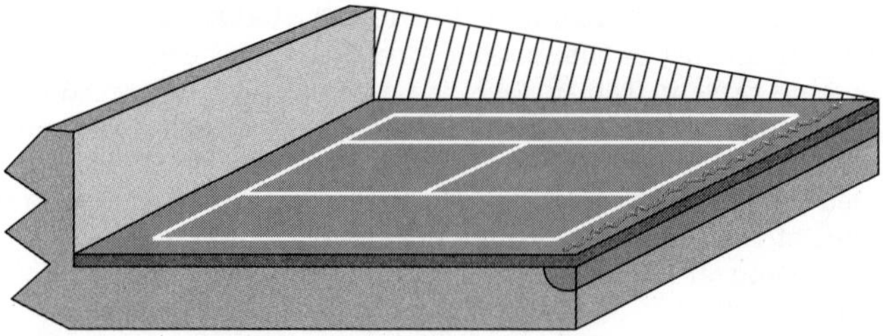

▨	Urban Land and Ferris Soil	▨	Concrete retaining wall
▨	Portion of soil with erosion problems	▨	Asphalt court foundation
▨	Indicates excavated soil when extruded along length of court	▨	Court surface
		▨	Crack due to soil erosion

Figure 2: 3-D Cross-Section View of Current Court

on the hill. The soil at the northern edge has tendencies to erode and slide because the soil was not compacted as well here. When the soil slides it creates a void spot below the court's northern side (Figure 2). These problems have caused long parallel cracks along the entire northern edge and have destroyed the structural integrity of this side.

Purpose, Scope, and Methodology

The purpose of this project was to recommend the best foundation reconstruction alternative for English Aire Apartments' tennis court while considering the soil shifting problems. The foundation alternatives considered for reconstruction of the court are asphalt and concrete.

The alternative solutions were evaluated according to a set criteria provided by English Aire Apartments management: overall project cost, safety, material properties, and maintenance.

In order to evaluate asphalt and concrete as suitable foundation material, RCG acquired information using textbook materials, site research, phone interviews with sport court companies, Internet searches, and library research. RCG determined the problem of expansive soils and foundation erosion through site inspection and soil survey maps. Cost estimates were obtained from Dobb's Sport Court and Sport Court of Austin. Please note that the cost figures given are to be considered estimates and should not be considered actual prices. Safety was determined through phone interviews with Dobb's Sport Court, Austin Sport Court, and by direct connection with the material properties. According to Vick Sanchez of Austin Sport Court, "A safe court is one with no cracking and graveling problems." The material properties of asphalt and concrete were collected from textbooks, library research, and Internet searches. Maintenance requirements were acquired from Internet searches and interviews.

2

This explicit statement of purpose and approach tells readers what to expect and provides a framework for comprehension. Note that the four criteria are listed in the order in which they are discussed in the report.

This section identifies methods. They are appropriately described in the introduction because there are multiple methods that apply to various criteria.

(continued)

EXAMPLE 14–5 (Continued)

Organization of Report

This report is divided into five sections. The first section, Alternative Solutions, explains in detail the two options considered by RCG. The second section, Solution Criteria, explicates the criteria used to evaluate concrete and asphalt. The third section, Evaluating Foundation Alternatives, assesses each option according to the criteria. The fourth section, Conclusions, shows how concrete best meets the criteria. The report concludes with recommendations for implementing the conclusions.

Alternative Solutions

Ruddy's Consulting Group decided to evaluate asphalt and concrete because asphalt and concrete are the most commonly used foundation types for tennis courts. Other options were not considered based on English Aire Apartments management's desire for relatively low maintenance and overall cost.

Asphalt

Asphalt is a mixture of hydrocarbons that is used in various foundation applications. It is a durable material that is often used for tennis and sport courts. Proper construction using asphalt as the foundation can lead to a long lasting tennis court.

The basic construction of a properly designed asphalt court involves four components (see Figure 3 for a detailed cross section of an asphalt foundation):

1. A stable, compacted, and well-drained soil sub base,
2. A 6-to-8-inch compacted crushed stone base course,
3. A 3-to-4-inch-thick compacted hot asphalt pavement usually laid in two courses of a minimum thickness of 1½" each and sloped from side to side of the court (not diagonally or end to end) at a slope of 0.8% to 1.0%, and
4. A liquid applied acrylic playing surface system squeegeed on top of the cured asphalt pavement (usually between 3 and 10 layers of surfacing)

If constructed properly, an asphalt tennis court has an expected life of 18 to 20 years, requiring repainting every 3 years, depending upon seasonal usage.

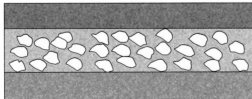

Figure 3: Typical Asphalt Cross Section

3

The report's structure reflects the investigation process: identification of a problem, consideration of alternative solutions, identification of criteria, and answering the research question.

This section clarifies that the writers have responded to the client's directions and have thought through the options rather than selecting one and trying to justify the choice.

The implicit research question is: Which is better, asphalt or concrete?

(continued)

EXAMPLE 14–5 (Continued)

Concrete

Concrete is used in construction when strength is a main concern. Post-tensioned concrete is commonly used for the reconstruction of tennis courts on expansive soils. Post-tensioned concrete is cured concrete that has been reinforced with steel rods (Figure 4). Also embedded in the cured concrete are tubes and channels where cables or rods are inserted and stressed. The concrete bonds to the stressed rods, which then places the concrete in compression generating built-in resistance to tensile forces caused by ground movement.

Post-tensioned concrete is used in the majority of new tennis court construction. It is also the reconstruction method of choice for badly heaved or cracked courts. This method involves the following:

1. A pad of 4-inch thick concrete with ⅛" sheathed tension cables are centered in the concrete and anchored in a perimeter beam.
2. The fence and net posts are integrated into the slab when the concrete is poured. While the concrete is curing the cables are tensioned in both directions to pull the concrete together.
3. The surfacing system and lines are applied to create the playing boundaries.

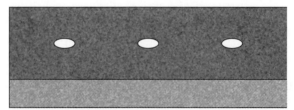

Figure 4: Typical Concrete Cross Section

Solution Criteria

English Aire Apartments management provided RCG with set criteria in order to evaluate and recommend a suitable foundation material for the tennis court. The criteria are listed in order of importance.

Overall Project Cost

The overall project cost of the court includes the reconstruction cost and year-round maintenance costs over the life span of the court. The cost per year ratio is the overall project cost divided by the suspected life span of the court.

Safety

Unsafe conditions can result in increased liability concerns and costly lawsuits. Safety is defined as the ability of the foundation

4

Defining the criteria in a separate section emphasizes the reasons that will lead to the conclusions. It also eliminates the need to repeat the definitions in the two main sections on asphalt and concrete.

(continued)

EXAMPLE 14–5 (Continued)

material to resist cracking and graveling. Graveling occurs when pieces of aggregate loosen from the surface. Increased numbers of cracks and loosened gravel could cause players to trip or slip and injure themselves.

Material Properties

Material properties include durability, resistance to cracking and graveling, and reactions to soil swelling and ground shifting. Durability of the material is determined using the court's life span. Resistance to cracking, graveling, and reactions of the materials under soil swelling and expansion conditions are determined by the strength of the material. These material properties will help gauge how well the foundation material will perform at the site's location.

Maintenance

The condition of the court will be determined by the original construction of the court and regular maintenance. Poorly constructed courts will need more maintenance over the life span, which increases the overall cost. Repainting and cleaning are the requirements for general maintenance. Maintenance is calculated on a per year basis. Other repairs are not included in the maintenance section due to the inability to forecast possible problems.

Evaluation of Foundation Alternatives

This section evaluates asphalt and concrete according to the criteria defined by English Aire Apartments management.

Asphalt

OVERALL PROJECT COST

The cost of reconstructing English Aire's tennis court can be broken into three parts: reconstruction, maintenance per year, and repainting. The cost of the reconstruction is one lump sum at the time of construction. Cleaning will be done weekly and repainting will occur every 3 years. Repairs made to the court vary in cost and are not included in the overall project cost.

According to Austin Sport Court, the reconstruction of the tennis court using asphalt costs $25,000. The maintenance cost is $350 per year for cleaning and $2,500 every 3 years for resurfacing. This brings the total cost to $46,300 over an 18-year life expectancy period (Table 1). The cost is $2,572 per year.

SAFETY

If the court is not constructed properly, the soil shifting could create cracks, bumps, and uneven surfaces and make the court unusable in as little as a few months. According to Avi Mor, asphalt can withstand heavy loads, like a car, due to its bending properties, but the forces caused by soil shifting tear the asphalt, or place it in tension, and crack it from the bottom to the top due to its low tensile

5

Because the research question foregrounded the two alternatives, readers are thinking in those terms, and the organization by alternatives will make sense. Some comparative analyses are organized according to criteria.

The description and table show how the writers calculated the costs.

(continued)

EXAMPLE 14–5 (Continued)

Table 1: Overall Project Cost Items for Asphalt

Description	Individual Costs	18-year Cost
Reconstruction	$25,000	$25,000
Maintenance Per Year	@ $350 × 18 years	$6,300
Repainting	@ $2,500 × 6 times	$15,000
Total Cost (18 years)		$46,300
Cost per year	$46,300/18	$2,572

strength. The tensile strength of asphalt cannot be compared in a quantitative analysis to that of concrete due to different testing methods, but asphalt has a low tensile strength in a qualitative analysis compared to that of concrete. Asphalt has a very high tendency to bend and become very brittle causing asphalt to have a low resistance to graveling. The surface of the court becomes shifted with these forces and the aggregates break apart and become loose on the surface. Overall, asphalt generates a low safety rating because of its high ability to crack and gravel.

MATERIAL PROPERTIES

The properties that are relevant to constructing a long lasting court on expansive soils are durability, resistance to cracking and graveling, and reactions to soil swelling. The life expectancy for an asphalt tennis court, according to Xsports and S and S Developers, can be as much as 15 to 18 years. The life expectancy depends solely on proper construction and maintenance of the court. Asphalt has a low tolerance to cracking and graveling due to its low tensile strength; therefore, asphalt will fail and crack when placed under the soil swelling forces present at the proposed site, as stated by all of our sources.

MAINTENANCE

Court maintenance can extend the life of the tennis court. Maintenance of an asphalt court necessitates regular cleaning and periodic repainting. The court must be blown and swept of debris regularly. If debris remains on the court it can act as an abrasive. When a player steps on the debris, the court surface may be scratched. RGC anticipates that an asphalt court will require cleaning and sweeping very frequently because of the surface's propensity to gravel. According to Xsports, weekly court cleaning and sweeping will be required in order to preserve the life of the court's surface. An asphalt tennis court must also be repainted. Xsports estimates that for a safe playable asphalt court to be maintained it must be repainted every three years. An asphalt court must be repainted more often than concrete due to the inevitable debris abrasion.

6

(continued)

EXAMPLE 14–5 (Continued)

Table 2: **Overall Project Cost Items for Concrete**

Description	Individual Costs	18-year Cost
Reconstruction	$50,250	$50,250
Maintenance per year	@ $175 × 30 years	$5,250
Repainting	@ $2,500 × 6 times	$15,000
Total cost (30 years)		$70,500
Cost per year	$70,500/30	$2,350

Concrete

OVERALL PROJECT COST

The cost of reconstructing the English Aire Apartments' tennis court with concrete can be broken into three parts: construction, cleaning, and repainting. The cost of the reconstruction will be one lump sum at the time of construction. Cleaning will be done once a year and repainting will occur every five years.

Dobb's Sport Court estimates the reconstruction of the tennis court using concrete will cost $70,500. The maintenance cost of a concrete court is $175 per year for cleaning and $2,500 every 5 years for resurfacing. This brings the total cost to $70,500 over a 30-year life expectancy period (Table 2). This gives a cost per year ratio of $2,350.

SAFETY

According to Prestressed Concrete Design, concrete normally has a high resistance to cracking due to its high compressive strength, 5,000 to 8,000 psi. the ground movement and poor soil conditions, however, will place tensile forces on the foundation. According to Avi Mor, concrete's tensile strength is one-tenth its compressive strength. Post-tensioned concrete increases its tensile strength to approximately 3,000 to 5,000 psi.

Post-tensioned concrete as the reconstruction method will reduce the chance of cracking under ground movement and soil swelling conditions according to all of our sources. In addition to cracking, Vick Sanchez states that graveling is not likely to occur with post-tensioned concrete. Overall, concrete has a superior safety rating compared to asphalt.

MATERIAL PROPERTIES

Constructing a long lasting court on expansive soils requires high durability, high resistence to cracking and graveling, and minimal reactions to soil swelling. A post-tensioned court on average lasts 30 years, according to S & S Developers and Xsports. Post-tensioned concrete also has a high resistance to cracking and graveling. According to Prestressed Concrete Design, concrete has a high resistance to cracking due to its high compressive strength. The post-tensioned

In addition to facts, the writers offer an explicit interpretation: concrete is safer than asphalt.

Note the similar pattern of development in each section: definition, data, interpretation.

7

(continued)

EXAMPLE 14–5 (Continued)

concrete is designed to resist the tensile forces present in conditions such as ground shifting, so cracking should not be a concern with this method. Overall, post-tensioned concrete performs well under ground shifting conditions, according to all of our sources.

MAINTENANCE

The maintenance required for a concrete court is similar to maintenance required for an asphalt surface, but the cleaning can be less frequent. According to Xsports, a concrete surface must be swept and cleaned once every two weeks to remove any dust, dirt, and eroded surface particles. Xsports also states that a concrete court must also be repainted every five years. Intervals between paintings are longer with concrete than with asphalt because a concrete surface is tougher and is less likely to have serious abrasion problems.

Conclusions

The conclusions address each of the criteria for each of the options.

Based on the valuation of foundation alternatives using the criteria defined by English Aire Apartments management, RCG has reached the following conclusions concerning the reconstruction of the English Aire Apartments' tennis court.

1. Constructing an asphalt tennis court had the lower overall project cost of the two alternatives at $46,300. It has a life expectancy of approximately 18 years with regular maintenance. Asphalt, however, had a cost per year ratio of $2,573. Cracking and graveling will occur more quickly on an asphalt court due to the soil shifting properties causing unsafe conditions. The asphalt foundation requires weekly maintenance and repainting every 3 years.

2. Constructing a post-tensioned concrete court had the higher overall project cost at $70,500. The post-tensioned court has a life expectancy of 30 years giving a cost per year ratio of $2,350. On the post-tensioned court, cracking is rare and graveling is virtually nonexistent under the soil swelling conditions. Without cracking and graveling, the court will be safe for the players. Also, a post-tensioned concrete court requires cleaning every two weeks and repainting every 5 years.

The research question (which is better?) is answered directly.

In spite of the higher initial cost, concrete will provide a more satisfactory long-term choice. Choosing to construct a post-tensioned tennis court, English Aire Apartments management will benefit for several reasons. The advantages of post-tensioned concrete courts over conventional asphalt courts are as follows:

- Lower cost per ratio of $2,350
- Resistance to cracking and graveling
- Increased safety
- Greater durability and longer life span
- Less maintenance which leads to lower maintenance costs

8

(continued)

EXAMPLE 14–5 (Continued)

Recommendations

In order to increase usage of the courts, attract new tenants, and decrease liability concerns, Ruddy's Consulting Group recommends that English Aire Apartments management reconstruct the present tennis court with post-tensioned concrete. In order to implement this recommendation, RCG suggests the following:

1. Remove current asphalt foundation and sublayers.
2. Compact and level the soil starting from the bottom up with a roller to create a firm base for the next layers of construction.
3. Create a frame or form the size of the outside diameter of the finished court.
4. Pour a pad of 4" thick concrete with ⅛" sheathed tensioned cables centered in the concrete and anchored in a perimeter beam.
5. Apply the surfacing system (a ⅛" of cold asphaltic emulsion, or cushion) and paint the lines after the concrete has cured.

References

Asphalt Institute. 1978. Soils Manual No. 10. College Park, MD: The Asphalt Institute.

Asphalt Institute. 1978. Asphalt Overlays and Pavement Rehabilitation. College Park, MD: The Asphalt Institute.

Derucher, Kenneth N., Korfaitis, George P., and Ezeldin, A. Samer. 1994. Materials for Civil and Highway Engineers. Englewood Cliffs, NJ: Prentice Hall.

Dobbs, Tyra. Dobb's Sport Court Salesperson. Personal Interview. Austin, TX. 30 October 1997.

Hurst, M. K. 1988. Prestressed Concrete Design. New York: Chapman & Hall.

Lester, Michael L., and Tucker, Richard L. 1972. Movement of Residential Foundations Founded on Expansive Clay Soils. Arlington, TX: Construction Research Center.

Mor, Avi. Doctor Mor and Associates, Head Consultant. e-mail. November 5, 1997.

Pollack, Herman W. 1988. Materials Science and Metallurgy. Englewood Cliffs, NJ: Prentice Hall.

Sanchez, Vick. Austin Sports Court Salesperson. Personal Interview. Austin, TX. October 26, 1997.

S&S Developers. 1996. "Types of cracks and what they mean" and "Court Maintenance." Computer-Mediated Communication Magazine. http://www.tenniscourts.com

United States Department of Agriculture. 1974. Soil Survey of Travis County, Texas. Washington, D.C.: U.S. Government Printing Office.

Xsports 1996. "Rehab: Asphalt Tennis Courts" and "Post Tensioned Concrete Court Construction." Computer-Mediated Communication Magazine. http://www.xsports.com

9

ACTIVITIES AND EXERCISES

1. Many of the stories you read in the local or campus newspaper have a report for decision making in their background. For example, a city government considers whether to build an all-purpose arena at a cost of $36 million. The newspaper reports that the issue has been raised, and it presents some of the reasons for building the arena as well as some of the drawbacks. A feasibility study must be completed before the city can formally propose the center to the citizens and ask contractors for bids. Scan your local and campus newspapers, and identify possible studies in the background of the stories.

    Bring a selection of your findings to class and with your teacher and classmates discuss what some of the reports would need to contain.

2. Summarize the problem in the following situations by creating *A* but *B* statements (see Example 14–1). Be prepared to discuss your answers in class.

    a. A student who desires part-time work to make money for living expenses considers whether to open her own home-based pet care service. Although she has a tight schedule already, she needs the extra money.

    b. The new owners of a feed yard need to organize as a business for legal and tax purposes. The options include a general partnership, S corporation, and C corporation.

    c. A student in mechanical engineering must choose a design project for a senior class. He considers a diver propulsion vehicle that is complex and expensive to assemble.

    d. The superheater at Acme, Inc., produces steam to heat the seven buildings at the Acme plant. High-temperature, high-pressure steam exits the refrigeration cycle's steam turbine at the plant and enters the desuperheater, where water is added to the steam to drop the temperature and pressure of the steam. The large drop in pressure and temperature produces energy, which is currently wasted.

    e. Livestock graze on 95 million acres of permanent rangeland in Texas. An estimated 88 million of these acres are infested with woody plants that suppress the grass on which the livestock graze.

3. In the following situations (or those you identified in activity 1), identify the research question and brainstorm the criteria for decision making. Work in pairs or small groups for brainstorming. Then join another group to compare the research questions and criteria you have identified.

    a. Moving from a dorm to an off-campus apartment.

    b. Preparing for a career in technical communication.

    c. Purchasing a laptop computer.

    d. Choosing between two brands of pickup truck.

    e. Investigating whether to change from styrofoam containers to paper wraps in a fast-food restaurant.

    f. Determining why enrollment has dropped in two high schools and the impact of the drop on staffing and construction in a city school system.

    g. Determining whether to pave a section of highway with concrete or asphalt.

    h. Determining why the teenage pregnancy rate in a city is higher than the national average in order to decide what to do to lower the rate.

4. Discuss in class why the lists of criteria might differ for two different people investigating the feasibility of moving from the dorm to an off-campus apartment. What might some points of difference be?

5. People base decisions on hunches, experience, personal preferences, self-interest, and the desire to please others as well as on the procedure(s) of investigation and reasoning that this chapter describes. For each of the claims stated below, determine the basis for the decision. In what circumstances would a decision based on the claim be a sound decision? When might the claim result in a flawed decision? If the claim might result in a flawed decision, what would be a better method of arriving at the decision? Discuss your findings in class.

   a. It worked last time; therefore, it will work again.
   b. The choice worked for her, and therefore it will work for me.
   c. You can't solve social problems by throwing money at them; therefore, we will not fund your project.
   d. I will become a lawyer because my father-in-law urges me to do so.
   e. I like rock music, so I will recommend that the Activities Committee schedule a rock concert for fall.

6. Discussion: What are the differences in meaning of these words: *evaluation, opinion, judgment, interpretation, perception, assumption, bias?* How, in particular, does judgment differ from opinion? Why must some technical communication include judgment and evaluation? Discuss your answers in class.

7. Assume that you are a paid staff member of the Minnesota AIDS Project (MAP). Review all the documents available through this and other organizations in Appendix A, Case Documents 1. Your supervisor has asked you to write a report reviewing the type of information on HIV and AIDS that is being sent to the public. You have collected samples, as represented in the case documents. Your supervisor would like you to evaluate the documents designed for the public and to recommend where MAP money and efforts should go next. Your supervisor is particularly interested in how effective the documents seem to be in helping the public make wise decisions about changing their lifestyles to include safer sex. In a brief report, complete the evaluation and offer your recommendations. In particular, you must help your supervisor decide how to make MAP money and volunteer efforts be more effective in Minnesota communities.

8. In a brief report to your supervisor, Dean Karen Wartala (see Appendix A, Case Documents 3), recommend what means of future communication should take place between the university and the public (particularly the Coalition of Citizens Concerned with Animal Rights). Some promises and demands were made in the letters between the university and CCCAR, but so far no one has set up a schedule to implement these promises and meet these demands. That's your task. You will need to be as concrete as possible in your recommendations.

   As you complete your report to Dean Wartala, keep a list of the criteria that affect your decisions and recommendations. Identify which are social, technical, and managerial criteria. If you see a solution that was not discussed in the series of letters, you can also make that a recommendation. In an attached memo to your instructor and classmates, explain your criteria and your ranking of those criteria.

WORKS CITED   Carter, Michael. "Problem Solving Reconsidered: A Pluralistic Theory of Problems." *College English* 50, no. 5 (September 1988), pp. 551-565.

Hayes, John R. *The Complete Problem Solver.* 2nd ed. Hillsdale, NJ: Lawrence Erlbaum, 1989.

Rude, Carolyn D. "The Report for Decision Making: Genre and Inquiry." *Journal of Business and Technical Communication* 9, no. 2 (April 1995), pp. 170-205.

Souther, James W. "Identifying the Informational Needs of Readers: A Management Responsibility." *IEEE Transactions on Professional Communication* PC-28.3 (September 1985), pp. 9-12.

_____. "What to Report." *IEEE Transactions on Professional Communication* PC-28.3 (September 1985), pp. 5-8.

Winsor, Dorothy. "Communication Failures Contributing to the *Challenger* Accident: An Example for Technical Communicators." *IEEE Transactions on Professional Communication* 31, no. 3 (1988), pp. 101-7.

_____. "The Construction of Knowledge in Organizations: Asking the Right Questions About the *Challenger*." *Journal of Business and Technical Communication* 4, no. 2 (1990), pp. 7-20.

Proposals

Proposals serve three purposes. First, they're really important because we get money to support our programs from them. Second, when we write them, we get clearer about what it is we're doing. The more we have to write to get someone to buy into our idea, the more we're forced to articulate clearly our goals, objectives, and outcomes. And finally, for every proposal we don't get funded, we still raise the funder's level of knowledge and understanding of the problem we are trying to solve.

Mary Hartmann, Executive Director, New Foundations, Inc.

Introduction

Proposals, like reports, originate with the need to solve problems. A proposal sets forth the solution or the way to meet a need or take advantage of an opportunity; it also requests support for implementing a proposed plan. Proposals within an organization establish a plan of action for solving a problem and claim the support of the organization (in personnel, time, money) to pursue the plan. Proposals may also seek support from other organizations: They link organizations or individuals who have problems to solve with the organizations or individuals who can solve the problems or provide financial assistance.

For example, all organizations using computer systems had to consider the Year 2000 (Y2K) problem, but the method of solving the problem varied. If an organization had a programming staff in-house, the technology specialist probably wrote a proposal to upper-level management to use staff and time for this purpose and to hire new staff if necessary. The proposal included a schedule of tasks with completion dates and a budget along with a statement of the problem and need. An in-house proposal defines priorities and establishes commitments so that the organization can get its work done well.

If the organization could not solve the Y2K problem in-house, it might have solicited proposals from companies that specialize in Y2K solutions. These companies would write a proposal to offer their plan for solving the organization's problem. The organization seeking help would select the company with the best offer as represented by the proposal.

A proposal is also the tool for seeking funds to support charitable or educational activities or research. The organization that can deliver services or conduct research lacks the funds to do so, and foundations have funds but not the staff and expertise to achieve their goals directly. The proposal links the two groups.

In this chapter we focus on how to write persuasive proposals, no matter whether the proposals are written in-house or whether they seek support from another organization. We also discuss sources of funding and the review and selection process in case you seek support outside your organization.

Proposals Solve Problems

Proposals are so pervasive in the workplace that it is useful to consider the different purposes that define them. Not all proposals fit into the same mold. In this section we discuss the management proposal (usually an internal document) and the research, project, or service delivery proposal (usually one that seeks support from an external source or offers a service to another organization).

MANAGEMENT PROPOSAL

Sometimes people look around them and think that a procedure could be more efficient, safer, or fairer if it were changed or if different equipment or facilities were available. Or there is a need for a change imposed from outside, such as the Y2K situation or changes in laws, but they do not have the authority to make the changes themselves. Instead, they must persuade someone with power to make the changes by means of a **management proposal.** The internal management proposal points to action within the organization that should be taken to solve the problem. The proposal writer says, "*We* should do these things to solve the problems I have noted."

management proposal
proposal designed to convince someone in authority to make a change in a procedure or policy

The proposal to commit resources to solving the Y2K problem is a reflection that a company has a problem that requires action. The proposal argues that management priorities should be adjusted to reflect this need. Likewise, probably all the classes you are taking were at one time proposed to a committee and university administrators. The faculty had to convince upper-level administrators that the course would be a useful addition to the curriculum and that faculty should be assigned to teach it.

Occasionally, management proposals appeal to organizations outside the organization. The person who observes a need or an opportunity for improvement offers a solution to be implemented by someone with power and authority. In this type of proposal, the writer says, "If *you* will do these things, then the problems I have pointed out will be solved." A student, for example, who rides her bike to campus is worried about car-bicycle accidents on campus. She proposes to the director of facilities construction that bicycle paths be built so that bikes and cars won't have to share the same street. She doesn't plan to build the bicycle paths herself; rather, she wants to convince the university to invest in the paths. The benefit to her, and to others, is a safer place to ride a bike.

To persuade the facilities director to act on her proposal, the student will research evidence of the problem beyond her own experience as a bicyclist and develop a plan of campus paths to show how to solve the problem. She might survey bicyclists and motorists to identify their gripes and fears about each other. She might use statistics about bicycle-car accidents from the traffic office. And she might cite some statistics from other universities that had similar problems, installed bicycle paths, and reduced the number of bicycle-car accidents. Finally, she would research the national standards set for width and materials of bicycle paths and plan some routes using a campus map. This information shows how the proposed paths could be implemented.

It is hard to convince people to change procedures that are well established or to purchase costly new equipment. Even when people agree that a change is needed, they may think that there is no way to accomplish the change. Details of *how* the change might occur are more persuasive than simply telling *what* might be changed. The plan for implementation makes the proposal more substantial and convincing than an idea in a suggestion box or even the recommendation following a study.

RESEARCH, PROJECT, AND SERVICE DELIVERY PROPOSALS

research, project, or service delivery proposal proposal that offers a service in exchange for support, usually including financial support

A **research, project,** or **service delivery proposal** offers a service in exchange for support, usually including financial support. The service could be research, design, installation, education, health care, or any other service that someone is willing to buy.

Your professor might ask you to propose a project for your feasibility study assignment or for your senior design assignment. This proposal is a kind of research proposal in which you seek approval for the project by demonstrating that you have identified a significant project and a workable method and that you have the qualifications for completing it. The proposal in Example 15-5 at the end of this chapter is a project proposal: David Silk and his teammates propose to build a prototype ISDN-compatible terminal adapter. The proposal seeks approval of the professor, but it also seeks the cooperation of Motorola. Both the professor and Motorola will be convinced by the careful plans for the design and the plan for building the adapter.

Other examples of this type of proposal would include one in which a Y2K-solution company offers to provide the service of reprogramming a company's computers in exchange for a fee to cover salaries and equipment or another in which a social service agency offers to provide affordable housing and support services to low-income families.

When developing a research, design, or service delivery proposal, you expect to participate in implementing the solution. In effect, you say to a potential supporter, "If you will give us money and time, we will solve the problem that concerns you by completing the tasks described in this document."

Grant proposals offer to provide research or service. A scientist offers to continue the research to find out which defective gene causes Alzheimer's disease in exchange for funding for salaries and equipment from the National Science Foundation (NSF). A consulting firm offers to find the best manager for a city to hire in response to the city's request for help with this task. A home health service appeals for support from a government fund created by legislation to provide respite care for caregivers of homebound individuals. Research and service delivery proposals are usually written to external organizations. Many people make their living by winning contracts to deliver a service or perform research for someone else.

The Request for Proposals (RFP)

When two organizations are involved in problem solving, the one writing the proposal needs to know of the supporting organization's interests and needs. This information is often communicated in a **Request for Proposals (RFP).** An RFP identifies a need for a service or research and invites proposals for completing the necessary work. Then the organization proposing to offer the service or research knows what to offer. Your instructor's assignment for a proposal is a kind of RFP.

Example 15-1 illustrates an RFP issued by a city seeking consultants to determine the feasibility of building an all-purpose arena. The RFP guides proposal planning and development by defining the problem and goals in specific terms. It gives you many clues about objectives, schedule, and budget as well as specifications for equipment and supplies. A consulting firm responding to the RFP in Example 15-1 can use the

request for proposals (RFP)
a document identifying a need for a service or research and inviting proposals for the necessary work

---

EXAMPLE 15–1   Request for Proposals

REQUEST FOR PROPOSALS FOR PROFESSIONAL SERVICES
FOR THE LUBBOCK CHAMBER OF COMMERCE, LUBBOCK, TEXAS

I.   General:
The Lubbock Chamber of Commerce is seeking professional services for a consultant, architect, or engineer experienced in determining the feasibility of an all-purpose event facility for the Lubbock community.

II.   Scope of Services
The consultant shall determine what size of an all-purpose event facility the Lubbock community can support and what unique features must be included to make the facility feasible in the Lubbock market. The information provided by the consultant shall include but not be limited to the following:

1.   *Market Analysis*
   a.   Review historical event statistics of the Civic Center and Auditorium/ Coliseum for the past three years (FY 1995–96, 1996–97, 1997–98).
   b.   Define market area (primary and secondary geographic area).
   c.   Project usage by type of event (i.e., entertainment, sports, conventions), quantity of events, attendance, and timing.
   d.   Demographics by population trends and spending profiles.
   e.   Review competitive facilities by location and projected attendance to determine market impact.

2.   *Projected Size*
   a.   Determine what size facility can be supported by the Lubbock community (number of seats in arena, total square footage).
   b.   Determine unique features (e.g., arena, exhibit area, dining, meeting rooms, skating arena) that need to be included to make facility economically feasible.
   c.   Determine number of parking spaces required for projected facility.
   d.   Determine number of acres of land needed for projected facility.

3.   *Financial Feasibility*
   a.   Prepare estimates of projected revenue by income source (e.g., rent, concessions) for first 10 years of operation of facility.
   b.   Prepare estimate of projected operating and maintenance expenses for first 10 years of operation of facility.
   c.   Project total cost of construction to include architectural fees, engineering fees, construction costs, furnishings, and so on, excluding the cost of land acquisition.

(continued)

EXAMPLE 15–1 (Continued)

    d. Provide estimated unit cost per square footage, per arena seat, and so on.

    e. Determine economic impact on community (additional revenue generated in hotel/motel tax, sales tax, estimated dollars spent by increased attendance).

III. **Experience and References**

Professional consultants will be selected on the basis of a combination of proven experience, ability to complete the project in a timely manner, and recommendations from recent clients with similar projects. Submit the following information:

1. A listing of similar projects completed or in progress, including a list of references by name, address, and telephone number for each project listed.

2. A listing of the individuals, with professional qualifications, who will be assigned to this project.

IV. **Schedule and Methodology**

The consultant shall be capable of performing the services in the shortest time possible without sacrificing the quality.

1. Include a projected timetable for the work to be performed in the scope of services.

2. Describe the methodology to be used to complete the project.

V. **Estimated Cost for Services and Basis of Payment**

The Chamber of Commerce desires to obtain the services in Section II at its actual cost not to exceed a maximum fee.

1. List cost for services to be provided to include professional services and expenses.

2. Provide anticipated payment schedule.

VI. **Contract Negotiations**

Upon selection and approval of a consultant, negotiations will be scheduled to determine final project fees and contract terms.

VII. **Submission of Proposal**

To be considered, 10 copies of the proposal should be submitted to:

    All-Purpose Facility Planning Committee

    c/o Chamber of Commerce

    P.O. Box 561

    Lubbock, Texas 79408

    Phone 806 763-4666

Submissions must be received no later than 2 P.M. Friday, January 8, 1999.

Source: Reprinted with the permission of the Lubbock Chamber of Commerce.

specific tasks defined in it to establish project objectives. The RFP also specifies that the city wants to see work and payment schedules. However, this RFP does not tell everything you might want to know; it is vague about money, for example. If the RFP specified an amount, chances are that all proposals would request that amount or a few dollars less,

EXAMPLE 15–2 Announcement in the *Commerce Business Daily* (Wednesday, August 18, 1993) of a Contracting Opportunity

---

F Natural Resources and Conservation Services

Contracting Officer 90C,
V.A. Medical Center/Outpatient Clinic,
150 South Huntington Avenue,
Boston, MA 02130

F — SERVICE FOR REMOVAL AND DECONTAMINATION OF REGULATED MEDICAL WASTE SOL RFP 523–38–93 DUE 091793 POC Contracting Officer, Edward Camey, (617) 232–9500x5536 Contractor shall provide all tools, equipment and labor to provide for the removal and decontamination of regulated medical waste in accordance with specifications at the VA Medical Center/Outpatient Clinics, 150 South Huntington Avenue, Boston, MA 02130. The contract period is from October 1, 1993 through September 30, 1994 with three option years. Contract award is subject to the availability of funds. Solicitation packages will be available on or about August 18, 1993. Requests for solicitation must be in writing. No phone or faxed requests will be accepted. Proposals are due by the close of business on September 17, 1993. SIC 4959 applies to this solicitation which is a 100% small business set-aside. (0228)

Source: *Commence Business Daily*, August 18, 1993.

eliminating the competition that the proposal process aims for. But it is very specific about the outcomes it would like. The RFP serves both parties in the contracting process: The city is more likely to get what it wants if it defines what it wants; the proposing organization does not have to guess at the expectations nor waste time developing proposals for projects that it is not prepared to do.

Example 15-1 illustrates a specific, one-time need, but organizations that regularly distribute money or contract for services may publish their needs and interests in handbooks, on Web sites, or in periodicals. Example 15-2 illustrates a call from the *Commerce Business Daily*. More details on these publications are provided later in the chapter in the section, "Sources of Project Support."

The Competitive Nature of Proposals

There is always competition for the support that you seek as a proposal writer. The researcher who wants to study genes and Alzheimer's disease competes with a dozen or more researchers who want to study the same or a similar thing. The National Science Foundation has limited funds and can fund only the projects that have the best chance of achieving the objectives and that identify the most significant problems. The proposal for bicycle paths competes with other campus projects to provide safe and convenient means of transportation. Perhaps the proposal for new parking lots will argue more effectively. Even when an employee proposes to work on the Y2K problem, the proposed use of her time competes with other uses.

All writing is persuasive. In some genres, however, you try to delay committing to a point of view until research is complete. The effort to remain open to a range of possible results is especially important for genres that report research, such as the scientific article or the report for decision making. In a proposal situation, you commit to a project from the outset. Your overt purpose is to persuade.

Being persuasive means that you have to establish the need for the management change, service, or research; show that what you propose will serve the goals of the sponsor or organization, not just your own; and demonstrate your qualifications.

plan of implementation
a central and defining
component of a proposal
that often suggests change

A proposal persuades in part by spelling out how the change or project will be accomplished, with details of who will do what and when. The **plan of implementation** is the central part of the proposal. This information distinguishes a proposal from a simple suggestion for a change or even from a recommendation at the end of a report. It's hard to convince people to support your project if they don't understand exactly what you will do. You have to propose a sound and feasible plan for solving a problem. You also have to anticipate and answer reviewers' questions satisfactorily. And you have to convince the potential project supporters that you can provide the service or research better than anyone else or, in the case of management proposals, that the change is worth the investment of time, retraining, or new equipment.

Competitive proposals reflect awareness of what the potential sponsors are seeking combined with high quality project planning. Persuasion also requires attention to arrangement, style, and presentation of the proposal.

Parts and Structure of Proposals

Your proposals may be complete in a one- or two-page letter, or they may require several volumes of text and supplements. Proposals to external organizations are likely to be formal and detailed because you and the proposal reviewer do not share the same knowledge of the problem or even of each other. An internal proposal may be formal and detailed when the change proposed will be substantial or dramatic.

We discuss some standard parts of proposals in this section (Example 15–5 at the end of this chapter shows some of the standard parts). Your proposals may not contain all of these parts, or the parts may carry different names. Before you write a proposal, you should find out the expectations for the proposals: parts required, labels for parts, and arrangement of parts. Proposal guidelines or the RFP from the sponsoring organization will list the requirements. One of your first strategies of persuasion is to offer the proposal reviewers the information they seek in the way that they expect to see it. You are offering to provide a service for them, so you should demonstrate that you are in tune with their wishes by conforming to their requirements for documents. Be careful even with details of wording. If guidelines specify a part called "Statement of the Problem," don't label your corresponding part "Problem Statement" or "Needs Assessment"; if it asks for an "Overview," don't provide a "Summary."

Each part of your proposal presents significant information that answers questions the reviewers will have. Anticipating and answering the questions satisfactorily is a key to persuasion. Example 15–3 names standard proposal parts and the questions they answer.

The proposal not only wins support for a project but also serves as a planning and management tool for the project. Details of the plan of implementation, which include dates, tasks, and specific outcomes, provide direction for carrying out the project. Objectives, schedule, and budget also define standards for evaluating progress. When you write a proposal, you make a commitment to carry out the tasks you propose in the time and for the budget you indicate. Your follow-up progress reports measure what has happened according to what was planned.

COVER LETTER (OR LETTER OF TRANSMITTAL)

A **cover letter** for your proposal for outside funds may be written by the executive officer of your organization to establish that the organization supports the project. The cover letter for a management proposal might be written by an immediate supervisor. This support assures the funding source that if the grant, contract, or support is awarded, the completion of the project according to the stated plan will be a high priority of the organization and not something that gets lost among other projects. For example, if you were proposing support for a design project from an engineering firm, you might ask your faculty advisor to write a cover letter affirming that the university approves the project. The letter would confirm the availability of facilities and faculty advisement for you and for the proposed project. When there is no executive officer involved, you may write the cover letter, perhaps over your supervisor's signature.

cover letter
a letter indicating support for a project written by the executive officer of the organization submitting a proposal

EXAMPLE 15–3   Proposal Parts and Questions They Answer

Proposal Part	Reviewers' Questions
Cover Letter	Do the division managers or executive officer and board of directors of the proposing organization support the proposed project?
Summary or Overview	What is this project about, and what do the writers propose to do? How does this project relate to our interests? How will this project be better than similar ones we might fund?
Introduction	Who is proposing this project? What are the organization's interests and activities? What kind of track record does it have with similar projects?
Problem Statement or Needs Assessment	What is the problem? What is the need? Is the problem important? Is this the type of problem in which our organization is interested? Does the writer have a grasp of the problem? Does the writer's concept of the problem match ours?
Objectives	Exactly what will be the outcomes of the proposed work?
Plan	What will be done? In what order? What methods will be used? Are these the right methods to achieve the objectives? When will the project be complete? Can the project be finished in the specified time, or can we predict delays and budget overruns?
Evaluation or Quality Control	What measures will be used to establish that the plan is achieving the objectives? How will we know that the plan is working?
Budget	How much will the project cost? How much will each component cost? Are the costs fair and reasonable for the work performed?
Personnel and Qualifications	Who will do the work? Do they have the qualifications (education, experience, interest) to do the work well? Are they the best qualified of the various people who have applied to do the job? Do they have a record of success in similar projects?
Facilities and Equipment	Do the proposers have the necessary facilities and equipment to do the job they say they will do?
Other Sources of Funding; Future Plans	When our grant ends, will the project die, or is the support broad enough to ensure its continuation? Have others committed to the project?

SUMMARY OR OVERVIEW OF THE PROPOSAL

The summary or overview provides a capsule of the whole project, including the problem, objectives of the project, plan, budget, personnel, and facilities. It may highlight special features of the proposal that distinguish it from competing proposals. Such features might include the following:

- An unusually well-qualified staff.
- A new, efficient method of performing the proposed service.
- Outcomes that will exceed the minimum requirements.

Thus, your summary may go beyond answering the main question, "What will this project accomplish and how?" to answer, briefly, "How will this proposer do a better job than others who have applied?" All the information appears in more detail in the rest of your proposal, but your summary gives reviewers their first impression of the project and establishes the ethos of you and your company.

INTRODUCTION TO THE PROPOSAL

"Introduction" has different meanings for different organizations, but it may be simply a problem statement that introduces the project. This type of introduction is typical in a management proposal within an organization. If a client requests an introduction separate from a problem statement, you should make the introduction focus on the organization that offers services or requests support rather than on the project itself. This type of introduction is valuable when the writer and audience are in different organizations. It is typical in a research or service delivery proposal. Use it to provide a brief history and to state the organization's purposes. It provides an opportunity to establish some credibility. You may also review achievements in activities related to the proposed project.

PROBLEM STATEMENT OR NEEDS ASSESSMENT

The problem or need that you have identified is the reason for proposing some corrective action. Therefore, you should use the problem statement to describe a discrepancy between the way things are now and the way they might be and to explain the reason for the proposed project. It is similar to the problem statement for a report (see Chapter 14), and it should tell

- What the problem is.
- When and where the problem takes place.
- Why the problem happens.
- Who is affected by the problem.
- Why the problem is significant—for example, the short-term and long-term costs (financial and otherwise) of not solving it or the benefits of pursuing a solution.

Some proposal guidelines call for a *needs assessment* instead of a problem statement; a needs assessment tells what is needed, who needs it, when and where it is needed, and why it is needed.

If you are responding to a request for proposals (RFP), you may see the problem already described there and assume that a problem statement is redundant in the proposal itself. One part of your task in developing a problem statement, however, is to convince the reviewers that you understand the problem well enough to plan and carry out a solution. Your own statement of the problem, which draws on the content and even phrases of the problem statement in the RFP, establishes a sense that you have a firm grasp of the problem.

OBJECTIVES

objectives
statement in concrete terms of what one plans to accomplish if the proposal is funded

The **objectives** state in concrete terms what the outcome of your work will be if you receive the support you request. Objectives answer the question, "What will happen if the proposed project is implemented?" They identify intended results. The next part, the plan, will tell what you will do to achieve the results. If there are multiple objectives, you should list them, but you may also need to explain or elaborate upon them.

The following objectives identify project outcomes in specific, even measurable, terms. The first three objectives name specific results of actions. The others use verbs with the implied subject of the proposers. This phrasing is fine because the outcomes are identified, but verb phrases are risky for objectives if they emphasize what you will do rather than what you will achieve. Note that the verbs *determine* and *characterize* indicate a research proposal (a study) whereas the other verbs suggest service delivery.

All company computers will make the transition to the year 2000 without data problems.

Car-bicycle accidents on the central campus will decline from an average of seven per year to zero by 2001.

As a result of efforts to increase awareness of AIDS transmission and prevention among teenagers, 75 percent of all high school students in Minneapolis–St. Paul will achieve a score of 85 percent or better on a paper-and-pencil test covering AIDS transmission and prevention.

To determine the optimum seating capacity for an all-purpose events arena in Lubbock.

To reduce oil contamination on 80 square miles of Alaska beach by an average of 50 percent.

To determine the degree to which fertilizer amendments enhance the biodegradation of surface and subsurface oil.

To characterize ecological risks associated with fertilizer amendments.

To develop a computerized advisement system for juniors and seniors in mechanical engineering that will be at least 95 percent accurate in its advice.

To recruit, train, and place 100 teenage volunteers in local service agencies.

PLAN: METHODS, DATES

Your plan of implementation will make or break the proposal. The reviewers will probably select the proposal with the best plan for solving the problem, assuming the budget is reasonable. Often the biggest barrier to finding workable solutions to problems is knowing *how* to make them happen. *What* will be done has already been established by the problem statement and objectives; the plan specifies *how*.

The following example illustrates the power of a plan to persuade. Several engineering students were frustrated by long waits to see their advisor at registration time. They also felt that the advice they received was redundant anyway because it was so similar for all students in their major. The engineers thought a computerized advisement system might save time for both students and faculty advisors. Knowing *what* they wanted, they could have simply suggested the computerized system to the dean. Nevertheless, the dean, who might have responded positively to the idea, would probably have defaulted to the standard response to a suggestion for change: It's a good idea, but we don't have the means to implement it. The engineering students' suggestion would have died a natural death.

Instead, the students developed a plan for creating software and a database that incorporated all the degree requirements and prerequisites so that students could tell what courses they needed in any given semester. They recruited an expert computer student to do the programming, established the contents and structure of the database, and included pilot tests of the program with representative students. By showing how they could achieve the benefits of a computerized system at a reasonable cost and offering the people to do the job, they created a solution that would have been hard for the dean to refuse. The engineering students created temporary jobs for themselves as well. Thus, by shifting the focus from what the recipient could do to what they could do for the recipient—the *how* of their proposal—they made it easier for the recipient to agree to the proposal.

The specific methods of proposal implementation depend on the objectives and the nature of your project. David Silk's adapter (see Example 15–5) required the building and testing of a prototype. Solving the Y2K problem requires coding and testing. A public information campaign requires development of materials. Whatever your plan for achieving the stated objectives, be specific about what will be done, when, where, by whom, and in what quantities. *Don't leave any doubts about your ability to get started and to complete the task.*

The description of methods lets reviewers envision your plan as it will be implemented. A schedule of tasks and completion dates shows them when the work will be done. A **Gantt chart** offers even more information; in addition to showing completion dates, it shows when the tasks are being accomplished in relation to other tasks. The proposal in Example 15–5 uses such a chart in the "Project Schedule" section. The tasks listed on the chart have already been described in narrative form, but this chart shows at a glance where the project should be at a given point in time. It also shows that some tasks are simultaneous rather than sequential.

Gantt chart
schedule of tasks and their relationships and completion dates

EVALUATION AND QUALITY CONTROL

A good proposal includes a section on evaluation or quality control. The evaluation methods tell whether your plan has achieved its stated objectives. Alternatively, you promise in this part evidence of quality control and progress according to your established schedule. From the variety of evaluation methods available, your task is to choose and develop those methods that measure what your project aims to do. When your objectives involve education, your evaluation methods often include a test, perhaps a pre- and post-test that compares what the students knew before the project began and what they knew after. If the test is reliable, rising scores indicate success. The objectives may establish how much the scores will rise.

If your project is an ongoing construction or service delivery project, your sponsors may request periodic progress reports that indicate whether the project is on schedule. Your progress reports compare what has happened with what was supposed to happen. If the objectives state, for example, that 500 high school students in Minnesota will participate in AIDS discussion groups by May 15, but 600 students have participated by the target date, you have support for a positive evaluation of the project.

If you find it hard to write the evaluation methods, your objectives may not be specific enough and may require revision.

BUDGET

You can be sure that proposal reviewers will study your budget carefully, looking at the individual items as well as the bottom line. An external funding organization may restrict how you can spend the money you may be awarded; some grants will not provide salaries or equipment, for example. Instead of guessing and using round figures, you should take the time to get specific estimates for materials, printing, travel, and any other expense. Also follow the suggestions in Writing Strategies 15–1.

The proposal with the lowest bid does not always get the job. Funding organizations are looking for quality, safety, timeliness, attention to detail, and sound arguments, as well as low prices.

WRITING STRATEGIES 15–1   Creating Proposal Budgets

1. Make sure that the plan explains or at least prepares for each item in the budget. Don't introduce surprises at this point. If your budget includes postage, for example, be sure the plan specifies what will be mailed.
2. Check your figures to avoid errors in addition.
3. Don't inflate the budget, even if you haven't spent all the money that has been offered. Estimate reasonable and fair costs. Don't try to sneak in a new piece of equipment that you want even though the project doesn't require it. Sponsors don't want to be cheated.
4. Display your figures in a table so that the reviewers can easily see the item, the formula for calculating the cost (e.g., per piece cost times number of pieces), and the total expense for the item as well as the bottom line.

PERSONNEL AND QUALIFICATIONS

For an internal management proposal, you will need to identify staff needs—positions plus time required, but as employees you do not have to demonstrate qualifications unless the proposed project is outside your usual job duties. For a proposal to another organization that does not necessarily know your qualifications, you provide additional information. You can attach résumés of the key project staff to establish an overall record of qualifications and achievement, but your information may be easier to interpret if you offer a paragraph description that focuses on the project staff's specific qualifications. Your paragraph need not be comprehensive; instead, highlight certifications and previous experience with related projects and their outcomes. You might also include testimonials from your clients and customers.

FACILITIES AND EQUIPMENT

You may require specialized facilities, such as a dust-free manufacturing area, or equipment, such as an electron microscope, to complete your proposed project. A description of available facilities and equipment assures the reviewers that the project is feasible in terms of these requirements. If the facilities belong to someone else, you'll need a letter from the owner to confirm their availability for the proposed project.

OTHER SOURCES OF FUNDING AND ONGOING FUNDING

In an external proposal, a potential funding source may be interested in other sources of funds you have obtained as a way of determining need and assessing the support the project has already gained. If your proposed project will last longer than the support requested, the funding source may want to know your plans for continued funding. Perhaps other agencies will help to support the project if it proves successful, or perhaps there will

---

WRITING STRATEGIES 15–2   Working with Proposal Supplements

When your proposal allows you to add supplemental materials in appendixes, consider whether the following supplements would strengthen your document:

- A map or architectural drawing showing the location of the project.
- Letters of endorsement from experts who think the proposed project has value.
- Résumés of the key people who will implement the project as a way of confirming their qualifications.
- Detailed budget analyses that are only summarized in the text.
- Brochures or other informative literature describing the organization proposing funds.
- Other evidence that supports the proposal but is not central to its arguments.

---

be fund-raising activities or client fees. This information helps your funding source assess whether it is investing in a long-term project or one that may fail after the initial support period.

SUPPLEMENTS TO THE PROPOSAL

Supplements, often presented in appendixes, support the proposal but are not essential to understanding your plan. If you prepare such supplements, consult Writing Strategies 15–2 for what you might need to include.

For a nonprofit organization, you may need to submit evidence of its tax-exempt status, audited financial statements, and the operating budget. You may bind these supplements with the proposal as appendixes or, in the case of a large map, enclose them as separate items. But don't stuff the appendix full of every piece of paper you can find that may be related to the project or your organization. Reviewers are impatient with having to wade through excessive material to find what is really important.

Some organizations require you to complete application forms. The form might be in addition to the proposal narrative, or the proposal itself may have to be typed into the form. Academic proposals that seek the approval of faculty advisors for research projects often include literature reviews.

VISUAL DISPLAYS

Visual displays are not a separate part of the proposal, but they are mentioned here as a reminder that visual displays may have a greater impact and be easier to understand than words. Thus, they support the persuasive and informative purposes of your proposals. In addition to presenting the budget in a table, consider drawings to illustrate key concepts or processes and graphs to demonstrate the significance of the problem.

CONCLUSION TO THE PROPOSAL

Many proposals simply end without a formal conclusion once all the requested information has been provided. Your whole proposal has been a request for support and to restate it may seem redundant. If you need a formal conclusion, use it to summarize the benefits to the funding source and request support. You will leave the reviewers with a good image of the outcome of the proposed project.

There are good reasons for collaborating with other writers on proposal writing. The project may require different types of expertise, and the people who are expert in a project requirement should be the ones to write about how it will be fulfilled. Involving people who will help to implement the project in developing the proposal gives these people a personal and professional stake in the project that can carry over into their work on the project. Moreover, project deadlines may be so tight that one person could not possibly complete the entire proposal within the allotted time. Note that the announcement in Example 15-2 of the need to remove and decontaminate medical waste was published August 18, but the completed proposals were due September 17. In the intervening month the contractor also had to request the RFP (here called a "solicitation package") in writing. That schedule gave less than a month to develop the proposal, including research, planning, and writing.

Proposals Written Collaboratively

In the late 1980s the federal government issued an RFP to locate, construct, and manage a particle accelerator known as the superconducting supercollider (SSC) to support scientific research. The SSC would allow physicists to explore the building blocks of nature by smashing atoms. That RFP specified requirements for space (a tunnel with a 50-mile circumference was required), details on environmental impact, and even cultural opportunities for the people who would work on the supercollider project. The engineers who could conduct soil tests as part of the environmental impact study were the best people to write the part of the proposal that evaluated environmental impact, but the chamber of commerce and civic leaders were better suited to describe the cultural and educational opportunities in the city closest to the prospective site. Thus, the scope of the project required a division of labor and the contributions of multiple people. Furthermore, the deadlines were tight: Proposers had little more than a month to gather all the necessary information and to organize and present it. The tasks had to be completed simultaneously rather than sequentially. Collaboration was the only way to complete a credible proposal. The project was indeed awarded to Texas, over competitors such as Illinois.

Collaboration will probably require a division of responsibilities, but your whole team will need to review the proposal parts as they develop. All team members should have a stake in the outcome, not just in their own part. Members responsible for one part can be good reviewers of the parts of others' because they can spot confusion or unanswered questions that are hard for writers to see in their own work.

A project coordinator keeps the work on schedule and facilitates communication between the different work areas. Each work area may also have a leader to keep its part of proposal development focused and on schedule and to serve as a liaison to the project coordinator. Review meetings provide opportunities to resolve any unanticipated findings or problems, to refine project goals and methods, to make necessary adjustments in the project or in the document, and to edit the sections as they are written.

Proposal Management

Because of the competitive nature of proposals, the likelihood of collaboration, and the likelihood of deadlines that are earlier than you would like, you need to use all your best management and writing strategies when you develop proposals. The main tasks are summarized here:

1. Define the problem or need; describe the proposed solution in broad terms. Study the RFP if one is available. Define the specifics of the problem and the criteria for a solution.

2. Conduct an audience analysis: learn about the potential sponsor's interests, needs, and constraints on awards. Screen the RFP, directories, proposal guidelines, Web site, and annual reports. Interview the sponsor for more information on the problem or constraints on the solution. Learn who will review the proposal. Most funding agencies welcome phone calls about the project—they want good proposals, and helping you get it right is a good investment of their time. Determine that there is a match between what the potential sponsor wants to have done and what you want to do.

3. Plan a solution: get information that will help you devise a sound and feasible plan; refine the details of the plan; develop the required parts of the proposal.

4. Develop a presentation: apply the funding source's guidelines for format and content to this proposal; answer the questions directly and clearly. Create proposal sections using the terms in the proposal guidelines and in the order specified. Use topics and phrases from the RFP as a way of establishing mutual understanding of the problem.

5. Manage the task: Organize collaborators; define tasks; set a schedule for completing tasks. If there are collaborators, schedule team meetings throughout the project as well as at the beginning (to plan it) and at the end (to review the results and make last-minute changes). The meetings will keep collaborators on schedule and give you all a chance to assess progress and goals and to work out unanticipated difficulties.

Writing Strategies 15–3 will prompt you for each of these tasks.

WRITING STRATEGIES 15–3   Proposal Management Worksheet

1. **Define the problem or need**
   Who
   What
   Where
   When
   Why
   Significance

   Plan summary: State briefly what you plan to do about the problem (or what someone else should do if this is a management proposal)

   Type of proposal:
   _____ research or project
   _____ service delivery
   _____ management

2. **Analyze the audience**
   - Who will review the proposal? Who will make the decision about accepting or rejecting the proposal?
   - Are the reviewers specialists or do they have a general knowledge?
   - What are the interests of the audience in your proposal?
   - How will the proposal benefit the audience?
   - How well does the audience understand the problem?
   - What constraints will influence the audience's response to the proposal? (Consider time, money, competing projects, limits of facilities or equipment, limits in the type of project that the audience will support)
   - Methods of gaining information about the reviewers and decision maker
     _____ Study an RFP
     _____ Locate proposal guidelines from the organization
     _____ Read descriptions in directories
     _____ Interview
   - Implications of the responses: What will make this proposal attractive to the audience?

3. **Plan a solution**
   Objectives: What results will describe an effective solution? What do you want to achieve? What does the audience want to be achieved?

   Plan: What methods or facilities will achieve these results?

   Schedule: On what schedule can the plan be completed?
   Task   Time required

   Evaluation: How will you be able to demonstrate that the objectives have been achieved?

   Budget: How much will the plan cost?
   Staff
   Equipment and supplies
   Postage and printing

(continued)

WRITING STRATEGIES 15–3 (Continued)

4. **Presentation: Imagine the document itself, and choose the best strategies of organization, format, style, and visual displays to make it effective.**
   What parts will be required?

_____ Cover letter	_____ Evaluation
_____ Summary	_____ Budget
_____ Introduction	_____ Personnel
_____ Problem statement	_____ Facilities and equipment
_____ Objectives	_____ Ongoing funding
_____ Plan	_____ Supplements
_____ Time/task schedule	

   What information should be displayed in tables, graphs, or other displays in order to increase comprehension or to persuade?

5. **Management: Identify research and proposal preparation tasks, and set a schedule for completing them.**

Research tasks	Completion date	Person responsible
a.		
b.		
c.		
d.		

Writing, proposal preparation tasks	Completion date	Person responsible
Writing		
Revision and editing		
Preparation of visual displays		
Preparation of preliminary pages		
Preparation of supplements		
Final proofreading		

Sources of Project Support

Because proposals often link organizations with problems to solve with organizations or individuals who can either solve them or who can provide the funds for solving them, the first step in linking is to understand the organizations that may support projects and the goals that they have for the support they can give. Successful proposals match their project goals to the goals of the supporting organization.

The discussion in this section applies primarily to research and service delivery projects that are funded externally, that is, by organizations other than the one that provides the service. The support these organizations provide does not have to include money, but typically it does. Foundations, government agencies, professional societies, civic groups, and corporations sponsor research and service delivery projects.

FOUNDATIONS

Some organizations exist not to make money but to distribute it. Public (local, state, and federal government), corporate, or private foundations are examples of funding sources. A **foundation** is a nonprofit entity that uses its funds to support certain research, educational, arts, or charitable projects. Foundations may be funded by donations, taxes (in the case of public foundations), a certain percentage of earnings (in the case of corporate foundations), or by permanent endowments. An **endowment** is a sum of money that generates interest or investment income to fund particular kinds of projects. The purpose of foundations is to distribute funds to individuals or groups who can achieve the objectives the foundations have set for the use of their funds.

> **foundations**
> nonprofit entities that use their funds to support research, educational, artistic, or charitable projects

> **endowments**
> sums of money that generate interest or investment income to fund particular projects

A foundation sets goals but doesn't necessarily have the expertise to achieve the goals with its own staff and facilities. For example, the National Science Foundation (NSF) uses tax money to fund scientific research. The problems the NSF considers include lack of knowledge about the physical and natural world and very specific problems such as ozone depletion. There would be no way to conduct all the scientific research that NSF sponsors in one place because of the scope and variety of the research, nor would it be desirable to do so. Researchers must maintain a measure of independence from their sponsors and from one another. At the same time, the researcher with the expertise to conduct the investigation doesn't always have the funds that it takes to conduct the research—funds for salaries, equipment, supplies, travel, and **overhead,** such as electricity and rent for office space.

> **overhead**
> research costs including such things as space and utilities

Foundations establish their own goals for uses of their funds. Some foundations are especially interested in education while others are concerned with supporting projects in health or agriculture. Some define their interests geographically; that is, they may be willing to fund a range of projects in a certain city or state. Some are willing to pay for equipment and salaries, but others will provide only travel or printing expenses. A private, nonprofit foundation with a substantial amount of money to distribute is probably listed in *The Foundation Directory*. The entries identify the foundation's geographic and subject matter interests.

GOVERNMENT PROGRAMS

The U.S. government contracts, by means of competitive proposals, for many of the services it needs to have performed. The *Commerce Business Daily,* published by the Department of Commerce, announces projects each day in multiple categories, including research and development, installation of equipment, education and training, transportation, data processing, and social services. Example 15-2 shows an announcement of a need to remove medical waste. Various companies that specialize in medical waste disposal will submit proposals to do the job; after a review, one will be selected to complete the job and awarded the funds to do it. State and local governments may also seek services or research and award contracts to the organization that presents the best proposal.

Information About Funding Sources

A persuasive proposal not only identifies a workable solution for a significant problem but also targets a potential sponsor's interests. A perfectly good proposal won't get funded if it goes to an organization without an interest in the problem or with restrictions that exclude the type of support being requested. Another way to doom a proposal is to define the problem in a different way than the potential sponsor defines it. One important first step in proposal development, then, is for you to learn about the potential project sponsor. You should explore printed sources of information including requests for proposals, directories of funding sources, annual reports, and guidelines for proposals available from potential project sponsors. You can make inquiries by phone or in person. This information is essential for you to gauge your audience's interests and expectations.

According to Janis Cavin, "the most common reason for rejection of a proposal is a mismatch between the purposes of the proposal and the goals of the funding source" (7). Researching the funding source's goals is a good investment of your time and energy.

FOUNDATION DIRECTORIES AND OTHER SOURCES OF INFORMATION

If the problem that the proposal addresses originates with an organization that needs funds to provide services instead of with a funding organization, there will be no specific RFP. Instead, proposal developers search for potential sponsors who are interested in the type of project. A basic source of this information is *The Foundation Directory* or the *Directory of Corporate Philanthropy,* or a specialized directory such as *Funding in Aging* or *Private Funding for Rural Programs.*

The descriptions in the directory will help you to screen potential sponsors and find one suited to your project. The foundations themselves may have fuller statements of their interests in print or on the Web. By studying the descriptions you can easily rule out organizations that are unsuitable for the project you are proposing, or you can adjust your project so that it will be attractive to the foundation. If you do this research, you can be reasonably confident of a match between your project and the potential sponsor.

Funding organizations may also have detailed guidelines that specify award criteria and methods of selection. Like an RFP, proposal guidelines provide specific information for audience analysis. Annual reports of the potential sponsor also reveal information about its interests by showing the type of projects it has funded previously.

In addition to studying an RFP, guidelines, and annual reports, take the initiative to talk with potential sponsors about the proposed project. These conversations can help you determine whether the project is suitable for a particular organization and also provide the opportunity to inquire about the funding schedule and deadlines that must be met. Representatives may be available to talk with you to clarify their expectations, to answer questions, and even to offer suggestions for proposal development. It is in their best interest to receive excellent proposals because they want to spend their money well. They also learn about you in a more personal way than the proposal itself allows. These conversations establish relationships as well as provide an opportunity for information sharing. Such conversations are more likely to occur with private than with government funding sources. Government regulations may require that all applicants receive the same information to ensure a fair award process.

Proposal Review and Selection

You will screen potential projects and sponsors at the beginning of the proposal development process, and the sponsor will screen the proposals you submit. Reviewers read and evaluate the submitted proposals and select the one or ones most likely to accomplish project goals within the limits of time and budgets. The reviewers are people who understand the goals of the funding organization. They may or may not have expertise in the methods of the project being proposed. To write for the audience, you need to know whether the reviewers will be specialists or generalists.

peer reviewers people with expertise in a research area or discipline who judge the relative merits of proposals

Peer reviewers are people with expertise in a field of research or service delivery. If the NSF were reviewing a proposal for gene research related to Alzheimer's disease, it would select reviewers who themselves could conduct gene research rather than, say, psychologists or technical communicators. These reviewers can evaluate research methods and determine whether the methods are likely to yield the intended results. Likewise, evaluators of the proposals to dispose of medical waste must be expert in methods of waste disposal or in the administration of waste disposal projects. If the proposal will be reviewed by peers, you may assume your audience has some knowledge about the problem and its significance as well as technical knowledge. You need to include details of the methods, but you may be able to describe the problem in an abbreviated way.

On the other hand, if the proposal will be read by people with general rather than specific backgrounds in the problem, you need to explain the problem and its significance in terms that a generalist can comprehend.

You may describe methodology with less detail about procedures and more explanation of what the methods will accomplish. For example, a university may have funds to support student research for which students from throughout the university submit proposals. The reviewers read proposals for projects outside their own disciplines. A biology professor may need to compare a proposal from engineering with a proposal from English. The proposal writers will have to explain what the project is and why it is important in terms that the reviewers can understand.

Proposals are often read and ranked by a team of reviewers, especially when a lot of money is at stake. Different reviewers will see different merits and limitations in the proposals. The team approach to review also neutralizes possible personal bias in an individual reviewer. Some proposals are reviewed "blind"; that is, the name of the proposer does not appear on the proposal that the reviewers see. Thus, they can focus on the merits of the proposal itself rather than letting opinions about an individual intrude in their decision.

The reviewers evaluate according to specific technical, managerial, and social criteria (remember the discussion of some of these criteria in Chapter 14). Upon request, you may be able to obtain these review criteria from the funding agency. The criteria would provide you with a good checklist for proposal development. The reviewers will consider the significance of the problem, responsiveness to the RFP or to the goals and limits set for the use of available funds, the merit of the proposal (including methods), qualifications of the staff, adequacy of facilities and other essential resources, and record of past performance.

Example 15–4 illustrates the process of linking the organization with the problem to the organization that can solve the problem by means of an RFP and competitive proposals. In response to the RFP that describes the problem, need, or opportunity, several organizations develop proposals. After a review, one proposal is selected and the developing organization is funded.

When an organization seeks funds for its own projects rather than responding to an RFP, the RFP stage would be replaced by a search for an appro-

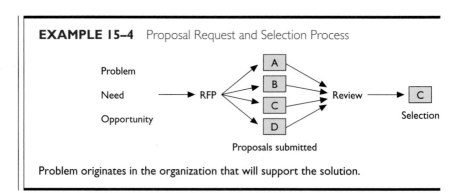

EXAMPLE 15–4   Proposal Request and Selection Process

Problem originates in the organization that will support the solution.

priate funding source. The proposal would still be one of multiple proposals submitted to the funding source, only one or some of which would be funded.

UNSOLICITED PROPOSALS

In some cases you might submit a proposal without this explicit or implicit invitation. Management proposals, in particular, may be unsolicited. The manager or potential project sponsor may not even have recognized a problem, let alone imagined a solution. The proposal may be a surprise—but as a shrewd writer you will prepare the sponsor before submitting the proposal.

An unsolicited proposal is more difficult to write and less likely to succeed than a proposal that responds to an RFP or an announcement of interests in funding. You must first convince the people who review and decide upon the proposal that there is a problem, that the problem is significant, and that the problem concerns the potential sponsor. In addition, you must convince the reviewers that this problem and solution are more worthy of available funds than other projects to which the sponsor might commit. Convincing a possibly skeptical reviewer that there is a problem is only half your task, because many reviewers who agree there is a problem don't necessarily believe that anything can be done about it. You can help the potential sponsor to imagine the solution by giving complete details of the way the project will unfold and what results would do to solve the problems.

When you write an unsolicited proposal, you may have only general guidelines about expectations for proposals, such as the ones in this chapter, rather than specific guidelines. Nevertheless, you should make every effort to get some direction from the potential project sponsor. Communication before the proposal is submitted can prepare the audience for the proposal as well as yield ideas about ways to appeal to the reviewer.

- Proposals solve problems by claiming new priorities within an organization or by linking organizations that have problems to solve with the organizations that can solve the problems. Many individuals and companies make their living by providing services or research for other companies. They get their work by developing proposals, usually in response to stated needs for the type of services that they provide. Many charitable and educational organizations, which have limited ability to generate their own income, are supported by grants from foundations and government programs. They, too, develop proposals for the funding they need. Proposals are also the means by which individuals or groups may encourage changes in procedures or facilities to improve efficiency, safety, or fairness. The stakes are high: Developing a good proposal requires a significant investment of time and other resources, and there is no sure reward for the investment.

SUMMARY

In this chapter we discussed two kinds of proposals: research or project and service delivery proposals and management proposals. We also discussed several sources of project support, such as foundations and government programs. We gave you advice on how to get information about funding sources and introduced you to *The Foundation Directory* and the *Commerce Business Daily*. Often proposals are solicited through Requests for Proposals (RFPs). Proposals are then reviewed by peer reviewers or reviewers with general knowledge.

■ To increase the chances that your proposal will succeed, you should look for a match between your goals and capabilities and the goals of the potential funding source. You must develop proposals that meet the guidelines, interests, and expectations of your potential funding source. Your proposal must answer questions about the objectives, plan, budget, and evaluation specifically and clearly.

■ Common parts of proposals include a cover letter, summary or overview, introduction, problem statement or needs assessment, objective, plan, evaluation or quality control, budget, personnel and qualifications, facilities and equipment, and other sources of funding. The planning and review required for proposal development also facilitate good management when the plan is implemented.

ACTIVITIES AND EXERCISES

1. Study Example 15-5 to determine how it illustrates the principles of proposal development and uses the various proposal parts to achieve its goal of persuasion.

   a. Who is the audience for the report? What purpose does the writer hope to accomplish in writing the proposal?

   b. What type of proposal is it—research or project/service delivery or management? (The distinction relates to whether the proposers will implement the solution to the problem or whether they are trying to convince managers to implement a solution.)

   c. What are the most persuasive features of the proposal? What is the most detailed section, and why? Use the reviewers' questions in Writing Strategies 15-3 as you consider the ways that this proposal persuades. How does the proposal answer the reviewers' questions? Do you think the proposal will win the support it seeks? That is, will the project be approved? What will convince the Iowa State University engineering faculty to approve the project? What will convince Motorola to provide supplies?

   d. How do the visual displays aid the task of persuasion?

   e. Compare the parts of the proposal with the list of proposal parts in Example 15-2. What variations do you notice? What explanations can you suggest for the variations? Is the proposal complete? What parts are omitted, and why?

EXAMPLE 15–5   ISU/MOTOROLA ISDN SENIOR DESIGN
PROJECT PROPOSAL

Submitted To:
Dr. David Morgan
Director of Corporate and Computer Motorola Research and Development

Professor Doug Jacobson
ISU Faculty Project Monitor

Submitted By:
S. David Silk
Team Leader

August 20, 1990

TABLE OF CONTENTS

1

(continued)

EXAMPLE 15–5 (Continued)

1. INTRODUCTION

In order for ISDN to take a foothold in tomorrow's telecommunication market, it must accommodate existing non-ISDN equipment as well as current ISDN compatible terminal equipment. Although ISDN standards are still evolving, both the technology and the emerging implementation strategy of ISDN are well understood. A team of electrical engineering and computer engineering students (named in Appendix B) wishes to take advantage of this available technology and strategy in order to pursue a senior project. In pursuing this senior project, we seek not only to educate ourselves in this technology, but also to acquire practical application by building a prototype ISDN-compatible terminal adapter (TA).

The purpose of this proposal is to explain in detail the project we wish to pursue demonstrating Motorola's chip set. The following sections discuss project objectives and plan of implementation, design constraints, additional resources, and the project schedule.

2. OBJECTIVES AND PLAN

2.1 Project Overview

Using the Motorola chip set, we will develop our own board layout implementing IDL to link the chips together to provide a powerful Layer 1/Layer 2 ISDN foundation. Please refer to Appendix C for a preliminary layout. (Note: the layout is subject to change.) Beyond this point, we will specialize by incorporating our own version of Layer 2 and 3, three data ports, a keypad, handset, feature keys, and a LCD with a user interface. All of these features will be incorporated into a prototype that, when finished, will have the basic configuration as illustrated in Figure 1.

2.2 Voice Functions

A primary concern in the TA design is the versatility our device will provide the user. Our device will emphasize voice functions and data functions equally. The voice functions listed below will be implemented using either the ISDN technology or individual components.

Auto-dial:	Perform one-touch dialing of stored numbers from a built-in phone number library.
Flash:	Drop caller and await new number to be dialed.
Hold:	Place calls on hold.
Memory:	Store phone numbers to memory for auto-dial.
Mute:	Provide privacy by preventing the caller from hearing your conversation with others in the room.
Program:	Select distinctive ringing pattern to identify incoming caller.
Redial:	Redial the last number dialed manually.
Speakerphone:	Provide two-way speakerphone.

2

(continued)

EXAMPLE 15–5 (Continued)

FIGURE I
Projected configuration of the prototype terminal adapter

2.3 Data Functions

We will incorporate three data ports into our design to enhance the data features but also to increase the versatility of our device. The first port will provide asynchronous and synchronous data access to the B-Channel with in-band signaling for call control. The second port will provide data access to the D-Channel. This port will support both ASCII communications and generic LAP-B. Our TA will do a LAP-D to LAP-B conversion to allow a linkup to computers with existing X.25 hardware/software. Finally, the third port will provide an out-of-band signaling facility to control the ISDN channels as well as the other two ports. Figure 2 illustrates the three data ports. For simplicity, they will be labeled as they were described above: Port 1, Port 2, and Port 3 respectively .

Here is a brief listing of the port characteristics.

- Port 1:
    1. B–Channel access with in-band signaling for call control.
    2. Support of asynchronous and synchronous data.

3

(continued)

EXAMPLE 15–5 (Continued)

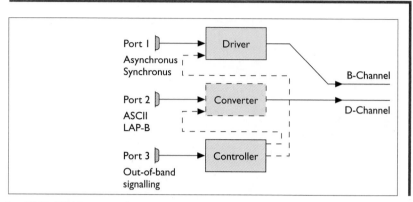

FIGURE 2
Port Configurations

- Port 2:
    1. D–Channel access with in-band signaling.
    2. Support of both ASCII communications (contain a PAD-packet assembler and disassembler) and generic LAP-B.
    3. LAP-D to LAP-B conversion to incorporate X.25 hardware/software to provide a linkup with a host. This capability opens up numerous software related features for advanced support of personalized services from the host.
- Port 3:
    1. Support of an out-of-band signaling facility to control the ISDN channels as well as the other two ports.
    2. Contain a EIA-232 interface for a PC/Mac linkup.
        a. Enable a user to control, from the PC, call setup and tear down features.
        b. Enable information to be down loaded from the host by controlling Port 2.
        c. Initiate phone calls by controlling Port 1.
        d. Perform reprogramming of the TA's personalized features.

2.4 Screen Layout and User Interface
Our device will incorporate an LCD capable of displaying several lines of information at once. In addition to the LCD, our device will also incorporate a user interface to facilitate in the operation of the TA. The combination of these two features, an LCD and a user interface, will provide an efficient method for on-screen editing of all functions built into the TA.

4

(continued)

EXAMPLE 15–5 (Continued)

2.5 Prototype Testing

Due to the lack of a Central Office in the area that supports ISDN, we will build two prototypes to facilitate the testing of our device. With the two prototypes connected together, we will then be able to test and refine our design fully.

3. DESIGN CONSTRAINTS

3.1 Power Consumption

Due to the ISDN technology involved, our device will not operate on the conventional power levels currently supplied by the telephone companies. Our TA will incorporate a power supply supporting ±5 volts and ±12 volts.

As a result of our TA's increased power consumption, it will be designed to be turned off when it is not in use for extended periods of time (e.g., overnight or when a user is on vacation). Also, when it is not in use for short periods of time, it will be designed to "sleep." The ability for our device to be completely turned off will prevent the waste of power. Furthermore, the ability to sleep will allow some nonessential components to be powered down and remain "asleep" until some event "awakens" the device (e.g., the user touches any of the buttons or picks up the handset to make a phone call, or an incoming phone call is received). Since our TA will have its own power supply and be able to be completely turned on and off at the user's will, it will not be dependent on other devices for startup.

3.2 The Ports

Although we have identified and described three data ports, implementing Port 2 is not within the scope of this project. Since the required software used to operate Port 2 is proprietary, we will have to develop complementary software to operate the port ourselves. Accomplishing this task will not be feasible within the thirty weeks allocated for our project. Software for Port 1 and Port 3 will be developed and implemented.

We will, however, complete a well-rounded device that is versatile in design. Section 3.2.1 of this proposal discusses our short run approach that will actually make a useful long run solution in the future evolution of our device. Therefore, we intend to design and to incorporate the hardware for Port 2 into our device in the event that our device is recirculated through the university for another session with a new senior design team. A future senior design team will then be able to continue where we left off on Port 2. Guidelines for future implementation of Port 2 are described in Section 3.2.2.

3.2.1 Short Run Solution for Port 2

As described in Section 3.2, the inability to complete Port 2 to the desired level imposes some limitations to the versatility of our device. Port 2 provides a link to a host via X.25 software/hardware to support an advanced array of personalized services. The major consequence of its absence is the lack of available memory to store the user's personal preferences (i.e. assignment of desired functions to buttons on the TA which can be saved on a per-user basis).

5

(continued)

EXAMPLE 15–5 (Continued)

It is important that our device can function as a stand-alone without being dependent on other devices to perform routine operations such as startup. The best solution is to design the TA to read, from internal memory, a generic setup during startup. As it turns out, this short run solution also proves to be a useful feature to have in the long run. In the future when a host is connected to our device, upon device startup the host can simply download the personal preferences superseding the generic setup. This feature will prevent our device from being dependent on the host to start up. Thus, in the event that the host is not functioning properly, the TA will still be able to power-up with its preset generic features and provide immediate service.

In addition to the integration of a generic setup of the features, additional memory will be integrated into our device to enable the user to store alterations to the generic setup. This will eliminate the inconvenience of requiring the user to modify the features each time the device is started up.

3.2.2 Long Run Solution for Port 2

In order to take full advantage of Port 2, when it is fully operational, the additional features listed below should be incorporated into the device.

1. The TA should be designed to register with the host with its appropriate user identification information upon startup. Once a user is registered, the user's personal preferences would automatically load on the registered TA. When a user notifies the host of a temporary station change, the host can download the custom phone features to the new TA. Therefore, a user can use any TA in the office complex; the custom TA features follow the user around. This feature would also allow the host to reroute incoming phone calls to the appropriate user if a change in location has occurred. In such a situation, the host can act like a PABX, a Private Address Branch Exchange.

2. The TA should be designed to assist the host computer with a feature called call accounting. The TA can be set up to keep track of all the information pertaining to the calls made on that TA, for instance, date, calling phone number, call starting time, call length, etc. At regular intervals, the host computer can poll this information to acquire all the necessary statistics on how the telephones are being used and provide detailed information for billing purposes.

3. The TA should be designed to notify the host computer each time it is turned off. Hence, if the TA is off, the host will execute the necessary procedures to take care of incoming phone calls whether the user is unavailable or at another station.

(continued)

EXAMPLE 15–5 (Continued)

3.3 User Interface

All features of our device, whether it be call setup/teardown or personal phone preferences, will be manipulated via an LCD and a user interface. The integration of an LCD, with the ability to display several lines of information at once, will enable the user to view and/or edit a group of related settings. Also integrated into our device are programmable "soft" keys. The soft keys will be mounted directly under the LCD below their individual captions on the LCD.

As mentioned, there are two modes for the interface, view and edit. When the interface is in the view mode, the user can view listings of the selected features, check a phone number in the call library, or view the standard screen layout to make a phone call. For a depiction of the standard screen layout used to make a phone call, please refer to Figure 3.

Conversely, while in the edit mode the user will be able to scroll through the features and modify options, edit the call library, or alter the information displayed in the standard screen layout by adding or deleting features.

3.4 System Software

The core operations of our TA will be controlled by the first three layers of OSI, Layer 1, Layer 2, and Layer 3. As a result of Layer 2 and Layer 3 being classified as proprietary material, we will have to develop and implement complementary versions in order to run our device.

4.   FACILITIES AND EQUIPMENT

In addition to the Motorola ISDN chip set, we will need the ADS302 Development System to provide a platform for software development for the MC68302. The university will supply us with the necessary host computer to facilitate in the development of all necessary software. For a detailed listing of what is essential for our project, please refer to Appendix A for a parts listing.

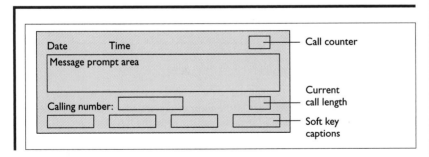

FIGURE 3
Screen Layout

7

(continued)

EXAMPLE 15–5 (Continued)

5. PROJECT SCHEDULE

Thirty weeks have been allocated for our project. The schedule in Figure 4 illustrates how we will utilize the thirty weeks. These dates are subject to change.

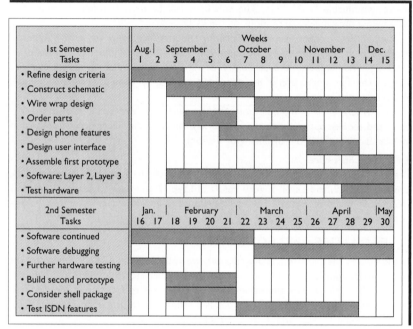

1st Semester Tasks	Aug.	September				October			November			Dec.	
	1	2	3	4	5	6	7	8	9	10	11	12	13	14	15
• Refine design criteria
• Construct schematic
• Wire wrap design
• Order parts
• Design phone features
• Design user interface
• Assemble first prototype
• Software: Layer 2, Layer 3
• Test hardware

2nd Semester Tasks	Jan.	February			March			April		May
	16	17	18	19	20	21	22	23	24	25	26	27	28	29	30
• Software continued
• Software debugging
• Further hardware testing
• Build second prototype
• Consider shell package
• Test ISDN features

FIGURE 4
Project Schedule

6. CONCLUSION

We respectfully request the approval of this project by the engineering design faculty and Motorola as well as support in the form of project advice, use of facilities and equipment, and supply of the parts needed to build the TA. This interdisciplinary project will allow electrical and computer engineering students to use the information, concepts, and skills from their backgrounds to study the complexities of ISDN. This project will allow the students to interact as a team and to get firsthand experience at problem solving.

8

(continued)

EXAMPLE 15–5 (Continued)

APPENDIX A: Parts

From Motorola:

Quantity	Part Name
3	MC145554 PCM Codec Filter
3	MC145474 ISDN S/T Interface Transceiver
3	MC68302 Integrated Multi-Protocol Processor
1	ADS302 Development System
3	15.36 MHz Crystal
3	16.67 MHz Crystal

From Iowa State University:
This is a list of parts from the preliminary design of the schematic. As we finalize the design, more parts will be added to this list.

Quantity	Part Name	Quantity	Part Name
3	MC1455 Timing Circuit	6	30pF Capacitors
6	MC1488 Quad Line Drivers	6	25 pF Capacitors
6	MC1489 Quad Line Receivers	6	0.1µF Capacitors
6	MC6821 Peripheral Interface Adapters	3	0.47 µF Capacitors
3	MC6850 Asysnc. Comm. Interface	12	Miscellaneous Cap.
	Adapters	27	10 kΩ Resistors
6	SN74LS04 Hex Inverters	3	10 mΩ Resistors
3	SN74LS32 Quadruple Input Positive	3	700 kΩ Resistors
	OR Gates	3	0.43 kΩ Resistors
3	1N4001 Diodes	3	29.4 kΩ Resistors
33	Miscellaneous Diodes	18	500 Ω Resistors
6	1:1 Ratio Power Transformers	9	4.7 kΩ Resistors
6	1 mH Inductors	21	Miscellaneous Resist.
6	MCM6206 32KX8 Static RAM		
6	32×8 EPROM		
6	DB-25 Connectors		
2	Handsets		
2	Keypads		
2	LCD		
2	Power Supplies		

9

(continued)

EXAMPLE 15–5 (Continued)

APPENDIX B: Personnel

For this project to be successful, it will need a group of approximately eight students: four electrical engineering students and four computer engineering students. In the fall, when the project is officially formed, I will have a commitment from the people I have solicited. Below is a listing of the current members of the design team.

- Design Team Composition
  1. Roger Najera   E.E.   Communications / Power
  2. John O'Brien   E.E.   Analog Electronics / Math Minor
  3. Bill Ryan   Computer E.   Software
  4. S. David Silk   E.E.   Communications / Telecommunications
  5. Eric Welch   E.E.   Communications / Control Systems

  At least three more computer engineers will be added in the fall.
- Management
  1. Faculty Project Monitor   Professor Doug W. Jacobson
  2. Technical Writing Advisor   Professor Patricia Goubil-Gambrell
  3. Group Leader   S. David Silk

Source: Reprinted with the permission of the author, S. David Silk.

10

2. Using *The Foundation Directory* or *Directory of Corporate Philanthropy* in your library, identify two potential sources of funds for the Minnesota AIDS Project, either for client services or for prevention and education. Use the geographic and subject indexes. Briefly explain the criteria you used (e.g., the number and size of awards and interests of the foundation) to select the two sources.

3. Assume that you would like to gain corporate support for a student project—research, design, service delivery. (Alternatively, if you happen to volunteer or work for a nonprofit agency, consider seeking support for one of its program objectives.) Use the *Directory of Corporate Philanthropy* or *The Foundation Directory* to identify a potential sponsor for your project. In a brief report, state why you think the sponsor might be interested in your project.

4. A home health care agency wrote a proposal for government funds to provide respite care for caregivers of people with handicaps. They offered to relieve the caregivers, who were often stressed by their 24-hour responsibilities, by providing basic nursing care and living assistance. The RFP required them to describe the project's plan to target services to socially and economically disadvantaged participants, particularly low-income minority individuals. Here was their response:

   Services available will be advertised at the following locations:

   All senior citizen centers

   Adult Day Care Center

Resources United

Coalition on Aging meetings

All hospital senior programs

Civic groups and organizations

Public school counselors

Hospital discharge planners and agency social workers

Physicians, residents, and pharmacists

Caseworkers, MHMR [Mental Health/Mental Retardation] center

Alcohol and drug abuse counselors

Clinics in low-income areas

As a reviewer of the proposals submitted, evaluate their response to the question. What unanswered questions might you have? How could the agency be more persuasive in this section of the proposal?

5. According to your instructor's directions, write a proposal for your writing instructor or an instructor in your major field proposing a research and report project that you will conduct for course credit. Your broad goal is to convince the instructor that you have identified a significant and feasible project, that your methods will achieve the stated objectives, and that you can complete the project within the time and budget constraints.

6. Assume that your local school district has requested proposals from individuals and organizations to head an AIDS task force for education in the high schools. The district has $250,000 to spend on AIDS education over the next two years.

Using the Minnesota AIDS Project (MAP—see Appendix A, Case Documents 1) as a guideline, you have been asked by your local AIDS organization to write a proposal to the district requesting that the organization be granted the funding and authority to head the task force. This goal is slightly different for your organization, as it has depended on funds to maintain itself over the years. Now the organization feels it's in a position to take over specific and discrete projects. You have selected a number of staff members and volunteers to work with you (your instructor will place you and others in a collaborative writing team).

You will need to write a detailed proposal on what your organization can offer, how much it will cost (estimate this cost as realistically as possible), what time period it will cover, when you'll be ready to get started, what expertise you'll bring, how high school students will benefit, why yours is the best organization to head such an effort, how the organization might draw on other educational and community resources to support the effort, and all the aspects of an effective proposal that were covered in this chapter. In some cases, you can draw inspiration from the case documents in Appendix A, but in others you may have to do some secondary or primary research to supplement your knowledge about AIDS education.

7. Assume that the Coalition of Citizens Concerned with Animal Rights (CCCAR) has decided that it wants to submit a proposal to the University of the Midwest suggesting alternative ways of reducing the trapping of birds on the experimental crop fields (see Appendix A, Case Documents 3). This proposal will be unsolicited and might be met by some hostility on the part of the university. Therefore, it will require careful and tactful development. Moreover, CCCAR wants the university to bear the cost of funding the proposal so it will have to be very attractive in terms of the solutions it offers.

   Write a detailed proposal to the university that describes the problem, defines what alternative approaches can be undertaken to stop bird trapping on the fields, what their costs are likely to be, and what benefits are to be gained from these alternative plans. In addition to looking at the case documents carefully for alternative approaches, you will need to do some work at the library and perhaps make some calls in the community to determine what some cost-effective solutions might be.

WORKS CITED   Cavin, Janis I. *Understanding the Federal Proposal Review Process.* Washington, DC: American Association of State Colleges and Universities, 1984.

Hartmann, Mary. Executive Director of New Foundations, Inc. Personal interview, September 1998.

Developing and Maintaining a Professional Edge

Professional Communication

Introduction ▮ Creating Effective Letters and Memos ▮ Following the Conventions of Workplace Correspondence ▮ Creating Effective Letter and Memo Reports ▮ Creating Effective Letters of Transmittal and Cover Letters ▮ Writing Effective Letters of Recommendation ▮ Communicating Effectively Online ▮ Developing Job Search Materials

"Ourselves among others" captures something of today's burgeoning internationalism. But equally important, it can serve as a metaphor for that relationship between the writer of business/technical documents and his or her diversified audience; it is a metaphor of the changing contexts of business/technical writing in which today, more than ever before, writers write to "others" who are "different" from them in some significant way.

Timothy Weiss, "Ourselves Among Others: A New Metaphor for Business and Technical Writing," *Technical Communication Quarterly* 1, no. 3 (1992), p. 23.

The problem was that prospective employers were leery of a person whose entire career had been spent with one employer, with one kind of service. He had no specific training in technical writing, having learned all he knew about it on the job. He knew his subject matter well enough and was proficient enough to hold his job, but his résumé was flat: no evidence of attempts at self-improvement, no civic involvement, no volunteer activities, no STC membership. Nothing about specific interests, hobbies, or participation in sports gave employment managers any hint of his being able to fill their needs . . . His problem was that he had really had only *one* year of experience, repeated 14 times, and there was nothing to show that he had tried to improve or advance in his trade.

Dart G. Peterson, Jr., "The Business of Technical Communication," *Technical Communication* 38, no.4 (1991), pp. 594–97.

Introduction   Correspondence is a broad category covering letters, memoranda, messages, and e-mail. This chapter teaches you how to use the conventions and accepted patterns of correspondence to send your audience information that will help them solve problems or to persuade your audience to take certain actions. This chapter will also help you understand how you can modify the standard conventions of workplace correspondence to accommodate the newer communications media—fax (facsimile) machines, mail merge software, e-mail, discussion lists, and company Web sites and home pages.

Knowing how to write correspondence—and being able to do so quickly and effectively—is one of the most important skills you will need in the workplace. The computer makes it easier and faster to put thoughts into writing and has resulted in more communications being sent both in print and online. Mail merge software, for example, allows us to send hundreds or thousands of "personalized" letters in the time it would have taken us to type only three or four. Fax machines allow us to send our correspondence immediately to anyone with a fax machine or a fax modem on their computer anywhere in the world. Moreover, computer networks, which allow us to send e-mail and documents, have also enlarged and extended our audiences.

This chapter suggests ways to write letters and memos that get the job done without overwhelming either you or your audience. Despite the growing popularity of computer technology and the communication modes it allows, traditional conventions governing workplace correspondence still prevail. Therefore, we discuss some general principles about writing effective letters and suggest ways to write some of the letters and memos often composed by technical communicators: short reports in letter and memo form, transmittal letters, cover letters, and letters of recommendation.

This chapter also covers the job search, which usually begins with a letter and résumé sent to prospective employers. Regardless of the career you choose, one of the most challenging and exciting activities you will engage in is searching for a job. You may perform this task a number of times in your professional career, and in every case you will have a good deal of time, energy, and ego invested in a good outcome. In this chapter we also offer suggestions about this important task: how to prepare yourself and find job opportunities, how to analyze what you have to offer prospective employers, and how to contact employers successfully. We discuss the typical documents you write for this task: the letter of application, résumé, and follow-up letter.

Two of the most common kinds of correspondence you will create are letters and memoranda (memos). All that you've learned about analyzing audience and purpose, arguing with good reasons, and gathering sufficient evidence applies to correspondence. Additionally, workplace correspondence has special conventions governing its format. You should consider these formats flexible **templates,** which you adapt to your audience and situation.

In some cases, your audience for correspondence will be quite small, perhaps one or two specific people. In those cases, you'll know your audience personally. Letters often respond to earlier correspondence: a telephone call or an e-mail message. Memos have audiences internal to an organization and generally deal with subject matter known well to both sender and receiver. In other cases, you might be sending a mass mailing, such as a cover letter for a survey. Or your memo or letter may become part of a paper trail in a legal case or an extended consumer complaint, although you will originally send it to only a few readers.

Except for short reports that are sent in the form of letters and memos, correspondence doesn't have a very long life. Correspondence is used to respond to a given issue at a particular time. You may not have much time to prepare correspondence. You can't take a week to write a memo informing the factory that a shipment of a needed part will be delayed. Correspondence must be timely to be effective, and it must also be correct. Your and the company's reputation can be damaged by errors in content as well as grammar, spelling, or mechanics.

Because you will be very busy at work, writing a memo or letter may seem time consuming. Before writing correspondence, ask yourself whether this particular matter can be handled by telephone, a quick note, or an e-mail message. Generally speaking, if you think that there will ever be a need for a written record, write the letter or memo. For example, if a complex issue needs clarification, if there are data to be interpreted, if you are providing cost information, or if you need to document the content of a phone call, use correspondence. If all you want to do is remind your colleague that the room for the meeting has been changed, use the phone or send an e-mail.

Creating Effective Letters and Memos

templates
formats for creating documents such as memoranda, transmittal letters, cover letters, and letters of recommendation

CORRESPONDENCE IN COMMUNICATION LINKS

Because correspondence is a quick way to request or offer information, letters and memos commonly create and sustain many **communication links.** One letter or one memo leads to another, which in turn leads to another, and each becomes a part of a pattern or paper trail of requests and decisions. Before writing correspondence, you need to be aware of the context

communication links
links created by the paper trail resulting from interconnected requests and decisions

into which you are writing and of the documents that have preceded and may follow yours. The three examples that follow are part of one of those paper trails or communication links.

Let's say that you have received the following request from a consumer, Ms. Patrick, to replace a computer chip for a personal computer that your company, Systems Tech, produces (see Example 16–1). This letter will begin a chain of communication, some oral but much written.

EXAMPLE 16–1 Initial Request

July 27, 1999

Systems Tech
1150 Pearson Ave.
Suite 54
St. Paul, MN 55108

Dear Systems Tech,

On January 12, 1998, I purchased a Systems Tech 175 personal computer with hard drive (160 K memory, operating system 6.1) for my place of business. This week the computer malfunctioned because a chip behind the hard drive cracked and began to smoke. I took the computer to a technician located in Denver and was told that the entire mother board would have to be replaced at the cost of $322 (not including labor).

Although the computer is no longer under the year's warranty available with the systems Tech 175, I believe that Systems Tech should replace the mother board without cost. The computer is less than two years old and was not damaged by mishandling. Moreover, the computer technician stated that this kind of malfunction is highly unusual. I believe that the computer must have been improperly assembled in the first place or that the chip itself was faulty.

I will not have the computer repaired until I hear from you. I would prefer to have the computer repaired locally, as I need a quick repair, so I would like to be reimbursed for the repair.

Sincerely,

Janet Patrick

Ms. Janet Patrick
Executive Consumer Surveys
135 West 15th St.
Denver, CO 80210

Obviously, the next link in the chain will be your reply to Ms. Patrick. If you cannot meet her entire request, you still want to retain her good will, to explain why you cannot reimburse her for the repair, and to compensate as much as possible for not being able to reimburse her.

Before you write the letter, you need to gather as much information as possible on similar reported malfunctions on this computer model and to check with your own repair shop. You need to determine whether the problem Ms. Patrick reports is a common one and, if so, you will probably take some responsibility for it. If the problem is an unusual one, you will want more information about how it might have happened and how it can be avoided in other machines.

Only by repairing the computer yourself can you control the situation, gather the information you need for your own future manufacturing concerns, and be assured that indeed Ms. Patrick's local computer technician has assessed the problem accurately. Moreover, you need to check on Systems Tech's previous correspondence with Ms. Patrick—how often has she ordered from you? Has she had any previous complaints? Has your association with her been positive, neutral, or negative in the past? See Example 16-2 for one way you could respond to Ms. Patrick.

In the meantime, you will send a memo to shipping and the repair shop so that they can anticipate Ms. Patrick's call and the arrival of the computer. You'll also request that your team of computer engineers investigate the problem (see Example 16-3). In another memo you'll ask the repair shop to alert computer engineering when the computer arrives. Computer engineering will then send you a memo about their findings; the repair shop will confirm when they send the computer back to Ms. Patrick; you will write to her if computer engineering discovers any manufacturing problems with the computer (and if so, you'll write to other owners of the same model to recall or alert them); and so on. You can probably think of other links in the communication chain yourself, most of which would be filled by correspondence. It's this chain or context you must understand as you generate and respond to correspondence.

CORRESPONDENCE STYLE

Correspondence often has a more conversational tone than do other forms of technical communication. For example, you are likely to find the use of contractions, personal pronouns, and first names. Correspondence also makes good use of reader-based principles, addressing the reader in the second person, as "you" is seen in Examples 16-1 to 16-3:

- The repair shop will call you . . .
- Although indeed the warranty on your computer has expired, we want to help you as much as possible.
- I will not have the computer repaired until I hear from you . . .

EXAMPLE 16–2 One Possible Response

Systems Tech
1150 Pearson Ave.
Suite 54
St. Paul, MN 55108

August 1, 1999

Ms. Janet Patrick
Executive Consumer Surveys
135 West 15th St.
Denver, CO 80210

Dear Ms. Patrick:

We are sorry to hear about your problems with our Systems Tech 175 personal computer. Because each machine is tested by hardware engineers before being shipped to our consumers, we find hardware failure is uncommon in our computers. We are particularly concerned when a problem occurs with one of our customers whom we have served as often as we have served Executive Consumer Surveys.

Although indeed the warranty on your computer has expired, we want to help you as much as possible. We can do so only if we service the computer ourselves. This procedure would not only allow us to replace any necessary parts with Systems Tech parts but would also help us to determine what exactly caused the problem and to prevent it from happening again. Therefore, we are able to offer you the following:

We will repair the machine at no labor cost.

We will replace the mother board at a parts cost of no more than $250.

We will pay shipping costs to and from Systems Tech.

We will repair the computer within two weeks of its arrival at Systems Tech.

Finally, we will guarantee these repairs and the new mother board for five years from the date of repair.

We value the association we have had with Executive Consumer Surveys and hope that you will find our offer satisfactory. Place call our shipping department at 1-800-555-2233 to arrange for UPS packing and pickup of your computer.

Sincerely,

John Smitzer

John Smitzer
Marketing Supervisor

EXAMPLE 16–3   Memo to Computer Engineers

Systems Tech
1150 Pearson Ave.
Suite 54
St. Paul, MN 55108

MEMORANDUM

DATE: August 2, 1999
  TO: Jane Smalley, Computer Engineering Supervisor
FROM: John Smitzer, Marketing Supervisor   *JS*
  RE: Possible Problems with 175 Mother Board

Please be aware that a customer has reported a cracked and smoking chip in the Systems Tech 175 some-where "behind the hard drive," which appears to have damaged the entire mother board. I have checked with the repair shop, and they told me that although the problem is unusual, three such complaints have come in on the 175.

The repair shop will call you when the computer comes in and will copy the other two complaints for you. Please give the computer a quick check to make sure that we do not need to change the 175 design in some way. I promised the customer a fast repair so I hope that we can get someone on this as soon as the machine arrives.

As the writer, you can speak about yourself in the first person: "I believe that the computer must have been improperly assembled in the first place . . .". This use of first and second person in correspondence makes for a less formal tone, a more conversational exchange.

However, because of this conversation tone, you need to resist any tendency to say more than might be needed. People who need information want it fast. Strange things seem to happen when people begin to write correspondence. They put on paper things they would not ordinarily write. For example, some writers make use of archaic phrases and clichés because they think that's what they're supposed to use in correspondence:

Please be advised . . .

Enclosed please find . . .

As per our agreement . . .

We are in receipt of . . .

Please don't hesitate to call . . .

Or they lapse into jargon without considering whether the audience can understand their special technical terms and abbreviations. They write:

The RFP for the YETP has been issued by RETC, an agency funded through CETA.

Instead of writing:

The Request for Proposal for the Youth Employment and Training Program has been issued by the Regional Employment and Training Consortium, an agency funded through the Comprehensive Employment Training Act.

In the example above the passive voice still adds to the confusion. The active voice—"the Regional Employment and Training Consortium has issued the Request for Proposal"—states the main idea more forcefully, placing the actor and action at the beginning of the sentence.

Writers of correspondence sometimes lapse into imprecise and inaccurate prose in an attempt to sound impressive. Simple, direct language is more effective:

Instead of	*Use*
approximately	about
apprise	tell
initiate	begin
utilize	use
terminate	stop, end

Other symptoms of the bureaucratic prose frequently found in correspondence include abstract, rather than simple, concrete words. Letter writers talk about "equipment" when they mean "a tachometer" or call something a "facility" when they are referring to "an office." Redundancies add more words as well: "a new beginner" instead of simply a beginner, "general consensus of opinion" instead of consensus, and "round in shape" instead of round. Look at the effect of accumulating all these stylistic problems in Example 16-4.

What do you suppose is the message of this memo? How many times would the state director have to read the memo to get the message? Look at the fourth paragraph. Should memos sound like legislative bills? Some people do think lawyers or solicitors are supposed to talk like this, but is it appropriate in a memo to a busy person who needs a straightforward answer? If state directors issue special-use permits, they know they cannot assign the property rights of the U.S. government to someone else. And even if the lawyer wanted (just to have it in writing) to remind the state director that this information should be spelled out in the special-use permit, the reminder could be worded in a much clearer and more concise way. Consider the revised memo in Example 16-5.

EXAMPLE 16–4   Roland Trespass Memorandum

MEMORANDUM

DATE: January 12, 1999
TO: State Director
FROM: John Lawbook, Solicitor   *JL*
SUBJECT: Roland Occupancy Trespass

This responds to your memorandum dated December 21, 1998, requesting that we review and comment concerning the subject Roland trespass on certain lands under reclamation withdrawal.

We appreciate your apprising us of this matter and we certainly concur that appropriate action is in order to protect the interests of the United States.

We readily recognize the difficult problem presented by this situation, and if it can be otherwise satisfactorily resolved, we would prefer to avoid trespass action. If you determine it permissible to legalize the Roland occupancy and hay production by issuance of a special use permit, as suggested in your memorandum, we have no objection to that procedure.

Any such permit should be subject to cancellation when the lands are actively required for reclamation purposes and should provide for the right of the officers, agents, and employees of the United States at all times to have unrestricted access and ingress to, passage over, and egress from all said lands, to make investigations of all kinds, dig test pits and drill test holes, to survey for reclamation and irrigation works, and to perform any and all necessary soil and moisture conservation work.

If we can be of any further assistance in this matter, please advise. We would appreciate being informed of the disposition of this problem.

Source: John O'Hayre, *Gobbledygook Has Gotta Go* (Washington, DC: U.S. Government Printing Office, 1973), pp. 11–12.

EXAMPLE 16–5   Revised Roland Trespass Memorandum

MEMORANDUM

DATE: January 21, 1999
TO: Jonathan Leigh, State Director   *JL*
FROM: John Lawbook, Legal Services
SUBJECT: ROLAND TRESPASS OCCUPANCY

I received your memo on the Roland trespass case. You're right—action is needed. The problem is tough, and we'd like to avoid trespass action if we can. So, if you can settle this case by issuing Roland a special-use permit, go ahead. Please spell out the Government's cancellation rights and right-to-use provisions in the permit as usual.

If we can be of further help, please call.

Source: John O'Hayre, *Gobbledygook Has Gotta Go* (Washington, DC: U.S. Government Printing Office, 1973), pp. 14–15.

CORRESPONDENCE ORGANIZATION

As with other writing tasks, the way you structure your message strongly affects how well your reader responds to it. To achieve the conciseness that a busy reader needs, consider a direct plan in your workplace correspondence, a deductive or top-down approach. This direct plan is especially useful if you have good news to tell your reader, and this plan follows the typical organization of technical communication—getting to the point in the beginning.

In some cases, you must report bad news or turn down a request, announce a price increase, or deny an application, so you might begin this type of correspondence with some buffers to your bad news. These buffers will help you keep the goodwill of your reader: maintain a business relationship, help a reader save face, or ensure some future relationship. You can begin by offering the reader another choice than the one requested, assure the reader that the applications were all outstanding even though the job went to someone else, or offer a discount on the reader's next purchase. Bad news needs to be de-emphasized, so placing it more toward the middle of your correspondence will give you a chance to soften the blow your reader might otherwise feel. Our best advice then is to state the good news first—the bad news early, but only after you have had a chance to prepare the reader.

To test this advice look at Example 16–4 again. The good news is buried. The answer to the state director's question, "Is it legal for me to issue Roland a special use permit?" is yes, although it's a qualified yes. Burying the answer in the middle of the memo forces the reader to search for it, and even though the news is good, the reader might be alienated by that search. Now look at Example 16–5 again. Here the good news is immediately perceived. You probably realize now that it's not just direct language that makes this message clear; it's also the direct plan.

Following the Conventions of Workplace Correspondence	Letters and memos require conventional formats. To deviate from these formats distracts or even irritates readers, causing them to question your knowledge and background. Make use of what people expect but use your creativity and originality in organizing your arguments and articulating your ideas.

Businesses generally find value in these conventions, although some firms are more flexible than others. Always know the accepted form for correspondence within your company and follow it carefully. If you were not given a corporate style guide, find some example of other work done in the company. Even though you may not be responsible for actually preparing final copies of the letters and memos, your name will be on them so the responsibility for their appropriateness is yours.

CONVENTIONS FOR WRITING BUSINESS LETTERS

First impressions are important. Letters should look like a picture in a frame. Most letters have seven main parts: letterhead or heading, inside address, salutation, body, complimentary close, signature, and initials. Example 16-6 illustrates these main parts. Read through all the parts of this letter carefully to understand the conventions of the business letter.

Corporations usually have printed **letterhead** for their correspondence. The letterhead includes the name of the company, its address, and other information such as telephone and fax numbers and department identification. The date of the letter is usually two spaces below the last line of the letterhead. If there is no letterhead, the **heading** includes the sender's street address, the city, state and zip code, and the date. Sometimes you will find the sender's address after the sender's name in the signature block.

The **inside address** names the recipient of the letter. It includes the recipient's full name, job title, company name, address, and the city, state, and zip code. An appropriate form of address always precedes a person's name; for example: Ms. Andrea Bayard Smith; Professor William T. Anderson; Representative Thomas D. Henderson; or The Reverend Michael Duin. If you do not know the name of the person receiving your correspondence, the first line of the inside address may be a job title such as purchasing manager, the name of the firm, or a department within the firm. In writing the company name, follow the way the company writes its name. For example, some companies do not abbreviate the word *company* (Co.) in their name.

The **salutation** is the traditional hello used in letter writing. What salutation you use depends on two things: the first line of the inside address and how well you know the receiver. If the first line of a salutation is a company name, you can use Gentlemen, Ladies, or Ladies and Gentlemen, although repeating the company or department name instead will eliminate the need to guess the gender of the reader. The title Ms. eliminates having to decide whether to address a woman as Miss or Mrs., but a better solution is to find out what title the woman you are writing prefers and use it. This strategy avoids the impersonality of "Dear Executive" or "Dear Purchasing Agent."

The **body of the letter** contains the message. Notice in Example 16-6 that the body of a letter is single spaced with double spacing between paragraphs. Bulleted items work in letters as well as reports, and topic headings can improve the readability of letters and memos.

The **complimentary close** is the "good-bye" in a letter. Its level of formality matches the level of formality in the salutation. In the **signature block,** if the company name appears two spaces after the complimentary close, the letter writer is legally established as a representative of a company. Otherwise, the company name follows that of the letter writer. **Initials** tell who dictated and who typed the letter.

letterhead
stationery that includes the company name, logo, address, telephone and fax numbers, and Web address

heading
the sender's name, address, and the date at the top of a letter not printed on stationery

inside address
receiver of a letter

salutation
greeting or "hello" found at the beginning of a letter

body of the letter
the message or content section of a letter

complimentary close
the "good-bye" found at the end of a letter

signature block
signature and company name at the end of a letter

initials
identification of who typed a letter

EXAMPLE 16–6   Main Parts of a Letter

At least 1½"

Letterhead

Coalition of Citizens
Concerned with
Animal Rights
CCCAR

August 2, 1999

3–12 blank lines

Inside address
(single spaced)
1 blank line

Dr. Karen Wartala
Dean, College of Agriculture
University of the Midwest

Salutation
1 blank line

Dear Dr. Wartala:

Body (single spaced with double
spacing between paragraphs)

The Coalition of Citizens Concerned with Animal Rights (CCCAR) was alerted by a concerned person two weeks ago that the University is trapping and killing large numbers of birds in the experimental crop fields on its campus. In our surveillance and investigation we have learned that:

- Birds are lured to the traps by bait placed in them, and the number of birds in these large traps has been observed to be at least 50 at times.
- The reported method of killing the birds was suffocation, by placing them, 400–500 at a time, in a bag.
- Approximately 500 birds are killed every day on the campus; 10,000 were killed in a three-month period.

We object to this for the following reasons:

- It is a waste of animal life.
- It is ineffective. Trapping and killing birds will not permanently reduce the bird population in the area. Even temporarily it will have no more than a minimal effect.
- The trapping and suffocation of these birds is cruel in the extreme.

For these reasons, we insist that the University immediately stop the killing of these birds. If you wish to discuss this matter, please contact me at 555-5537 or 555-5856.

1 blank line
Complimentary close
Signature block
3 blank lines

Yours very truly,

Ed Younger

Ed Younger
Vice President, CCCAR

Initials

EY:sl

Professional letters can also have other parts (see Example 16-7), depending on whether they are needed or not. **A special handling line** at the beginning of the letter, usually after the dateline, tells about a letter's transmission. If you are sending a letter by certified mail, registered mail, overnight express mail, or with a return receipt requested, you will include

special handling line
line at the beginning of a letter, following the dateline, indicating the letter's transmission as certified, registered, or overnight mail

---

EXAMPLE 16–7 Other Letter Parts

The ABC Manufacturing Company
2121 Evans Road
San Diego, CA 92021
Fax 619-667-5588   *Special handling line*

Attention: Mr. H. John Bryan   *Attention line (considered part of inside address so used on the envelope also)*

Dear ABC Manufacturing Company:

SUBJECT: Other Letter Parts   *Subject line*

You will need to use special parts of a business letter when they are appropriate. Notice the spacing and punctuation used with the attention line.

The subject line can appear even with the left margin, centered two spaces below the salutation, or centered on the same line as the salutation.

The additional letter parts illustrated in this letter are the enclosure notations, found below the initials, copy notations, and the postscript. The postscript appears a double space below the last item on a letter. You can use the abbreviation P.S. although you don't have to.

Sincerely yours,

Frances Burgus

(Ms.) Frances Burgus

sl
Enclosure 2   *Enclosure notation (1 or 2 spaces below initials)*
   1.   Bill of materials
   2.   Check for $102.01
cc: Juan Martinez   *Copy notation (1 or 2 spaces below enclosure notation or initials if no enclosure)*

The postscript is either indented or even with the margin, determined by how you arrange the rest of the letter.

          Postscript (always last item; can be handwritten)

attention line
used as the salutation if no name appears on the first line of the inside address, e.g. Attention: Marketing Department

subject line
line at the beginning of a letter summarizing in four or five words what a letter will discuss

enclosure notation
an indication that enclosures are attached to a letter

copy notation
an indication at the end of a letter that copies were sent and who received them

this information. If you use a fax machine to send your letter over regular telephone lines, you can indicate this mode of transmission by adding the fax number to the inside address. Use an **attention line** only if no name appears on the first line of the inside address. It directs a letter to a particular department or person, but it also directs the letter to someone else if that person is no longer in that department or is on vacation.

The **subject line** saves words in the beginning of a letter by summarizing what the letter is going to discuss, usually in four or five words. You can use it to refer to an earlier letter or to a file reference number. Sometimes it is indicated by RE: or Ref or by the word "SUBJECT" followed by a colon.

If a letter includes an enclosure, mention the enclosure in the body of the letter and after the initials in the **enclosure notation.** You can use the word "Enclosure" or the abbreviations, Enc. or Encl. List the items enclosed if they are important, and if your letter has more than one enclosure, let the reader know how many items to expect. If you are sending copies of the letter to other people, use the **copy notation.** The notation c: or cc: (standing for "carbon copy," a holdover from the era before photocopying and computers) is followed by the names of the people to receive copies. Also, pc stands for photocopy and bc for blind copy, marked only on copies that are sent to other readers without the main recipient knowing it.

Use the abbreviation P.S. to indicate a postscript, or you can merely add an additional paragraph after the last item in a letter. Handwriting the postscript gives a personal touch to a letter. The postscript is rarely an afterthought left out of the body of the letter, but instead can emphasize a main point.

continuation page heading
an indication at the top of additional pages of a letter that they are part of a single document

If a letter is long, more than 250–300 words, continue it on additional pages so the letter is not crowded on one page. Indicate additional pages with a **continuation page heading** that keeps the letter intact in case pages become separated. Do not use letterhead paper for continuation pages, but the paper should be of the same quality and color as in the letterhead. It's a good idea to have at least two lines of a paragraph to continue onto the next page and never hyphenate a word at the end of one page to carry over to the next page.

CONVENTIONS FOR WRITING MEMOS

You use memos to correspond with people within your company whether they are sitting in the office next to yours or in a division office across the country. You can use letters for this purpose too, especially if you are writing to a superior in top management. Generally, however, you will use memos for in-house correspondence.

Many firms use printed forms for communication within the company. These memos display much more variety than letters in their styles and format conventions. Example 16–8 illustrates a typical memo form. Notice the

EXAMPLE 16–8   Memo Form

MEMORANDUM

DATE: July 31, 2000

TO: All Departments

FROM: H. Watanabi, President   *HW*

SUBJECT: Interoffice Memo Forms

Most interoffice memos discuss only one, well-defined subject. The style is usually slightly less formal than most business letters. Often a file or reference number placed on the subject line or in a spot especially designated for it makes the memo easy to file away or helps the reader to locate preliminary information to which the memo refers.

Memo formats vary, although all include lines at the top that identify it as a memo, give the date, and indicate who will receive it, who is writing it, and what its subject is. Compare some of the sample memos in this chapter to see the different ways you can arrange this information.

heading calls for a date, the writer's name, the receiver's name, and a subject. Job titles usually follow names, and if you are sending the memo to someone outside your department, you would identify both your department and that of the recipient, especially if all who receive copies of the memo would not know this information.

The body of the memo begins two or three line spaces after the subject line, and it is usually single spaced with double spacing between paragraphs. Memos do not have closing lines, although you should initial or sign the memo where your name appears in the heading or at the end of the memo. The typist's initials appear below the body, and other notations, such as enclosures and copies, appear as they do on letters. If your memo is more than one page, you can either use the continuation page headings used for letters or merely number the following pages as you would in a report, usually in the upper right-hand corner of the page.

Memo forms can vary from simple, handwritten notes to highly structured forms with attached copies of papers and complicated, predetermined routing slips (Example 16-9). Very large businesses may have a variety of memo forms, used according to the type of communication being sent.

Networking systems that provide e-mail capabilities are popular for in-house correspondence. Although electronic communication may help cut down on the paper load, this off-the-record and ephemeral communication poses problems, both ethical and legal, that its users have just begun to contemplate. Often electronic messages do not become permanent records of

---

EXAMPLE 16–9   Memo Pack

	Date:
	I.D.#
	REF#
FROM:	SUBJECT:

MESSAGE

Blue--Originator

Yellow--Purchasing

Green--Accounting

---

correspondence, creating gaps in the information that documents the work history of projects. When people do print out e-mail messages so they have a record of the material covered, the sender sometimes discovers that a hasty, informal, and maybe even ungrammatical and misspelled document is put into the permanent file and identifies him or her as the originator. As a consequence, when you use e-mail for other than run-of-the-mill material, approach your document as an electronic memo and follow the guidelines for effective memos. Example 16–10 shows a typical e-mail memo and reply, a version of the memo example shown earlier in this chapter (Example 16–3).

Notice that the content does not change much, but the memo is much more informal. Smalley just "hit reply" to send a message, assuring Smitzer that she would follow through on his request, and her message appears before the copy of his. Both parties can tell exactly—to the second—when the messages were sent, and they each have a complete copy of the correspondence.

EXAMPLE 16–10   E-mail Memo

> From: Jsmalley@SystemsTech.ComEng.com
> Date: 12 Aug 2000 15:02:16
> To: Jsmitzer@SystemsTech.Marketing.com
> Subject: Possible Problems with 175 Mother Board
>
> Jack: Sure thing! The repair shop just called, and I have already asked my team to look over the computer. Will get back to you by Tuesday.
>
> From: Jsmitzer@SystemsTech.Marketing.com
> Date: 12 Aug 2000 13:59:28
> To: Jsmalley@SystemsTech.ComEng.com
> Subject: Possible Problems with 175 Mother Board
>
> A customer has reported a cracked and smoking chip in the Systems Tech 175 somewhere "behind the hard drive," which appears to have damaged the entire mother board. I have checked with the repair shop, and they told me that although the problem is unusual, three such complaints have come in on this model. The repair shop will call you when the computer comes in and will copy the other two complaints for you. Please give the computer a quick check to make sure that we do not need to change the model design in some way. I promised the customer a fast repair so I hope that we can get someone on this as soon as the machine arrives.

Although you write memos with a more informal style, what is true in the writing of letters is also true in the writing of memos, and you will want to prepare them carefully. Many times the only identity you have in the workplace is based on your memos.

Creating Effective Letter and Memo Reports

Often you will use the letter formats discussed in this chapter for your short reports for people outside a company. Short reports are usually defined as those reports under 10 pages. Documents of more than 10 pages need the help of the conventions associated with longer reports (see Chapters 13 and 14).

You can use many of the conventions of effective documentation within the letter formats: headings to chunk your prose, items in a list, and graphics to complement your prose. Example 16-11 shows a letter report. (See Appendix A, Case Documents 3 for the background of this letter report.) Notice that after the traditional salutation, a short introductory paragraph begins the letter. This paragraph functions in much the same way as the letter of transmittal or cover letter. Do not label it as "Introduction," however. This heading sounds out of place in letter reports. The report begins immediately after this paragraph and follows the pattern of organization most appropriate to the subject. After the conclusion and recommendations, if any, the conventional complimentary close and other letter parts appear.

How is Example 16–11 part of a communication chain? How did the letter report writer establish *ethos* and use *logos*? Do you see any examples of *pathos*?

Using memo format is common practice for short periodic reports such as progress reports, field reports, accident reports, and trip reports. Such reports are usually under 10 pages and written for an in-house audience.

EXAMPLE 16–11   Letter Report

August 10, 1999

Mr. Ed Younger
Vice President, CCCAR
67 East Main
St. Paul, MN 67492

Dear Mr. Younger:

I appreciate receiving your letter stressing your objections to our bird control program. To clarify the program, I am sending you this short report that explains the purpose of the program, details the procedures followed in controlling the birds, and provides information about the trapped birds and what happens to them.

Purpose of the Bird Control Program

The purpose of our bird control program on the campus plots is to protect the experimental plantings that may start new crop varieties. The seed produced on these plots is extremely valuable because it represents the product of genetic crosses carried out over a period of years. With each plant selection, only a few seeds may be produced so it is essential that they be protected. The new plant varieties produced may increase food production worldwide and, thus, may reduce world hunger. We believe the benefits of these plant breeding experiments are important enough that we must protect these experimental plants from being destroyed by pest birds.

Procedures for Controlling Birds

The campus plots occupy approximately 100 acres. The comprehensive bird control program on these plots involves several approaches that we are continually improving as we learn about new techniques. These approaches include the following:

- Plastic strings stretched across the plots vibrate to create a noise that frightens birds.
- Brightly colored, large balloons with metallic designs (scare-eyes) suspended above the plots deter birds.
- A hired, licensed falconer uses trained hawks to frighten birds away from the plots daily.
- We are encouraging sparrow hawks to nest in the plot area to help control the birds naturally.
- Eight traps in the plot area, baited with bread, trap birds. Trapped birds receive food and water until they are removed from the cages.

Information on Trapped Birds

The cages trap approximately 150 to 200 birds daily, with the number of birds trapped varying greatly. On unusual days as many as 500 birds may be trapped. Table 1 shows the kind and number of birds trapped from 1995 through 1998.

(continued)

EXAMPLE 16–11 (Continued)

Table I	Kind and Number of Birds Trapped, 1995–1998				
	Kind of Bird				
Year	**Sparrow**	**Blackbird**	**Starling**	**Grackle**	**Total/Year**
1995	6305	581	1086	163	8135
1996	2317	1468	400	131	4316
1997	3843	502	1104	325	5574
1998	3396	395	2160	412	6363
Total/Kind	15861	2946	4750	1031	

The program never permanently reduces the bird population. Our bird control program attempts to reduce bird population in the plot areas only during the growing season. All songbirds are released. The only pest birds destroyed are blackbirds, starlings, and English sparrows. Destroyed birds are given to the Raptor Center to feed birds housed there. The Raptor Center uses all birds supplied to them.

New Method Used to Euthanize Pest Birds

Beginning August 6, trapped birds will be euthanized with carbon dioxide. Birds collected in cloth bags will be kept in a cool place until they are euthanized with high concentrations of carbon dioxide in a closed chamber. Collecting and transporting the birds in a cloth bag is more humane than using a wire cage, because the birds remain quiet in the dark bag and do not damage themselves by fluttering about, as occurs in a small cage. Euthanizing with carbon dioxide is an accepted, approved method used for small animals that results in loss of consciousness within 45 seconds and respiratory arrest within five minutes. Information from the 1986 Report of the American Veterinary Medical Association Panel on Euthanasia indicates that carbon dioxide is an effective euthanizing agent. Inhaling carbon dioxide causes little distress in birds, suppressing nervous activity and inducing death quickly. Euthanized birds can still be used for food in the Raptor Center.

Alternate Methods for Handling Pest Birds

Transporting and releasing birds is not possible because a suitable release site is not available that would be acceptable to the general public. Releasing significant concentrations of pest birds is not a desirable outcome to those who would be living in a potential release area and would be irresponsible on our part.

Placing nets over the plots to prevent bird damage is impractical because of the size of the plot area (100 acres). It is also impractical because it would restrict management and evaluation of the plots. Nets would also be subject to damage by storms.

This report provides information about the bird control program on experimental plots at University of the Midwest. If you have further questions about our bird control practices, please call Dr. Pat Lennox, Department Head, Department of Plant Pathology, or Mr. David Obi, Research Plot Coordinator, Department of Plant Pathology.

Yours very truly,

Karen Wartala

Karen Wartala
Dean, College of Agriculture

They begin with typical memo headings. Two or three lines after the subject line, the introductory paragraph explains what is to follow, as in the letter report. At this point, the writer presents the report, following the pattern of organization appropriate for the topic. As in all memos, no signature lines or complimentary closes appear at the end. Topic and audience determine the level of formality of memo and letter reports, but letter and especially memo formats would be inappropriate for a formal technical report.

Creating Effective Letters of Transmittal and Cover Letters

letter of transmittal
letter accompanying the transmission of a report to the recipient informing recipient about the authorization for the report and its title

letter of authorization
letter authorizing someone to begin a project and providing that person with task specifications

Letters of transmittal "transmit" reports to recipients. Usually, projects begin as a result of a **letter of authorization,** which authorizes you to work on a project, gives specifications for the project, and tells you exactly what to do. Reporting becomes the record of the progress of projects, and a final report ends the documentation on most projects. Many companies hire industrial librarians to archive these histories of their projects.

Basically the letter of transmittal tells the reader, "Here's the report you asked for," although not as informally as this. At the beginning of the letter or memo, you refer to the authorization and the request for the report along with its topic or project title. Usually the letter of transmittal appears at the beginning of the report, placed either before the title page or after it. If it is after the title page, it directly precedes the table of contents. Example 16–12 illustrates a letter of transmittal.

You may include other information in the letter of transmittal, which acts much like a preface. Remember that the letter of transmittal is usually brief. Some of the following items would likely appear in the letter:

- Significant information you wish to stress to the reader, perhaps sections of the report the reader will be particularly interested in or that other people in the organization would find important.
- Very brief summary of the report (more descriptive than informative, especially if an abstract follows within the next page or two).
- Conclusions and recommendations (depending on audience, some writers prefer not to report bad news this early in the report).
- Possible action to take as a result of the information in the report.
- Relevant background information not found elsewhere in the report (funding enabling your research, limitations of your results—but be careful not to undermine your report).
- Thanks and acknowledgments for help not mentioned elsewhere in the report (reference librarians, members of your staff, and so forth).

Letters of transmittal frequently end with a statement of appreciation for working on the report (especially if you are reporting to a client on the work your company has done and hopes to continue to do), an indication of what you may have learned from working on the report or project, an offer to answer questions or provide additional information the reader may need, or any combination of these comments.

EXAMPLE 16–12   Letter of Transmittal

September 1, 2000

Ms. Megan Dodd, President
Information Services, Inc.
P.O. Box 4750
Denver, Co 50765

Dear Ms. Dodd:

Here is the report on the results of the survey you asked me to conduct last month on methods being used in the Sales Department to eliminate sexism in our business letters.

As you can see from the report, most of our letter writers favor using Ms. at all times, although several of the supervisors surveyed reported some negative responses to this choice. You will find these comments listed on page 15 of the report.

When I give my report at the meeting next week, I suggest we go over the alternatives listed on page 19 of the report. Although this problem is a sensitive issue to most of the people responding to the questionnaire I sent out, 87 percent indicated they would like to see a policy established for all departments to follow.

I would be happy to answer any questions you might have about the results of the survey.

Sincerely,

Benjamin Evans

Benjamin Evans
Vice President, Sales

Enc.

Cover letters also may transmit reports, especially ones that are not formally solicited by a letter of authorization. These letters provide the same kind of information as letters of transmittal except at the beginning you will need to explain more directly why you are sending the report. Frequently cover letters accompany the résumé, questionnaires for surveys, and any other items you are sending to people who will need an explanation about the information they are receiving. Depending on the audience, you can use either memo or letter formats.

Cover letters can be extremely brief, especially if someone has requested information informally. Depending on the information requested, your cover letter may merely indicate "Here's the information you asked

cover letter
resembles a letter of transmission, but explains more directly why the report is being sent to the recipient

for." In many offices, handwritten Post-it memos suffice. Specialized self-adhesive paper can also be used as a cover sheet for fax transmissions. These adhesive notes indicate source and receiver and allow you to send a fax without attaching an additional transmittal sheet.

In some circumstances, your cover letter will need to be extremely persuasive, especially if you are asking your reader to take time to do something for you, such as fill out a questionnaire you are enclosing with your letter. As in any correspondence in which you make requests, tell the readers the benefits to them in granting your request before you make the request. Example 16–13 shows that you can also offer inducements to the reader, such as an offer to provide the results of the survey. Assurances that completing the questionnaire will take only a short time may result in more returns.

EXAMPLE 16–13   Cover Letter

October 22, 1999

Dear Fellow Student:

Finding a parking place on campus has become an increasingly vexing problem. Many of us arrive on campus hours before our classes to ensure we will find a place to park. Lines of waiting cars queue up in parking lots with students offering to take departing students to their cars so they can have their parking slots.

I am exploring solutions to the parking problem on our campus. Would you be willing to take a few minutes of your time to answer the enclosed questionnaire that asks about alternative solutions to the parking problem on campus? The questionnaire asks you to respond to questions about car pooling and other possible solutions to the problem of finding a parking place on campus.

The report of my survey will be sent to the Department of Physical Facilities with recommendations on how to resolve this problem. If you would like to receive the results of this survey, please fill in the separate form attached to the enclosed questionnaire, and I will mail you the data I collect. Your answers will remain confidential.

Your responses will provide valuable information about solving this problem. Please respond by October 31 by mailing your completed questionnaire in the enclosed return envelope.

Sincerely,

Spenser Reid

Spenser Reid

Enc.

People are frequently asked to write **letters of recommendation,** especially as they move into management positions. Some firms, because of legal considerations, refuse to provide any information beyond a person's employment history with the company. Letters of recommendation, however, are still frequently written either as a result of company policy or a desire to help someone in the job search. Letters of recommendation are also written for promotions, appointments, and admissions to special programs and schools.

Many letters of recommendation include glowing, but vague abstract commendations that are of little value to the person you are recommending. Try always to supply concrete, specific detail. Rather than using vague adjectives such as "conscientious," "dependable," "hardworking," "motivated," and "responsible," which have little specific meaning, try "completed a six-month-long project on time," "initiated a newsletter within the department," or "supervised a crew of seven electricians." In other words, supply good reasons for your argument.

Begin the letter by indicating your relationship to the person you are recommending: Did you work with this person as a supervisor? As a co-worker? Usually you will mention the length of time you are reporting on, provide a brief record of employment, and give an overall appraisal of your recommendation.

As Example 16–14 illustrates, after the opening paragraph, you provide the specific and concrete description of what you know about this person's activities. Follow this description with your concrete and specific evaluation of this person's performance. The closing paragraph emphasizes how strongly you are recommending this person. It might include a statement of your willingness to answer any questions, with your phone number provided.

Be careful in the language you use to describe the person you are recommending. What might have been intended as a compliment might in fact be seen as inappropriate. Do not make any comments about a person's race, age, sex, national origin, or disability as qualifier—either negative or positive. For example, don't say, "Mary Smith is a highly skilled technician and a very attractive presence in the laboratory" or "Despite the fact that he is over 50 [or only 20], Jamil Polanski is a diligent and hard worker."

Whether to include negative comments is a tough decision. Some companies are fearful that if the person does not get the position, negative comments in recommendation letters may be used in lawsuits. For example, in a grievance, the person about whom you made a negative comment might require you (and your company) to prove what to you seemed a matter of professional judgment or interpretation. When you are asked to write a letter of recommendation by a person about whom you cannot write favorably, suggest that the person ask someone else, if you can. In a competitive job market, lukewarm recommendations have

Writing Effective Letters of Recommendation

letter of recommendation
letter written at the request of the person seeking a position; describes writer's knowledge of a job applicant's abilities

EXAMPLE 16–14   Letter of Recommendation

June 6, 1999

Mr. Benjamin Evans
Vice President, Sales
Information Services, Inc.
P.O. Box 4750
Denver, CO 50765

Dear Mr. Evans:

I am writing in enthusiastic support of the application of Keith Grey for a position in your department. Mr. Grey worked for me for over a year as a student assistant in the Validation Research Center at San Diego State University. My impressions of his work record are overwhelmingly positive.

Mr. Grey assisted at every stage of the usability tests we performed. He helped arrange for test subjects, maintaining a pool of applicants for this position and corresponding with them. He prepared questionnaire forms, test booklets, sample documentation to be tested, and reports of the tests. He also entered data collected into a series of databases in *Lotus 1–2–3* and *dBaseIII*. He is also proficient on several word processing programs, *WordPerfect, Microsoft Word,* and *Microsoft Word for Windows.* He has designed graphic displays of data for our reports using a number of different graphic programs.

In all he did, he was scrupulously accurate. The only times he redid something were when I changed my mind, and these changes he did cheerfully. He works well without constant supervision, for much of his work involved staffing a one-person office. I often could present a problem I was working on, for example, data collected during a test session that I wanted to display in some way, and he would present to me, very quickly in many cases, several possible solutions. He has an eye for knowing what needs to be done and takes it upon himself to do it.

In addition to these exceptional attributes, he is a congenial, friendly worker. He seems to enjoy his work, is always on time, and has missed not one appointed work time. He works well with the people who call or came into the Validation Research Center, is usually able to answer their questions, and, if not, passes on careful records of their questions to me.

Mr. Grey is obviously an important asset to our Validation Research Center, and I regret to see him leave. I would consider him an asset to any organization and recommend him without hesitation.

Sincerely,

Sherry Burgus Little

Sherry Burgus Little, Director

extremely negative results. At times you may receive a request for a letter of recommendation when the person has not asked you first if you will provide one. If you have not agreed to supply a letter of recommendation as the result of a personal request, you are under no obligation to write one, unless you are that person's immediate supervisor and company policy dictates this function as part of your job. It is always risky to give someone's name as a reference without first asking that person's permission, but some people do it.

Communicating Effectively Online

We have already discussed sending memos by e-mail, but you might also be invited at some time to participate in an online discussion group or some other system of computer-mediated communication (CMC). For example, you and your collaborative team might find it useful to set up a discussion list to discuss the problems detected with the Systems Tech 175 personal computer, the subject of the examples at the beginning of this chapter. If so, you will probably find the immediacy of discussion lists attractive, but you might be fooled by what appears to be informal and even anonymous communication. Some people even develop a different persona or *ethos* as they participate in the discussion conducted online in chats or discussion groups. Also, "flaming" (hostile replies) sometimes gives online discussion groups a character much different than that of face-to-face communication within a meeting. Participants sometimes find it easier to attack directly an opinion or even a person, attacks they would suppress or mask in a face-to-face situation.

Your participation in CMC should at all times be professional. Your messages can become part of a paper trail somewhere if they are printed out, and your professional relationships can become strained when and if you meet other CMC participants in person. Try carefully to think through and edit all your responses—and wait at least a few minutes before firing back a response to give yourself time to rethink your strategies.

Developing Job Search Materials

One important process usually begun through correspondence is the job search. As is true for all writing tasks, your written documents will be more successful if you do some planning before you begin to write. Many people slight this important part of the writing process and end up with inadequate job search materials that get them no job offers. You'll be spending most of your life in your chosen career, and your level of satisfaction is a result of one of the most serious decisions you will make.

IDENTIFYING JOBS

What are some things you can do to make sure your decisions are the best ones? Being sure you understand what you will be doing on the job can help ensure a successful job search. Talking to people working at what you think you want to do is probably one of the most effective ways to make the right decisions. For example, conducting job information interviews can provide you with valuable insights about how companies function, how their employees are treated, what kind of conditions they work under, what goals and values the company has, and what opportunities for advancement, relocation, and further education exist. Most people enjoy an opportunity to talk about themselves and their jobs. Making plans about what you want to talk about, realizing people are busy so that you don't waste their time, and being courteous before, during, and after the interview ensure successful encounters.

If you don't know someone you can interview, you can meet people by becoming active in the professional organizations of your field. Most professional organizations have local branches that meet monthly, providing not only contacts but also professional programs or guest speakers of interest to practitioners in your field. The Institute of Electrical and Electronic Engineers (IEEE), American Society of Mechanical Engineers (ASME), the American Business Communication Association (ABC), and the Society for Technical Communication (STC) are just a few examples of these professional organizations. Most professional organizations post jobs online for their members and have student chapters in which you can participate—an important step to take well before you begin thinking about your job search. It's true that most job interviews result from personal contacts or referrals, not from answering job advertisements or sending out unsolicited job applications. The **networking** important for such referrals should begin as soon as possible and continue throughout your professional life.

networking
in the context of a job search, a group of personal contacts who can assist you in identifying jobs and opportunities

If you live where national organizations have no local chapters, you can use other resources. Another important source of information about careers is the *Occupational Outlook Handbook,* published periodically by the Department of Labor. Its companion volume, *The Dictionary of Occupational Titles,* identifies job clusters or "occupations" and gives you job titles within these clusters, as well as short descriptions of occupations. It then provides a decimal number you can use to find more detailed information about job opportunities within the clusters in the *Occupational Outlook Handbook.* You will find information about what kinds of jobs are available, where such opportunities exist in the United States, background and skills needed, and information about salary and future job opportunities. You can also visit a company's home page on the World Wide Web. Often an interviewer will expect you to write a job letter or come to an interview already familiar with the information posted on the company's home page.

Your academic advisor or the career counselor at your college or university can also help you explore not only general opportunities for graduates in your major, but also specific companies that hire such graduates. You also will want to talk to the reference librarian at your school and look at the references listed at the end of this chapter before you start making decisions about your job search. You will want to start with the following suggestions months and perhaps years before you search for a job.

Many people look for job advertisements in newspapers, trade or professional journals, placement services, or employment agencies. Again, the most successful source of jobs is referrals by people you know, such as contacts you've made before looking for a job, your professors, and your professional colleagues. Take advantage of internships offered by your school, summer job experiences, and volunteer work to find contacts and information about available jobs. Participating in job or career fairs sponsored by school placement services or professional organizations can also help you to identify job opportunities.

Once you have targeted jobs to apply for, start gathering as much information as you can about the companies you've identified. Read information about the company in newspapers, annual reports, brochures, advertisements, and other sources of information. Trade journals and business publications such as *The Wall Street Journal* or *Forbes* publish information about the activities and accomplishments of companies. *F & S Index of Corporations and Industries* will direct you to publications about companies. Your school library may have information files on local companies, including recent news releases, to help you tailor your job search to the needs, interest, and philosophy of your targeted companies. In each state, the Office of the Secretary of State, Corporate Division, will have information about the status of corporations, the names of the corporate officers, and the address of the corporation. The *CPC Annual,* published by the College Placement Council and available in school placement centers and most libraries, provides names, addresses, and general descriptions of companies as well as suggestions for conducting successful job searches. *Thomas' Register of Manufacturers* and Dun and Bradstreet's *Reference Book of Manufacturers,* available in most libraries, are two more sources of job information.

Again, school placement centers and university career counselors provide information about all parts of the job search. Most offer workshops on writing letters of application and résumés. Some provide help with preparing for job interviews and videotapes of practice interview sessions to suggest ways to handle typical interview questions. Some provide an electronic résumé review that allows employers to review résumés by computer. State offices of economic development offer similar help, if you do not have a school placement center available. Employment agencies can also help in your job search; however, some charge you for such services, although others charge the company that hires you. Be sure to understand exactly how payment will be handled before signing any agreements because some agencies can charge as much as your first month's salary.

These sources can make your job application stand out from all the rest. Being thorough is time well spent.

WRITING THE EFFECTIVE RÉSUMÉ

The résumé will be your prospective employer's first glimpse of your experience and accomplishments. You should always have an up-to-date résumé on file. Even though you may not be considering changing jobs or looking for one, résumés are frequently part of proposals and reports. *Always* include the following information:

- Name, address, phone numbers, and e-mail address (if you have one).
- Education, including major field and degree obtained.
- Experience.

Résumés usually include the following information:

- Career objective.
- Skills, licenses, certificates.
- Honors and awards.
- Professional activities.
- Information on how to obtain references.

You include in your résumé those things you want to advertise or highlight, based on your careful self-analysis and the position for which you are applying. Your résumé headings emphasize what you have to offer or have accomplished, such as "Computer Programming Languages" or "Community Services."

Although résumés alone seldom get jobs, they can determine whether you have a chance at a position. Because a résumé sells you to potential employees, its main purpose is to get you an interview. Competition for jobs is fierce; often the hiring process, in its earliest stage, is not one of choosing the best applicant but of screening out those who don't promise the best fit for the job. Hundreds of résumés may result from a single job advertisement; personnel directors, with only a general knowledge of what is needed, may do the initial screening of applicants, spending only seconds on each résumé. Then managers will receive the résumés that have made it past the personnel directors and will decide whether to schedule an interview.

Some job candidates make their résumés available online. If you decide to do this, remember that your résumé must always look professional, but you will also have to consider how much information your audience will scroll through. Your online résumé should be short and concise, with the major categories of your education, experience, and skills easily identifiable.

What are some of the things that can help ensure that your résumé will go in the yes pile? Generally you will want to think about the overall appearance of your résumé, content, persuasive strategies, and format.

Appearance

A résumé should invite your reader to read it. It should be placed attractively on the page, with careful use of white space (wide margins and space between headings) and easy-to-read type. It should be carefully proofread and edited; some personnel directors and prospective employers discard résumés with typos and misspelled words. They might infer, "If something as important as a résumé is done so carelessly or hastily, what does that tell me about the quality of work this person will produce?" Use all your knowledge about document design when you plan your résumé.

It's usually not a good idea to have a fancy résumé (that is, one that deviates too much from the more traditional formats) or one that has been done by a professional résumé service, as people who look at many résumés recognize the distinctive styles of most résumé services. For professional positions you should design your résumé and then have it prepared on a word processor to look typeset. Print your résumé on a laser printer and make photocopies on quality white or cream-colored paper.

Remember that your résumé must be easy to skim, and the important details of your education, experience, and accomplishments must stand out. Use headings, columns, lists, and boldface or italics to organize it.

Content

Most résumés are a single page, although they increase in length with the level of your education and the length of your work experience. Your résumé is not a complete history of everything you have done, but rather a careful selection of your qualifications and skills. However, leave out information such as your marital status, your age, your health, and other characteristics that might invite discrimination. Include personal interests and hobbies, if this information relates to your qualifications for a position. Perhaps you are an excellent speaker, work with disadvantaged children, or have organized a literacy campaign in your community. These personal activities can highlight your talent and motivation.

References are not always listed on résumés. You may choose to indicate that references will be furnished upon request, although it is safe to assume prospective employers will know this, and you may wish to save space on your résumé for other information or provide different references for different job applications. You should prepare a list of references separately to give to prospective employers before or during the job interview. This list must include names, titles (showing their work relationship to you), addresses, phone numbers, and even fax or e-mail addresses (always make it easy for the prospective employer to reach your references).

Sometimes letters of reference can be part of a portfolio you bring to the interview, although prospective employers are likely to find confidential recommendations more convincing.

Persuasive Strategies

In résumés for technical positions especially, you need to be sure your readers can understand what you are talking about and that they can grasp the information quickly and easily. Avoid acronyms, abbreviations, and technical jargon. Both the technical supervisor and the nontechnical reader must understand your résumé, although you should use the *terminology* of your career field. Keep paragraphs and sentences short, and use phrases when possible. Leave the subject "I" out of sentences so that sentences begin with dynamic action verbs. Again, listing allows the reader to scan through your qualifications quickly without missing any of the important details.

Pay particular attention to your writing style as you describe what you have to offer prospective employers. The following guideline will help you select words and phrases you need to advertise your skills and persuade your readers:

1. Use action words—powerful, dynamic verbs:

adapted	designed	installed	presented
advised	developed	invented	produced
analyzed	directed	maintained	programmed
applied	edited	managed	sold
assessed	established	operated	supervised
conducted	evaluated	organized	supported
coordinated	identified	oversaw	trained
created	improved	planned	wrote
cut	inspected	prepared	

   Try to stay away from phrases like "Responsibilities included" or "Duties were."

2. Emphasize accomplishments and achievements instead of describing what you did. Try to make these quantifiable, such as "Saved $20,000," "Improved sales by 20 percent," or "Supervised 10 people with a budget of $450,000." In choosing what to describe, always remember you are emphasizing what you have to offer your employer.

3. Keep information simple. Don't try to pack too much information into each phrase or you will dilute your key selling points.

4. If possible, include all experience and skills you can offer whether they are from education, work experience, volunteer experience, or social activities.

5. Stress important information in the beginning of your résumé.

Finally, if you do not have much relevant experience, you will probably want to highlight another aspect, such as your education. However, don't list all the relevant courses you have taken, but concentrate instead on the skills you have acquired as a result of your education that are of value to your prospective employer. For example, you might refer to your ability to write programs in several different computer languages or to your document design skills.

Because the format of your résumé will help you emphasize your most impressive qualifications, we discuss the three most common formats below.

The Chronological Résumé

The **chronological résumé** is most common in the workplace. As in all résumés, your name, address, and phone number should be displayed in the beginning (see Example 16-15). In this résumé, the most recent information is considered the most relevant and important. Experience and education are listed in reverse chronology, with the most *recent* listed first. Dates (both month and year) are usually listed first, followed by the name of the organization, the job title, and the job duties. You may wish to highlight the work you have done instead of the length of time you did it. In that case, place dates in parentheses after the job title and organization name. You can also list these actions with bullets rather than use paragraph form. Bullets also make the résumé easier to scan, but don't use so many bullets that they become distracting.

chronological résumé
résumé listing experience and education in reverse chronology, with the most recent information listed first

Again, because most people have experience with the chronological style of résumé, it provides them with a familiar guide for conducting the job interview. Because this style follows a chronological order, you'll find it easy to prepare. You simply structure it by providing dates, company names, job titles, and other information about your education. This format allows you to highlight a steady employment and education record.

Unfortunately, this format also draws attention to any time gaps that might appear, gaps that might be interpreted negatively. If you are a student or are seeking a career change, this format also emphasizes your recent jobs that may not be directly related to the position for which you are applying. Putting yourself through college waiting tables, for example, will show your determination but it doesn't relate directly to a position as computer programmer.

The Functional Résumé

The **functional résumé** highlights your qualifications without emphasizing specific dates or work history (see Example 16-16). As you can see, this style de-emphasizes a work history with gaps and downplays unrelated work experience and education, which might best be handled in the job interview. Instead, the functional résumé emphasizes professional growth

functional résumé
résumé highlighting qualifications without emphasizing specific dates or work history

EXAMPLE 16–15   Chronological Résumé Using Paragraphs

RÉSUMÉ OF KAHUKU V. OADES

46480 Madrid Court
San Diego, California 92124
(619) 888-2319

Education
Sept. 1994–present

San Diego State University, San Diego, California. Bachelor of Science degree in Biology expected in May 2000.

Experience
Jan. 1998–present

Undergraduate researcher, Department of Biochemistry, San Diego State University. Developed and performed smooth muscle cell isolation protocols, conducted spectrophotometric enzyme assays, performed protein purification, and attended local, national, and international professional meetings.

Sept. 1996–Jan. 1998

Research Assistant I, Center for Research on Aging, Department of Medicine, University of California at San Diego. Maintained cell lines in primary culture, performed assays including ELIZAS, cytotoxicity, RIA protein assays, and radioisotope assays, cultured bacterial cells including thermophilic bacteria, partially purified proteins, maintained a cryostorage facility.

June 1996–Sept. 1996

Lab Helper, Clinical Immunology Department, Scripps Clinic and Research Foundation. Prepared buffers and reagents, ordered laboratory supplies, maintained cryostorage tanks.

Honors
1998

Received National Institute of Health: Minority Biomedical Research Support (MBRS) Grant.

1994

Received Samuel Gompers Secondary School, Math/Science/Computer Science Magnet Program Certificate.

Publications
Muscle Membrane Plasma MgATPase is an Ecto-MgATPase. K. Oades, P. J. Yazaki, H. B. Cunningham, and A. S. Dahms. Abstract, *MARC/MBRS Symposium Proceedings,* October 1998.

The Skeletal Muscle Transverse Tubual MG-ATPase Identity with Mg-ATPases of Smooth Muscle and Brain. H. Brad Cunningham, Paul J. Yazaki, Kahuku V. Oades, Ron Domingo, Rodger A. Sabbadini, and A. Stephen Dahms (submitted for publication).

References
Available upon request.

EXAMPLE 16–16 The Functional Résumé

CANDIS L. CONDO

4566 Polk Avenue, San Diego, California 92105   714-283-9668

OBJECTIVE

TECHNICAL WRITER in an organization that needs precise, clearly written documents.

WRITING EXPERIENCE

As *Job Analyst* at San Diego Trust and Savings Bank, developed precise job descriptions based on information obtained through questionnaires and interviews. Have prepared more job descriptions in the first quarter of this year than were prepared in the entire previous year.

As Technical Writer at the same bank, wrote and revised entire Personnel Policy Manual, finishing project four months ahead of schedule.

As Technical Writer for RAIR, Inc. a computer timesharing company, wrote custom software documentation for custom programs.

As College Instructor, taught basic composition, technical writing, and business writing courses at Ohio State University, San Diego State University, American Institute of Banking, and San Diego City College.

RESEARCH AND ANALYTICAL SKILLS

Analyze both exempt and nonexempt jobs and develop written descriptions containing sufficient detail for recruiting, selecting, placing, training, and evaluating bank staff.

Research, prepare findings, and make recommendations on sensitive legal issues, such as sexual harassment, solicitation/nonsolicitation, and age discrimination.

Conduct salary surveys, prepare results, and make recommendations on the appropriate "slotting" of both exempt and nonexempt jobs within bank.

VERBAL COMMUNICATION SKILLS

Schedule and conduct interviews at all levels up through senior management to obtain information necessary to describe accurately job content.

Solicit information from employers and survey services by letter, telephone, and personal visits.

PROFESSIONAL AFFILIATION

Member of the Society for Technical Communication.

PERSONAL DATA

Avid scuba diver and jogger, enjoy classical music, successful public speaker.

and accomplishments that relate directly to the position for which you are applying. The tasks or functions you can perform receive primary attention.

On the other hand, this format does not provide the conventional kinds of information employers normally expect on a résumé, and some potential employers may be uncomfortable with what they consider omissions.

The functional résumé usually begins with the job objective, which also emphasizes what the job applicant has to offer—a persuasive strategy in itself. Making the objective general rather than listing a job title avoids limiting the position for which you may be considered.

Finally, you can use capital letters, italics, bold, and different-sized type to emphasize information. Achievements can then be highlighted by concrete accomplishments and quantifiable details.

The Combination Résumé

combination résumé
résumé combining chronological and functional résumés, listing qualifications and providing company names and dates in a separate section to minimize gaps in work history or temporary jobs unrelated to career goal

The **combination résumé** allows job applicants to take advantage of both the chronological and the functional styles. It is similar to the functional résumé, but it includes company names and dates in a separate section. It allows you to emphasize the skills and capabilities that are relevant, while placing much less emphasis on gaps in your work history or temporary jobs unrelated to your career goal. Both the combination and functional résumé are harder to write than the chronological résumé, demanding careful and critical analysis of the information you have compiled, the job position for which you are applying, and your assessment of the organization to which you are applying.

Study Examples 16-17, 16-18, and 16-19. These résumés illustrate a writer's search for an acceptable format to allow him to discuss his long work history, which has given him excellent qualifications. This writer believed that providing a detailed history, one that spans 12 years of different positions, would prevent his highlighting skills and qualifications, making it difficult for the reader to assess what he has to offer. Listing all these positions and providing a detailed description of each job would take two pages and mean repeating activities performed on different jobs, burying significant information about his performing increasingly more sophisticated tasks. Thus, the chronological résumé format (Example 16-17) prevents readers from easily understanding what he has to offer. How else could this résumé have been improved by following guidelines about persuasive strategies and appearance?

On the other hand, the writer identified those skills he wished to advertise in a combination résumé (see Example 16-18). However, these skills are listed with no relation to each other so that the reader is still left trying to make sense of the long list.

In Example 16-19, however, the writer was been able to classify his skills into easily grasped categories, while still providing a work history and educational information. When this writer interviews for a job, he can provide a complete list of all the jobs with dates, titles, and job descriptions if the interviewer wants to see it.

EXAMPLE 16–17   One Case: A Chronological Résumé

This résumé buries important information.

BRENT MOSBARGER
4266 RIVER DRIVE
SAN DIEGO, CA 92224
789-7234

EMPLOYMENT HISTORY

Date: 3/89 to 1/91 J. S. King Electric, San Diego, California
Project: South Elementary School
Remarks: I supervised this .5 million dollar electrical project which consisted of 5
             new buildings and the remodel of 3 existing buildings.

Date: 2/88 to 3/89 DDD, Escondido, California
Project: North County Hospital
Remarks: Journeyman, I worked on roughing the shell and almost all of the me-
             chanical and backup generator controls.

Date: 12/91 to 1/88 J.S. King Electric, San Diego, California
Project: Midway Junior High School
Remarks: Lead journeyman, I supervised all underground, consisting of large con-
             crete encased duct banks and setting all handholes and transformer pads
             to finished grade. I also installed all of mechanical controls and most of
             the interior pendent mounted lighting.

Project: Blue Onion Restaurant
Remarks: Lead journeyman, I laid out and installed the dance floor lighting system,
             consisting of mounting the unistrut grid, line and control circuits, mount-
             ing of trusses and fixtures. I also installed the class A F.A. system and
             ended up taking over the project for the 4 weeks to completion.

Project: GE Building 99
Remarks: I supervised this project which consisted of service change from 1200
             amps to 1600 amps. Several sub-panels, 225 kva x-former, and miles of
             G-4000, 1500, and 2600 wiremold. I had 11 electricians working for me.

Project: Software, Inc.
Remarks: I supervised this project. It was the addition of 4 class 10 clean rooms.
             Consisted of remote ballast panels, variable speed drives for 100 hp 3
             phase fan motors, a complete pneumatic control system, toxic gas sensors
             with master and slave monitors. This was a very complicated control job.

Date: 5/85 to 12/85 Jones Electric, Denver, Colorado
Project: State Prison Complex
Remarks: Journeyman/foreman. I did the auto-body building and worked in the
             other 12 buildings on their contract.

(continued)

EXAMPLE 16–17 (Continued)

Date:	3/89 to 5/91 New Mexico Electric, Albuquerque, New Mexico
Project:	Federal Building
Remarks:	Journeyman, computer facility for GSA payroll, consisting of a large UPS system, 400 Hz motor generator sets, and some 15 kv splicing. Most of the conduit was G.R.C.

Project:	Department of Energy
Remarks:	Journeyman, installed wiring systems, UPS systems and computers designed to measure and correct air quality. All of the conduit used on this job was G.R.C. up to 4 inch.

Project:	Lees Southwest
Remarks:	Journeyman, 150,000 sq. ft department store. A commercial job with a great deal of lighting and an elaborate energy management system.

Date:	9/88 to 2/88 Fairbanks Electric, Fairbanks, Colorado
Project:	Tunnel Lighting
Remarks:	Journeyman, installed HPS fixtures in 2 highway tunnels between Denver and Central City, Colorado.

My apprenticeship was spent working for Gleen Electric and McLeod Industries in Denver, Colorado, and Spring Electric in Omaha, Nebraska.

Education:	San Diego State University, Electrical Engineering
Graduation:	May 2000

Experience Highlights: My experience with most of the projects was from ground breaking to the final inspection. I spent 8 months estimating for J. S. King Electric Company. I hold a C-10 license through the state of California. I am comfortable with prints and specs and have a good knowledge of code and construction.

Obviously, your audience and your content will dictate the résumé format you choose. You may want to have more than one résumé on hand, using different styles depending on the way you choose to present yourself. As you can see, the information you generate on the self-analysis worksheet helps you select the skills and accomplishments you choose to highlight. You can vary this emphasis quite easily once you've completed the worksheet.

Finally, you might want to develop your own professional Web page to promote your skills and job interests. Your résumé should offer the URL for your Web page, and all the materials contained on the Web page should supplement the skills, education, and experience you listed in your résumé. For example, if you write a feasibility study within a technical communication course or an analytical report for a summer position, you can make

EXAMPLE 16–18   One Case: Combination Résumé

This résumé leaves too much interpretation up to reader.

RÉSUMÉ OF BRENT MOSBARGER

4266 River Drive
San Diego, California 92224
789-7234

EDUCATION

Candidate for Bachelor of Science degree in Electrical Engineering from San Diego State University in May 2000.

WORK ACHIEVEMENTS

Administered change orders.

Supervised and motivated up to 15 people.

Managed electrical projects with budgets of over one million dollars.

Possess a California State Electrical Contractors license.

Ensured that all systems installed were properly operating.

Possess a working knowledge of the National Electrical Code Book.

Qualified as industrial and commercial journeyman electrician.

Resolved conflicts and discrepancies in plans and specifications.

Responsible for ensuring that projects meet plans, specifications, and deadlines.

Participated in estimating and submitting bids on large electrical projects.

Coordinated time schedules between different trades and activities of my trade.

WORK EXPERIENCE

12/91–1/97	J. S. King Electric, San Diego, California. Foreman/Journeyman electrician.
5/91–12/91	Jones Electric, Denver, Colorado Journeyman electrician.
3/89–5/91	New Mexico Electric, Albuquerque, New Mexico Journeyman electrician

REFERENCES

References available upon request.

EXAMPLE 16–19   One Case: Revised Combination Résumé

This résumé is the most persuasive choice.

RÉSUMÉ OF BRENT MOSBARGER

4266 River Road
San Diego, California 9222
(714) 789-7234

EDUCATION

Candidate for Bachelor of Science degree in Electrical Engineering from San Diego State University in May 2000.

TECHNICAL

Possess California State Electrical Contractor's license.

Ensure that all systems installed are properly operating.

Possess a working knowledge of the National Electrical Code Book.

Qualified as industrial and commercial journeyman electrician.

MANAGEMENT

Administered change orders.

Supervised and motivated up to 15 people.

Managed electrical project with budgets over one million dollars.

Resolved conflicts and discrepancies in plans and specifications.

Responsible for ensuring that projects meet plans, specifications, and deadlines.

Participated in estimating and submitting bids on large electrical projects.

Coordinated time schedules between different trades and activities of my trade.

WORK EXPERIENCE

J. S. King Electric, San Diego, California. Foreman/Journeyman electrician.	(12/91–1/97)
Jones Electric, Denver, Colorado Journeyman electrician.	(5/91–2/91)
New Mexico Electric, Albuquerque, New Mexico Journeyman electrician	(3/89–5/91)

REFERENCES

References available upon request.

these writing samples available through your Web page if you own the copyright to them. One important note of caution: Don't mix the message and purpose of your Web page. Don't post jokes or your pet's picture for your friends and family if you expect prospective employers to visit your Web page. Don't fill your page with vivid colors and graphics. Keep the materials professional in content and appearance.

WRITING THE LETTER OF APPLICATION

The **letter of application,** sometimes called a cover letter, accompanies your résumé. It is best written after you have written your résumé. The letter of application, like the letter of transmittal, serves as a tool to transmit the résumé to the prospective employers and to emphasize to readers significant details on the résumé—those skills and accomplishments that relate to the specific position and company addressed. Remember, the letter of application is asking for something—in this case, a job interview—so you will want to be persuasive. You'll need to establish an appropriate *ethos* and use good reasons as to why the reader would benefit from hiring you (try some *because-clauses* as explained in Chapter 4). As in the letter of recommendation, you are evaluating—yourself this time—and you want to be concrete and specific. The content and organization of your application are designed to:

letter of application
cover letter accompanying a résumé that summarizes significant details of the résumé

- Create a successful link to the reader.
- Emphasize throughout your letter the skills, experiences, and accomplishments you have to offer the prospective employer.
- Request action—the job interview.

Creating Successful Links

To create an immediate connection with your reader, address the letter to a particular person. You may need to call the company and talk to a receptionist or operator to identify who will be receiving the letter. While you are at it, be sure to ask about pronunciation, spelling, and preferred title. Often, the job advertisement itself will include a contact name. If so, use it in the address and the salutation, and spell it correctly. If the ad lists an S. Baker as the contact and you do not know which title to use in the salutation, use the full name or call the company to ask what the S. stands for.

In the first paragraph, often in the first sentence, mention how you learned about the position. If someone has told you to apply, mention this referral. If you are answering an ad, name the source of the ad (remember: names of newspapers and trade journals are in italics or underlined, months should be spelled out, and dates require special punctuation). You want to make clear in this beginning that you are applying for a specific position. The following example gets the job done directly and simply:

> I read your ad for the position of a nutritionist in the May 3, 1999, issue of the *San Diego Tribune.* I would like to be considered an applicant for this position.

This opening sentence is typical, however, and the person receiving letters of application may see hundreds of similar opening sentences. Try writing a more interesting beginning, one that still does the job but grabs the attention of the reader, setting your letter and you off from all the others, as in the following:

> Your company's advertisement for a nutritionist in the May 3, 1999, issue of the *San Diego Tribune* calls for someone with a combination of scientific and interpersonal skills. My B.S. degree with a double major in nutrition and biology and my experience working with children and older clients provide me with the qualifications you are seeking.

Notice also that this opening leads the reader smoothly into the next section of the letter, a demonstration that the education and experience do indeed match those advertised. This paragraph also demonstrates the "you" approach, addressing the reader's needs rather than your own. For example, the last example especially emphasizes the "here's what I can offer you" rather than what the first example suggests: "Here's what I want."

If you are not responding to an ad or able to mention a referral in the beginning of your letter, you may consider mentioning a fact about the company you have learned as a result of preparing for the job search. Did you read a recent report about a product or process the company is working on? Did you read about a new service that requires the experience, interest, or education that you can offer the company? Mention the information and where you read it. Most readers will be impressed with your thoroughness, especially if you can demonstrate sincerely and enthusiastically how your background and this information are linked, as in the following example:

> A recent article in *HNS News* detailed the advancements Hughes Network Systems is making in digital cellular communications. I have a career interest in this technology and believe that both my education and work experience qualify me for a position as engineering technician in your department.

Such openings encourage the reader to pay more attention to what you say in the rest of your letter.

Emphasizing What You Have to Offer

The "you" approach is especially critical in the letter of application. Here you must discuss your skills, talents, experience, and education so that your reader will invite you to an interview. It is difficult to write about yourself with a confident tone without appearing to be bragging, but still you must discuss what sets you apart from all the other applicants and how your skills and interests will benefit your potential employers. Remember that you need proof—concrete, detailed, good reasons—to explain why you should be hired (or interviewed). For example, rather than saying "other people love to work with me," you can state, "On our last seven international proposal projects, I was specifically added to six of the groups at the committee leaders' requests."

EXAMPLE 16–20   Letter of Application

4810 Campanile Drive
San Diego, CA 92115
August 9, 1999

Ms. Cheryl Herbers
Computer Sciences Corporation
Applied Technology Division
4045 Hancock Street
San Diego, CA 92110

Dear Ms. Herbers:

In Sunday's issue of the *Union-Tribune,* I found your job advertisement for UNIX system programmers. My experience in "C" programming and with the UNIX operating system has prepared me for the position at Computer Sciences Corporation.

As you can see from the enclosed résumé, I gained this experience at SDSU. I will be graduating in December 1999 with a Bachelor of Science degree in Electrical Engineering. During my studies, I have taken seven classes requiring both "C" programming and using the UNIX operating system. In one class, I wrote a program that broke down enrollment figures of various courses into categories. The program then calculated grade point averages and unit totals in each category, printing the results to the screen in an attractive format using graphics routines.

In addition to "C" programming experience, I have used IBM compatible personal computers both on the job and at home. Since December 1995, I have run income tax programs and entered data into an income tax database. For several years, I have also used a personal computer at home. On this computer, I program in Quick C as a leisure activity.

I believe my interest and experience qualify me for a position as an UNIX system programmer. May I have an interview to discuss my qualifications with you in more detail? Please call me at 619-555-4848. I am available mornings and afternoons until August 30; after that I can most easily be reached in the early mornings.

Sincerely,

Brent W. Warren

Enclosure: Resume

In two or three short paragraphs, you must select concrete and specific details that illustrate the skills, experience, and education you have to offer. Notice in Example 16–20 that the writer has managed to provide persuasive detail, emphasizing his qualifications that match the job mentioned. The writer directs the reader's attention to the résumé and the additional details offered there early in the letter. Rather than simply stating that the résumé is enclosed, the writer points to the good reasons in that résumé for considering him. He is experienced in "C" programming both at school and on the job, skills required for the Computer Sciences Corporation position. Also, the writer carefully establishes his *ethos*. He demonstrates his interest and professionalism by researching the position and analyzing his own qualifications.

Finally, remember that instead of repeating in your cover letter what is on the résumé, you want, in essence, to "teach" your reader how to understand the information provided. For example, you might add up the years of experience from the jobs you have listed in your résumé and call the reader's attention to that impressive length of time in your letter. In doing so, you are persuading the reader that your qualifications match the reader's needs.

Asking for Action

The last paragraph of your cover letter should be a call for action—a request for the interview in which you can demonstrate that you are as impressive in person as you are on paper. Some writers may feel comfortable being assertive here and saying, "I will call you next week to set up an appointment for an interview." Be aware that just one word can change the tone of the ending. Notice the difference in these two sentences:

- *If* we arrange an interview, I can give you more information about my qualifications. [This leaves the action up to the reader.]
- *When* we arrange an interview, I can give you more information about my qualifications. [This assumes the action will take place.]

Whatever your approach, end by making it easy for the reader to contact you. Even though your telephone number is on the résumé, provide it again in the letter. Let the reader know a good time to reach you. If you are applying for a position out of town, you may want to let the reader know when you will be in the vicinity. Some companies may not want to pay your fare to come in for an interview, but if you will be in their location at a particular time, they may be willing to see you. It's better not to thank the reader in advance for anything, and beware of using stereotyped expressions found often in letters like "at your convenience," "please do not hesitate to call," or "feel free." Finally, as the letter in Example 16–20 illustrates, the enclosure notation after the signature block indicates your résumé.

We will cover the next step of the job search process, the job interview, in Chapter 17.

SUMMARY

- In this chapter we have looked at the wide range of correspondence you are going to encounter in the workplace. As with all writing, you will need to consider both the rhetorical dimensions of the project— audience, purpose, and organization—as well as the characteristics of the genre. Correspondence often has a smaller audience, often one or two people known to you, and it has a kind of immediacy lacking in other documents you create. You must write correspondence quickly for it to be useful, but you still must follow both the conventions of the genre and requirements of effective communication.

■ Correspondence, despite its short length and transitory nature, is important both in your career and in the life of the corporation for which you work. Correspondence is part of a paper trail or communication link of any number of documents that address a particular issue or problem. Correspondence has to be clear and concise. These documents require special attention to style because they are too short for you to waste space with vague or imprecise sentences or inadequate organization.

■ Correspondence has clearly defined formats that are well accepted throughout the workplace, and you need to be familiar with those conventions. Therefore, you must know the main parts of letters and memos and how to lay these parts out in the document.

■ Often short reports appear in the form of letters and memos and follow the conventions of correspondence at the same time they solve problems or help people make decisions. Letters of transmittal and cover letters introduce longer documents, such as reports and questionnaires. Letters of recommendation, on the other hand, offer professional assessments of people's performance.

■ In this chapter we've looked at the job search process and the materials you will have to prepare to carry it out successfully. A successful job search begins with careful preparation on your part, some of which involves analyzing your own strengths and weaknesses as well as your experience and background. Another part of the preparation is finding the jobs to apply for.

■ After you have identified a possible job lead, you will need to make successful contact with a potential employer by creating an effective résumé and letter of application. You have several résumé formats to choose from and are now more aware of the advantages and disadvantages of those choices. Always keep a current résumé. Your letter of application should be persuasive and be "you" or reader oriented. The letter of application "teaches" the reader how to read your résumé in order to highlight your most impressive and relevant qualifications.

■ Many job applicants need to remember that finding the right fit is a two-way street. Not only is the company trying to find the right applicant, but applicants are trying to find the right job for themselves. You will receive negative responses. You may even find yourself realizing that a job may not be the right one for you. Most important is not to become discouraged. Be aware that you have much to offer a prospective employer and applying for a position means that you will be interviewing the company as well, as we'll cover in the next chapter.

1. Study the memo in Example 16–4, and without doing any reorganizing, eliminate all unnecessary words. How many words can you eliminate and still convey the message?

2. Collect any business or advertising letters that you receive in the mail. Do you find unnecessary words, redundancies, jargon, and stereotypical phrases in your letters? How did the letter writer establish *ethos* and use *logos* in the letters? Are there examples of *pathos*? Was the letter writing helping the reader solve a problem or make a decision? What other features do you find in these collected letters? As your instructor directs, bring your collection to class for discussion. Working in groups of three, discuss what you find in the collected letters. Report your findings to the class.

3. Explore the ways you can use your word processing software to assist you in creating effective and consistent correspondence. Create a style guide using the style selection features of your software. Experiment with the mail merge feature of your software and use it to send a letter in a form you have selected to your instructor and your classmates. In the letter discuss how mail merge programs can improve and simplify the correspondence process.

4. If you have access to e-mail on your campus, consider joining an online bulletin board, chat, or discussion group to engage in a "correspondence" with someone who shares your interests. In a memo to your instructor and classmates, describe the nature of this correspondence and how it might differ from more traditional means.

5. Look back at the "communication link" between Systems Tech and Executive Consumer Surveys. Complete the following assignments:

   a. Assume that you did indeed find a faulty chip in the Systems Tech 175 computer. Write a letter to Ms. Patrick explaining your finding and offering what you consider adequate compensation.

   b. Assume that you did indeed find a faulty chip in the Systems Tech 175 computer. Write a letter to the other 175 owners to recall the machine or do whatever action you think appropriate.

   c. Assume that when Ms. Patrick's 175 computer arrives and the repair shop checks the machine, they find evidence that a liquid has seeped into the machine. Write a letter to Ms. Patrick explaining your findings and taking whatever action you think appropriate.

6. Select and study one of the sets of case documents in Appendix A. With a partner, describe the communication links that have been generated and will probably be generated by the documents you study.

   Then individually write a letter or memo that might be generated by the documents. For example, you could assume that you work for the Seattle Treatment Education Project and have just written the AZT Step Fact Sheet (Case Documents 1). You need to alert such groups as the Minnesota AIDS project of the availability of the Fact Sheet in a letter that contains a brief summary of the purpose, audience, and content of the Fact Sheet. Or, you might need to alert the university community by memo about the communication link generated by the complaints about bird trapping on experimental fields (Case Documents 2) in case the faculty and staff receive calls or letters about the situation. You would want to summarize the previous correspondence and set a policy as to how to respond to inquiries from the public, press, or special interest groups.

Finally, exchange letters or memos with your partner. Your partner will now reply to your letter or memo. This reply might take some imagination, but remember there's usually another link in the communication to be written.

7. Locate an advertisement for a job for which you would like to apply. Look in the newspaper, in your professional organization's newsletter or online job posting, or on an announcement at a job fair. Prepare a résumé and write a letter of application for the job as advertised.

8. Assume that you are the Public Relations Director for the university involved in the controversy depicted in Appendix A, Case Documents 3. You are relocating and applying for a similar job in another university. Write an appropriate cover letter to explain your role in this controversy in the best possible light. Decide how to turn this experience into a way of describing your skills and abilities.

WORKS CITED

Bolles, Richard Nelson. *What Color Is Your Parachute? A Practical Manual for Job-Hunters & Career Changers.* Berkeley, CA: Ten Speed Press, 1988.

Jackson, Thomas. *Guerrilla Tactics in the Job Search.* 2nd ed. New York: Bantam, 1991.

Lathrop, Richard. *Who's Hiring Who: How to Find the Best Job Fast.* 12th ed. Berkeley, CA: Ten Speed Press, 1989.

Peterson, Dart G., Jr. "The Business of Technical Communication." *Technical Communication* 38, no.4 (1991), pp. 594–97.

Weiss, Timothy. "Ourselves Among Others: A New Metaphor for Business and Technical Writing." *Technical Communication Quarterly* 1, no.3 (1992), pp. 23–36.

CHAPTER

17

Oral Presentations in the Workplace

Introduction ❚ Preparing Oral Presentations in the Workplace
❚ Using Visual Aids and Presentation Software ❚ Developing Support
for Your Presentations ❚ Delivering Oral Presentations ❚ Preparing
for the Job Interview ❚ Succeeding at the Job Interview
❚ Employing New Technologies

As much time as college-educated workers spend writing, they generally spend more in oral communication.

> Paul Anderson, based on a meta-analysis of survey research conducted in workplaces in "What Research Tells Us About Writing at Work," in *Writing in Nonacademic Settings,* ed. Lee Odell and Dixie Goswami (New York: Guilford Press, 1985).

Don't tell me the details of how you got the data, just tell me what the data means.

> A request from a project manager at Hewlett-Packard Labs upon listening to an extemporaneous presentation, as reported by Frederick Gilbert, "The Technical Presentation," *Technical Communication* 39, no. 2 (1992), pp. 200–1.

Like many people, you may dread the times when you have to stand up before your colleagues or strangers and make presentations. By giving a lot of attention to audience analysis, preparation, practice, and your own *ethos* as a speaker, you can do a great deal to reduce your anxiety and improve the quality of your oral presentations in the workplace.

Introduction

Although the majority of this book has provided instruction on how to design written documents, this chapter helps you communicate technical information through oral presentations suitable to your listeners' needs.

Keep in mind that oral presentations encompass an enormous variety of experiences, including:

- A two-minute pitch to sell yourself to a potential employer.
- A technical presentation to a local society concerning the complexities of composting.
- A 10-minute slide show to upper-level managers about a proposed product.
- A 20-minute presentation providing information to management.
- A scripted talk detailing research results at a professional conference.
- An impromptu explanation of networking systems to interested colleagues.

To prepare for these varied experiences, you apply the same rhetorical principles that you used in developing written documents; that is, you analyze your purpose and audience, and then you develop your presentation. You need to consider your own knowledge, attitudes, and needs or purpose as well as those of your listeners. Once you complete that analysis, you can use the information you gathered to construct and deliver your presentation.

We begin with a guide for preparing oral presentations (see Writing Strategies 17-1). As you might guess, most people focus immediately on the "Text and Delivery" column, determining what content they will include and how they will present this content to their listeners. However, focusing first on the delivery is equivalent to drafting a document without taking time to plan it. Therefore, we recommend that you begin your development process first by considering:

Preparing Oral Presentations in the Workplace

- *Knowledge*—what you and your listeners know about the subject.
- *Attitudes*—how you and your listeners feel about the subject.
- *Needs*—why you need to communicate this information orally or your purpose in doing so.

Responding to these questions helps you to determine what content you should include and how you will present this information. When you

WRITING STRATEGIES 17–1   Oral Presentation Development Grid

	Speaker	**Listeners**	**Text and Delivery**
Knowledge	What do I know about this subject? Where does my knowledge come from?	What do my listeners know about this subject? What limitations are there to their knowledge?	What content should I include? In what order? How much detail?
Attitudes	How do I feel about this subject? How do I feel about addressing this particular audience?	How do my listeners feel about this subject? Will their reactions be mixed?	How should I present this information— formally or informally?
Needs	Why do I need to communicate this orally? Am I informing, teaching, or persuading?	Why do my listeners need to hear this? How will they put the information to use? What kinds of approaches would appeal to them?	What kind of visual aids should I use? How will I meet or exceed the listeners' needs and expectations for this presentation?

give an oral presentation, you are often persuading; responding to these questions will also help you present good reasons for your argument, establish your *ethos,* and use *pathos* when appropriate.

Completing this grid also helps you to determine exactly what you want to accomplish in your presentation. That is, it helps you to clarify your purpose, your listeners' attitudes and expectations, and the type of oral presentation you wish to deliver. As an example, one speaker responded to these questions as a way to brainstorm how she would design and deliver a presentation on AIDS to a local community group (see Example 17–1). In this case, after brainstorming responses to the questions in the grid, she highlighted specific responses and then used these to develop an outline for her presentation.

CHOOSING A TYPE OF ORAL PRESENTATION

Research indicates that you will be making presentations involving scientific or technical data frequently in the workplace. Only about 2 percent of the respondents to a recent survey of scientific, technical, and managerial professionals said they never made presentations, and these individuals were new on the job (Scheiber 161). Most audiences are internal and presentations are directed toward colleagues within the organization, technical managers, senior managers, and general managers (Scheiber 166). The ma-

EXAMPLE 17–1   A Speaker's Completed Oral Presentation Development Grid

	Speaker	**Listeners**	**Text and Delivery**
Knowledge	My own knowledge comes from: 　Experience 　Magazine articles 　Newspapers 　Pamphlets 　Medical reports 　Financial reports	They may know about AIDS in general, but I doubt they are aware of the emotional scarring and financial debts that the disease leaves behind.	I should include excerpts from my sources as well as my own experience. I should use facts about the rising medical costs that do not increase the quality of care. I should tell how painful it is to watch people become ill and not be able to do anything for them, not to be able to comfort them in any way.
Attitudes	I feel there should be more financial support of AIDS patients as well as free or low-cost counseling available to family members. I want to reach this audience so I will have to make a personal appeal to them.	Many people feel that AIDS is brought about by carelessness, although many times it is not. These people feel that the patient is at fault and cannot relate to my point of view. Other people may support my point of view, but regardless of the method of exposure, the financial impact is extreme.	Although I might begin by writing a script, I want to practice so that I won't read the presentation. During my personal stories especially, I want to stop and just talk to the people. For facts, I want to display actual statistics and graphs (using an overhead projector).
Needs	I feel I need to communicate this because otherwise people will continue their biased way of thinking and nothing will be done to help the victim and his or her family. Just as the number of victims is increasing steadily, so in turn are the number of families affected by AIDS.	I think my listeners will take notice because these facts are too strong to ignore. Anyone can experience AIDS in one way or another. Most can relate to the illness and death of a loved one. Most can relate to being in debt. Probably not as much as I am referring to, but in debt nonetheless.	I think I will meet the expectations if I can find the statistics to back me up. I want to put some of these statistics in a handout and leave this with people, so that they will remember this information and know how easily it could happen to them. I want the listeners to share this information with others and be made aware of the consequences of AIDS.

jority of these presentations are to inform, instruct, and share data. Given the nature and frequency of the presentations you will give in the workplace, you need to have strategies for preparing these presentations quickly and effectively.

You will be called upon to construct formal or informal presentations to meet your needs and those of your listeners. We're going to look at three different types of presentations you can use in the workplace: the impromptu presentation, the extemporaneous presentation, and the scripted presentation.

Impromptu Presentations

impromptu presentations unrehearsed and unprepared presentations delivered at a moment's notice

The **impromptu presentation** is delivered at a moment's notice, unprepared and unrehearsed. Impromptu speeches are, by their nature, unplanned. You have no notes and give little thought to organization.

You make impromptu presentations daily. For example, perhaps a group stops by and asks you to explain why you have chosen a particular piece of software. Or a project team decides to brainstorm on a product related to one that you are currently developing and asks you to offer advice.

You will generally use an impromptu presentation when your listeners need immediate feedback from you or when you need immediate feedback from your listeners. Thus, you mainly use these presentations to respond to questions. With an impromptu presentation, you need to "think on your feet" and quickly provide a snapshot of what your listeners need to know.

Listeners want responses to questions, although keep in mind that they may not be asking a precise question because of their lack of knowledge on the subject. Respond with shorter speeches (keep them under two minutes) and follow up with questions to determine your listeners' understanding of the information. Let listeners ask questions to determine the direction of your responses.

Do not use impromptu presentations to deliver complex information, even if the subject is familiar and comfortable for you. Use them to respond to questions, give quick updates, describe a problem, and brainstorm new ideas. Think of an impromptu presentation as a dialogue with your listeners. When you are called upon to give an impromptu presentation in your workplace, take a few seconds to glance around the room and decide what you really want to say. Be brief and be gracious.

Impromptu presentations take little preparation time, but this carries the risk of treating your subject in a disorganized or incomplete manner. Generally speaking, your audience understands the circumstances of your presentation and wants to hear information in your own words. They may prompt you with their questions, and you will have to shape what you say in response to those probes.

Extemporaneous Presentations

extemporaneous presentations presentations planned thoroughly in advance yet delivered in a spontaneous manner

Many people confuse extemporaneous and impromptu presentations, but they are very different. An **extemporaneous presentation** is one that you plan carefully in advance, prepare thoroughly, and then deliver in a spontaneous manner. The best lecturers you heard in college probably used this method of presentation. And this is often the method used by motivational speakers, trainers, and CEOs who want to energize their employees with an annual presentation.

The extemporaneous presentation is most preferred by speakers and is the one you are most often called upon to deliver. It allows you the best opportunity to shine as a speaker because it gives you the time to prepare thoroughly and through careful practice deliver a relaxed, conversational presentation. When you give this kind of presentation, you'll do so from a set of notes containing your exact ideas, the sequence to be used when presenting the ideas, and the supportive material that you will need. You do not, however, prepare the exact language of the presentation. Some people call the extemporaneous presentation an outlined talk.

You are most likely to design an extemporaneous presentation when you want to have solid control of your material but also want to keep things on a comfortable, face-to-face level with the audience. Imagine the following scenarios:

- A group of technicians asks you to explain differences in asbestos removal processes.
- A local club asks you to present the current status of drug control efforts in your company.
- Your board of directors asks you to analyze and present current marketing efforts.
- The marketing division asks you to discuss the differences between your company's current software product and the newer version to be released so they can prepare new advertising.

In such situations, your listeners want well-prepared information, but they want to "hear your voice" rather than a stilted, read speech. Respond by using Writing Strategies 17–1 to acquire a thorough knowledge of the subject and your listeners' attitudes and needs.

We suggest that you prepare extemporaneous presentations in three steps:

1. In an outline or list, write down all of your ideas in the order of their introduction.
2. In a set of brief notes, write down no more than a dozen key words that will act as reminders during your presentation.
3. Practice to work out your general treatment for each part of your talk, decide how to emphasize your main points, and develop transitions that are clear and concise. In addition, use visual aids to guide both you and your listeners.

Extemporaneous talks can be created rather quickly and delivered with a natural speaking voice. Based upon listeners' reactions, you can speed up or slow down, eliminate unnecessary material, or add something that you discover is needed. However, you can easily run over your time limit, leave out crucial information, or encounter difficulty in finding the appropriate phrasing that will explain what you mean accurately and clearly. If you get tongue-tied or tend to forget your message, you may want to choose instead a scripted presentation. Keep a watch handy to make sure you stay within the time allotted.

Scripted Presentations

A **scripted presentation** is delivered verbatim from a written copy of what you will present. Although initial preparation is similar to that of an extemporaneous speech (i.e., you begin by using Writing Strategies 17–1), the similarity ends there. For a scripted presentation, you determine every word, all graphics, and many gestures in advance, and when you present your information, you read your script to your listeners. Even though you will be reading your paper, it's essential you remember that hearing a written paper is very different from reading one, and you need to take listeners' needs into consideration by providing a clear road map of your points.

You need to design a scripted presentation when you must deliver technical information clearly and accurately and when you are in a timed setting, such as a presentation during a teleconference. Use it for situations when you must be precise and when small slips in phrasing could be embarrassing or damaging, such as in the following scenarios:

- A professional association accepts your paper on desktop videoconferencing for presentation at a national meeting.
- An NCAA committee asks you to make a presentation applying nutritional concepts to the use of training tables before collegiate games.
- The Alaska Department of Environmental Conservation (ADEC) asks you to present your firm's research findings on the effectiveness and safety of bioremediation.
- A government commission asks you to testify regarding occurrences of sexual harassment in the workplace.
- The management of your company asks you for a 12-minute presentation on your company's Year 2000 (Y2K) compliance measures over a teleconference broadcast to company offices in three countries.

Listeners want well-prepared, accurate, and clear information on which to take notes and use in the future. Again, respond by using the grid in Writing Strategies 17–1 for a thorough knowledge of the subject and your listeners' needs. Remember that listeners need more help following an oral argument than readers need following one in print. Prepare a scripted presentation by writing the text much as you would any technical document; however, avoid multisyllabic terms and remember to define new information. Use examples and visual displays whenever possible.

If you expect to be nervous during your presentation or if you are speaking to government or corporate officials, a scripted presentation ensures that you will communicate clearly and concisely. However, this kind of presentation takes a long time to prepare, and when you are delivering it, you cannot easily adjust it in light of the reactions you get from your listeners.

FOUR ESSENTIAL GUIDELINES FOR ORAL PRESENTATIONS

Whenever you need to prepare an oral presentation in the workplace, the following guidelines will help you be more effective and comfortable:

1. Make it short.
   - It takes twice as long to listen as to read.
   - Stick to a few main points.
   - Time your presentation a bit short so that you can *talk slowly.*

   Let's say that you need to give a presentation to your supervisors at the Environmental Protection Agency (EPA) about bioremediation in the cleanup efforts in Prince William Sound after the *Exxon Valdez* oil spill. You have 30 minutes, so obviously you will have to choose the most important points to present. Because your audience is knowledgeable about environmental issues and probably somewhat familiar with bioremediation, you can probably focus on their main concern: the safety and effectiveness of the process in this case. Your audience is also concerned about whether it should promote bioremediation in any other situations. In your 30-minute slot, you decide to speak for 20 minutes and leave 10 minutes for questions. When you practice your presentation several times, you find you generally speak for about 18 minutes—a time you are comfortable with because you can then slow down, emphasize, or add information or definitions if your audience appears confused at any point in the presentation. Running a few seconds short on your speech is seldom a problem; running over time generally irritates your audience and damages your *ethos.*

2. Make the organization obvious.
   - Acknowledge your introduction (if you've been introduced). If you haven't been introduced, take a moment to let people know who you are or in what capacity you are presenting ("I'm here today as the communications department representative to speak about document guidelines . . . ").
   - Tell your audience what you're going to tell them—capture their attention, define any essential terms, give a background preview.

- Then tell them—inform them, persuade them, teach them (entertain them too, if appropriate).
- Then tell them what you told them—summarize, conclude, recommend.

In your presentation on bioremediation to the EPA, remind your audience of the definition of bioremediation, preview the points you will cover, and prepare the audience for your conclusions and recommendations. You also want to capture your audience's attention and motivate them to listen to you, a task that should not be too difficult with this knowledgeable and interested audience. Although you are informing them about bioremediation, you want to persuade them to accept your recommendations. Thus, you might begin your presentation as follows:

> Today, I want to address the following question: Do we have an effective and safe method for cleaning up shoreline oil spills? Is it a method that we can recommend or implement with confidence?
>
> I believe that with bioremediation we can say yes to the first question. Bioremediation is effective. Bioremediation is safe. But our yes must remain a cautious one. Bioremediation is safe and effective only in certain circumstances. And bioremediation must be backed up by other cleanup procedures. Any EPA endorsement of bioremediation must include those other procedures.
>
> I will briefly review the bioremediation process for you, using the cleanup efforts in Prince William Sound following the *Exxon Valdez* oil spill as a test case. First, I will describe the environment of the sound and the extent of pollution in that environment. Second, I will discuss the safety of bioremediation from two points of view: the precautions necessary to protect the workers involved in the bioremediation process and the environmental consequences of bioremediation. Third, I will assess the effectiveness of bioremediation in removing oil hydrocarbons from the spill site. Finally, I will detail when and to what extent the EPA can feel comfortable recommending and endorsing the bioremediation process.
>
> In reviewing the bioremediation process, let me remind you that bioremediation is basically biodegradation, a natural process . . .

Your audience will immediately understand your approach to the topic of bioremediation and be looking for the three parts of your presentation. Also the audience will be expecting you to end with a cautious endorsement of bioremediation and will listen for the benefits you highlight and the qualifications to those benefits. In your introduction then you have prepared your audience to *hear* you.

3. Make the ideas simple and vivid.
   - Add minipreviews and summaries throughout the presentation to help your audience keep track of where you are and remember what you have said.
   - Use obvious and clear transitions. Design your presentation much like a road map with all the turns, curves, and changes in direction and speed obviously marked.

- Repeat your main terms often. Don't rely on pronouns—your audience might lose track of what your "it," "their," "they," and such refer to.
- Try to show your audience how things *look*.
- Explain ideas in plain language before doing so using mathematics or formulas.
- Explain the purpose of any procedure or idea.
- Incorporate rhetorical questions to keep your listeners' attention (e.g., "So I asked, what are the real needs of our clients?").

As you precede with your presentation, at times you will be teaching your audience about the bioremediation process in Prince William Sound. You will need to describe vividly the environment and the process, such as:

> Typically about *three million oil-eating micro-organisms* exist in every ounce of sediment in Prince William Sound. After bioremediation, *thirty million such organisms* were present in each ounce of sediment. And, the oil-eating ability of *those thirty million micro-organisms* increased *tenfold*.

Changes in your tone or voice level and repeating phrases or using similar structure help emphasize the essential words in your presentation (italicized above).

Transitions, minisummaries, and minipreviews again help you announce changes in direction within your presentation. For example, after you finish explaining the bioremediation process, you might conclude that part and announce the next part of your presentation as follows:

> Thus, we see that bioremediation in Prince William Sound relied on a greatly increased number of microorganisms to break down oil hydrocarbons into carbon dioxide, water, and microbial biomass. But how safe was this process to the environment and to the workers?
>
> First, the workers—what application procedures and protective clothing were designed to ensure worker safety . . .

Transitions such as "thus," "but," and "first" in the part of your presentation above and your preview of the next part of your presentation help your audience follow your presentation. And the brief summary of your description of bioremediation will help your audience remember your presentation.

4. Summarize and be ready for questions.
   - Repeat your main points in your conclusion.
   - Repeat each question for the benefit of the audience.
   - Reword clumsy questions.
   - Wait until after the speech to give handouts to listeners.

In your presentation on bioremediation, you will conclude with a summary of the three main parts of your presentation and your recommendations concerning when bioremediation will be effective, perhaps as follows:

> I've reviewed with you the process of bioremediation, as used in the Prince William Sound cleanup, as a natural biodegradation process which relies on microorganisms to break down oil hydrocarbons. As we saw, the plant, bird, and animal life, the overall environment of sound, was conducive to this process. Worker safety was ensured by protective clothing. Environmental safety was ensured by birds and mammals' instinctive avoidance of unnatural substances. Finally, I detailed this most effective aspect of bioremediation—the speed with which the oil disappears and the environment returns. The EPA can then be confident in endorsing bioremediation with the five precautions I detailed earlier:
>
> 1. Bioremediation must be backed up by hot water spray and the use of absorbents.
> 2. Bioremediation works best after bulk oil is removed by manual and natural processes . . .

Following these four guidelines—make it short, make the organization obvious, make the ideas simple and vivid, and summarize and be ready for questions—will help you to prepare and deliver more effective oral presentations in your workplace and elsewhere.

Using Visual Aids and Presentation Software

visual aids
anything provided by the speaker for the audience to look at during a presentation, including but not limited to handouts, charts, videotapes, posters, overheads, and slides

CREATING VISUAL AIDS

Oral presentations nearly always benefit from **visual aids** (see Chapter 10 for specific suggestions on how to develop visual displays). A visual aid is anything that you give your listeners to look at during your presentation. It might be something they can touch and hold such as a handout, or it might be a table, graph, drawing, illustration, map, or photograph presented via slides, overheads, blackboards, posters, videotape, PowerPoint™ graphics, or computer simulations.

Remember that there's always the risk of overdoing visual aids. Frederick Gilbert, president of a technical communication consulting firm, describes an all too familiar scene: "A technical expert stands next to the overhead projector with a huge pile of transparencies, reading the hard-to-read text to a half-sleeping audience. This has been called the 'talking overhead projector'" (201). Gilbert's solution is not to eliminate visual aids but to hold your listeners' attention by:

- Using fewer rather than more visual aids.
- Using color graphs and charts rather than words.
- Never beginning or ending with visuals.
- Keeping visuals big and bold. (201)

Never overcrowd your visual aids, a common problem on overhead transparencies, and label clearly all the lines, bars, and other parts of any visual display of data.

Despite the possibility of overdoing it, you run a greater risk by not using visuals. Raised on MTV, Nintendo™, and computers, today's audiences increasingly demand a visual-oriented, fast-paced presentation. Visual aids can also increase the effectiveness of your presentations by:

- Helping you attract and hold your listeners' attention (you are giving people a second place to gaze that is directly related to your message).
- Helping your listeners follow the organization of your presentation and keep track of your main points (you can design a poster that displays the main points of your talk).
- Helping you explain your material (drawings, graphs, charts, and other visual aids communicate material with an economy and effect that you cannot achieve with words alone).
- Helping you remember what you want to say (especially in an extemporaneous presentation, visual aids remind you of what you intend to say next and help you not to forget relevant material).

USING PRESENTATION SOFTWARE

You may worry about how to create high-quality, professional-looking visuals to accompany your presentations. In the past many presentations were weakened because of the amateurish visuals that went with them. Now, however, if you follow the basic design principles we include in this book and you learn to use **presentation software** such as PowerPoint™, you can create polished-looking work.

Presentation software takes some time to learn well, but it's basically not very difficult to use; and it permits you to create a wide range of supporting materials for your workplace presentations. Before you get started, you have to decide what kinds of visual displays you will need, and then you can generate them directly from the presentation software. Power-Point™, for example, allows you to create slides that can be displayed electronically using a computer or developed into 35mm slides to be displayed using a slide projector. The same slides can also be printed in color on paper or as overhead transparencies. Presentation software also allows you to create printed handouts for your audience that include your graphics and any notes you wish your audience to take with them.

One important feature of presentation software for you to consider is the ease with which you can transfer information from other applications to it. You can avoid the hassle of retyping and insert graphs and charts from spreadsheets or text from word processing programs directly into your presentation. Remember, however, that you will have to make some adjustments to formatting. All text and numbers must be readable to your entire audience.

Finally, think about how else your presentation can be distributed. Most presentation software allows you to distribute your visual aids online

presentation software
graphics programs and other desktop publishing software used to develop visual aids for oral presentations

so you can get feedback from your colleagues. Some presentation software has a comment function that allows your colleagues to insert small electronic "notes" right on the graphic. You can use their suggestions to revise your graphics and your presentation. When you're finished with your presentation, you can add **hyperlinks** to it (perhaps linking to an annual report or a sales summary) and post it on the company's Web page.

Because of the increased emphasis on high-quality visual aids, we recommend that you use a **storyboard** to integrate visual aids into your presentation. You can also use a storyboard to integrate gestures and supporting materials into your presentation. A storyboard is a two-column tool developed by people who write film, video, and multimedia scripts. In the right-hand column, you write a general outline or list (for an extemporaneous presentation) or the script of your talk. In the left-hand column, you indicate the visuals your listeners will see as those words are spoken. Example 17–2 shows a rather elaborate storyboard for a manuscript presentation on collaboration via

hyperlinks
embedded addresses found on a Web document; a click on them puts you in direct connection with another Web page

storyboard
two-column script used to prepare oral presentations in which the left-hand column is used for the visuals and the right-hand column for the text

---

EXAMPLE 17–2   An Example Storyboard for a Manuscript Presentation

Collaboration via Desktop Videoconferencing: Implications for Technical Communication

Visual	Oral
Title slide—"Collaboration via Desktop Videoconferencing."	Are we satisfied with the ways in which we use technology to collaborate? Are we satisfied that we have enough access to information in our world? Could we link ourselves with other companies?
Slide—*purposes* for the presentation.	I do not think that we should be satisfied; in fact, one purpose for this conference is to examine the past and design the future of technical communication. The purpose for my presentation is to promote innovative ways of using technology for collaborating, to describe one means for linking collaborators and providing access to numerous companies . . .
Illustration—the desktop videoconferencing system.	We have taken a configuration designed for business applications—Cameo™ (for audio and video) over an Integrated Services Digital Network (ISDN™ or a broadband services phone line)—and
30-second video—two collaborators exchanging information and collaborating.	have applied it to our company's needs. This short video shows the basic configuration of the multimedia system [discuss this extemporaneously], and the following video shows how collaborators exchange and view similar texts, all the while seeing each other and verbally discussing their responses . . .
	[the storyboard continues]

desktop videoconferencing, but a storyboard doesn't have to be this elaborate to be effective. You can simply write a few notes in the margins to help you integrate your visual aids into the presentation.

As in any writing or speaking situation, consider your listeners' expectations. If they expect you to present a slick, packaged slide show or PowerPoint™ presentation and you draw lines on an overhead, assume that no matter how excellent your content may be, you are likely to be judged otherwise. Or if you have a set amount of time to present, your listeners may become disturbed if you take time to draw diagrams on a chalkboard or overhead. In the information age in which we live, listeners expect you to come prepared (except, of course, in the case of impromptu presentations) and to use appropriate communication technologies to develop and present your oral presentations.

Developing Support for Your Presentations

In Chapters 6 and 7 we discussed how to conduct primary and secondary research to gather data for your documents. When you prepare workplace presentations, you will also have to develop support for the arguments you make. Secondary research can provide you with the examples and data you need to make your workplace presentation interesting and effective, and primary research can provide you with expert testimony and the data to support your claims.

USING EXAMPLES IN ORAL PRESENTATIONS

Examples are powerful tools for making what you say memorable and interesting in workplace presentations. They are also useful tools for adding human interest to your presentations. Examples can be short or extended, depending on the point you wish to make. They help clarify and reinforce the ideas that you are presenting.

Imagine, for example, that you were asked to give a presentation on the availability of high-yielding varieties of soybeans that are resistant to soybean cyst nematode (SCN). You can and should talk about the numbers of varieties available, how they are best used, and some of the concerns about using these varieties. You may be especially concerned that many farmers are choosing to grow SCN resistant varieties without confirming the presence of SCN in their fields. This practice is a concern because it can set up fields for resistant weeds. To make your point, it would be effective to use a specific example. You could discuss a particular farmer who planted all SCN resistant varieties without doing a nematode count. When such a verification was finally completed, this farmer's fields were found to be virtually without SCN, and so the planting of resistant varieties was uncalled for. You could begin your example by saying, "Let me tell you about a farmer in southwest Iowa . . . .".

RELIABLE DATA

To make claims in your workplace presentations and not back them up with reliable data can damage your credibility. You can gather data from experts, library sources, and the Internet. Once you're convinced about the quality of your data, use them to give your presentation some punch. If you are trying to convince your audience that Americans eat a lot of peanut butter, it's much more effective to say "Americans eat 700 million pounds of peanut butter each year" than to say "Americans eat a lot of peanut butter." The first statement sounds as if you did your homework. The second statement sounds like an opinion. Wherever possible be specific. If you are talking about high-yielding varieties of soybeans resistant to the soybean cyst nematode and want people to know there are various sources of resistance (each of which is identified by a specific number) and they are being used in different places, get the facts straight and build them into your presentation. For example, to an expert audience of plant pathologists, you might say, "In Iowa, for example, 98 percent of the SCN varieties are from PI88788. Three varieties have Peking resistance. The Hartwig resistance is available only in Hartwig, and Fairbault has a PI209332 resistance."

USING EXPERT TESTIMONY

Your listeners are often interested in hearing who provided you with information and who the experts were you consulted in your research on your topic. When you use expert testimony, state others' information in your own voice. If you have chosen your experts with care, their voices add to the credibility of your own voice. If you were using expert testimony to support your presentation on soybean cyst nematodes, you might say, "According to Frank Pfleger, head of plant pathology at the University of Minnesota, there are seven sources of resistance being used in breeding programs. Four of those sources are available in today's varieties."

CITING YOUR SOURCES

When you give a workplace presentation that has a great deal of data or when you use testimony from experts not everyone knows, you may be asked for the sources of that information. There are two simple ways to address these requests but they can work only if you prepare carefully ahead of time. Just as you took care to gather information about your sources and put those into a "Works Cited" section in your reports and other print documents, so too you must take care with your sources when you prepare oral presentations. If you are using slides or handouts, make sure that you provide full citations on those materials. Follow the standard citation format for your profession. If you are giving a presentation in which you use many sources, it is often helpful and a real courtesy to provide a reference page for your audience.

When you deliver an oral presentation, you have to be like a stage performer, with words to deliver and visual aids to manage. For your delivery to go smoothly, you need to prepare your stage and listeners, recognize special needs of your listeners, and cope with anxiety. Apprehension about speaking, called **communication anxiety (CA)** by researchers, is a very real thing, and if you are one of those people who has more than the common amount of anxiety, you will need to address it. People will often tell you that the way to cope with anxiety is to practice again and again. That's not bad advice, but it's limited. Research has found that people with high CA can reduce their anxiety and do a better job with their presentations if they spend a significant portion of their time on planning and audience analysis rather than simply on rehearsing.

When delivering oral presentations, consider the following:

1. Set the stage and the audience.
    - Get there early and get ready—try out all audiovisual equipment.
    - Remove distractions if you can (when you can't, acknowledge them and go on).
    - Cool off the audience (open windows if you need to).
    - Choose a room that is somewhat small.
    - Choose a room with the entrance in the rear.
2. Carry an "Insurance Policy" (for extemporaneous presentations).
    - Put notes on cards and number them clearly in case you drop them.
    - List your main points.
    - Write on one side only.
    - Talk from visual aids and other materials.
    - Have a plan in mind if the overhead projector breaks, the bulb in the slide projector burns out, or your PowerPoint™ presentation won't load.
3. Use visual aids.
    - Make the print on overheads or any visuals large and easily readable (practice the document design techniques in Chapter 9 and the visual display principles in Chapter 10).
    - Make a photocopy of all color visuals to see if there is enough contrast for them to be easily visible to people who are color blind or have difficulty seeing.
    - Label all overheads and slides and mark your script with when to display and remove them.
    - Practice with your visual aids prior to the presentation so that you don't have to fiddle with them during your presentation.
    - If you are in a large room, ask someone to turn your overheads if necessary so you can stay at the lectern.
    - If you are working from a PowerPoint™ or other display program, make sure that the computer from which you are directing is close by and facing you so that you can easily and unobtrusively change slides.

Delivering Oral Presentations

communication anxiety (CA) apprehension many people have in making an oral presentation

4. Talk enthusiastically and slowly.
   - Keep in mind that you are talking to people, not reciting words.
   - Use your voice to clarify your message.
   - Show enthusiasm and interest.
   - Keep good posture.
   - Keep your eyes on your listeners' noses or foreheads.
   - Place your hands at your sides except when highlighting something on a visual aid or making gestures. If you're nervous, write in your gestures on your script (and practice them so they don't look wooden).
   - Move on transitions and make purposeful gestures.
   - Do not go over your allotted time!

PREPARING YOUR STAGE AND LISTENERS

Your presentation will be more effective if you think deliberately about the arrangement of your stage and the needs of your listeners. If you are going to speak in an unfamiliar location, visit the room beforehand and arrange your presentation area. If you are speaking in a room that is larger than you had expected, make sure that your visual aids will be visible by all listeners and that everyone in your audience can hear you. If you are using slides as visual aids, be sure that you will have a source of light for reading your notes when the room is dark. Most importantly, make sure that all of your props or materials are available and easy to access. If possible, place them on a table in front of you in the order that you will be using them.

One of the best ways to prepare your listeners is simply to look at them before you start to speak. This helps you to create a personal bond with them, gets their attention, and helps establish your *ethos* as a sincere and interested speaker. If you have difficulty looking straight at your listeners' eyes, look at their foreheads or noses. Your peripheral vision will still allow you to recognize their reactions and needs. Throughout your presentation you can continue to look at your listeners (even with a manuscript presentation) in order to see if you need to change your rate of speaking, speak more loudly, or give your listeners more time to study a visual aid. It's sometimes a good idea to ask audience members in the back rows if they can hear you and see your visual aids. When you are working from a manuscript, leave your finger at the line you just completed as you survey the room. A quick glance can bring you right back to your place. Avoid any disruptive behaviors you might have—fiddling with rings or paper clips, for example. If you're not sure what those behaviors might be, arrange to have yourself videotaped during a rehearsal or give your presentation to a friend or two and ask for specific feedback.

One way to help your audience is to anticipate the traps they might fall into when listening. Sometimes listeners prejudge ideas and decide to "tune out" even though they do not yet know the content of your message. Listeners also misplace their attention by concentrating on remembering all the facts rather than important ideas, taking too many notes, or pretending to be listening. Or listeners let trigger words elicit their emotional reaction to your content rather than continuing to listen to your view toward the subject.

To help listeners avoid these traps, you can provide handouts that include your major facts, stating that you will distribute these after your presentation. You can also keep listeners' attention by being truly enthusiastic about your content. In addition, if you limit your major points to no more than seven (plus or minus two), you will be keeping within the limits of the human mind to store and process incoming information. When distractions occur—a baby starts to cry, traffic noises increase, someone mows the lawn outside your window, air-conditioning breaks down—acknowledge that these conditions are affecting the audience, and enlist their good will in your endeavor.

RESPONDING TO QUESTIONS

You also help listeners by responding to their questions either during your presentation (if you have stated that you will allow questions any time during your presentation) or after your presentation concludes. Handling questions can be tricky, so remember to do the following:

- Greet each question politely and listen carefully. If the question has several parts, jot a quick note about what the questioner is asking.
- Repeat the question for the other listeners (this allows you to rephrase the question to see if you have heard it correctly, and it gives you additional time to develop your response).
- Always be accurate with your answer. If you don't know the answer, say so.
- Answer each question completely while being as brief and specific as possible.
- When you don't understand a question, say so and ask the person to repeat or rephrase it.
- When a question seems truly absurd, don't say so; clarify it and continue.
- Don't counter one question with another question.
- Never answer by simply saying yes or no; this response offends listeners.
- When questions come in the middle of your presentation, use them as transitions if possible.

Answering questions completely and honestly will maintain the *ethos* you have established while delivering your oral presentation.

RECOGNIZING LISTENERS' SPECIAL NEEDS

More than 40 million people in the United States have some kind of disability. Some of these disabilities may make it difficult for people in your audience to follow your presentation. **Hearing impaired people** will be able to see your visual aids and view the signing of your presentation (provided that you have planned for a professional in American Sign Language to be available).

It is often assumed that because listeners who are visually challenged can hear, they will have little trouble understanding your presentation. Quite to the contrary, your visual aids present unique difficulties for **visually challenged people.** To illustrate this fact, turn back to back with a partner, and have one person describe the flow chart of the human information processing system (Example 17–3) while the other person draws it. How much difficulty did you and your partner have?

Parkin and Aldrich studied home readers or people who tape-record materials for people who are visually challenged. They found that the most frequent problem for such readers was communicating information given in maps, diagrams, and other visual displays. They often found that home readers simply gave up trying to describe visual aids, hoping that the text alone gave sufficient information. To illustrate how difficult it is to describe a visual display of data, Example 17–4 presents a complex graph and its oral description.

Although there are few guidelines for developing oral presentations to be delivered to an audience comprised of the visually challenged, the National Braille Association's instructions for reading a table are as follows (this same manual includes suggestions for presenting illustrations, graphs, and diagrams to the visually challenged):

1. Read the title, source, captions, and any explanatory keys.
2. Describe the physical structure of the table. Include the number of columns, the headings of each column and any associated subcolumns, reading from left to right. If column heads have footnotes, read them following each heading.
3. Explain whether the table will be read by rows (horizontally) or by columns (vertically). The horizontal reading better conveys the content. On rare occasions it is necessary to read a table both ways.
4. Repeat the column headings with the figures under them for the first two rows. If the table is long, repeat the headings every fifth row. Always repeat them during the reading of the last row.
5. Indicate the last row by saying "and finally . . ." or "last row . . .".
6. At the completion of the reading say "End Table 3."

In any oral presentation, listeners have varied skills. Those who have become visually challenged later in life will have conceptions of what figures, tables, diagrams, and flow charts look like; however, those who have been visually challenged from birth will have much greater difficulty

---

hearing impaired people
audience members whose inability to hear may require special development and placement of audio aids and the support of a sign-language professional

visually challenged people
audience members who may require special placement of visual aids as well as alternative media

EXAMPLE 17–3   Flow Chart of the Human Information Processing System

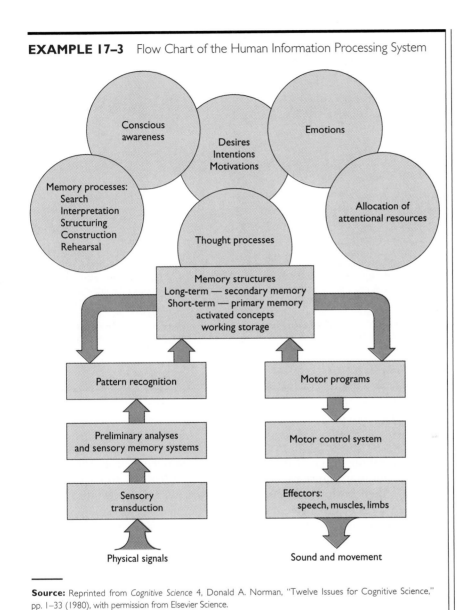

Source: Reprinted from *Cognitive Science* 4, Donald A. Norman, "Twelve Issues for Cognitive Science," pp. 1–33 (1980), with permission from Elsevier Science.

with visual arrangements and the visual imagery often used to describe them (Hartley). Of course, you'll also need to consider how cultural background and ability to read and understand English might make a difference in how you deliver your oral presentation. Therefore, be aware of these listeners' needs, plan for *all* members of your audience, and your attention to these needs will be appreciated.

EXAMPLE 17–4 A Complex Graph and its Oral Description

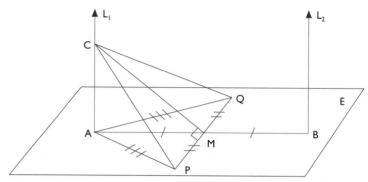

Description: *The figure illustrating theorem 8-7 appears on page 254, and is depicted by a parallelogram representing a horizontal plane E. Segment AB occurs horizontally in the plane. A at the left, B at the right, and its midpoint M. The ray L₁ is shown drawn vertically upward from point A, and a ray L₂ is shown vertically upward from B. The segment PQ is drawn in the plane perpendicular to AB at point M. Tick marks indicate that AM is equal to BM, and PM is equal to QM. Additionally, the segments AP and AQ are drawn, with tick marks indicating that they are equal in length, and line segments are also drawn to a point C on ray L₁ above the plane from points P, M and Q. A small right angle symbol indicates that MC is perpendicular to PQ.*

Source: *Recording for the Blind News,* Winter 1990.

Preparing for the Job Interview

Think back to Chapter 7 on primary research, where you learned how to prepare questions to elicit the information you needed from the person you were going to interview. For example, you learned about the value of asking open rather than closed questions when you wanted to get more than a yes or no answer. With the job interview, the situation is reversed and you are being interviewed by someone who has given lots of thought to the kinds of information he or she wants to gather about you. It's critical to remember that when you are being interviewed for a position someone has spent considerable time preparing. To be successful, you will need to spend a good deal of time preparing for the interview as well.

PUTTING YOUR BEST SELF FORWARD

We have referred to *ethos* a number of times in various chapters. You were told to consider your *ethos* when you gathered data for your oral presentations and when you carefully proofread your documents before sending them to your boss. *Ethos* is especially important when doing your job interview. You want to be and come across as competent, interested, intelligent, and as a good colleague.

The first thing to consider is how you look. A survey of top business executives found that 84 percent said their companies do not hire people who appear at job interviews improperly attired (Malloy 36). If you are going to make an error in dressing, make it in dressing on the conservative side. Business suits and professional-looking shoes are always your best bet. If you go out to buy a new suit and shoes for your interviews, it's sometimes best if you wear them a few times before you wear them to the interview. You don't want to look so uncomfortable in your new attire because your shoes pinch, for example, that the person interviewing you thinks you never dress up or you are not happy with the interview.

GATHERING THE RIGHT MATERIAL FOR THE JOB INTERVIEW

Once you've settled on the look that will enhance your *ethos,* it's time to begin gathering the data you need for your successful interview. The first thing you must do is gather the necessary material about yourself:

- Names, addresses, and dates of your work history.
- Names, titles and addresses for your references.
- Dates and details about your education and training, including copies of all certificates and licenses.

You should outline specifically what you have to offer, especially those points you want to be sure to cover in the interview. Develop a list of appropriate questions to ask during the interview, but avoid salary or fringe benefits questions at this point. You can take your list of questions to the interview.

Learn About the Company

It is critical to a successful interview that you gather information about the companies to which you have applied. You want to learn as much as you can about the firm interviewing you (and remember it's also the firm you are interviewing): Know well what kinds of jobs you can do for your prospective employer and what products or services the firm offers. A company wants to see that you take the initiative. Use the skills you developed doing library and Internet research to find out all you can about the company interviewing you. You will especially want to visit the company's Web pages to see what they say about themselves, what they value, and what's new. The person interviewing you will be impressed if you have done your homework and he or she doesn't have to use interview time telling you something you could have found out on your own. Getting information about a company can also give you the opportunity to explore what tasks or projects they need people for and what your particular interests in the company might be.

Anticipate Questions

Once you know who in the company is interviewing you, you can turn your attention to anticipating the questions he or she will be asking you.

Anticipate questions you may be asked and how you will respond. Some likely questions include the following:

- Tell me something about yourself (usually at the beginning of the interview to set you at ease). Remember to avoid supplying illegal or irrelevant information in response to this question.
- Why do you want to work for us? (A great opportunity to show you have done your homework.)
- What are your strengths?
- What are your weaknesses? This is a tough question to handle. Do you have a weakness that is really a strength? For example, are you a perfectionist about your work, wanting to do the best job possible at all times? When discussing a perceived weakness, it's always a good idea to discuss as well what you are doing to remedy the situation or how a weakness may somehow work in your favor.
- What were your favorite undergraduate, graduate, or continuing education courses? In which ones did you make your best grades or learn the most?
- What work experience have you had that is particularly meaningful?
- What interesting projects did you complete? What skills have you learned?
- Where do you see yourself in five years? Ten years?

Most of these typical questions don't have "right" answers. In most cases, the interviewer just wants to see what you will do with them. Usually, the résumé acts as an outline for the interview. If your résumé discusses a particular job you have had, anticipate such questions as "I see you worked for four years with General Dynamics, Electronic Division, as a technician. What were some of your major activities? What accomplishments are you most proud of? Why did you leave?"

Remember that you have rights as an interviewee. Companies are not allowed to discriminate against candidates. Decisions about whether or not to hire you must be made on the basis of your qualifications, not personal items. Therefore, you should be wary of any company that asks questions about your marital status, whether or not you plan to have children, your age, or your race, religion, or ethnicity. If you have special concerns about a spouse being able to find work, for example, you can go ahead and ask those questions. Once you have brought up this subject (or any related one) the company is then free to pursue that line of questioning to a certain extent. Companies often try to be sensitive to issues of family life and disability, so if you have an interview you might be asked if you require any special accommodations such as a wheelchair-accessible room or the like. Don't be afraid to ask for what you need.

If you have any questions about the most current equal opportunity and affirmative action guidelines in your state, ask the reference librarian for information on the subject as you plan for your interview. The following list demonstrates some questions that are lawful and unlawful for interviewers to ask, topics that interviewers may not raise, and documents that interviewers may not require legally:

Unlawful Questions

Are you naturalized or native-born?

When did you acquire citizenship?

Are your parents or spouse naturalized or native-born?

When did your parents or spouse acquire citizenship?

What is your first language?

How did you acquire your ability to speak, write, or read a foreign language?

Do you wish to be addressed as Mr.? Mrs.? Miss? Ms.?

Are you married? Single? Divorced? Separated?

Where does your spouse work?

What are the ages of your children, if any?

How old are you?

What is the date of your birth?

Do you have a disability?

Have you ever been treated for any of the following diseases?

Have you ever been arrested?

Unlawful Documents

Proof of age, birth certificate, or naturalization record.

Unlawful Topics

Birthplace of applicant and relatives.

Religious denomination, affiliation, church, parish, pastor, rabbi, or religious holidays observed.

Lineage, ancestry, ethnic group, descent, parentage, or nationality.

Nationality of applicant's parents or spouse.

Name, addresses, ages, number, or other information about spouse, children, or other relatives *not employed by the company.*

Applicant's *general* military experience.

Clubs, societies, or lodges of applicant (except for professional societies).

Lawful Questions

Are you a citizen of the United States?

Do you intend to become a citizen of the United States?

If you are not a citizen, have you the legal right to remain in the United States?

Do you intend to remain permanently in the United States?

Are you between 18 and 65 years of age? If not, state your age.

Do you have any impairments—physical, mental, or medical—which would *interfere* with your ability to perform the job for which you are applying?

Have you ever been convicted of a crime? (give details).

Lawful Topics

Languages applicant speaks and writes fluently.

Applicant's academic, vocational, or professional education and the public and private schools attended.

Applicant's work experience.

Name, addresses, ages, number, or other information about spouse, children, or other relatives *already employed by the company.*

Applicant's U.S. military experience or experience in the state militia.

Applicant's service in particular branch of the armed forces.

Applicant's membership in any organization which the applicant considers relevant to his or her ability to perform on the job.

If you are asked an unlawful question during an interview, seek the advice of your career counselor or major professor.

Take the Right Things to the Interview

Prepare a briefcase or a file folder with materials you plan to take to the interview. Include extra résumés; names, addresses, and telephone numbers of references and employers; and paper and pen. Although you probably will not take notes during the interview, you'll want to jot down as much as you can remember about the interview immediately afterward: names of people you met, topics discussed, and details you learned about the job, company, and interviewer. You may want to make up a portfolio to take with you to the interview. Include in it work samples, writing samples, letters of recommendation, copies of licenses or certificates—anything you think will help you show employers what you have done and can do. Arrange all your papers so you can find them easily.

Practice the Interview

Many placement centers have workshops for the job interview. Videotape yourself in a mock interview to identify mannerisms you may not be aware of. Do you fidget while you talk; for example, rubbing your nose or stroking your beard? Do you jingle change in your pocket, tap your finger or pen on the desktop, fuss with your purse, tie, or hair? Practicing an interview in front of a mirror can also be helpful. Interview for jobs every chance you have, but don't waste your time and others' by interviewing for jobs you do not want. The more experience you have, the more relaxed you will be. Talk to others who have been interviewed to find out about their experience.

Have an idea of what to expect by becoming familiar with interview routines. Some interviews are brief, lasting 20 to 30 minutes, especially first ones. Others may consume an entire day or two, in which you're being introduced to a number of different people. If your travel to the interview has been paid for by the company, don't make long-distance telephone calls at the company's expense or abuse the snack and drink refrigerator in your hotel room.

Succeeding at the Job Interview

With your preparation done, you will find the interview far less stressful than you might imagine. The most important rule for an interview is to be yourself. Acting a role is not going to fool an experienced interviewer. Of course, you will want to act professionally: You will want to arrive promptly, perhaps 5 or 10 minutes early, dress appropriately, be friendly but not too casual, and have your thoughts and questions carefully organized. Because you have done your homework, try to relax and enjoy the opportunity to tell the interviewer about yourself and what you can do.

Many say the first 20 seconds of an interview is crucial for making a positive first impression. Take cues from your interviewer: Wait until you are asked to sit down, for example. Generally it's best to take the initiative when shaking hands, especially if you are a woman. Some male interviewers may think it impolite for them to initiate a handshake. Don't allow awkwardness to intrude into these first few seconds of the interview process. When you shake hands, do so firmly. Greet the interviewer by name if you are sure of the pronunciation. Smile. Look your interviewer in the eye.

When the questions begin, answer questions fully and honestly. If you don't know an answer, say so. Try not to answer questions with either a yes or no, but also don't talk too much. Listen attentively and show your interest and enthusiasm. You'll want to make lots of eye contact with the interviewer, sit up straight, appear poised and at ease. Don't chew gum or smoke, even if invited to. If you are asked a question such as, "Tell me something about yourself," bring the conversation around to those things you know you want to talk about. Ask in return, "Would you like to hear how my educational background provides important skills for this position?" By countering open-ended questions like this with a question, you can often be sure the interview stays on your target: Tell the interviewer what you have to offer. Some interviewers are not terribly skilled at interviewing. They sometimes talk excessively about themselves, their jobs, or the company. By preparing for the interview, you know exactly what facts you want to be sure to cover. Use every opportunity to draw the conversation into these areas.

Try to be positive and courteous at all times. Don't whine or be disrespectful about other employers. Keep away from personal matters; don't discuss your family life, financial condition, or politics. Be alert to hints that the interview is over, and before leaving thank the interviewer for the interview.

Sometimes you will be interviewed by a panel of three or four interviewers. For management positions especially, some interviewers may try to test you in a stress interview by being antagonistic or deliberately placing you in a stressful situation to see how you will react. Sometimes panels assign members antagonistic roles while other members become rescuers when the applicant gets into trouble. Such tactics are unusual, but should you face situations that appear unnaturally stressful, realize you are being tested. Try to remain calm and in control of yourself.

FOLLOW UP ON YOUR INTERVIEW

Immediately after the interview, find an opportunity to write notes about the interview. Jot down important details, like names, titles, and job descriptions you will want to remember. If possible, ask for the cards of those people to whom you speak and use them for writing your follow-up notes.

follow-up letter
a letter sent after a job interview that emphasizes the interviewee's continued interest in the position

After an interview, especially one that went well, you will want to write a **follow-up letter** (see Example 17–5). A follow-up letter, usually not more than three paragraphs, thanks the interviewer for the interview, may make a specific reference to something that you and the interviewer might have discussed, and emphasizes again your interest in the position and what you can offer the company.

Sometimes during the interview, additional information is requested. Such a situation provides a perfect opportunity for writing a follow-up letter. Enclosing the information with the letter, you can make the letter a natural development of the interview. You can also use the follow-up letter to call to the interviewer's attention pertinent facts that you want to emphasize again. If you discovered during the interview a mutual interest in sports or hobbies, you can mention that fact (e.g., "Given that I just bought a new fly rod, I especially enjoyed hearing about the fishing trip you and others in the company took to Alaska"). The result of mentioning such one-of-a-kind information is to create in the interviewer's mind a personal picture of you that will make you stand out from all the competition. Establish courteous but personal *ethos*.

EXAMPLE 17–5   The Follow-Up Letter

143 South Camino del Rio
San Diego, CA 94623
September 21, 1999

Ms. Julie Wesson
Southwest Research Institute
P.O. Drawer 28510
San Antonio, TX 78228-0510

Dear Ms. Wesson:

Thank you for the time you spent last Wednesday discussing the instructional designer position with me. I appreciate the detailed description you gave me of the position and its responsibilities, especially because my interest and educational experience so closely parallel it.

As you asked during the interview, I am enclosing the instructional video that I scripted, storyboarded, and produced, transporting pictures, sound, and animation from one medium to another. I believe this video reflects the experience I have had in analyzing, designing, developing, and evaluating instructional design projects for the past two years.

Once again, thank you for your interest in me and my qualifications and for taking the time to talk with me. If you would like any further information, please call me. I sincerely hope to hear from you soon.

Very truly yours,

Jesse Wilson

Enclosure: Video

New technologies are constantly being created to aid face-to-face presentations and presentations across distances. Keep in mind that the minute this textbook hits the press, new technologies will have replaced or augmented the ones we describe. However, these descriptions will give you an idea of how you might use these technologies in oral presentations. Even if your school does not yet have these technologies, your workplace most likely will.

As we have discussed already in this chapter in relation to visual aids, numerous software programs can help you present your information effectively. Presentation software often includes graphics programs and other

Employing New Technologies

desktop publishing software. If you have a computer projection system available for your presentation, you can use presentation software to supplement your presentation or to create self-standing visual displays that can be shown at a kiosk at a conference or professional meeting. The value of this computer technology is that the electronic text and visual aids you have developed for your oral presentation can now be directly used in other forms of presentation.

Increasingly, companies are relying on video to support or capture oral presentations. Some companies prefer to have prospective employees submit a videotape along with their résumés. In that way companies can judge a person's communication abilities. Some schools now have small studios where students can go to prepare these presentations. If you are asked to do such a presentation, you will have to practice looking at the camera and addressing the viewers "directly." This takes some doing, so if you need to create an interview tape, plan to allow plenty of time to script your presentation, deliver it, critique the tape, and do the whole thing over again until you have a final version you are satisfied with.

Because many companies are transnational or global, they are relying on videoconferencing to cut travel costs but still keep in contact with staff at distant locations. **Videoconferences** provide speakers and listeners with simultaneous audio and video across distances. Videoconferencing systems combine television monitors and cameras, special coder/decoder systems, and digital transmission services from long-distance carriers and local telephone companies to provide full interactive video and audio meetings for speakers and listeners. Thus, as a speaker you might be located in Minneapolis while your listeners are in Orlando. With the price of these systems dropping, advice for delivering presentations will change in the future as speakers view listeners on television monitors.

videoconferences
use of television cameras and monitors and digital transmission devices to provide interactive video and audio for meetings at distant sites

Videoconferencing systems have also been developed for use from your desktop computer. Desktop videoconferencing units such as Connectix™ integrate camera and microphone systems that run on computers. The Connectix camera is an inexpensive video camera about the size of a fat golf ball that sits on top of your computer monitor and transmits your image to others' computers to which you are linked. This means that you can deliver your presentation live from your computer to anyone else who has a computer equipped with videoconferencing equipment and connected to digital transmission services. You can make a presentation, for example, to 8 or 10 others whose faces will be displayed on your computer monitor as you address them.

The University of Minnesota, for example, has used the Cameo system and the Integrated Services Digital Network (ISDN™) phone service to allow university students to mentor high school students across distances (see Duin et al.). Essentially, the university students have acted as speakers in highly interactive impromptu presentations with the high school students.

The combination of desktop video and the World Wide Web is leading to the development of a whole generation of new ways for people to interact with each others. A few years ago, all you might be able to do would be to download your PowerPoint™ presentation to your Web page. Now, with developments in streaming audio and streaming video, we can place audio and video presentations on Web sites so that others can watch them on their television monitors. Streaming audio and video differ from the jerky compressed video with which you might be familiar. Because they are run from a distant server and you don't have to download them onto your computer before you watch them, you can watch high-quality video or listen to high-quality sound at the desktop. More and more, given the technological changes occurring in the workplace, you will be called upon to make presentations in a variety of media.

Before we leave this discussion of communication technologies in the workplace, we want to touch on the growing importance of asynchronous Web conferencing. Web conferencing can be local or global, and it is used by companies to keep people in touch who are working on developing products, coordinating writing projects, or pursuing some other corporate goal. Whenever group participation and collaboration are needed, companies are relying more often on Web conferencing products such as Web-Board™. These products allow people to have virtual meetings when schedules and locations won't allow face-to-face interactions. Their information management functions and archiving abilities allow information to be organized locally logically or chronologically. Moreover, most Web conferencing products are designed to facilitate foreign language conferences, an important feature in a global marketplace. Buttons and menu bars can display the same commands in a variety of languages.

If you think that you will never use these technologies, think again. Technologies will continue to be designed to bring speakers and listeners closer together across distances, and as we increase our use of desktop videoconferencing, the demand for clear impromptu presentations and accurate extemporaneous presentations will also increase. Therefore, remain aware of new technologies and plan to use them throughout your professional career.

SUMMARY

- This chapter provides you with a basic oral presentation grid (Writing Strategies 17–1) to analyze your and your listeners' knowledge, attitudes, and needs before designing the content of your presentation and how you will deliver that content.
- Oral presentations include impromptu, extemporaneous, and scripted presentations. The impromptu presentation is delivered at a moment's notice, unprepared and unrehearsed. The extemporaneous presentation is delivered from a set of notes and any supporting materials you need. For a scripted presentation, you

determine every word and gesture in advance, and when you present it, you simply read to your listeners. Speakers largely prefer the extemporaneous presentation because it allows them to respond to their listeners and speak in a normal tone of voice.

■ When preparing your presentation, remember to keep it short, make the organization obvious, make the ideas simple and vivid, and summarize and be ready for questions. Because oral presentations nearly always benefit from visual aids, use them but don't rely totally on them. Also, keep your listeners' special needs in mind, and if listeners are visually challenged, plan for complete oral descriptions of any visuals used in your presentation. To coordinate your visuals with your presentation, you can create a storyboard that designates when each visual should be used.

■ When delivering your presentation, remember to set the stage, carry notes on cards (for extemporaneous presentations), use your visual aids, and talk slowly yet with enthusiasm. Think deliberately about the arrangement of the room and the auditory and visual needs of your listeners. When responding to questions, repeat each question and answer it completely while being as brief and specific as possible.

■ When you are preparing for a job interview, do your homework: Gather information on the company and rehearse your answers. After the interview, write an appropriate follow-up letter.

■ Finally, be on the lookout for new technologies that will aid both the preparation and delivery of your presentations.

ACTIVITIES AND EXERCISES

1. Look back on the examples from the EPA presentation on the bioremediation cleanup process in Prince William Sound and review Appendix A, Case Documents 2. Using note cards for an extemporaneous oral presentation, prepare the complete presentation and deliver it to your classmates and instructor.

2. *Preparing for an Oral Presentation.* You can use the oral presentation development grid in Writing Strategies 17–1 to begin brainstorming for a presentation that you need to design and deliver. If you don't have a specific subject in mind, use any of the case studies in this textbook. For example, you could brainstorm for:

   ■ A presentation to the Coalition of Citizens Concerned with Animal Rights (CCCAR) in which you describe Midwest University's efforts to control birds in the experimental crop fields on its campus (or change the listeners to a group of professors whose raw data comes from the experimental crop fields).

   ■ A presentation to colleagues in your workplace about the impact of product liability suits on the development of documents accompanying your products.

   ■ A presentation to colleagues in your workplace about the need to develop a policy on sexual harassment.

   ■ A presentation to managers in your company about anticipating software problems such as the Y2K bug.

After you have determined your understanding and attitudes, and those of your listeners toward the subject, as well as your needs, you can then concentrate on determining what content to include and how to deliver the presentation.

3. *Preparing for the Visually Challenged.* Locate any visual aid (graph, table, flow chart) and following the guidelines from the National Braille Association, write out how you would present this visual aid orally to the visually challenged. Bring your description to class and compare it with those of other students.

4. *Determining Type of Presentation.* To gain an understanding of how you might mix and match oral presentation types, work with a partner and determine the type of presentation that you would deliver in each of the scenarios in Example 17-6.

EXAMPLE 17–6 Scenarios for Exercise 4

Context for Speaking	Listeners' Need 1. To learn 2. To do 3. To learn to do something	Type of Presentation 1. Impromptu 2. Extemporaneous 3. Scripted 4. Combination
Instructing clients how to fill out a purchase agreement form		
Teaching a local community group the best ways to avoid exposure to HIV		
Describing your own rights in a legal contract with a landlord		
Encouraging customers to purchase exercise equipment		
Describing product liability laws to corporate officials		
Encouraging local residents to vote against the building of a nuclear power plant in your county		
Explaining how to solve a telecommunications problem to a staff member		

EXAMPLE 17–7   Room Layout for a Computer Laboratory

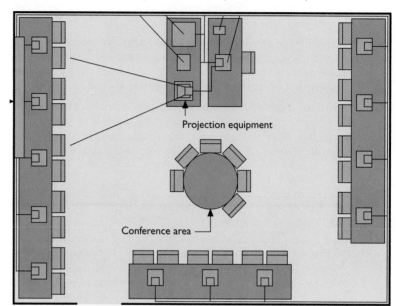

Projection equipment

Conference area

Source: Mary M. Lay and William M. Karis. *Collaborative Writing in Industry: Investigations in Theory and Practice* (Amityville, NY: Baywood Press, 1991). Reprinted with permission.

5. *Delivering Presentations in Difficult Rooms.* Note the layout for a computer laboratory in Example 17-7. Imagine that you are to deliver an oral presentation in this room. Describe the strategies that you would use to arrange your stage and hold your listeners' attention.

6. *Offering Feedback to Other Speakers.* The feedback form shown in Example 17-8 was created as a way to have students give each other feedback on their presentations. Critique this feedback form, and based on your critique and the suggestions in this chapter, design a new feedback form.

7. Note the document titled "What Are Universal Precautions?" from the case study materials on AIDS in Appendix A, Case Documents 1. Read through this document and prepare three different types of presentations:

    ■   A one-minute impromptu speech in response to the question, "What are universal precautions one might take to avoid exposure to HIV?"
    ■   A three-minute extemporaneous speech
    ■   A five-minute scripted speech.

    What are the differences in your preparation and delivery processes in these three types of presentations? Which type of presentation was most effective? Easiest? Most difficult? Best for a classroom audience?

EXAMPLE 17–8   Oral Presentation Feedback Form

Speaker _____   Person giving feedback _____

	Poor	Satisfactory	Good	Excellent

Does the speaker appear:

1. Well prepared and knowledgeable   ____   ____   ____   ____
2. At ease, yet professional   ____   ____   ____   ____

Has the speaker:

3. Determined how the *audience* and   ____   ____   ____   ____
    purpose affect the *content* of the
    presentation
4. Developed an effective *strategy*   ____   ____   ____   ____
    for the presentation
5. Designed effective *visuals/handouts*   ____   ____   ____   ____

Rate the speaker's:

6. Volume, pitch, clarity, and speed   ____   ____   ____   ____
7. Verbal fillers (uh, um, okay)   ____   ____   ____   ____
8. Gestures   ____   ____   ____   ____

Additional Comments—

8. Using the guidelines for visual display of data from Chapter 10, design three graphics that you would use to support a five-minute scripted presentation that looks at the effects of the oil spill caused by the *Exxon Valdez* in Prince William Sound after 10 years. You will have to use information gained from library and Internet searches to describe Prince William Sound today.

WORKS CITED

Allen, M., J. Hunter, and W. A. Donohue. "Meta-Analysis of Self-Report Data on the Effectiveness of Public Speaking Anxiety Treatment Techniques." *Communication Education* 38 (1989), pp. 54–76.

Anderson, Paul V. "What Survey Research Tells Us About Writing at Work." Lee Odell and Dixie Goswami, eds. New York: Guilford Press, 1985.

Connolly, James E. *Making More Effective Technical Presentations.* Minneapolis: University of Minnesota, 1982.

Duin, Ann Hill; Linda A. Jorn; Craig J. Hansen; Elizabeth Lammers; Lisa Mason; and Sandra Becker. *Mentoring via Telecommunications/Multimedia.* Narrative report submitted as part of an EDUCOM award application, 1993.

Gilbert, Frederick. "The Technical Presentation." *Technical Communication* 39, no. 2 (1992), pp. 200–1.

Hartley, James. "Presenting Visual Information Orally." *Information Design Journal* 6, no. 3 (1991), pp. 211-20.

Malloy, John T. *Dress for Success.* New York: Warner, 1975.

National Braille Association. *Tape Recording Manual.* 3rd ed. Washington D.C.: National Library Services for the Blind and Physically Handicapped, Library of Congress, 1979.

Parkin, A. J., and F. K. Aldrich. "How Can Studying from Tape Be Made Easier?" *New Beacon* 71 (1987), pp. 340-41.

Robertson, Doug, and Rowan Carroll. "Videoconferencing: Matching Systems and Peripherals with Applications." *Presentation* 27, no. 5 (1993), pp. 32, 35-36.

Scheiber, H. J., and Peter J. Hager. "Oral Communication in Business and Industry: Results of a Survey on Scientific, Technical, and Managerial Presentation." *Journal of Technical Writing and Communication* 24, no. 2 (1994), pp. 161-80.

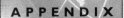

Case Documents 1

Case Documents 1: MAP brochure used with permission of the Minnesota AIDS Project; Taking Care: used with permission of the author, James Rothenberger, School of Public Health, University of Minnesota; Monthly Surveillance Report used with permission of the Minnesota Department of Health AIDS epidemiology Unit; Tuberculosis and AIDS Project; AZT Fact Sheet used with permission of the Exxon Company, USA.

Case Documents 2: all documents used with permission of the Exxon Company, USA

659

Taking Care

A Manual for
AIDS Caregivers

In the next decade, AIDS will affect all Americans. Many of us will be caring for a patient or a loved one who has AIDS.

There are no rules for taking care of a person with AIDS, but there is information in this manual that can help. It was produced by people with years of experience in health care, AIDS education and counseling.

1989 Midwest AIDS Training and Education Center (MATEC)
School of Public Health, University of Minnesota. Funded by the U.S. Public Health Service, Health Resources and Services Administration, Education and Training Center Program, Grant # HHS BRT 000033-01-0.

MATEC Coordinator:
James Rothenberger
Publication Coordinator:
Patricia Ohmans

First, the facts

With AIDS, it's what you do that matters.

- AIDS stands for Acquired Immune Deficiency Syndrome. This serious infection, caused by the Human Immune Deficiency Virus (HIV) destroys the body's natural ability to resist infections. People with AIDS become seriously ill and often die of diseases that most people's bodies easily resist. Some of these illnesses can be treated, but there is no cure for HIV infection itself. (This brochure uses "AIDS" when referring to a condition, such as "AIDS patient," and "HIV" when referring to the virus that causes AIDS.)

- Infection with HIV can, by itself, be a serious health condition, but not everyone infected with HIV has AIDS. Some people seem unaffected by the virus. Others develop only limited symptoms. However, anyone infected with HIV can pass the virus on to others.

- People infected with HIV can be male or female. That is why this manual refers to a person with AIDS alternately as "he" or "she." HIV infects white people and people of color, heterosexuals, and homosexuals. With AIDS, it's not who you are, it's what you do that matters.

- HIV is spread through close sexual contact, in which blood, semen or vaginal secretions are exchanged. It is also spread through the shared use of intravenous needles, and perinatally, from mother to child around the time of birth. Some cases of HIV transmission through non-intact skin in contact with HIV-infected blood and through transplantation of infected organs and tissues have also been documented.

- People risk becoming infected with HIV when they have unprotected anal or vaginal intercourse with an infected person, or when they share needles with someone who is infected. A pregnant woman infected with HIV can pass the virus on to her unborn child through shared blood, or to an infant through breastmilk. Before 1985, many people received blood transfusions contaminated with HIV. Now the blood supply is screened and it is safe to receive blood.

What are universal precautions?

They are the best way to avoid exposure to HIV.

What are universal precautions? They are the simplest way to ensure that you will not be infected with HIV when you care for a person with AIDS.

- **Wear protective clothing.**
  All health care workers should routinely use appropriate barrier precautions when in contact with blood or other body fluids.

  Wear gloves for touching blood and body fluids, mucous membranes, or non-intact skin of all patients, for handling items or surfaces soiled with blood or body fluids, and for performing venipuncture and other vascular access procedures. Change gloves after contact with each patient.

  Wear masks and protective eyewear or face shields during procedures that are likely to generate droplets of blood or other body fluids to prevent exposure of mucous membranes of the mouth, nose and eyes. Wear gowns or aprons during procedures that are likely to generate splashes of blood or other body fluids.

- **Wash your hands.**
  Wash your hands and other skin surfaces immediately and thoroughly if they are contaminated with blood or other body fluids. Wash your hands immediately after gloves are removed.

- **Be careful with needles and sharps.**
  Take precautions to prevent injuries caused by needles, scalpels and other sharp instruments or devices during procedures; when cleaning instruments; during disposal of used needles; and when handling sharp instruments after procedures.

  To prevent needlestick injuries, do not recap, bend or break needles. Do not remove them from disposable syringes or otherwise manipulate needles by hand. After they are used, disposable needles and syringes, scalpel blades and other sharp items should be placed in puncture-resistant containers for disposal; the puncture-resistant containers should be located as close as practical to the use area. Large-bore reuseable needles should be placed in a puncture-resistant container.

- **Keep resuscitation devices handy.**
  Although saliva has never been implicated in the spread of HIV, try to minimize the need for direct mouth-to-mouth resuscitation. Keep mouthpieces, resuscitation bags or other ventilation devices available in areas in which the need for resuscitation is predictable.

- **Protect chapped or irritated skin.**
  If you have oozing lesions or dermatitis, wear gloves or avoid direct patient care and patient care equipment until the condition resolves.

- **Be careful if you're pregnant.**
  Pregnant health care workers are not known to be at greater risk of contracting HIV infection than health care workers who are not pregnant. However, if you develop HIV infection during pregnancy, the infant is at risk of infection from perinatal transmission. Because of this risk, pregnant health care workers should strictly adhere to precautions to minimize the risk of HIV transmission.

Wear gloves.
Wash your hands.
Be careful with needles.

Minnesota Department of Health AIDS Epidemiology Unit Pediatric/Adolescent HIV/AIDS Monthly Surveillance Report November 1, 1992

The following five tables summarize information on transmission category, race, and vital status for cases of HIV infection and AIDS in children and adolescents which have been reported to the Minnesota Department of Health. In order to relate the age of the person as closely as possible to age at time of HIV infection, the information reflects age at time of first positive test for HIV infection or time of diagnosis of AIDS. In accordance with the U.S. Centers for Disease Control criteria for pediatric AIDS surveillance, we have defined "pediatric" to include those less than 13 years of age at time of diagnosis or test, and "adolescent" to include those 13 to 19 years of age at the time of diagnosis or test.

Table 1 Pediatric and Adolescent (0–19 years of age) AIDS Cases by Exposure Category

Exposure Category	Male Number (%)	Female Number (%)	Total Number (%)
Men who have sex with men	0 (0)	0 (0)	0 (0)
Injecting drug use (IDU)	0 (0)	0 (0)	0 (0)
Men who have sex with men and IDU	0 (0)	0 (0)	0 (0)
Hemophilia/Coagulation disorder	2 (33)	0 (0)	2 (14)
Heterosexual	0 (0)	1 (13)	1 (7)
Transfusion, blood/components	1 (17)	1 (13)	2 (14)
Mother with/at risk for HIV infect.	3 (50)	5 (63)	8 (57)
Other/Undetermined	0 (0)	1 (13)	1 (7)
TOTAL	6 (100)	8 (100)	14 (100)

Table 2 Pediatric (< 13 years of age) and Adolescent (13–19 years of age) Cases of HIV Infection (Including AIDS) by Race/Ethnicity

Race/Ethnicity	<13 Years Number (%)	13–19 Years Number (%)	Total Number (%)
White (not Hispanic)	18 (56)	37 (59)	55 (59)
Black (not Hispanic)	11 (34)	21 (34)	32 (34)
Hispanic	2 (6)	2 (3)	4 (4)
Asian/Pacific Islander	1 (3)	0 (0)	1 (1)
American Indian/Alaskan Native	0 (0)	2 (3)	2 (2)
TOTAL	32 (100)	62 (100)	94 (100)

Table 3 Pediatric (<13 years of age) Cases of HIV Infection (Including AIDS) by Exposure Category

Exposure Category	Male Number (%)	Female Number (%)	Total Number (%)
Hemophilia/Coagulation disorder	7 (37)	0 (0)	7 (22)
Transfusion, blood/components	3 (16)	1 (8)	4 (13)
Mother with/at risk for HIV infect.	9 (47)	11 (85)	20 (63)
Other/Undetermined	0 (0)	1 (8)	1 (3)
TOTAL	19 (100)	13 (100)	32 (100)

Table 4 Adolescent (13–19 years of age) Cases of HIV Infection (Including AIDS) by Exposure Category

Exposure Category*	Male Number (%)	Female Number (%)	Total Number (%)
Men who have sex with men	28 (60)	0 (0)	28 (48)
Injecting drug use (IDU)	2 (4)	1 (9)	3 (5)
Men who have sex with men and IDU	3 (6)	0 (0)	3 (5)
Hemophilia/Coagulation disorder	13 (28)	0 (0)	13 (22)
Heterosexual	0 (0)	5 (45)	5 (9)
Transfusion, blood/components	0 (0)	1 (9)	1 (2)
Other/Undetermined	1 (2)	4 (36)	5 (9)
TOTAL	47 (100)	11 (100)	58 (100)

*Exposure category data are pending for 3 males and 3 females

Table 5 Pediatric (<13 years of age) and Adolescent (13–19 years of age) Cases of HIV Infection (Including AIDS) by Vital Status

Vital status	<13 Years Number (%)	13–19 Years Number (%)	Total Number (%)
Alive	27 (84)	58 (94)	85 (90)
Dead	5 (16)	4 (6)	9 (10)
TOTAL	32 (100)	62 (100)	94 (100)

AIDS PUBLIC POLICY IN MINNESOTA

A series of public policy position papers from the Minnesota AIDS Project

Number 5   August 1992

TUBERCULOSIS AND AIDS

In 1990 the Centers for Disease Control (CDC) received the first report of clusters of drug resistant TB cases among hospitalized AIDS patients.[1] Tuberculosis is an old disease staging a comeback from its recent obscurity. It poses an immediate threat to those infected with HIV and a longer-run threat to the entire public. TB is preventable and largely still curable, but aggressive public health programs must be fielded immediately.

BACKGROUND

Tuberculosis (TB) caused by Mycobacterium tuberculosis, is a contagious, airborne disease of the lungs. Sometimes TB affects other parts of the body. It is a major cause of death in many parts of the world today and was the leading cause of death in the early years of this century in the United States. As recently as 1950 Minnesota reported 5,000 new cases. TB has been called a "social disease with medical aspects" because it can be found wherever poverty, poor nutrition, homelessness, and substance abuse occur.

Robert Koch identified the cause of TB in 1882 in Germany. Control programs have existed for over a century. Clinics, case finding programs, isolation in TB sanitariums, and most important—improvements in the standard of living—helped bring tuberculosis under control. Antibiotics introduced in the 1940s and '50s (streptomycin and isoniazid) were very effective in curing the disease. By the 1960s TB had become relatively uncommon in the United States.

TB causes fatigue, weight loss, coughing, and hemorrhage. These and other symptoms become more severe as the disease progresses. Detection of TB can be done by mass screenings using the Mantoux skin test or chest X-rays. Sputum cultures provide definitive diagnosis but may take weeks to complete. Polymerase chain reaction (PCR) tests provide much faster results but are not ready for general clinic use.[2]

CURRENT STATUS OF TUBERCULOSIS CONTROL EFFORTS IN MINNESOTA

In 1990 there were 25,701 cases of TB in the United states, 9.4% more than in 1989.[3] While AIDS may be poised to make a comeback in many parts of the country, it is not as serious a threat in Minnesota at the present time.

	Number of New Tuberculosis Cases	Rate per 100,000 in 1991
United States	26,283	10.4
Minnesota	102	2.3
Metro Area	45	
Greater Minnesota	47	

In Minnesota, surveillance and monitoring of TB cases is a state responsibility, while TB treatment services are provided by county and city clinics located in the Twin Cities metro area and Rochester. Public Health nurses and physicians provide TB treatment services in most state counties. Federal grants currently supply $86,020 for 2.5 FTE Minnesota Department of Health TB surveillance staff (1992-1993 budget). $30,000 in state funds are available to supply free TB medicine to patients (1991-1992 budget).[4] Current federal spending on TB programs is $20,000,000. Prevention efforts are cost effective, repaying the money spent 4 to 1.[5]

HIV & THE REEMERGENCE OF TB

If tuberculosis has been brought under control in the United States, it has not been eliminated. Some metropolitan centers are seeing shocking numbers of new TB cases. Cities such as New York which combine high rates of IV drug use, alcoholism, homelessness, and poverty offer ideal conditions for TB to infect large numbers of people. These are the same cities that have the highest rates of HIV.

Persons living in substandard, crowded conditions are

more likely to be exposed to the airborne bacteria which causes TB and, owing to poor health, are more likely to develop active disease. The larger the pool of active TB cases, the more likely is transmission.

Infection by Mycobacterium tuberculosis does not automatically cause disease. TB often remains inactive, or latent, causing no symptoms. Experts believe that the lifetime chances of healthy people developing TB after exposure are about 10%. Tuberculosis tends to be much more aggressive in persons whose immune systems have been damaged by HIV or cancer treatments. HIV infection can activate latent TB and, in combination, TB and AIDS take more severe forms. Since Active TB is contagious, cities with large numbers of HIV cases see the most dramatic growth in TB.

While TB can generally be cured, the disease is becoming resistant to drugs which are inexpensive, easy to administer, and have few side effects. One reason for this is that patients most likely to be infected (chemically dependent homeless persons, for example) are the least likely to complete the lengthy course of treatment. Antibiotics help these patients feel better so they stop taking their medicine—long before all the TB germs have been destroyed. When they become ill again they return for more medicine. The result is hardier, more drug-resistant strains of TB. Multi-drug-resistant TB is difficult to treat, no matter how responsible the patient is.

Of the 102 TB cases in Minnesota, 3 were resistant to 1 drug and 1 case was resistant to two drugs. 2 of the 102 cases were HIV+ and neither were drug resistant.[6]

Effective hospital infection control procedures are a critical component of good medical care. In a recent study of a cluster of TB cases, it was found highly probable that TB had been acquired in hospital. The airborne TB germs were transmitted through ventilation ducts during hospitalization. In the same study it was reported that 18% of the health care workers in the hospital tested positive for TB.[7]

AIDS Public Policy in Minnesota
is produced by the Minnesota AIDS Project
2025 Nicollet Avenue South
Minneapolis, MN 55404
612/870-7773

Michael Jefferis, Series Editor
Tom Flynn, Director of Education
Lorraine Teel, Executive Director

Desktop Publishing: Brian Cockayne

Reviewed by: Mandy Carver, Susan Lentz, Scott Mayer, Mary Frances Skala

Supported through a grant from the
MINNESOTA AIDS FUNDING CONSORTIUM

The necessary lab work (culturing colonies of M. tuberculosis) and then testing various drugs on the colonies to determine drug resistance and find an effective treatment can take from a month and a half to 16 weeks. During this period of time ineffective treatments may be in use which further increase TB drug resistance.[8]

THE LAW

Persons with active cases of tuberculosis may become involved in legal action if their behavior can be shown to be a public health threat—for example, TB patients who are infectious and are failing to carry out prescribed treatment. MN Statutes sections 144.4171 - 144.4186 establishes a process beginning with a report to the Commissioner of Health and ending with remedial steps which the Commissioner can pursue in court, ranging from education to involuntary commitment for treatment.

The Commissioner may seek a court-ordered health directive regarding an individual who poses a health threat to others—that is, someone who appears either willing to transmit a communicable disease to another person, or unable to avoid transmission owing to mental incompetence. MN Statute 144.4182 permits emergency proceedings whereby a person posing a health threat can be taken into custody and held for up to 72 hours. The hold can be continued for up to 10 days.

RECOMMENDATIONS

TB is a dangerous disease, and in a mobile society local outbreaks of infection can become national epidemics in a short period of time. Such was the case with HIV. We are thus making recommendations for both local and national action.

In Minnesota:

• Meeting the housing, nutritional, and medical needs of poor and homeless people, chronic alcoholics, drug abusers and HIV infected persons is the first line of defense against the spread of tuberculosis, since these unmet needs will foster the continued spread of TB.

• Health education programs must be implemented to provide information about TB to the general public.

• Persons infected with HIV must receive education about their risk for TB

• The same data privacy standards protecting all Minnesotans should be applied to tuberculosis patients' records. The goal of insuring compliance with medication protocols can best be met by assuring the patient's confidentiality.

• HIV education and AIDS service organizations and TB surveillance and treatment programs should coordinate their efforts to prevent HIV and TB transmission.

RECOMMENDATIONS *cont.*

Nationally

• Tuberculosis surveillance activity should be increased rationally so that early treatment can be commenced before isolated cases develop into clusters and then into general outbreaks of infection. This includes making TB testing available at homeless shelters, community health clinics, etc.

• Immediate, effective, and complete treatment for tuberculosis must be made available through in-patient and/or out-patient treatment programs

• A continuum of intervention should be available for those with active TB who do not comply with treatment. This may include incentives—meals, transportation or pocket money. Court-ordered detention must be reserved as a last resort and only when the patient has adequate legal counsel.

• Infection control procedures which protect patients and health care workers alike must be followed in hospital and clinics. (For example, isolation in rooms with negative air flow; room air exhausted outside; UV radiation; barriers, etc.)

• Medical schools must re-establish tuberculosis as an active specialty area. There are few young researchers entering the field to replace those who have reached retirement age. New drugs, diagnostic tests, and vaccines need to be developed. Research into the genetic mechanisms by which Mycobacterium tuberculosis becomes resistant to drugs is needed.

REFERENCES

[1] B. R. Edlin, et al, New England Journal of Medicine, June 4, 1992, p. 1514; Vol. 326, No. 23.

[2] Lawrence K. Altman, New York Times, February 11, 1992, p. B6.

[3] AIDS Action Briefing: Tuberculosis and HIV: Challenges in Policy and Practice, May 1992.

[4] Charlotte Hagenmiller, Minnesota Department of Health.

[5] op. cit. AIDS Action.

[6] op. cit. Hagenmiller.

[7] op. cit., Edlin, et al.

[8] op. cit., Edlin et al p 1520.

RESOURCES

American Lung Association...........................612/227-8014
Minneapolis Health Department...................612/348-3045
Minnesota Department of Health.................612/623-5526
St. Paul Public Health Department................612/292-7731

MAP
Minnesota AIDS Project

AZT

What is it?

AZT (also known as zidovudine, azidothymidine, and Retrovir) is an antiviral that works to slow down the replication of HIV. This compound was developed in 1964 as an anti-cancer drug. AZT did not prove to be very effective as an anti-cancer agent, but the manufacturers continued to make small amounts for research purposes. In 1985, *in vitro* (test-tube) studies showed that AZT was a potent inhibitor of HIV. Human trials confirmed AZT's ability to inhibit HIV and it was soon approved for use in individuals with CD4 counts (T cell counts) less than 200 who had previously had an AIDS-defining opportunistic infection (such as *Pneumocystis carinii* pneumonia) at a dosage of 1,500 mg per day. Because AZT was only being used in people whose immune system was already severely compromised and because the dosage was so high, many individuals experienced severe adverse reactions to AZT, such as bone marrow toxicity. Further studies of AZT showed that lower doses of 500 to 600 mg were just as effective as the old dosage of 1,500 mg and the incidence of side effects was greatly reduced.

It has also been shown that individuals who begin AZT therapy early, with CD4 counts above 200, tolerate therapy much better with toxic side effects appearing in less than 3% of individuals.

How Does it Work?

AZT was the first antiviral approved for use in people with HIV. It does not kill the virus, but rather works by interfering with the virus's ability to integrate into the cell's genetic material (i.e. DNA). This has the effect of decreasing the rate of infection to other cells in the body. Under normal conditions, when HIV enters a cell, it is able to integrate itself into the cell's DNA. Once the virus accomplishes this, it is able to reproduce , making many HIV virions which in turn, can infect other cells. When an individual takes AZT, this process is interrupted. AZT, like ddI and ddC, is known as a nucleoside analog. It works by acting as a decoy for a newly forming chain of DNA. However, this decoy causes the premature termination of the DNA chain, leading to inhibition of reverse transcription. AZT has been shown to block HIV replication in CD4 cells (T cells), monocytes, and macroph-

ages -- all of which are cells that can be infected with HIV. In addition, AZT does cross the blood-brain barrier well which is essential in helping to prevent and fight HIV-related neurological complications.

Who Should Use AZT and How Much?

AZT is officially recommended for HIV infected individuals who have a CD4 count of less than 500, with or without symptoms. Many studies have confirmed added benefits for individuals who begin AZT before symptoms develop in addition to a decreased incidence of side effects in these individuals.
The current dosages recommended are either 500 mg a day or 600 mg a day. Many physicians recommend 500 mg for individuals who are asymptomatic and 600 mg for individuals experiencing HIV-related symptoms. There has been research which suggests a dosage of 300 mg is also effective. However, physicians are often reluctant to prescribe this dosage as it is the very minimum required for effectiveness. An individual who is only taking 300 mg a day absolutely could not miss a single dose or they would not derive any benefit from treatment that day.

What Does the Research Show?

Much research has been conducted on AZT in the past seven years. This section covers some of the larger more recent studies completed. The results presented represent statistically significant findings. For additional study information, you may call the STEP office.

The Multicenter AIDS Cohort Study

In April, 1992, results of the Multicenter AIDS Cohort Study (MACS) were published. This study assessed the survival of over 2,500 men in relation to AZT use and prophylaxis for *Pneumocystis carinii* pneumonia (PCP). The investigators found that the use of AZT before the development of AIDS, regardless of PCP prophylaxis, significantly reduced death in all follow-up periods. According to a press release from the U.S. National Institute of Allergy and Infectious Diseases, "Early treatment with AZT reduced the risk of death at six months by 57 percent, whether or

Seattle Treatment Education Project, 127 Broadway E, Suite 200, Seattle, WA 98102 (206) 329-4857 not it was combined with a drug that prevented PCP. Moreover, investigators found that, after 2 years, early treatment reduced the risk of death by 33 percent, when compared to those who did not take AZT before a diagnosis of AIDS."

European/Australian AZT Study

In February, 1992 the discontinuation of a large European/Australian study was announced[4]. The study was designed to determine the efficacy of AZT in people with HIV who are asymptomatic. The trial was stopped early after preliminary analysis showed that the incidence of disease progression in individuals receiving AZT was half that of the placebo group. In this study of nearly 1,000 individuals, disease progression was defined as a decrease in CD4 count below 350 mm^3 or the development of clinical symptoms. For those individuals with CD4 counts between 500 and 750, the probability of disease progression was 18% for the placebo group, versus 9% for the treatment group. Disease progression in individuals with CD4 counts between 400 and 500, was 38% for the placebo group versus 20% for those treated with AZT.

Italian AZT Study

In March, 1992, results from an Italian study of AZT were published[1]. This study examined the influence of long-term AZT therapy on survival in individuals with AIDS. 271 people with AIDS were enrolled, 159 of whom received AZT, 112 of whom did not. The overall median survival time for those receiving AZT was 22.1 months, and for those patients who were not treated, the median survival time was 10.6 months. The estimated 1 year survival rate for the AZT treated group was 85%, whereas the estimated 1 year survival rate for the untreated group was 46%. The estimated 2 year survival rate for the AZT treated group was more than twice that of the untreated group.

Australian AZT Study

In a multicenter study of AZT in Australia, the survival of 308 men with AIDS using AZT was compared with the survival of 482 historical controls who did not use AZT. The median survival time after diagnosis with AIDS in the group who received AZT was nearly 3 times higher than the historical control group.

AIDS Clinical Trials Group Studies

The ACTG 016 compared AZT with placebo in 711 mildly symptomatic individuals. 11 out of 260 individuals with CD4 counts between 200 and 500 mm^3 progressed to advanced ARC or AIDS versus 34 out of 253 individuals who received placebo.

The ACTG 019 study compared AZT (either 1,500 mg a day or 500 mg a day) with placebo in 1,338 asymptomatic individuals. In people with CD4 counts below 500 mm^3, 8% of the placebo group progressed to AIDS compared with only 2% of the group receiving 500 mg of AZT, and 3% of the group receiving 1,500 mg AZT.

Concorde Trial

The Concorde trail enrolling 1749 people from 73 sites in England, Ireland, and France, has generated an immense amount of controversy, deliberation, and confusion. The mean length of follow up in this large study was three to four years. Its purpose was to determine the optimal time for initiating antiretroviral therapy with AZT in asymptomatic individuals. That is, is it better to give AZT to people when their CD4 counts fall below 500 (as is the current recommendation in the USA) or is it better to wait until the individual develops symptoms? The dose of AZT used in this study was 1200 mg/day, at least twice the current recommended dosage in the USA. The trail lasted from 10/88 to 10/91 and is the largest and longest trail of AZT use in asymptomatic individuals. Subjects were randomized to receive AZT (immediate group) or placebo (deferred group). The original intent was that individuals in the placebo arm would receive AZT if symptoms developed. However, in the fall of 1989 when the initial results of a portion of the ACTG 019 trail were released, a change was made in the protocol of Concorde. They now gave those in the placebo arm the option of receiving AZT when their CD4 count fell below 500. Of the 872 individuals in the placebo group, 282 (32%) chose to receive AZT at that time. However, due to the "intent-to-treat" analysis used by the Concorde researchers, these 282 individuals who began AZT were considered as belonging to the placebo group when the analysis was done. The interpretation of the data by the Concorde investigators was that there was no difference in the disease progression or survival between the immediate and deferred groups. When CD4 counts instead of clinical end points were used to evaluate the results, the immediate group showed an advantage over the deferred group. The investigators go on to say that CD4 cell counts as marker of clinical disease progression in asymptomatic individuals on AZT monotherapy is unreliable. Their caveat on CD4 counts is that they are probably valuable in natural history studies, but not so reliable in evaluating drug treatment trails, where markers of viral load would be more helpful. The investigators also stated that the results of their trail do not indicate that "early intervention" is bad or harmful, but suggest that they are not sure there is a benefit either. They go on to say that some individuals will benefit from early intervention and others will not. The difficult task is deciding which course any single individual will follow with regards to progression. There has been a large amount of discussion generated from this study. Although additional data was presented in Berlin, there are immunologic and virologic analyses still being performed. The entire study data will appear in publication at some future point. Most major HIV/AIDS treatment newsletters throughout the country have written on this topic including *Treatment Issues, AIDS Treatment News*, and others. Therefore the reader is invited to search out other commentaries on this issue. It was not the purpose of the Concorde to ascertain the efficacy of AZT in symptomatic individuals. The benefit of AZT therapy in this group has clearly been previously established. Also, it was not the purview of Concorde to ascertain the efficacy of AZT in combination with other antivirals in any patient population with regard to disease progression or survival.

Some concerns about this study are as follows. The "intent-to treat" analysis skews the data in favor of no difference between the two groups : any potential benefit which the 282 individuals who received AZT were considered in the placebo group. The dosage of AZT used in this trail could possibly have cytotoxic effect, thereby decreasing a perceived benefit. This point is further made in the following report on ACTG 019 when the effects of 500 mg/day of AZT is compared with 1500 mg/day. It is generally acknowledged that the benefit of AZT monotherapy is limited and decreases over time. Therefore, any benefit from treatment with AZT is going to be seen early on. As the length of time on treatment in creases, either a change in monotherapy or the institution of combination therapy has suggested benefit and has been advocated by some for many years. Seen in this way, the results of Concorde are not all that surprising. Keeping individuals on single nucleoside antiretroviral will no doubt lead to decreased effectiveness of that drug (no matter what it is) over a prolonged period of time such as three to four years. However, the conclusion drawn from the study investigators are in contrast to the five other studies of early AZT use in asymptomatic persons. Studies from Italy, EACG 020 (Europe/Australia), the MACS study in the USA, ACTG 019, and from John Hopkins have concluded that early treatment with AZT is advantageous in survival and/or disease progression. Recent accounts if the "non-latency" of the virus even during the asymptomatic stage, offer the rationale that intervention with drugs which inhibit the infection of new cells (like nucleoside analogues do) are going to be more effective at earlier stages of disease in contrast to later ones. The clinicians I speak with or have heard from , are not making any changes in their recommendations regarding treatment strategies based on Concorde.

In an on going follow up of disease progression in individuals previously enrolled in ACTG 019 with less than 500 CD4 cells, the duration of AZT's efficacy was studied. These are people who initially were randomized to one of three treatment arms: placebo, 500 mg/day AZT, 1500 mg/day AZT. The trail for this subset of participants lasted almost two years until it was stopped in 8/89 and those in the placebo group were offered AZT. Looking at disease progression for a subsequent follow up period of 2.6 years, 232 of 1556 people progressed or died. When considering all 1556 individuals, decreased disease progression is seen in those initially randomized to receiving 500 mg/day of AZT compared to 150 mg/day AZT or placebo. This benefit has not been shown for survival. When these 1556 individuals are broken down into two groups with either more or less 300 CD4 cells at entry, in those with less than 300 CD4 cells, no difference in progression among the three groups is discernible. However, in those with greater than 300 CD4 cells at entry, decreased progression was seen in the 500 mg/day group compared with the 1500 mg/day or placebo groups. These results argue favorably for instituting antiretroviral therapy at early stages of infection rather than later. Dr. Paul Volberding, protocol chair and presenter of this study put forth two hypotheses: in patients with less than 300 CD4 cells, the benefit of AZT may be around 18 months; in patients with 300-500 CD4 cells, the benefit of AZT use could be greater than two years.

Conclusions Drawn From Studies

Conclusions that can be drawn from the above mentioned studies as well as countless others not mentioned are:

*AZT prolongs survival in symptomatic individuals.

*AZT decreases frequency and severity of opportunistic infections.

*AZT delays progression to AIDS.

*AZT delays progression to symptomatic disease.

*AZT improves cognitive or neurologic function.

*AZT improves weight gain.

*AZT increases CD4 cell numbers.

*AZT increase CD8 cell numbers.

Seattle Treatment Education Project, 127 Broadway E, Suite 200, Seattle, WA 98102 (206) 329-4857

*AZT decreases serum and cerebrospinal fluid (CSF) p24 antigen levels.

*AZT increase skin-test reactivity.

*AZT increases platelet counts.

*AZT decreases the incidence of AIDS Dementia Complex

*AZT increases quality of life.

Many researchers have proposed that AZT's antiviral actions only comprise about one third of AZT's total benefits. This is why many physicians will not discontinue AZT therapy in individuals who no longer appear to be deriving antiviral benefits from the drug. These physicians will add another antiviral, such as ddI, to the treatment regimen, yet still keep the person on AZT to derive other benefits. For example, AZT improves the response to the pneumococcal vaccine in individuals with HIV infection. *In vitro* studies have also shown that AZT can inhibit some bacteria such as *Salmonella, Escherichia coli, Enterobacter, Shigella,* and others. It has also exhibited inhibitory activity against the Epstein-Barr virus *in vitro*.

What About Side Effects?

Although AZT has been shown to have a lot of benefits for the HIV infected person, it is by no means the perfect drug. AZT's most serious side effects are anemia (decreased red blood cell counts) and neutropenia (decreased number of neutrophils, one type of white blood cell) due to suppression of the bone marrow. In early studies of seriously ill individuals and extremely large doses of AZT (1,500 mg), anemia and/or neutropenia was seen in nearly half of the study participants. Even with the high doses, these side effects were seen much less frequently in individuals without an AIDS diagnosis. Now that individuals are beginning AZT therapy before they are seriously ill and using a much lower dose, the incidence of anemia and neutropenia have drastically decreased to around 3%. For the small percentage of individuals who do develop bone marrow suppression, there are other options available. Many individuals choose to continue taking AZT and take a drug called EPO (erythropoietin) to combat the anemia. Other individuals choose to try decreasing the dosage of AZT. Some individuals will decide to discontinue AZT therapy and try another antiviral approved for individuals who can not tolerate AZT called ddI.

AZT also has other, less serious (but still bothersome) side effects. Many individuals will experience headaches, nausea, hypertension, and a general sense of not feeling well during the first weeks of therapy. These effects are probably a combination of the body adjusting to the drug and the person's anxiety over beginning therapy. However, these effects will gradually subside in nearly all individuals.

After long-term therapy with AZT, a very small percentage of individuals may develop muscle weakness or pain exacerbated by exercise (myopathy or myalgia). The incidence of these side effects has also drastically decreased with the reduction in dose from 1,500 mg to 500 or 600 mg. If myopathy or myalgia does occur it usually goes away once AZT therapy is discontinued or the dosage is reduced. Additionally, myopathies and myalgias are not always caused by AZT and can be due to HIV itself. The underlying cause should first be determined by your health care provider before changes in your dosage are initiated.

A rumor that long-term AZT use causes lymphoma has caused undo worry in some individuals considering AZT therapy. The study which sparked this rumor showed that 8 out of 55 individuals (14.5%) who had used AZT for a median of two years developed lymphoma. Further analysis of this study showed that of the eight individuals with lymphoma, the median CD4 count when they *started* AZT therapy was 26. Their median CD4 count when they developed lymphoma was 6. Even before AIDS, it was a well known fact that immune suppression due to other causes increased the risk of lymphoma. The vast majority of researchers agree that the severe, prolonged immune suppression of the individuals who developed lymphoma in this study is what caused the lymphoma, not long-term AZT use.

Interestingly, in the ACTG 019 study mentioned in the research section, the placebo group reported just as many side effects as the AZT treatment group. This brings up the psycho-logical factor. If a person is reluctant to begin AZT therapy because they believe it is a dangerous drug with horrendous side effects, that person is likely to develop side effects. Likewise, if an individual believes that a placebo is actually a wonder drug, it is likely that person will experience short-term benefits, even increases in CD4 counts. If an individual is considering AZT therapy, try to keep an open mind. Don't anticipate side effects or you will be more likely to experience them.

What about Viral Resistance?

The matter of viral resistance has raised concern in some individuals as to when to start AZT. Laboratory studies have shown that HIV can mutate and new strains of HIV can evolve that are resistant to AZT. Based on these findings, some individuals have argued that AZT therapy should not be instigated early, but rather saved for more advanced disease when the person needs it most. When looked at closely, this view really doesn't hold up

Seattle Treatment Education Project, 127 Broadway E, Suite 200, Seattle, WA 98102 (206) 329-4857

and may actually increase the chances of developing resistant strains. First of all, AZT is not the only drug available. If resistance does occur, an individual can always consider using ddI or ddC. Studies have shown that HIV is not "cross-resistant" to these antivirals, meaning that if the HIV strain is resistant to AZT, it will not be resistant to ddI and vice versa. Also, studies have shown that the mutation of HIV which causes AZT resistance is directly proportional to the rate of viral activity. Therefore, if AZT is initiated earlier in the course of HIV, when the viral load is less, the virus may have a more difficult time mutating. Studies have been done to corroborate this theory, reporting that individuals who begin AZT when they are asymptomatic, develop far fewer AZT-resistant strains of HIV than individuals who initiated therapy after symptoms began.

AZT in Combination

For further information on combination therapies, one may call the STEP office and request the combination therapy fact sheet.

*Acyclovir: The combination of AZT and high-dose acyclovir has been studied by a number of different researchers. Unfortunately, the majority of the studies did not show any increased antiviral benefits from the addition of acyclovir. However, if individuals suffer from recurrent herpes outbreaks, acyclovir should be considered to decrease the incidence of outbreaks, thereby decreasing additional immune suppression caused by the herpes viruses.

*Alpha Interferon: Some researchers believe the combination of AZT and alpha interferon may increase antiviral effects. Studies have been done, but there is no definitive answer yet. It appears the added antiviral benefits are minimal, if any, and the combination appears to have synergistic toxicity. Side effects included liver dysfunction, decreased neutrophils and platelets, and fatigue.

*ddC: The combination of AZT and ddC looks very promising. The two drugs appear to work synergistically. This means that the benefits of the drugs used together are greater than simply the sum of their effects. Sort of like adding two plus two and coming up with six instead of four. Because the drugs do not have overlapping side effects, the combination is well tolerated and does not result in additional toxicity. The FDA advisory board has recommended that the combination of AZT and ddC be approved.

*ddI: The combination of AZT and ddI works similar to the combination of AZT and ddC. AZT and ddI also do not have overlapping side effects. ddI is already approved for use in individuals who have failed AZT therapy. Although combination therapy is not one of the official uses for ddI, many physicians are prescribing ddI for use in combination with AZT and insurance companies have been paying for it in many cases.

References:
1. Vella S, et al. Survival of zidovudine-treated patients with AIDS compared with that of contemporary untreated patients. JAMA 1992;267:1232-1236.
2. Swanson CE, Cooper DA. Factors influencing outcome of treatment with zidovudine of patients with AIDS in Australia. AIDS 1990;4(8):749-756.
3. Graham N, et al. The effects on survival of early treatment of human immunodeficiency virus infection. N Engl J Med 1992;326:1037-1041.
4. Burroughs Wellcome News Release, February 3, 1992.

July 1993

Seattle Treatment Education Project, 127 Broadway E, Suite 200, Seattle, WA 98102 (206) 329-4857

MAP Services
for HIV Positive Persons

A brief guide to services for HIV positive persons in both the Twin Cities area and throughout Greater Minnesota.

Minnesota AIDS Project

Minnesota AIDS Project

Serving Minnesota with offices located in:

Minneapolis
1400 Park Ave.
Minneapolis, MN 55404
612-341-2060
800-243-7321

Duluth
205 W. 2nd St.
Suite 306
Duluth, MN 55802
218-727-2437
800-731-2437

St. Cloud
810 West St. Germain
Suite 104
St. Cloud, MN 56301
320-259-1909
800-761-2437

AIDSLine
Metro Area
373-AIDS (2437)
(TTY: 612-373-2465)
Greater Minnesota
800/248-AIDS
(TTY: 888-820-2437)

The Minnesota AIDS Project (MAP) is a statewide, non-profit, volunteer-based organization serving Minnesotans since 1983. The mission of MAP is to reduce the impact of HIV and AIDS in Minnesota.

MAP Rating Level 1 — General Audience
© 1996 Minnesota AIDS Project

Greater Minnesota Services

Through the Duluth, Rochester, and St. Cloud offices, MAP meets the needs of Minnesotans affected by HIV. For more information, call the office nearest you.

HIV Case Management

Regional case managers can help HIV positive individuals identify a broad range of services that may be beneficial. Case managers can help people find resources to meet financial, medical, legal, and personal support needs.

Education

Staff and volunteers provide HIV and AIDS education to schools, businesses, and organizations throughout the state. MAP representatives staff tables at community events, county fairs, and the Minnesota State Fair, distributing information and answering questions about HIV and AIDS.

Transportation Reimbursement

Receive reimbursement for the cost of gas or the cost of public transportation needed to travel to medical appointments or support groups. Verification of HIV status is required and receipts must be submitted before reimbursement is paid. Payment for other transportation needs will be considered if there is sufficient funding.

Short-Term Housing Support

To help people avoid a housing crisis, payment can be made to cover the costs of moving expenses, utilities, rent and mortgage costs. There is an annual limit for each person. Payment can only be made directly to a moving company, landlord, utility company, etc.

Additional Services

The Minnesota AIDSLine, Positive Link, and Legal Program can also provide assistance to persons in Greater Minnesota. Additionally, if you will be visiting the Twin Cities find out about the free or discounted events available through the Life Enhancement Program.

675

A History of Care

Since 1983, the Minnesota AIDS Project has been a leader in the fight against HIV and AIDS. Founded by a small group of dedicated volunteers, MAP has grown to a staff of more than 55 and a volunteer base of more than 1,400 to meet the challenges of this epidemic.

MAP has combatted ignorance and fear through education, forcefully advocated for the rights of those affected by HIV, and provided compassionate care for those living with HIV and AIDS.

With offices located in Minneapolis, Duluth, Rochester, and St. Cloud, MAP has grown to meet the changing face of AIDS. The ever-increasing number of people infected with HIV, combined with limited funds, makes MAP's programs and services more critical than ever.

MAP Services for HIV Positive Persons

Through MAP's offices in Minneapolis and throughout the state, Minnesotans living with HIV and AIDS are provided confidential and non-discriminatory services, including practical, emotional, and social support.

The Minnesota AIDSLine

The Minnesota AIDSLine is a toll free phone line that offers prevention education, information, assessment, and referral in a supportive environment. Trained personnel are available to talk with you about your concerns and answer your questions **Mon–Thurs 9a.m.–9p.m.** and **Fri 9a.m.–6p.m.** *All contacts are confidential.* In the metropolitan area, call **612-373-AIDS (2437)**. From Greater Minnesota, call **1-800-248-AIDS**.

Positive Link/Family Link

Volunteers provide peer support and education for those who are infected or affected by HIV/AIDS. Available to persons in the metropolitan area and Greater Minnesota by calling the Minnesota AIDSLine.

Client Support Services

Staff provide support, assessment, technical assistance and referral for HIV positive and HIV affected individuals who want to meet face to face with a service provider. Appointments or more information are available by calling the Minnesota AIDSLine.

Women and Families Program

The program provides resource information to women, families, and service providers. It works to make current services for women and families more easily accessible. Staff promote awareness of the needs of HIV women and families.

Transitional Housing Program

Adults living with HIV and AIDS and their families who are at risk of homelessness are assisted through the Transitional Housing Program. The program provides short-term housing and rental assistance to persons who are case managed by an HIV social worker from any agency and live within the Twin Cities metro area.

Transportation

Rides for people with HIV are provided through the Transportation Program. Volunteers drive eligible participants to doctor's appointments, support groups, and other health-related visits.

Legal Program

Legal assistance is provided free or on a sliding-scale basis by volunteer attorneys and a staff legal coordinator. Assistance may range from estate planning and drafting living wills to fighting discrimination in housing, employment, and other areas.

Life Enhancement Program

Opportunities for education, socializing, and recreation are coordinated through Life Enhancement. Free or discounted tickets are provided for movies, sports, theater, and other events. Other special services may include income tax assistance, veterinary services and others as available.

HIV Case Management

In addition to the services previously listed, some clients of MAP may be eligible for Case Management. Case Management is for persons who have several needs that require ongoing assistance. Case managers can help HIV positive individuals identify a broad range of services that may be beneficial. They also can help people find resources to meet financial, medical, legal, and personal support needs.

Buddies and Home Helpers

Case managed clients may obtain the assistance of a buddy/home helper. Practical and emotional support is offered to clients who are in need of help or want social support. Volunteers help with household chores such as cleaning, cooking, and shopping. Volunteer "buddies" develop supportive on-going relationships with those living with HIV.

Case Documents 2

How Tiny Organisms Helped Clean Prince William Sound

Alaska's Prince William Sound today is essentially clean. There are several reasons for the rapid cleanup of the area, but among the least known and most important factors in the recovery process were microscopic petroleum-eating organisms that were assisted by a process known as bioremediation.

Prince William Sound is home to naturally-occurring single-celled creatures that continuously feed on hydrocarbons from natural petroleum seepages and evergreen droppings. Immediately after the 1989 *Valdez* accident, these micro-organisms increased their numbers to respond to the oil from the spill. However, their ability to "eat" this oil was limited by the available amounts of nitrogens and phosphorus—the basic nutrition the organisms need to multiply and biodegrade oil.

To help the microbes, and enhance their natural cleaning action, scientists from the US Environmental Protection Agency and Exxon applied special fertilizers to oiled shorelines in Prince William Sound. These fertilizers provided the microbes with more nitrogen and phosphorus, allowing them to reproduce and consume oil at a quicker pace. This process of stimulating the microbes is known as bioremediation.

Under typical conditions there are about three million oil-eating microorganisms in each ounce of sediment in Prince William Sound. After bioremediation the number of microbes rose to as high as 30 million and the oil-eating ability of the organisms increased approximately 10-fold. Repeated tests have shown that in Prince William Sound bioremediation was both effective and safe.

First Large-Scale Use   Never before had such a large-scale use been made of the bioremediation process in an oil spill. And it was undertaken in Alaska with extreme care because of concern over potential side effects from the fertilizers, such as growth of algae.

The Alaska project began in 1989 when a joint scientific panel formed by the US Environmental Protection Agency (EPA) and Exxon recommended that fertilizers containing nitrogen and phosphorus be applied to test areas to determine the feasibility of large-scale bioremediation work. Ordinary garden fertilizers could not be used because tidal action would wash the nitrogen and phosphorus away too quickly, rendering them ineffective. To meet this challenge, the scientists applied two patented products—Inipol EAP22 and Customblen—developed especially for the slow release of nutrients. These compounds are designed to stick to oil and remain there until consumed by microbes.

In 1989, tests on 74 miles of shoreline were so encouraging that bioremediation was used again in 1990 on some 400 patches of oil on shores in Prince William Sound and the Gulf of Alaska. Scientists found that two summers of bioremediation achieved a level of biodegradation that would otherwise have taken many more years.

Accomplishments in Alaska

The Exxon-EPA project provided scientists with new insights and knowledge about bioremediation:

- First, there is a better understanding about the processes through which microbes convert oil and other hydrocarbons into substances such as carbon dioxide, water, and microbial biomass. Scientific data showed that the pace of natural degradation can be accelerated, within limits, by controlling the amount of nutrients.
- Second, laboratory and field tests showed conclusively that bioremediation can be used safely, and that the microbes reduce their number as the oil is eaten.
- Third, the operational techniques pioneered in Alaska have significantly advanced knowledge about how to use bioremediation safely and effectively to treat future oil spills. The Alaska project demonstrated for the first time on an actual spill that accelerated microbial activity can be a major factor in cleansing oil from beaches. In future oil-spill cleanup operations, government and corporate officials can now consider bioremediation as a proven tool to supplement more traditional processes, such as spraying with hot water or the use of sorbents.
- Finally, the Alaska project, which took bioremediation from the experimental stage to proven application for the right conditions, demonstrated how a large corporation can join with Federal and State agencies to find an effective solution to a complex problem. The US Environmental Protection Agency, Exxon and the Alaska Department of Environmental Conservation managed to launch the Alaska bioremediation project quickly and efficiently.

THE PROCESS

What Is Bioremediation?

To understand bioremediation one must first understand biodegradation. Biodegradation is a natural process in which oil hydrocarbons are broken down, over time, by micro-organisms that then produce simple substances such as carbon dioxide, water, and microbial biomass. Bioremediation is a human effort that quickens this natural process. In some instances, bioremediation means transporting oil-eating microbes to a spill site. In Alaska, however, bioremediation consisted only of feeding nitrogen and phosphorus to the environment's own oil-hungry micro-organisms.

Where Do the Microbes Come from and How Do They Work?

The Gulf of Alaska region, including Prince William Sound, is rich in native oil-eating organisms, which have evolved over millions of years to degrade hydrocarbons produced by evergreen forests and natural petroleum seepage. These organisms are always at work reproducing and converting hydrocarbons into carbon dioxide, water, and microbial biomass.

SAFE SUBSTANCES

Is Bioremediation Safe?

Yes. Extensive studies indicate that bioremediation in Alaska posed no adverse environmental consequences. Although birds and mammals instinctively avoid unnatural substances, scientists made doubly sure that they remained away from newly treated beaches by placing brightly colored balloons and flagged ropes in bioremediation areas. In addition, because application procedures were carefully designed and workers wore protective clothing, there were no health problems for workers involved in the bioremediation process. As expected, field monitoring results show that once the oil is eaten, microbe populations return to pre-spill levels and there are no long-term effects from bioremediation.

What Are the Substances Employed in the Alaska Bioremediation?

The fertilizers used in Alaska are essentially the same as two of the common nutrients used by the home gardener, nitrogen and phosphorus. However, these nutrients were especially prepared to ensure that there would be no environmental damage to the adjacent waters. Inipol EAP22, a sticky liquid fertilizer with the viscosity of honey, contains a water soluble form of nitrogen with an oleophilic form of phosphorus in a biodegradable base. Inipol EAP22 was applied to oiled rock and gravel surfaces. Customblen, a slow-release formulation of soluble nutrients encased in a polymerized vegetable oil with a nitrogen to phosphorus ratio of 28:8, was applied in some areas to reach subsurface deposits through rain and tidal action.

If a Liquid Substance Was Applied, How Do We Know It Wasn't Washing the Oil into the Water and onto the Ground?

Extensive laboratory tests proved conclusively that the liquid fertilizers were sticking to the oil and accelerating the biodegradation process, not washing oil off the rocks and other shoreline surfaces.

THE ROLE OF BIOREMEDIATION

If Bioremediation Is So Safe and Effective, Why Use Anything Else?

Bioremediation works best after bulk oil has been removed by manual and natural processes. Traditional removal methods are more effective for the heavy concentrations of crude oil that are encountered immediately after a spill. Thereafter, bioremediation is effective in removing oil both on the surface and in subsurface sediments.

OUTLOOK FOR 1991

Will Bioremediation Be Used in Prince William Sound This Year?

This spring, Prince William Sound will be examined by a team of scientists representing the US Coast Guard, the Alaska Department of Environmental Conservation, and Exxon to determine shoreline conditions. If the team locates segments that can benefit from bioremediation, nutrients may be applied to those areas. Even without bioremediation, the biodegradation process will continue as a normal and natural part of the ecology of Prince William Sound

---

Scientific Summary

TEST FINDINGS

A comprehensive US Environmental Protection Agency/Alaska Department of Environmental Conservation/Exxon monitoring program produced the following conclusions in December 1990:

> Bioremediation proved an effective way to accelerate biodegradation of oil both on shoreline surfaces and in those limited areas where oil penetrated beneath the surface. The process was effective in removing residual oil that had been weathering for almost one and one-half years.

> Initial bioremediation accelerated the rate of biodegradation by three of four times for both surface oil and subsurface sediments sampled to a depth of 12 inches. Second applications replenished nutrients and accelerated oil removal five to ten fold. Fertilizer reapplication was desirable every 3–5 weeks to sustain and enhance microbial action.

> Bioremediation produced no adverse environmental consequences. Even immediately after application, nutrients applied in recommended strength pose no risk.

Even without bioremediation, biodegradation would have proved an effective means of removing oil. But the cleansing process would have taken much longer.

Indigenous Organisms of Alaska

In Prince William Sound, an active community of microorganisms (or heterotrophs) is present at all times. This community has made a significant contribution to the ecology of the region over millions of years.

These organisms are active in cold climates and in both fresh and sea water. Before the spill, scientists estimate that hydrocarbon-degrading microbes constituted one tenth of one percent of the total heterotroph population in Prince William Sound. The introduction of the oil and the fertilizers increased their number so rapidly that they soon represented more than 10 percent of the microbe population.

Special Fertilizers

Two products especially useful for beach areas were used in the Alaska bioremediation program:

Inipol EAP22, manufactured by Elf Aquitaine of France, is classed as an oleophilic or oil-soluble compound which sticks to oil and remains there until consumed by biodegradation, thus preventing nutrients from entering the water. Inipol EAP22 is a microemulsion which contains a water soluble form of nitrogen with an oleophilic form of phosphorus in a biodegradable base. It was developed following the 1978 grounding of the tanker *Amoco Cadiz* on the coast of France and is especially effective in the removal of surface oil.

Customblen, produced by Sierra Chemicals of Milpitas, California, is a granular slow-release formula which encases the nutrients in polymerized vegetable oil. It is similar to fertilizers sold to home gardeners as Osmocote and used on many golf courses. Customblen's nutrients are released gradually during storms and tidal action, making it especially effective feeding oil-eating microbes in subsurface deposits.

Project Monitors

The 1989 bioremediation study was launched by the US Environmental Protection Agency's Office of Research and Development and Exxon under the Federal Technology Transfer Act of 1986. This act encourages collaborative efforts between public agencies and private companies for the economic, social and environmental benefit of the nation.

The EPA and Exxon tested the capacity of bioremediation to treat oiled sections of shoreline. EPA was responsible for oversight and management of the study, while Exxon funded logistical, laboratory, and field activities. In addition, the Alaska Department of Environmental Conservation participated in the planning, operation, and monitoring of the project. Independent organizations were employed to measure the results of bioremediation and ascertain whether there were any adverse environmental consequences.

A Report Detailing the Effectiveness and Safety of Bioremediation in Prince William Sound

In December 1990, The Alaska Department of Environmental Conservation (ADEC), the US Environmental Protection Agency (USEPA) and Exxon jointly submitted to the US Coast Guard a report detailing the effectiveness and safety of bioremediation in Prince William Sound. A reprint of the Executive Summary of this report follows. For further information about this report or Exxon's bioremediation work in Alaska, contact R. C. Prince, Ph.D., or R. R. Chianelli, Ph.D., Exxon Research and Engineering Company (201) 730-0100.

Executive Summary*

The joint ADEC/USEPA/Exxon biodegradation monitoring team successfully organized and implemented a comprehensive program for assessing the utility of fertilizer amendments for enhancing the biodegradation of surface and subsurface oil, and for characterizing the associated ecological risks. Interim reports, with analyses of the data then available, were presented in July and September. The Final Report presents all the monitoring data, and reinforces the earlier conclusion that biodegradation is an important mechanism in the removal of oil from shorelines in Prince William

*Reprinted as originally released.

Sound, and that the application of fertilizers is a safe and effective means to enhance this natural process.

- Chemical analyses of the remaining oil on the beaches indicates that biodegradation has already removed 10–70% of the oil in individual samples.
- By employing ratios of degradable and undegradable fractions of the oil components, we have derived an estimate of the baseline oil degradation rate. This is approximately
     2.2 g oil/Kg sediment/year on the surface
     1.1 g oil/Kg sediment/year in the subsurface
- The activity of oil-degrading bacteria in surface sediments and subsurface sediments sampled at a depth of 30 cm was enhanced three to four fold relative to the unfertilized sediments. This enhancement was sustained for 32 days after the initial fertilizer application. A second application replenished nutrients and stimulated relative microbial activities five to ten fold.
- The microbial populations on the fertilized areas have consistently higher numbers of hydrocarbon degrading bacteria than the corresponding unfertilized areas. Adding fertilizer resulted in a five to ten fold increase in hydrocarbon degraders.
- Fertilizer application resulted in no adverse ecological effects.

This report provides evidence that fertilizer application is an effective means to enhance the activity of oil-degrading bacteria in surface and subsurface sediments with minimal environmental impact. The elevated activity of hydrocarbon-degrading bacteria and the measured increase in their populations together provided convincing evidence that fertilizer application indeed enhances oil degradation. Reapplication of fertilizers is warranted every 3–5 weeks.

Bioremediation Technology Development and Application to the Alaskan Spill

R. R. Chianelli, T. Aczel, R. E. Bare, G. N. George, M. W. Genowitz, M. J. Grossman,
C. E. Haith, F. J. Kaiser, R. R. Lessard, R. Liotta, R. L. Mastracchio, V. Minak-Bernero,
R. C. Prince, W. K. Robbins, E. I. Stiefel, J. B. Wilkinson, and S. M. Hinton
(Exxon Research and Engineering Company)

J. R. Bragg and S. J. McMillen
(Exxon Production Research Company)

R. M. Atlas
(University of Louisville, Louisville, Kentucky)

Introduction: Shortly after the *Exxon Valdez* ran aground on Bligh Reef (March, 1989), a task force was assembled at Exxon Research and Engineering Company to identify novel technologies for oil spill cleanup. One technique identified by this task force for potential application to the cleanup of oiled beaches in Prince William Sound (PWS) and the Gulf of Alaska (GOA) was enhanced bioremediation. Bioremediation is the process in which microorganisms are used to biodegrade oil, and enhanced bioremediation employs application of nitrogen and phosphorous or other nutrients that may be limiting the degradation process.

Biodegradation is a natural process in which bacteria consume petroleum and break it down to biomass and carbon dioxide. The bacteria that consume the petroleum are "hydrocarbon-oxidizers" that destroy petroleum molecules by adding oxygen to them. The newly formed oxygenated molecules are further consumed until only the final products, biomass and carbon dioxide, remain. The process of converting the oil to biomass and carbon dioxide is called mineralization.

Biodegradation, together with physical processes such as evaporation, is a major process by which petroleum is removed from the environment following an oil spill. The role of these processes for a large spill such as that of the *Amoco Cadiz* was described by Gundlach et al. [Gundlach, E. R., P. D. Boehm, M. Marchand, R. M. Atlas, D. M. Ward, D. A. Wolfe, 1983]. This study showed that biodegradation was acting as rapidly as evaporation even during the first days following the spill. It is thought that the high rate of biodegradation along the coast of Brittany was due to runoff of agricultural fertilizers from farms near the coast. The addition of nitrogen and phosphorus from this agricultural runoff enhanced the biodegradation on the beaches [Atlas, R. M., 1990] presumably by removing the nitrogen and phosphorous limitation resulting from low naturally occurring levels of these elements in seawater.

Although nutrient-enhanced bioremediation had been used extensively for land farming and ground water applications, its use in marine spills had been limited prior to the *Exxon Valdez* oil spill. However, since the large *Amoco Cadiz* oil spill in 1978, efforts have been made to develop enhanced bioremediation techniques. In 1985, an oleophilic, or "oil-loving", fertilizer INIPOL EAP-22 was applied to a small spill of diesel oil in Ny-Alesund, Spitsbergen, Norway. The results of this application were inconclusive due to the high rate of natural cleaning [Sveum, P., A. Ladousse, 1989]. Nevertheless, extensive laboratory studies suggested that enhanced bioremediation might be applicable to stranded oil on the beaches of PWS and the GOA. This paper reports on laboratory studies of nutrient enhanced bioremediation and the resulting application of nutrients to over seventy miles of beaches in PWS and the GOA in 1989, the largest shoreline bioremediation project ever attempted. Questions that arose in Summer of 1989 were answered in laboratory studies during the Winter of 1989–90, which contributed to the approval of nutrient reapplication in Spring 1990.

EXXON COMPANY, U.S.A.
POST OFFICE BOX 2180ÆHOUSTON, TEXAS 77252-2180

THE PRINCE WILLIAM SOUND
BIOREMEDIATION STORY

The Challenge: Effective cleanup following the 1989 *Valdez* spill.

The Response: After the spill, the **US Environmental Protection Agency (EPA)** approached **Exxon** to discuss ways to clean up the oil. A joint effort between the **EPA, Exxon,** and the **Alaska Department of Environmental Conservation** was organized to test the effectiveness and safety of bioremediation—a process in which nitrogen and phosphorus fertilizers are fed to naturally-occurring, oil-hungry microbes, allowing them to multiply and consume oil faster.

The Application: **Inipol EAP22,** a liquid based oleophilic ("oil-loving") fertilizer was sprayed on selected beach sites in Prince William Sound and the western Gulf of Alaska. Inipol EAP22 sticks to rocks and encourages surface cleaning.

Customblen, a slow-released fertilizer, was applied to encourage subsurface cleaning.

These fertilizers were applied to 74 miles of shoreline in 1989, and on some 400 patches of oil in Prince William Sound and the Gulf of Alaska in 1990. Laboratory and field tests showed conclusively that bioremediation is safe for the environment and effective for oil removal.

The Results: Bioremediation achieved a level of cleanup in Prince William Sound during a 24 month period that would have otherwise taken many more years to accomplish.

Bioremediation

for Shoreline Cleanup Following
the 1989 Alaskan Oil Spill

James R. Bragg
Roger C. Prince
John B. Wilkinson
Ronald M. Atlas

Background on Bioremediation

The Bioremediation Process

In recent years many biological products and application techniques have been proposed for oil spill cleanup and site remediation. Proposals have included a wide range of nutrient products, natural or "bioengineered" microorganisms, bioreactors, or *in-situ* treating conditions. The following terms define the processes as applied in Prince William Sound and discussed in this report:

- *Biodegradation*: the natural process by which microbes (bacteria and fungi) consume hydrocarbons and produce carbon dioxide, water, biomass, and partially oxidized, biologically inert by-products. The process is basically one of oxidation, wherein the bacterial enzymes catalyze the insertion of oxygen into the hydrocarbon so that the

molecule can subsequently be consumed by cellular metabolism. Some molecules are completely degraded to carbon dioxide and water, while others are altered and then incorporated into biomass. The process releases energy used by the microorganism for sustenance and for growth of additional microbial cells.

- *Bioremediation*: the process of accelerating the rate of natural biodegradation of hydrocarbons by adding fertilizer to provide nitrogen and phosphorus, which, following a spill, are usually limited in availability relative to the amount of hydrocarbons (see Figure 1.3).

Bioremediation as used in Prince William Sound and the Gulf of Alaska involved simply adding fertilizers to the beaches to supplement naturally available nitrogen and phosphorus. No trace metals or other micronutrients were included in the fertilizers, and no microorganisms were added. The bacteria involved were strictly native bacteria already present on the Alaskan shorelines.

Microbial Metabolism of Hydrocarbons

Hydrocarbon compounds are abundant in the environment, occurring both naturally and from anthropogenic (man-made) sources, and microorganisms have become adept at utilizing them for food. For example, as shown in Figure 1.4, total annual input of hydrocarbons to the oceans of the world is about 180 million metric tons. Most of this is biogenic (made by living organisms), such as alkanes produced by phytoplankton. The annual contribution to this total from all sources of petroleum hydrocarbons is estimated at 3.2 million tons (National Research Council 1985). Yet, almost regardless of their source, microbial biodegradation naturally consumes these hydrocarbons, maintaining an ecological balance (Atlas 1972, 1981, 1985; National Research Council 1975; Colwell and Walker 1977; Jordan and Payne 1980; Bossert and Bartha 1984; Floodgate 1984).

Hydrocarbon-degrading microbes have been studied extensively, and they have been found in all marine environments sampled (Robertson et al. 1973; Mulkins-Phillips and Stewart 1974; Bunch and Harland 1976; Atlas 1978; Roubal and Atlas 1978;

FIGURE 1.3 Bioremediation is the acceleration of natural biodegradation by addition of fertilizer to bring amounts of nitrogen and phosphorus into balance with the supply of carbon.

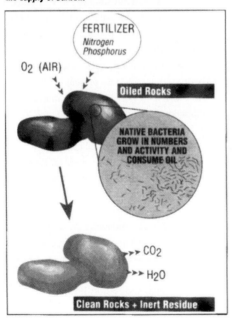

Gunkel et al. 1980). In general, researchers have found that in marine areas far from any source of petroleum hydrocarbons the numbers of oil-degrading organisms are lower than near sites of petroleum discharge. Hydrocarbon-degraders make up less than 0.1% of the total microbial population in areas removed from unusual hydrocarbon sources such as natural submarine seeps or municipal/industrial discharge sites (Atlas 1981; Azoulay et al. 1983), but increase their fraction of the total microbe population to 10% or greater near sources of oil discharge (Mulkins-Phillips and Stewart 1974).

Hundreds of different microbial species have been identified that are capable of metabolizing hydrocarbons, including bacteria, filamentous fungi, yeasts, cyanobacteria, and microalgae. However, each species has specific enzymes that give it the ability to degrade

FIGURE 1.4 *Most hydrocarbons that enter the marine environment are synthesized by living organisms such as phytoplankton.*

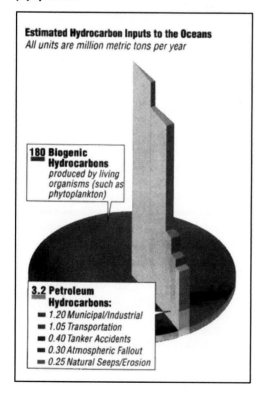

only a few of the many types of molecules in complex hydrocarbon mixtures, and this has important implications regarding how completely a given mixture of hydrocarbons can be consumed by a given microbial community.

The precise chemistry involved in the metabolic pathways by which microbes cleave the hydrocarbon molecules is varied, complex, and dependent on the specific chemical bonds to be broken and enzymes available to the microbes (Cerniglia 1984; Gibson 1971, 1977; McKenna and Kallio 1965; Perry 1977, 1979, 1984; Pirnik 1977; Ratledge 1978; Singer and Finnerty 1984; Van Der Linden and Thijsse 1965). However, all result in oxidation of all or part of the original hydrocarbon molecule. The original molecule may be broken into a smaller molecule containing one or more oxidized carbon atoms, be incorporated into cell structure, or be mineralized to carbon dioxide and water. Often the smaller molecule is an intermediate species that will be degraded further, but perhaps at a different rate from that of the original molecule.

One of the questions addressed in the Alaskan project was whether the more complex hydrocarbons such as polycyclic aromatic hydrocarbons (PAHs) could be broken down by the indigenous microorganisms. As described later in this report, these microbes demonstrated the ability to break down most types of hydrocarbon molecules present in North Slope crude oil. In large part, the breadth of hydrocarbon degradation realized following the *Exxon Valdez* spill was likely the result of the large diversity of microbial species found in the intertidal zones.

Even when conditions are ideal for microbial growth, not all hydrocarbon components of crude oil will be completely degraded to carbon dioxide and water. The very high molecular weight asphaltenes, resins, and complex aromatics biodegrade slowly (Wong et al. 1984). However, these materials are relatively harmless to the ecosystem because of their low solubility in water and their low biological activity (Hughes and Stafford 1983).

Role of Nitrogen and Phosphorus

Where does the need for nitrogen and phosphorus originate? These elements are vital for all living things. Nitrogen is an essential constituent of proteins and nucleic acids. Some bacteria can use atmospheric gaseous nitrogen, but most require the nitrogen in forms that are highly water soluble. Ammonia, nitrate, urea, and amino acids are among the most readily assimilable forms for most microorganisms. Phosphorus is an essential component of nucleic acids and the lipids of cell membranes and, as phosphate, plays a central role in biological energy transfer processes. The most readily assimilated form of phosphorus is phosphate, the form used in most fertilizers.

Normally, the beach environment is in relative equilibrium with respect to the amount of hydrocarbon to be degraded, the number of degrading microorganisms, and the amounts of nitrogen and phosphorus needed to balance microbial metabolism. However, after a spill, the large amounts of hydrocarbons introduced as food far outweigh the amounts of nitrogen that are available for balanced cell production. It is not clear whether the availability of phosphorus is also limiting, but it is prudent to assume phosphorus is also limited in supply. Unless more of these nutrients are added, microbial cell growth and the rate of biodegradation of oil are constrained. In the bioremediation process, more nitrogen and phosphorus are added, microbial growth and metabolism increase in response to the added food supply, and the rate of oil degradation increases. In effect, bioremediation is accelerating a natural process.

Under ideal conditions, the added nutrients should be applied so that they are in proximity to the target oil residue and remain there until needed by the microbes. However, in a marine beach environment, many physical processes such as tidal flushing and waves can dilute and diffuse added nutrients, reducing their concentrations below effective levels. The application strategy and fertilizer composition then become important elements for achieving success in bioremediation. For example, a slow-release solid fertilizer makes nutrients available to the microbes over a period of days and weeks rather than all being immediately water soluble and subject to rapid washout. An oleophilic liquid fertilizer adheres to oiled surfaces, providing proximity to the oil and slowly releasing nutrients.

Mass transport processes that control the rate of nutrient dissolution and dispersion are complex in the intertidal environment. Still, the applied nutrients must be able to remain in place and persist over time to be effective.

Role of Oxygen

The fundamental process of hydrocarbon biodegradation is oxidation. Unless adequate oxygen transport is maintained, oxygen availability can become the limiting component, even if sufficient nitrogen and phosphorus are available. In a beach environment, oxygen transport is affected by the permeability and porosity of the sediment and by the frequency and extent of flushing of shoreline sediment by seawater in tides and waves. During rising and falling tides, subsurface sediment receives oxygen dissolved in the seawater, providing the sediment has sufficient permeability to permit significant movement of seawater. During low tides, moist films of water and oil in the sediment are in contact with air, and oxygen diffuses into the films.

In the event that the beach sediment has low permeability, such as low-energy marsh mud flats, oxygen and nutrient transport into the subsurface may be low or negligible (Johnson 1970; Gibbs and Davis 1976). However, in such cases, penetration of oil may also have been insignificant.

For each potential bioremediation project, the availability of oxygen needs to be considered in advance of treatment. As described later, oxygen transport was investigated for the bioremediation project in Alaska and found not to be a rate limiting factor because of the generally large size of the sediment, its high permeability to seawater, and the ample content of dissolved oxygen in seawater flushed through the sediment during each of the two daily tide cycles.

Shoreline Bioremediation prior to the Exxon Valdez Spill

Prior to the *Exxon Valdez* accident, bioremediation of marine oil spills had been studied in laboratory tests and in a limited number of field trials (see, for instance, Lee and Levy 1986; 1989). Interest in using bioremediation to clean marine oil spills was stimulated following the *Amoco Cadiz* spill in 1978, when it became clear that biodegradation of hydrocarbons had played a major role in natural oil removal (Gundlach et al. 1983). There was evidence that fertilizers in water from agricultural runoff had accelerated biodegradation.

Even prior to the *Amoco Cadiz* spill, a pilot bioremediation project was initiated in 1976 at Spitsbergen, Norway, where small amounts of Forcados crude oil were spilled and microbial response monitored (Sendstad et al. 1984). After two years, agricultural fertilizer was added, and the rate of degradation with fertilizer was estimated to be increased about tenfold.

An experiment was conducted on Baffin Island (Sendstad et al. 1982) to determine the effect of mineral fertilizers on the biodegradation of Lago Medio crude oil. Oil was buried under gravel and sand. A rate enhancement of about fivefold was interpreted for fertilized crude samples versus those from untreated areas.

Recognizing the need for a liquid fertilizer that would allow slow release of nitrogen and phosphorus and that would tend to remain with the oil to be degraded, Elf Aquitaine developed Inipol EAP22 (Ladousse and Tramier 1991), which is an oil-external microemulsion that contains nitrogen and phosphorus nutrients. Because it contains an external oil phase, it readily sticks to oil residues and is therefore called an "oleophilic" fertilizer.

This fertilizer was first tested against a water-soluble agricultural fertilizer in an experimental spill in Norway (Sendstad et al. 1984; Halmo 1985). A threefold increase in biodegradation of alkanes was observed for Statfjord oil fertilized with either Inipol or agricultural fertilizer relative to oil from unfertilized areas, but with no statistical difference in rate between the two fertilizers.

Bioremediation was used following an actual spill in November 1985 of about 88,000 liters of marine gas oil on the shoreline in Ny Alesund, Spitsbergen (Sveum 1987; Sveum and Ladousse 1989). Inipol EAP22 was applied the next summer, and extensive monitoring of hydrocarbon levels was conducted. The increase in biodegradation rate of treated oil was about six to nine times higher than for untreated oil.

In summary, available experience suggested that addition of fertilizer might have significant potential to help eliminate oil stranded in intertidal zones. However, because prior tests were limited and results might be site specific, success of bioremediation following the *Exxon Valdez* spill could not be assured without field testing. ❖

Case Documents 3

```
┌─────────────────────────────┐
│ │ Coalition of Citizens    │ │
│ │ Concerned with           │ │
│ │ Animal Rights            │ │
│ │ CCCAR                    │ │
└─────────────────────────────┘
```

August 2, 1999

Dr. Karen Wartala
Dean, College of Agriculture
University of the Midwest

Dear Dr. Wartala:

The Coalition of Citizens Concerned with Animal Rights (CCCAR) was alerted by a concerned person two weeks ago that the University is trapping and killing large numbers of birds in the experimental crop fields on its campus. In our surveillance and investigation we have learned that:

- Birds are lured to the traps by bait placed in them, and the number of birds in these large traps has been observed to be at least 50 at times.
- The reported method of killing the birds was suffocation, by placing them, 400–500 at a time, in a bag.
- Approximately 500 birds are killed every day on the campus; 10,000 were killed in a three month period.

We object to this for the following reasons:

- It is a waste of animal life.
- It is ineffective. Trapping and killing birds will not permanently reduce the bird population in the area. Even *temporarily* it will have no more than a minimal effect.
- The trapping and suffocation of these birds is cruel in the extreme.

For these reasons, **we insist that the University immediately stop the killing of these birds.**

If you wish to discuss this matter, please contact me at 555-5537 or 555-5856.

Ed Younger

Ed Younger
Vice President, CCCAR

cc:
Dr. Pat Lennox, Head
Plant Pathology Dept.
University of the Midwest

President Hans Petersen
University of the Midwest

Lois Waseca, Head
Agronomy and Plant Genetics
University of the Midwest

F. Franklin Jefferson, Director
Agriculture Experiment Station
University of the Midwest

KTTP-TV
WDCO-TV
KNSP-TV
KURE-TV
WDCO-Radio
Minnesota Public Radio
KFCI-Radio
Associated Press
Minnesota Daily
Saint Paul Pioneer Press
Star Tribune
Twin Cities Reader

Bird Control
on the Experimental Plots
University of the Midwest

The purpose of our bird control program on the campus plots is to protect the experimental plantings that have been initiated for development of new crop varieties. The seed produced on these plots is extremely valuable because it represents the product of genetic crosses that have been carried out over a period of years. With each plant selection there may be only a few seeds produced, so it is essential that they be protected. The new plant varieties that are produced will be used to increase food production on a worldwide scale and, thus, may have a dramatic impact on reducing world hunger. We believe the benefits of these plant breeding experiments are important enough that we must protect these experimental plants from destruction by pest birds.

The campus plots occupy approximately 100 acres. The bird control program on these plots is a comprehensive one that involves several approaches that are being continually improved as we learn about new techniques. They include the following:

Plastic strings are used that when stretched across the plots vibrate to create a noise that frightens birds.

Brightly colored, large balloons with metallic designs (scare-eyes) are suspended above the plots which are known to have a deterrent effect.

A licensed falconer has been engaged who uses trained hawks to frighten birds away from the plots daily.

A program has been initiated to encourage the nesting of sparrow hawks in the plot area as a further means of biological control of birds.

Birds are trapped in large cages baited with bread. The trapped birds are supplied with water and food until they are removed from the cages.

There are eight traps located in the plot area. Approximately 150–200 birds in total will be trapped daily on the average. The number of birds per trap will vary a great deal. On unusual days as many as 500 birds may be trapped. The average number of birds trapped per year is usually about 5,000 with a range from year to year of 3,000–10,000. Bird populations are never permanently reduced as a result of the program. Our bird control program attempts to reduce bird populations in the plot areas only during the growing season. All songbirds are released. Only pest birds like blackbirds, starlings and English sparrows are destroyed.

The birds cannot be transported to another site and released because a suitable release site is not available that would be acceptable to the general public. The release of significant concentrations of pest birds is obviously not a desirable outcome to those who would be living in a potential release area and would be irresponsible on our part.

The birds that are destroyed are given to the Raptor Center for feed for birds that are housed there. The Raptor Center uses all the birds that are supplied to them.

Beginning the week of August 6, 1999, trapped pest birds will be euthanized with carbon dioxide. Birds will be collected in a cloth bag and kept in a cool place until they will be euthanized with high concentrations of carbon dioxide in a closed chamber. Collecting and transporting the birds in a cloth bag is more humane than using a small wire cage because the birds remain quiet in the dark bag and do not damage themselves by fluttering about as occurs in a small cage. Euthanization with carbon dioxide is an accepted, approved method that has been used extensively for small animals and results in loss of consciousness within 45 seconds and respiratory arrest within five minutes. Information obtained from the 1996 Report of the American Veterinary Medical Association Panel on Euthanasia indicates that carbon dioxide is an effective euthanizing agent. Inhalation of carbon dioxide causes little distress in birds, suppressing nervous activity and inducing death quickly. A further advantage in using carbon dioxide is that the euthanized birds can still be used for food in the Raptor Center.

Placing nets over the plots to prevent bird damage has been suggested. This approach is impractical because of the size of the plot area (100 acres). It would also be impractical because it would restrict management and evaluation of the plots and the nets would be subject to damage by storms.

If there are further questions about our bird control practices call:

Dr. Pat P. Lennox, Department Head
   Department of Plant Pathology   555-8200

Mr. Dave Obi, Research Plot Coordinator
   Department of Plant Pathology   555-3779

```
┌─────────────────────────────┐
│ ┌─────────────────────────┐ │
│ │  Coalition of Citizens  │ │
│ │     Concerned with      │ │
│ │     Animal Rights       │ │
│ │         CCCAR           │ │
│ └─────────────────────────┘ │
└─────────────────────────────┘
```

August 13, 1999

Dr. Karen Wartala
Dean, College of Agriculture
University of the Midwest

Dear Dr. Wartala,

Thank you for providing the Coalition of Citizens Concerned with Animal Rights (CCCAR) representatives with the opportunity to meet with you and other University members regarding the trapping and killing of birds in the experimental crop fields. We feel that a great deal was accomplished in clarifying the issues and look forward to a resolution that will satisfy all concerned.

I would like to review the following conclusions of that meeting. It was agreed that:

- The Department of Plan Pathology commits itself to researching bird deterrent methods with the ultimate goal of eliminating the current method of trapping and killing by next season of late June to mid-August, 2000.
- The Department of Plant Pathology will consult with the attached names and organizations, as well as any other resources at the department's disposal, for information and advice regarding alternatives. Please provide CCCAR with progress reports on or by December 1, March 1, and June 1.
- Several diverse alternative methods will be correctly applied and given an adequate length of time, and monitored and documented to assess their effectiveness.
- CCCAR will be provided with documentation of previous bird statistics from 1970 to present, and will be provided with new statistics at the end of the 2000 season.
- The Department of Plant Pathology will work with CCCAR member Jim Hamar during the remainder of the 1999 season to determine a means by which birds trapped can be released.

Please direct all questions, comments and correspondence to me. Again, please accept our sincere thanks for your cooperation in this matter.

Respectfully,

Malka Rothstein-Abrams

Malka Rothstein-Abrams
CCCAR
(655) 555-0304

cc:—President Hans Petersen, Vice President F. Franklin Jefferson, Dr. R. Lois Waseca, Dr. Pat Lennox, Dave Obi, KTTP-TV, WDCO-TV, KNSP-TV, KURE-TV, WDCO-RADIO, Minnesota Public Radio, KFCI-Radio, Associated Press, *Minnesota Daily, St. Paul Pioneer Press, Star Tribune, Twin Cities Reader*

Bird Control Resources

The Humane Society of the United States Guy Hodge, Naturalist 2100 L Street NW Washington, DC 20037	(202) 452-1100
International Alliance for Sustainable Agriculture Terry Gips, Executive Director Newman Center, University of Minnesota 1701 University Avenue SE Minneapolis, MN 55414	(612) 331-1099
Professor Ronald Johnson 202 Natural Resources Hall University of Nebraska Lincoln, Nebraska 68583	(402) 472-6823
David Tressemer (Bird Control Consultant) Boulder Colorado (has citations on USDA Department of Agricultural Ornithology, c. 1880–1930)	(303) 449-0486

August 22, 1999

Ms. Malka Rothstein-Abrams
Coalition of Citizens Concerned with Animal Rights

Dear Ms. Rothstein-Abrams:

I am pleased that Dave Obi, Richard Jones, Pat Lennox, Donna Harrigan, and I were able to meet with you and the others from the Coalition of Citizens Concerned with Animal Rights (CCCAR) on Friday, August 10 to discuss the bird control program in the crops research plots on the campus. I thought this was a productive meeting in that additional information was provided by both CCCAR and the University, that all agreed on the importance of the crops research and the necessity to control the damage done by the birds, and that some agreements on future actions were made.

I have read your letter of August 13 to me, and want to respond so that there will be no misunderstanding of what these agreements were.

- The CCCAR and the University agreed that the crops research is important, that it must be continued, that birds do damage to the research plots, and that this damage must be controlled.
- The University stated that the bird control measures currently being used would continue throughout the remainder of this season, but that between the end of this season and the start of the control period in 2000 the University will explore additional and alternative methods for controlling the bird damage. This does not mean that the University agrees to change its current program. It means that we will carefully explore valid, reliable, effective, feasible, and humane alternatives, and if one or more can be found that will meet our requirements, they will be adopted.
- The University is willing to work with CCCAR during the remainder of the current season to transport the trapped birds to a release site provided that an acceptable site can be found that meets all community and state regulations.
- The University will make available to CCCAR the bird trapping statistics from the 1970s to the present.

Although we did not agree to provide CCCAR with periodic progress reports, we will be happy to keep you informed on the alternative approaches that are being explored, and will inform you prior to the beginning of the 2000 control season of the methods that will be used.

Sincerely,

Karen Wartala

Karen Wartala
Acting Dean

Case Documents 4

The Year 2000 Problem on the Web

In 1998 and 1999, people around the world were worried about the "Year 2000 problem"—an error in computer programming that created problems for businesses, governments, and individuals. Computer software uses only the last two digits of a year for its internal dating purposes; therefore, for example, 1992 is programmed as 92. As a result of this internal machine shorthand, people at the end of the twentieth century were concerned that computer software would generate all sorts of errors when 1999 became 2000—when computers would interpret 00 not as 2000 but as 1900.

Although the problem was fairly easy to describe, it wasn't so easy to solve, for computers handled dates in a variety of ways and thus required a variety of expensive reprogrammings. Millions of lines of code written years ago had to be examined line by line and corrected. This problem with dating was anything but insignificant because matters of dating influence everything from employee records and financial transactions to how nuclear weapons are programmed. Dating errors can cause malfunctions in air traffic control operations and the workings of major utilities. People feared problems large and small: malfunctions in their VCRs and a shutdown in the Federal government. The computer problem was therefore a serious business problem and a legal problem as enterprises scrambled to cope with the job of correcting all their computer code.

Some people feared that the reprogramming would prove to be irremediable, at least in part, even for large organizations; others felt that small companies and organizations would prove especially vulnerable—that they would be unable to maintain their communications systems or share data with other systems. Thousands of lawsuits were expected from companies likely to attempt to lay blame onto software developers and service providers, and a number of people forecast business failures and other economic setbacks.

The Year 2000 Problem—also known as the Y2K Problem, or the Millennium Bug—inspired a great deal of communication. Major newsmagazines and newspapers devoted cover stories to the problem; conferences were held; articles, books, and newsletters devoted to the problem appeared everywhere; talk shows and news shows discussed the problem; e-mail and computer chatrooms buzzed about the issue. Because controlling the magnitude of the problem depended on preparation and planning, "Protection Guides" and "Resource Books" were drafted to advise companies, governments, and ordinary citizens about the matter, and magazines were full of how-to articles offering advice.

The tone in all this communication varied tremendously. Most people felt that the Y2K problem, while expensive to fix, would be handled by experts, so they wrote soberly and matter-of-factly about the technological problem. Others, however, in a mood of near panic responded with apocalyptic language and sensational predictions; they collected information into "Survival Plans," for instance, and even predicted the Second Coming and Apocalypse. Many of these communications used extensive religious references because the Year 2000 problem was associated with the new Millennium. Still others viewed the problem with amusement, and many resourceful types saw an opportunity for moneymaking.

A large amount of all communication about Y2K, not surprisingly, appeared on the World Wide Web. By 1998 the Web had already become the most frequently used medium for exchanging information in North America, and a great many people discussed the Year 2000 Problem by exchanging ideas and information electronically. This set of Case Documents, therefore, includes a few sites that illustrate some of the basic documents that appear frequently on the Web—Web pages, FAQ pages (FAQ="frequently asked questions"), sales documents, and a technical report—and that were developed during this extended conversation about the Year 2000 Problem. As you look over the documents, consider how the medium in which they originally appeared (the Web, not a book or magazine) affected their construction and composition.

Because no one can be sure how long any of these sites (or any Web sites, for that matter) will be maintained after January 1, 2000, we've included copies of the home pages of each site we reference. Those home pages can give you a feel for the language used, the layout, the audience addressed, and the purpose of the document. The first site in the series, <www.everything2000.com>, has the feel of a newspaper, for its job is to introduce the general public to various matters, serious and not so serious, about the Year 2000 Problem. The second, a FAQ page developed by the State of Wisconsin, <www.state.wi.us/y2k/faq.htm>, observes the general conventions of the FAQ genre in that it is organized around a series of queries. These are organized from most common and general to less common and more specific. Sometimes a FAQ site can go on for many screens, and indeed several Year 2000 FAQ pages did go on at some length—even to a hundred screens. This site seems to be an effort to reassure the citizens of Wisconsin (and other eavesdroppers) that the Year 2000 Problem was being managed in a satisfactory way.

The next two sites illustrate a familiar genre that has developed in new ways on the Web: the advertisement. In the first, a company called ENS— short for Enterprise Network Systems <ens.org>— attempts to attract new business by advertising its ability to solve Year 2000 complications for businesses and individuals. In the second, an individual from Austin, Texas, named Rob Campinell offered an "open letter to every CEO [chief executive officer] and CIO [chief information officer] regarding the Year 2000 Problem." The site <www.well.com/user/robcamp/y2k.html> offers information and advice that its author hopes also will translate into service opportunities for himself.

The final site included here is one created by the British Standards Institution definition of the Year 2000 conformity requirements <www.brainstorm.co.uk/disc/year2000.html> they developed. This site will give you a sense of the international scope of this issue.

home | site index | about us | guestbook 1999 | link | advertise | contact

Planet Prepare! Y2K PREPAREDNESS PACKS ✓water ✓food ✓stove ✓fuel *secure yours today!*

- Events
- Computer Y2K
- Destinations
- News
- Shopping
- Resources
- Countdown
- Movements
- Humor
- Life
- Politics
- Home

everything2000®

Your one-stop resource for celebrations, y2k, events, news, shopping, travel, links and more.

today's news today's news today's news today's news

White House Planning 24-hour Y2K Watch
Round-the-clock worldwide monitoring will begin on New Year's Eve
Full story...

Colorado Y2K Prepared
Contingency plans in place to move elderly, sick and prisoners in case of problems
Full story...

Millennium Caviar Shortage
Russian delicacies skyrocketing in price due to New Year's Eve demand
Full story...

Today's news archive

BUGS 2000 by o'Malley.
YOU TWO DISTRACT HIM, WHILE I SNEAK IN AROUND THE BACK..

year 2000 photo countdown

A customer looks over books on the table at Christian Publications Bookstore in midtown Manhattan that deal with the Year 2000 computer bug and millennial prophesy issues.
More...
Photo by Chris Hondros, Millennium Picture Group

everything weekly

Millennium Baby Updates
Our List of Millenni-moms is getting longer!
Full story...

New!
Everything Counts
Character Counts - Building character during every waking moment
Full story...

the year 2000

Email this site to a friend! ☺

New Year's Eve Entertainment Update
New!

Millennium Baby Updates

Bugs 2000

David Bowie in concert!
New Zealand 2000

Fiji for 2000!
Travel Package
GO

destinations
- United States
- Canada
- South America
- Europe
- Great Britain
- Africa
- Middle East
- Asia
- Down Under
- Antarctica

Motoring into the Millennium
Surviving Y2K in Hamilton Montana
More...

surveys
- surveys
- results

countdown clocks
- personal
- download

| SEARCH | FAQ | GLOSSARY | LINKS |

FREQUENTLY ASKED QUESTIONS

What is the Year 2000 problem?

In the past, virtually all computer programs used two-digits to designate years (e.g."98" instead of 1998). While this practice posed no special problems, a serious challenge lies ahead as the year "99" is followed by the year "00." In effect, computer applications that are not updated to become "Year 2000 compliant" may read "00" as "1900", instead of "2000," "01" as "1901" and so forth. This could result in either a malfunction or incorrect information. For example, because a computer may perceive "01" as coming long before "98", an employee hired in "01" could be treated as having more seniority than the one hired in "98."

Can we just switch Computer codes from two to four digits?

The conversion from two to four digits is only the tip of the Year 2000 iceberg: The problem begins with finding all the places where two digit year codes are used in computer applications and extends to careful testing of systems after conversion has been completed. Finding two digit year codes requires a line-by-line analysis of all applications, data bases and files. The state will also have to take steps to accommodate data imported from other sources (e.g. federal agencies) that may not be Year 2000 compliant.

What is state government doing about the problem?

Although it will be several months before most calendars roll over to the Year 2000, the State of Wisconsin entered its Fiscal Year 2000 on July 1 without any evidence of Y2K computer glitches. Because state government's financial records are maintained on a fiscal year that runs from July 1 to June 30, various accounting systems that track tax collections, payroll and payments are now operating in Fiscal Year "00." The fact that these programs continue to perform normally is viewed as further confirmation that state government can avoid significant Y2K computer problems on January 1.

All key computer applications that pose potential Y2K issues, including the state's central accounting system, have come though the transition to Fiscal Year 2000 with no unusual problems. This confirms the effectiveness of the State of Wisconsin's Year 2000 preparations.

Credit is due state managers and employes who continue to work on Year 2000 readiness. This is the first clear-cut confirmation we have that their efforts are paying off. While more work remains, we are increasingly confident that state programs and services will experience few, if any, Y2K problems. We are very pleased and proud that state agencies have responded so effectively to Governor Thompson's July 1998 directive making Year 2000 readiness our top business priority.

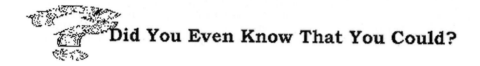

Did You Even Know That You Could?

Enterprise Network Solutions, Inc. welcomes you.

There are 150 days, 8 hours and 43 minutes left until the year 2000.

Have you started your Y2K Compliance project yet? Falling Behind? Project not going the way you thought? We can help you!

Who we are...and what we do....

Enterprise Network Solutions, Inc. is a single-source provider of Year 2000 services, Network Inventory, and Asset Management systems . We can provide businesses of PC based networks an array of services ranging from providing the tools to take advantage of current and emerging technologies, to developing a full program of assessment, design and implementation. In all cases, our services are delivered with the highest degree of professionalism using a personalized partnership approach.

ENS has partnered with industry leading companies to provide you with a solution to your Year 2000 and network problems. We realize that most businesses, large and small, are comprised of a variety of hardware and software environments. As a result of such diversity in the enterprise, we have vigorously tested each and every product we sell and implement.

Our years of experience have given us superior methodologies to help our clients get into the millenium while keeping control of their networks. Enterprise Network Solutions, Inc. strives to meet each clients business needs and objectives in every way. We have created a highly successful and repeatable approach towards Y2K remediation, asset management systems and automated network inventories.

Enterprise Network Solutions, Inc. can provide you with the software tools, training, project management, personnel, source remediation, and testing. Enterprise Network Solutions, Inc. possesses the tools and knowledge to perform Year 2000 assessments and conversions in a fraction of the time of in-house efforts or other consulting companies.

Please Bookmark this page so you may keep up to date with Y2K issues and products.

This site is best viewed with Internet Explorer 4.0 or Netscape Navigator 4.05 or higher

[Home] [Products] [Press] [FAQ] [Contact ENS] [Leasing] [Links]

Year 2000

Home

Life Story

Personal Details

Book Reviews

Year 2000

InterneTV

Streaming page

An open letter to every CEO and CIO regarding the Year 2000 problem.

Dear CEO and CIO:

By now, you have been confronted with the decision about a plan of action to confront the so-called Millennium Bomb. My main words of advice are:

TRUST YOUR INSTINCTS.

Apocalyptic predictions such as 90% of all computer systems will fail on January 1, 2000 are nothing more than scare tactics. The Year 2000 problem should be viewed as a new growth product for the computer, legal, and financial services industry. Although there are valid technical concerns regarding software and systems, the goal of the Year 2000 industry is to get you to part with your dear money. I urge you to stand strong against threats of systems meltdowns and litigation.

WE'VE SEEN THIS BEFORE

This is not the first time dire warnings about all your computer systems failing on an exact date in the future. The "Ides of March" virus scare in the 1980s predicted most of your PCs were infected and would fail on March 15. Looking back in retrospect, The predictions about the extent of the Ides of March virus were way off. The Ides of March turned out to be more of a hoax than an actual problem.

LIVING WITH BUGS

No software is 100% bug-free. We have been living in a software-center society for twenty year and have adapted to software bugs. Software bugs are like traffic jams, we don't like them, they are inconvenient, they get us frustrated. However, we have learned to live with them and find ways to work around them. They have become a part of our lifestyle.

In the year 2000, the entering college freshmen will have spent their entire lives living in a software-centered society. The software-centered society is here today, and we have to learn to live with it.

THE HYPE

It is a little more than two years away from that fateful day of January 1, 2000, and the hype about the Year 2000 problem keeps getting stronger. It has gone from the pesky "Millennium Bug" to the dangerous "Millennium Bomb." The majority of the public will be aware of the Year 2000 problem on January 1, 2000 just by the sheer shrill of the hype alone. They will able to tolerate minor inconveniences, but not failures which disrupt their daily routine.

PREDICTING THE FUTURE

No one can predict the future. Although they would like to think otherwise, the Year 2000 industry is no exception. Software Productivity Research out of Boston predicts the "Millennium Bomb" will cost $3.6 trillion. Technology Business Reports in California predicts $2 trillion. The Gartner Group and J.P. Morgan are a bit more modest. They predict $600 billion and $400 billion respectively. BZW in the U.K. estimates it will cost $52 billion dollars.

None of these groups know how many software programs worldwide are affected by the Year 2000 bug; nor do they know how many are worth saving. Even if you allocated resources in a preventive Year 2000 program, there is no guarantee that the solutions will work. Unlike previous information technology investments, investing resources in a Year 2000 program will not lead to increases in productivity. It only maintains the status quo on January 1, 2000, if you are lucky.

OPPORTUNITIES

If you undertake a Year 2000 program, it should include strategic planning to incorporate emerging information technologies into your present systems to give your company a competitive advantage. Current technologies such as relational databases and the desktop metaphor operating systems are 25 years old. Advanced database systems are replacing the relational data concepts. Operating systems are being developed based on your working process and not the furniture you use. The psychology of the millennium has a way of discarding conventional wisdom, and focusing on the need for progress. This will be true for the evolution of information technology.

CONCLUSION

I have worked as a consultant for a Fortune 500 company to implement and test a pilot Year 2000 compliance program. My experience has led me to believe that there will be some problems, but they will be minor in nature, and cause some inconveniences. I believe people will be able to adapt to the inconveniences. The hype generated by the media will alert the public to the problem.

The Year 2000 problem is an opportunity for people in the professional services business, which includes myself, to fleece their clients. Ultimately, you will have to make a decision on how much of your resources to commit to "solve" the Year 2000 problem. You have to judge the risks to decide whether to take preventative measures before January 1, 2000, reactive measures after January 1, 2000, or a little of both.

Rob Campanell
Austin, TX USA
robcamp@well.com

DISC PD2000-1:1998 *A Definition of Year 2000 Conformity Requirements*

Site last updated 24 June 1999

DISC PD2000-1:1998 *A Definition of Year 2000 Conformity Requirements*

This document addresses what is commonly known as "Year 2000 Conformity" (also sometimes known as Century or Millennium Compliance). It provides a definition of this expression and the requirements that must be satisfied in equipment and products which use dates and times.

BSI DISC originally published PD2000-1 in January 1997 and it has been widely adopted. A review of the document was conducted by the responsible committee (BDD/1/3) in the spring of 1998 taking into account comments received. The committee considered that amendments to the fundamental conformity requirements were neither necessary nor desirable. The Definition and the four Rules are unchanged but, to add value to the document and aid its interpretation, the Amplification sections have been amended. This document, PD2000 1:1998, replaces the previous version of PD2000-1 but does not change its requirements. An additional document PD2000-4, entitled "Guidance and Information on PD2000-1:1998" provides further information.

Please feel free to download these documents and distribute them to your clients, customers and trading partners so that together we can promote a common understanding of "Year 2000 Conformity" .

Document	Title	Further Information
PD2000-1:1998	*A Definition of Year 2000 Conformity Requirements*	Download a copy free (143kb) or Click to view a html version
PD2000-2	*A Code of Practice for Year 2000 Management*	For more information Click here
PD2000-3	*Understanding the Century Date Change Problem*	Download a copy free (37Kb)
PD2000-4	*Guidance and Information on PD2000-1:1998*	Download a copy free (223Kb)
PD2000-1:1998, PD2000-3, PD2000-4	*For your convenience these documents are available in a PKZIP Format*	Download a copy free (128Kb)
HB 10145	*PD2000-2 Plus "A business guide to the year 2000" as a package at a discount price*	For more information Click here

HB 10147	*The Millennium Time Bomb - A Practical guide to the Technical and Legal Issues*	For more information Click here

 STOP PRESS (24th June 1999)

Draft text of ISO/IEC 16509 Information technology - Year 2000 terminology, an international standard for Year 2000 has been approved. Watch this space for more details soon!

Other Relevent Publications avalible from BSI-DISC include

- *Year 2000 Guidance for the Blind and Disabled Communities (19Kb)*

- *Small Companies Feel the Y2K Bug Bite* - results of our Year 2000 survey.

- *A Year to Shutdown* - a report from our Year 2000 seminars.

(Note PD 2000-3 and PD 2000-4 are in Adobe PDF format - If you do not have a the Adobe reader Click here to visit the Adobe site)

 BSI DISC recently took a trip to the Turks & Caicos Island in the Caribbean to present Seminars on Year 2000 Awareness for the Government of the Islands. Your location may not be so exotic, but we will go anywhere to help raise the level of knowledge and understanding of the Year 2000 Problem. Send us an Email - we may be able to help you too!

Please leave your details to ensure that you are informed about our forthcoming Year 2000 guidance.

Do you require a copy of this definition?   No: ● Yes: ○ (Please select **YES** if you require a copy of PD2000-1:1998)

Name:

Job Title:

Address:

Company / department name:

Size of Company (*No. of employees*):

Building name and/or number:

Street:

Town:

County / region:

Tel number: [_____]

FAX number: [_____]

email address: [_____]

[Send to BSI-DISC]   [Reset form]

DISC is a part of the British Standards Institution
BSI, 389 Chiswick High Road, London W4 4AL
Tel: 0181 996 9000

DISC Home Page   BSI Homepage

Glossary

A

abscissa The horizontal axis that usually indicates time, items, or categories on a graph; also called the x-axis.

abstracts Standard parts of a scientific report that usually precede introductions, (2) reference texts that include listings of citations and brief summaries, and (3) condensed versions of longer documents that summarize main points.

accuracy A characteristic of technical writing; the validity and precision of the information in a technical document.

action verbs Strong verbs that convey precise, rather than general, meaning.

active voice A grammatical construction that clearly identifies the agent of the action, e.g., the police arrested the burglar. (*See also* passive voice.)

agonistic Descriptions of physical competition rather than cooperation; e.g., to fight it out over a point in a meeting.

analogies Figures of speech that extend a comparison for explanatory or persuasive purposes; often used to compare two dissimilar items to illustrate elements of similarity; one of the items must be familiar to the audience.

annotated Bibliographies that contain brief descriptions of the contents of each entry.

annual reports Company reports summarizing and evaluating annual activities.

ANSI Acronym for the American National Standards Institute, whose quality and reliability standards can be adopted by businesses and industry in the United States. Its Web site (http://www.ansi.org) offers the latest national and international standards.

antonyms Words that mean the opposite; e.g., the antonym of *dry* is *wet*.

appendixes Part of a report that includes material useful for reference but that is not essential to the argument of the report.

appositive A grammatical form that usually follows the statement of the key word in a definition.

argument In the rhetorical context, argument refers to the way that one's persuasive appeal (*see* persuasion) and arrangement of materials are used to convince or sway an audience.

Aristotle (384–322 B.C.) Greek philosopher and rhetorician considered, along with Plato and Socrates, one of the most famous thinkers of the ancient world. Wrote *Rhetoric* to describe the art of creating the most persuasive message on a given topic.

ascenders Parts of a letter that rise above the top of the letter (e.g., *b* and *b*).

attention line Used as the salutation in a letter if no name appears on the first line of the inside address; e.g., Attention: Marketing Department.

audience One component of the rhetorical situation; includes the readers, listeners, and viewers of a particular document who can be real, consisting of actual people, or imaginary, constructed in the writer's or speaker's mind; within the pages of documents they help writers communicate more effectively with readers as a kind of "character" who embody a role.

audience-based orientation Writing designed from the point of view of the audience and its needs rather than from the point of view of the writer.

audience roles The roles of transmitter, decision maker, advisor, or implementer that an audience takes after reading, viewing, or listening to a communication.

B

background information The additional information needed to help readers interpret a report; includes history, description of a mechanism or process, definitions, and needs assessments.

back matter Additional pieces of information that are useful but not essential to a report's argument.

benchmark A goal statement or a clearly defined set of expectations that a person can use to measure current achievements against or to determine what progress is being made on a given project.

because-clauses The good reasons supporting arguments or proposed ideal situations.

body The main section of a report describing the problem a project will solve; in a letter, the body is the message or content section.

boilerplate The standard sections of reports that can be stored as computer files and reused.

bookmarks Web pages visited and tagged by entering their address into a special file on the browser so that one can return to these sites directly.

Boole, George A mathematician (1815–1864) who invented a way to organize questions that is now called Boolean logic. Based on key operators AND, OR, and NOT, this logic is used in Boolean searches.

Boolean searches Database searches that follow Boolean logic by making use of three key operators: AND, OR, and NOT.

boxes Portion of a typeset text surrounded by a rule to emphasize a particular feature.

brainstorming An early stage in the writing process when one tries to create an extensive but unedited list of ideas or solutions.

brand A way for a business to show its corporate identity on all of its products and documents by means of a uniform and immediately identifiable image.

browsers User interfaces that provide access to the World Wide Web.

C

call numbers Coding devices used to identify library holdings; no two holdings will have the same call number, but related holdings may have related call numbers.

callouts Illustrations and definitions pulled from the main text and magnified to let the user see some aspect of the process or product in detail.

case Lowercase and uppercase (capital) letters.

categorical propositions Informal definitions that consist of a subject (word to be defined), a linking verb, and a predicate (something about the word to be defined).

CD-ROMs Compact disks that serve as storage devices for anything in digital form, including multimedia, encyclopedias, census and other demographic information.

Challenger U.S. space shuttle that exploded 73 seconds after takeoff on January 28, 1986, killing all seven crew members, including teacher-astronaut Christa McAuliffe, the first woman teacher to fly in space and one of the first civilians.

checklists Devices provided to help readers evaluate the outcome of their actions or decisions; for example, one consults a checklist of parts before assembling a device.

check-off questions Questions in a survey questionnaire that present a list of options from which a respondent may choose.

chroma The intensity or saturation of a particular hue.

chronological résumé Lists experience and education in reverse chronology, with the most recent information listed first.

citing sources Clear and accurate identification of the source of information.

clarity A cluster of elements in a document that includes freedom from ambiguity, logical development of ideas, unity, coherence, appropriate style and usage, clear transitions, and accurate word choice.

closed questions Questions in a survey questionnaire that force respondents to answer in specific ways; e.g., yes or no.

collaboration The act of a team or group working together on a common project such as a report, a Web page, feasibility study, or conference presentation.

collaboration process The three stages of successful collaboration are initiation, execution, and public presentation; each stage has a relationship and a task level.

collaborative levels The two levels of tasks necessary for a collaboration to succeed; the relationship level refers to the interpersonal activities through which collaborative teams get to know each other, share assumptions, maintain trust, resolve conflict, and acknowledge each other's work; the task level refers to the task completion activities in which teams clarify goals, generate ideas, establish procedures, draft, revise, edit, and present their work.

collaborative technologies The computer tools and software that support collaboration on the relationship and task levels.

collaborative writing Team or group writing on a project; *see also* divide and conquer model; specialization model; dialogic model; sequential model; and synthesis model.

color Used in a document to link or contrast objects, focus attention, serve as a navigational aid, or express values in charts and maps.

columns The vertical structures of information in a table.

combination résumé Combines chronological and functional résumés, listing qualifications and providing company names and dates in a separate section to minimize gaps in work history or temporary jobs unrelated to one's career goal.

communication anxiety (CA)   The anxiety many people feel when they have to make an oral presentation.

communication links   Created by the paper trail resulting from interconnected requests and decisions.

comparative terms   Expressions of what things are like or unlike a topic in a communication; used in argument by similitude.

complementary colors   The opposite of each primary color on a color wheel.

completeness   The internal integrity of the document; all parts of the assignment are present.

complimentary close   The "good-bye" found at the end of a letter.

computer-based training (CBT)   Consists of step-by-step training exercises completed at the computer.

computer-supported team   A writing group that uses computer technology to support the collaborative process.

computer tools   Computer applications designed to create a variety of visual displays based on information stored in data files.

conceptualization   The process in writing of conceiving of the purpose and the reader's need for information.

conciseness   Writing that is as clear and brief as it can be and still remain effective.

conclusions   Answers to the research questions explored in reports for decision making: the overall assessment of work in sections of progress reports; they can also be answers to a research question that go beyond the summary in providing evaluation, interpretation, and reconciling of contradictory results.

concurrent methods   Designed to collect data about how people think, feel, or respond as they use a document.

conditional searches   Library or Internet searches that use the Boolean key operator OR and specify that any included term is acceptable.

conflict   Part of the collaborative process that enhances decision making and can be managed effectively if understood; a number of different kinds of conflict occur in the collaborative setting: pseudo-conflict, a misunderstanding of the problem or solution; simple conflict, a disagreement over which course of action to pursue during a project; ego-conflict, results from the defense of a collaborative member's ego; and substantive conflict, disagreements about content, purpose, audience, organization, support, and design.

consistency   The quality of a document that enables a user to relate items and distinguish between them; includes consistency in capitalization, spelling, abbreviations, and use of numbers.

content analysis   A particular research approach that allows a researcher to determine the relative frequency of themes or actions in a collection of texts by counting the number of times they occur.

context sensitive help   Describes online help systems that present information on screen related to the procedure being performed at the time when the help is called up.

continuation page heading   Provides an indication at the top of additional pages of a letter that they are part of a single document.

conventional structure   The standard format for organizing a report that includes an executive summary, introduction, selection criteria, conclusion, recommendations, and appendixes in that order.

coordinate system   Describes the x and y axes on a graph.

coordination   The linking of two equal ideas, usually with a coordinating conjunction.

copy notation   A note at the end of a letter indicating that copies were sent and to whom.

corporate ethos   The way corporate values and beliefs are manifest in the persona the corporation adopts in its communications with people inside and outside the organization.

correctness   The level of correct grammar, punctuation, and spelling in a document.

cover letter   There are many kinds of cover letters, including that written by the executive officer of an organization submitting a proposal which indicates support for the project; other kinds include the letter of transmittal accompanying a report or résumé.

credentials   Experiences that give a communicator expertise.

credibility   Associated with ethos, the creation of a persona who appears knowledgeable and trustworthy.

criteria   The standards for making a good decision—e.g., when buying a car, selection criteria might include safety, gas mileage, dependability; used in arguments from value.

cross-sectional survey   A primary research tool that allows the researcher to ask questions of members of a heterogeneous group about current phenomena.

cultural background   The shared understanding of various symbols, shapes, and colors within a given culture.

cultural diversity   The differences among and between people due to their gender, ethnic background, religion, and so forth.

culture   Socially constructed and transmitted behaviors, beliefs, values, and institutions.

D

dangling modifiers   Modifiers not clearly linked to a referent.

data reduction   The process that allows a researcher to reduce raw scores into manageable summaries.

deductive organization   Common in technical writing; the document begins with the writer's conclusion or key point, and then presents supporting evidence or details.

definitions   Explanations of terms to the audience.

demographic information   Vital statistics of general populations that can be easily collected with closed questions and nominal scales.

descenders   Parts of a letter that drop below the base of the letter (e.g., *y* and *p*).

description   Information about the principles of operation or the working of a process.

descriptive abstracts   Description of the entire contents of a document.

descriptive statistics   A type of statistical analysis used to describe single variables or connections between variables.

design   The navigational devices that let readers find information: page numbers and running heads, title page and table of contents in a print document; menu of choices, topic list, and hyperlinks in a hypermedia document.

design features   Include many textual features such as white space, the use of color, and choice of binding.

diagram drafts   Start drafts created using designs such as circles, boxes, and lines to represent sections of a document.

dialogic model   A collaborative model in which each team member's responsibilities shift during the project; requires a high level of interaction.

digital images   Images captured and stored electronically, which increases their flexibility for modification to suit a variety of rhetorical contexts.

directed interviews   Structured interviews in which the interviewer asks a set of specific questions; the counterpart of a self-administered survey questionnaire.

direct measures   Direct measures are those that study how readers use, read, or rate a document and then judge that activity against a criterion.

discussion lists   E-mail discussion groups on specific topics that can be moderated and subscribed to by users.

distortion   The misrepresentation of data in a visual display.

divide-and-conquer model   A collaborative model in which each team member writes, revises, and edits one part of the project; assumes each team member is alike and replaceable.

dividers   Sheets of paper inserted between sections of a document.

document production process   The specified number of stages and checkpoints a technical writer must go through in creating documents; processes allow companies to control the quality of their documents, coordinate the writing time with product releases, and assure a consistent style and organization.

domain type   Part of an Internet address or URL, specifying the kind of organization sponsoring a Web site.

dramatizing   A tactic that includes emphasis and visual display, used with unmotivated audiences to involve them in the communication.

dyadic   Describes an exchange of information between two people.

E

economic contexts   Environments of cost, supply and demand, and systems of exchange in which terms can be placed.

editing   The process of making a document more accurate, effective, and readable by ensuring that the grammar, usage, style, and punctuation, as well as the organization, are accurate and appropriate.

E-mail   Electronic messages sent by computer over the Internet or an intranet.

enclosure notation   The indication at the end of a letter that enclosures are attached.

endnotes   The definitions placed (1) at the end of a chapter or section of text or (2) references to works cited in the text.

endowments   Sums of money that generate interest or investment income to fund particular projects.

English Renaissance   A period (1475-1640) marked by a resurgence of interest in science and the arts; it produced a large quantity of technical writing which recorded information on planting crops, curing diseases, cooking meals, navigating the seas, delivering babies, and catching fish.

environmental contexts   The environments or systems in which technical terms can be placed.

ethos   One of the three means of persuasion Aristotle describes; refers to the perceived trustworthiness, credibility, and reliability of a communicator. (*See also pathos* and *logos*.)

etymology   The history of words and their formation.

examples   Ways of extending definitions by including items that fall within the scope of the word or text being defined.

exclusive searches　Internet or library searches that use the Boolean key operator NOT, specifying that certain words are to be excluded from the search.

executive summaries　Abstracts that appear at the beginning of reports and give a condensed version of the entire document.

extemporaneous presentations　Oral presentations planned thoroughly in advance but delivered in a spontaneous manner.

extrapolate　A means of displaying trends in data by projecting trends. (*See also* interpolate.)

evaluative components　Part of an argument from value.

Exxon Valdez　A tanker that ran aground on March 24, 1989, releasing 11 million gallons of crude oil into Prince William Sound and the Gulf of Alaska.

F

facts　Information that can be verified or proven by personal observation or research.

feasibility　The determination that a project is both possible and desirable; documents that determine this are called feasibility studies.

field notes　Written records of observations.

field trials　A type of user testing of a document in an actual setting to determine what problems users face.

flow chart　Diagrams of processes that show the sequence of steps involved.

focus groups　A method of gathering data by bringing together groups of people around a specific topic or issue.

focus on the subject　In technical communication, refers to the fact that the main subject determines the kind of information and the form in which it must be conveyed.

folio　A two-page spread used in page design.

follow-up letters　Letters sent after a job interview thanking the interviewer and emphasizing the interviewee's interest in the position.

footers　Navigational aids set in small type at the bottom of a page which repeats information such as the title, topic, page number, or date.

footnotes　Notes that appear at the bottom of a page of text, containing definitions or references to citations in the text.

forecasting　A stage in report writing when one tries to select some possible solutions from the list of all possible solutions.

formal definitions　Definitions consisting of the term to be defined, the class/family/species to which the term belongs, and the special characteristics that distinguish the term from other terms in the same class/family/species.

format　The conventions required for a report; may include standards for typeface, margins, headings, spacing, length, binding, and the like; also refers to the way information is organized; e.g., the format of a letter differs from that of a memo by having a salutation.

foundations　Nonprofit entities that use their funds to support research, educational, artistic, or charitable projects.

four-color processing　The separation of hues and application of inks from four successive color runs to reproduce the original art.

frequency　The percentage of responses to a given question; frequency is the simplest descriptive statistic.

Frontinus Sextus Julius　Roman author and soldier (35-103) whose writing about aqueducts is an early example of technical communication.

FTP　File Transfer Protocol, defines how files are to be transferred from one computer to another; like World Wide Web servers, FTP servers are part of the Internet.

functional definitions　Definitions for processes, procedures, or phenomena that occur over time; constructed by using the name + the function or operation = special uses.

functional résumé　Highlights a job seeker's qualifications without emphasizing specific dates or work history.

functional specifications　The detailed descriptions of what the product will do and how it will perform.

G

Gantt chart　A schedule of tasks and their relationships and completion dates.

general population　All the people involved in the subject or topic being researched.

genre　A category of document in literature and academic writing characterized by a particular style, form, and content (e.g., a novel, a lab report, or cookbook); in technical communication, a genre is a form of communication created in response to a recurring situation, perhaps to solve a common problem or convey similar information (e.g., instructions, feasibility studies, annual reports, online help screens, or warning labels).

goal　Another name for the ideal situation.

glossary　A dictionary of terms that apply to the topic of a report; located at the beginning of a text or at the end of a document, depending on the needs of the readers.

grid layout The use of the page as a grid or blueprint upon which to make design decisions such as the number of columns and placement of graphics.

grid lines Used on a graph to help readers pick out data from given categories.

groupware Computer-supported tools to aid collaboration and revision efforts for people writing together in teams.

H

hanging heads Headings placed in the margins to facilitate a reader's skimming through a text to what he or she is seeking.

headings and subheadings Navigational devices that divide text into sections and subsections to help the reader's comprehension and selective reading; they indicate sections and subsections of a document and their relationship to each other. In letters, a heading consists of the sender's name, address, and date at the top of a letter that has no printed stationery.

hearing impaired people Audience members whose inability to hear may require special development and placement of audio aids and the support of a signer.

hierarchy The arrangement of positions and power within an organizational structure.

highlighting The identification of key words by the use of bold, italic, or underlined typeface.

historical information Elements that allow events to be placed in a temporal environment.

history Information on the origins of a problem in a technical document, previous attempts to solve it, and the results of those attempts.

hits Matches found by an Internet or library search engine to the names, phrases, or key words a user entered in a search; on Internet search engines, most hits are embedded links that allow a user to move directly from the search engine to the material discovered.

horizontal axis The *x*-axis on a graph that usually indicates time, items, or categories; also called the *abscissa*.

HTML HyperText Markup Language, a programming language used to create Web pages.

hue A synonym for color (e.g., red, blue, green, purple) that is often used to refer to shades of a color.

human subjects guidelines Written to protect the safety and rights of the human beings involved in primary research efforts; organizations require researchers to submit primary research proposals to an oversight committee for review, and also usually require researchers to obtain written consent from subjects indicating that they understand the research processes and voluntarily agree to participate.

hyperlinks Embedded addresses found on a Web document; when a user clicks on them, the user makes a direct connection to another Web page.

hypermedia Nonlinear digital documents that can be accessed online or over the Internet and consist of combinations of voice, data, text, still and moving images, and audio.

I

icons Symbols conveying abstract ideas or concepts.

ideological elements Components of a document that focus on the relationships between objects, phenomena, and concepts and the human beliefs that shape them.

imperative form of verbs Used when giving commands; e.g., "Unplug the machine before servicing;" the "you" is assumed.

impromptu presentations Unrehearsed and unprepared presentations delivered at a moment's notice.

IMRaD Acronym for Introduction, Methods, Results, and Discussion, an organizational structure of a scientific report.

inclusive language Does not exclude members of the audience.

inclusive searches Advanced searches using the Boolean key operator AND and requiring that all included terms be found in a given work.

indexes Navigational devices at the back of documents that help readers find information quickly.

indirect measures Designed to predict a reader's ability to comprehend or use a document.

inductive organization Writing in which the author lists ideas and evidence and then presents the conclusion or key point.

inferences Theories, hypotheses, and statements backed up by some observations or experiments but not completely verifiable.

inferential statistics A type of statistical analysis used to determine what a sample indicates about a larger population.

information maps Concise statements that map out the structure of a communication.

informative abstracts Abstracts providing an audience with an idea of the purpose and organization of a document and giving specific conclusions or findings covered in it.

initials Letters that identify who, other than the sender, typed a letter; placed at the bottom of the last page.

inoculation theory A persuasive strategy of mentioning counterarguments and refuting them before they occur to an audience.

inside address Address naming the recipient of a letter.

instructions Details of the specific steps an individual must take to complete a task; a common genre of technical communication.

intellectual property Creative work (text, image, music, objects, etc.) protected by copyright, trademark, and patent laws.

Internet An enormous network of computer networks.

interpolate A means of displaying trends in data by filling in gaps. (*See also* extrapolate.)

interval scale Closed questions that ask respondents to provide answers that are intrinsically numeric, such as age or income level.

interviews A primary research tool commonly used in the workplace, where the research instrument is the interviewer.

introduction draft A purpose or problem statement, a goals or outcome statement, and an information map at the beginning of a document.

introductions Early sections of a report that describe the problem the project will solve.

invention One of the five canons of classical rhetoric which refers to the strategies writers use to develop their content.

ISO Acronym for the International Standards Organization, which makes quality control of supplies and manufacturing processes possible. Its Web site is <http://www.iso.ch/>.

isotype Another term for icons used as numerical counting units in visual displays.

J

jargon The specialized vocabulary of a field or discipline.

K

kerning The space between letters in a printed document.

keyword search A way to examine Internet sites or library catalogs to find materials.

L

lab notes Description of methods and results of laboratory procedures in meticulous detail.

laboratory testing An effective way to test a lengthy document in the laboratory by video- or audiotaping users who are following the documentation while using a product.

lab reports Use laboratory or field notes to make a claim extending or challenging existing knowledge.

leading The space between lines of text.

left-justified margin All lines of text begin in a straight line at the left; readers are used to reading left-justified text.

legend Information included on a graph or map to explain devices such as color or patterns; also called a *key.*

legibility The design and formatting characteristics of a document, including typeface, type size, use of white space, and kerning.

letterhead The printed stationery used for correspondence that includes the company name, logo, address, telephone and fax numbers, and Web address.

letter of application A cover letter accompanying and summarizing significant details of a résumé.

letter of authorization Letter authorizing someone to begin a project and providing that person with task specifications.

letter of recommendation Letter written at the request of a person seeking a position and describing the writer's knowledge of the job applicant's abilities.

letter of transmittal A cover letter that is not part of the actual report but accompanies it to announce the attached document and project discussed.

line length The width of a column of text.

lists Chunks of information clearly separated from the rest of the text.

literature reviews Summaries and evaluations of published material on a topic; a genre of technical writing.

logos One of the three means of persuasion described by Aristotle; refers to the appeal to the evidence and the listener's or reader's reasoning process. (*See also pathos* and *ethos.*)

longitudinal survey A primary research tool that allows the researcher to ask questions of the same or similar group over a period of time.

M

management The research and writing tasks needed to produce a high-quality report in technical communication.

management proposal A proposal designed to convince someone in authority to make a change in a procedure or policy.

managerial structure A standard format for organizing a report that includes an executive summary, introduction, conclusion, recommendations, appendixes, and selection criteria in that order.

margins The white space at the top, bottom, and both sides of a page.

mean An arithmetic average obtained by dividing a sum by the number of items added.

median The middle number in a series containing an odd number of items or the number midway between the two middle numbers in a series containing an even number of items.

medium Part of the rhetorical situation technical communicators consider when they create a document. The medium for a document is the delivery mechanism in which it is created. For example, communications can be in any of the following media or in combinations of them: print, speech, video, graphics, online, etc.

metaphor A figure of speech that compares two unlike things without the use of the words like or as; e.g., "The wind was a bulldozer in the forest."

methodological notes Written records to remind the observer about additional research needed.

mode The number or value that appears most frequently in a series.

modification The process of adding words, phrases, clauses, or sentences to clarify or hone a definition.

modifiers Words, phrases, and clauses that describe other elements in a sentence.

MOO, MUD, IRC Three kinds of synchronous, or real-time, conversations in textual format or among multiple users online.

multiple editorial reviews The various stages of the editing process necessary to produce a complete, accurate, usable, and effective document.

N

natural work The work that happens naturally in a group, at a work site, or in an office.

needs assessment Information addressing the contingencies that might affect how a problem is solved; the part of a proposal that tells what is needed, who needs it, when and where it is needed, and why it is needed.

negation A technique used in creating descriptions; describes an object, concept, or term by what it is not or identifies characteristics it does not have.

networking A group of personal contacts who can assist a job seeker in identifying employment opportunities.

newsgroups Specialized electronic bulletin boards to which anyone can post or respond to a message.

NISO Acronym for the National Information Standards Organization.

nominalizations Verbs made into nouns; often identified by their endings: *-tion, -ment, -ance, -ing,* and *-ence.*

nominal scale questions Closed questions that ask respondents to place their responses in one of a number of categories provided for them.

nonverbal communication Communication signals and cues conveyed by means other than words; includes dress, posture, facial expressions, gestures, and tone of voice.

noun cluster One or more nouns used at the beginning of a definition.

O

objectives Expression in concrete terms of what one plans to accomplish if a proposal is funded.

observational notes Hastily written descriptions of what you see going on as you watch.

observations A primary research tool that gathers information about people and their activities as they go about their jobs or work with products and devices or in specific environments.

online help Information available on the computer; it can be built into a computer application, available from a Web site, or stored on a CD-ROM.

open interviews Unstructured interviews in which the interviewer imposes no limits on the questions asked or on the responses of the person being interviewed.

open questions Questions allowing respondents in a survey questionnaire to write anything they wish.

opinions Personal attitudes.

opposite values Positive/negative, gain/loss, increase/decrease values that can be displayed on a deviation bar graph.

optical character software Software allowing one to use the scanner to import text as text so that it can be edited.

ordinal scale Closed questions that ask respondents to place their answers in rank order.

organization The relationship between ideas and influences comprehension in technical writing.

organizational culture A corporation's sense of its own identity manifested in its mission statement, values, brand identity, documents, and in such informal things as how people dress and whether they bring their lunch or eat in the company cafeteria.

outlines Any of a variety of frameworks or skeletons for organizing data.

overhead The research costs for such things as space and utilities.

P

parallelism   Similar items that are expressed in a similar manner.

paraphrase   A restatement of a source that maintains the essence of the original.

passive voice   A grammatical construction that omits the agent, e.g., "The study was conducted." (*See also* active voice.)

pathos   One of the three means of persuasion described by Aristotle; refers to the appeal a communicator makes to the most basic and deeply held values and beliefs of people. (*See also logos* and *ethos.*)

PDF files   Portable Document Format files that can be printed from a Web site, preserving the look and all formatting of the original Web page.

peer reviewers   Experts in a research area or discipline who judge the relative merits of proposals.

perceptual task   Another name for the intellectual effort performed by an audience to retrieve and interpret data from a graph or other visual display.

performance   The use of the product or process being defined or described; often includes the tasks a user must perform. Performance is also the focus in writing instructions, specifications, and procedures.

periodic reports   Reports comparing the activities of the current reporting period with that of a previous period.

persona   The created narrator within a communication. (*See also ethos.*)

personal competencies   The qualities, skills, and abilities necessary in each collaborative team member for the team to function well.

persuasion   The ways in which written and oral discourse inform and motivate an audience, including rational appeal (*logos*), an emotional appeal (*pathos*), or an ethical appeal (*ethos*).

pilot surveys   Preliminary versions of a survey sent to a small, representative sample (about 3 percent) of a population to test a survey's usefulness and validity.

plagiarism   The failure to cite sources which can lead to dismissal at work, failure at school, and a lawsuit. Using someone else's words and calling them your own.

plan of implementation   A central and defining component of a proposal often suggesting change.

planning   The process a writer uses to develop a sense of his or her goals, determine the purpose for a document, and to decide how to handle timetables, organizational constraints, and sources needed for information. This activity takes place before drafting and redrafting as well as during the drafting and revision cycle.

Pliny the Elder   Roman soldier and administrator (23-79) who wrote and compiled information on natural history which provides early examples of technical communication.

preliminary pages   Pages preceding the text of a report, identifying the report and its contents, and giving credit to the people who created it.

presentation   In technical communication, the process of choosing the best strategies of organization, format, style, and visual display.

presentation software   Graphics programs and desktop publishing software used to develop visual aids for oral presentations.

primary colors   The dominant colors in a spectrum, such as red and blue, which provide contrast in a document.

primary research   The research methods designed to create or discover new knowledge, such as interviews, questionnaires, observations, and experiments.

procedures   A genre of technical communication that provides a description of how to coordinate an activity in which many people are involved.

product liability laws   Laws defining the responsibility of companies (including their writers) to protect consumers.

professional audiences   Highly trained professionals, including scientists, computer programmers, engineers, physicians, and systems analysts.

profile of users   The identifying factors gained from nominal questions.

progress reports   Reports describing the work accomplished on a project during a reporting period and identify the work remaining.

proofreading   The final check of a document for spelling and grammar errors and other surface mistakes; occurs before one finishes the last draft.

protocol-aided revision   A type of user testing in which users perform tasks and talk aloud as they do so to reveal how they are interpreting a text while they are performing an action; also known as "think-aloud protocol."

purpose   Indicates why a document is needed and what it is supposed to accomplish.

Q

quality metrics   A four-phase indirect measure of a reader's ability to use a document; the process ends with a regression analysis.

quick reference cards   A help document that lists important steps, hints, or points for readers to consult after they have used a document; they are single-page reference documents to get users started and provide basic information about a product.

quotations   Word-for-word statements of what a source said or wrote.

R

ragged right margin   A margin that is not right-justified; it guides readers to continue reading and avoids unnatural spacing between words in a line of text that often appears with right-justified margins.

random sample   A representative sample selected from a table of random numbers or a computer program.

raw scores   The final count of all answers gathered from a survey.

readability   How easily a document can be read, not necessarily how easily it can be understood; factors influencing readability include sentence length and complexity.

reading   An active process in which the reader interacts with a text to get from it what he or she needs; it is not a passive activity in which a reader "receives" a writer's message because it is presented to him or her.

recommendations   Advice designed in reports and feasibility studies to direct a particular action, unlike conclusions which answer a research question.

reference guides   A kind of glossary containing information arranged alphabetically that users will need to know to continue using a product.

reference librarians   Specially trained librarians who help researchers find needed information.

references   Sources used in a report and listed in the "Works Cited" section.

relevance   The appropriateness of evidence to the case at hand in a persuasive communication.

reliable questions   Survey questions that ask the same question every time without changing its meaning.

reports   Documents that contribute to significant decision making; along with business letters and memos, reports are the most common form of workplace writing.

report titles   Titles identifying the topic of a report and the approach taken toward it.

representative sample   A selection of some of the people in the general population of those involved in the subject or topic being researched.

Request for proposals (RFP)   A document identifying a need for a service or research and inviting proposals for the necessary work.

research   The process of finding credible and useful information through discovering what others have said about a topic or by conducting original experiments.

research, project, or service proposal   Proposal offering a service in exchange for support, usually including financial support.

research question   A step following forecasting in creating reports for decision making; focuses research by stating its objective.

retrospective methods   Methods designed to collect data as readers reflect on their experience when reading a document.

revision   A part of the writing process marked by a critical examination, reading, or rereading of a text by a writer and the changes that grow out of these considerations; revising (or re-visioning) takes place at many different levels, at many different times.

rhetoric   This term has a number of meanings, but in technical communication it refers to the use of language (written and spoken) to inform and persuade, and the relationship between language and knowledge; as defined by Aristotle, rhetoric is the facility of observing in any given case the available means of persuasion.

rhetorical function of a visual display   A description of the specific purpose of visual display, including the clarification of relationships among pieces of data and the trends they suggest.

right-justified margin   All lines of text line up along the right margin; such margins help readers identify discrete chunks of text or distinguish between them.

rows   The horizontal structure of information in a table; also known as lines.

rules   Lines used in the text to emphasize information or to draw the reader's attention.

running heads   A condensed version of the title of a report or major section of the report at the top of most pages of that section or report.

S

salutation   The greeting or "hello" found at the beginning of letters.

scaled questions   Closed questions that ask respondents to place their answers in a list of categories or to rate their response on a continuum of possibilities.

scanners   Electronic devices that allow users to convert text and images to digital files that can be stored on the computer and manipulated with image software.

scientific reports   Documents that present the results of experiments or other exact methods of research.

Scot, Reginald   Author of *A Perfite Platforme of a Hoppe Garden* (1574), which gives procedures for preparing the ground; planting, gathering, and drying hops; and preserving them from bad weather—an example of technical communication from the English Renaissance.

scripted presentations   Reports delivered verbatim from written copy.

search engines   Computer programs that examine information databases (e.g., library catalogs, Web sites) to find matches for the keywords, individual, or subjects.

secondary research   Locating and evaluating published or online scholarship to gather information for some task.

self-administered questionnaires   Printed surveys sent through the mail.

sequential model   A collaborative model in which each team member writes a draft and passes it to another for comment or sign-off, this team member in turn passes the draft to the next team member, and so on.

serif and sans serif   Two styles of type; serifs are the little "feet" attached to the ends of letters; sans serif means without serifs.

servers   Special computers devoted to maintaining such things as networks.

sidebars   The margins of a text where definitions can be placed.

signature block   The signature and company name at the end of a letter.

simulated work   Work that is re-created within a laboratory setting.

situated learning   The use of documents by readers who participate actively in the process of understanding a document.

social contexts   The human settings in which technological terms exist.

social process   In the workplace, the writing process is a social process in that many people are involved in helping a writer to plan, draft, and revise. The opinions, requirements, and needs of supervisors, coworkers, and clients influence the technical communicator's writing.

special features   Characteristics that distinguish a particular item from other items in a related classification group.

special handling line   A line placed at the beginning of a letter, following the dateline, which identifies the letter's transmission as certified mail, registered mail, overnight mail, and so forth.

specialization model   A collaborative model in which each team member is assigned a task according to his or her special expertise.

specifications   A genre of technical communication that document the standards that must be followed for a task to be completed safely and according to applicable codes.

spot color   The solid or tinted colors that do not overlap with each other to create new hues.

standing heads   Separate lines of text above the main text that are set off by a different typeface or graphic feature to identify different components of a document.

statistical analysis   A mathematical interpretation of the raw scores of a survey.

statistical significance   The result of inferential descriptive statistical analysis.

STC   Society for Technical Communication, the largest professional organization of technical communicators with more than 20,000 members worldwide. For more information contact the STC Web site at <http://www.stc-va.org>.

storyboard   A two-column script used to prepare presentations in which the left-hand column is used for the text and the right-hand column for the visuals that will accompany it.

structural headings   Headings identifying the largest chunks of information in a document.

stub heads   The categories or classification of line headings in the left-hand column of a table.

style   The writing choices made by an author including tone, point of view, level of formality, use of figures of speech, voice, emphasis.

style sheets   Devices used to ensure consistency in document design throughout a document; also can be used for style, usage, and grammar decisions.

subject line   A line placed at the beginning of a letter that summarizes in four or five words what the letter will discuss.

subordination   The linking of two unequal ideas so that the relationship between them is clear.

sufficiency   A judgment about the amount of evidence presented in a persuasive communication.

summaries   Reviews of the main points of a document; summaries and abstracts are ways to summarize or reduce information from many sources or from a single long source into a paragraph or two of notes.

superscript   A letter, number, or symbol (e.g., asterisk or dagger) placed above and to the right or left of a word.

suppressing the zero A distortion of data by starting graphs at other than zero.

survey questionnaires Primary research tools that ask sources to complete a question-answer form; often conducted through the mail or over the phone.

synonyms Words that mean the same thing.

synthesis model A collaborative model in which team members are brought together because of their different views or perspectives.

T

table of contents A list of the main divisions of a report and the page on which each begins; it appears at the beginning of the document.

tab markers Heavy markers attached to the right edge of a page that guide readers to appropriate sections of a document; also called *tabs.*

task analysis The systematic analysis of product performance.

task-oriented instructions User-friendly instructions designed to help users perform the tasks they wish with a product.

technical communication A type of applied communication designed to make scientific and technical information readily accessible to lay audiences; as a category includes technical writing, oral communication, visual display of information, and design and creation of online information.

technical knowledge The technical competencies required of each collaborative team member for the team to solve a problem, create something, or execute a specific task.

technical understanding The specialized knowledge about a product or process that an audience brings to a document.

technical writing A subset of technical communication, it is the broad umbrella that covers a variety of writing tasks, including everything from writing reports to writing speeches.

telephone surveys Surveys in which a caller asks respondents questions over the telephone and records answers by marking an answer sheet.

templates Overlays that fit on keyboards or control panels to remind users about the functions of a device; templates are also formats for creating documents such as memoranda, transmittal letters, cover letters, and letters of recommendation.

theoretical notes Statements that express one's opinion about or interpretation of what one sees.

think-aloud protocols An observation tool used when a person is asked to state his or her thoughts while performing a task. (*See also* protocol-aided revision.)

title page The page at the front of a report that identifies the subject as well as its recipient, author, and date.

transitions Words used to link thoughts; common transitions show relationships in time, place, cause and effect, contrast, and similarity.

tutorials Extended instructions that help beginning users understand a product's functions.

typeface or font A collection of type that has a unique name and set of distinctive characteristics (e.g., Times Roman).

type size, point and font size Terms often used interchangeably to refer to the size of the print in a text.

typography The style, size, and look of a printed page.

U

units of measurement The universal standards by which data are measured, e.g., inches, Hertz, meters, pints, or miles.

URL Uniform Resource Locator, an addressing system for the World Wide Web.

usability Describes how easily and accurately a document can be used.

usability study An investigation of how well a product meets its users' needs; these studies are conducted on-site or in a laboratory.

user edits A kind of user testing in which a "typical" user reads a document aloud while following the steps or points to see if it is possible to follow the instructions provided to carry out a process.

user friendly The concept of creating instruction manuals (and other documents) oriented toward the user's needs.

user guides A genre of technical communication that provides step-by-step instructions for completing all tasks associated with a product or process.

user's advocate The role a technical communicator should take in creating instructions, procedures, and specifications; the person writing has an obligation to represent the user's needs and interests.

V

vague pronouns The "it," "there," and other pronouns used to begin a sentence in place of the main idea.

valid questions   Survey questions that accurately ask what the researcher intends, without any misunderstanding between the researcher and the respondent.

value   In reference to color, the lightness or darkness of a hue.

vertical axis   The *y*-axis on a graph which begins at zero and usually indicates quantity; also called the *ordinate.*

videoconferences   Conferences held using television cameras and monitors and digital transmission devices to provide interactive video and audio for meetings at distant sites.

visual aids   Anything provided by the speaker for the audience to look at, including but not limited to handouts, charts, videotapes, posters, overheads, and slides.

visual design   The appearance of a document as created by choices of type, spacing and margins, use of headings, and illustrations.

visual display   The design or presentation of data in a graph, chart, or drawing on a page or screen.

visually challenged people   Audience members who may require special placement of visual aides as well as alternative media.

voice   The nature of a persona in a document; voice can be stern, friendly, condescending, formal, or gracious.

W

white space   The empty or blank areas on a page; guides readers through structure, prevents fatigue, and creates aesthetic appeal.

World Wide Web   The name for all computer files accessible through the Internet; the Web is not synonymous with the Internet, which contains many things in addition to the Web.

X

x-axis   The horizontal axis which usually indicates time, items, or categories on a graph; also called the *abscissa.*

Y

y-axis   The vertical axis on a graph which begins at zero and usually indicates quantity; also called the *ordinate.*

yes/no questions   Questions in a survey questionnaire that allow the respondent to answer only yes or no to a question.

Index